About Island Press

Since 1984, the nonprofit Island Press has been stimulating, shaping, and communicating the ideas that are essential for solving environmental problems worldwide. With more than 800 titles in print and some 40 new releases each year, we are the nation's leading publisher on environmental issues. We identify innovative thinkers and emerging trends in the environmental field. We work with world-renowned experts and authors to develop cross-disciplinary solutions to environmental challenges.

Island Press designs and implements coordinated book publication campaigns in order to communicate our critical messages in print, in person, and online using the latest technologies, programs, and the media. Our goal: to reach targeted audiences—scientists, policymakers, environmental advocates, the media, and concerned citizens—who can and will take action to protect the plants and animals that enrich our world, the ecosystems we need to survive, the water we drink, and the air we breathe.

Island Press gratefully acknowledges the support of its work by the Agua Fund, Inc., Annenberg Foundation, The Christensen Fund, The Nathan Cummings Foundation, The Geraldine R. Dodge Foundation, Doris Duke Charitable Foundation, The Educational Foundation of America, Betsy and Jesse Fink Foundation, The William and Flora Hewlett Foundation, The Kendeda Fund, The Andrew W. Mellon Foundation, The Curtis and Edith Munson Foundation, Oak Foundation, The Overbrook Foundation, the David and Lucile Packard Foundation, The Summit Fund of Washington, Trust for Architectural Easements, Wallace Global Fund, The Winslow Foundation, and other generous donors.

The opinions expressed in this book are those of the author(s) and do not necessarily reflect the views of our donors.

CLIMATE CHANGE SCIENCE AND POLICY

Climate Change Science and Policy

Stephen H. Schneider, Armin Rosencranz,
Michael D. Mastrandrea, and Kristin Kuntz-Duriseti

ISLANDPRESS

Washington | Covelo | London

Schneider, Stephen H.
  Climate change science and policy / Stephen H. Schneider . . . [et al.].
     p. cm.
  Includes bibliographical references and index.
  ISBN-13: 978-1-59726-566-9 (cloth : alk. paper)
  ISBN-10: 1-59726-566-7 (cloth : alk. paper)
  ISBN-13: 978-1-59726-567-6 (pbk. : alk. paper)
  ISBN-10: 1-59726-567-5 (pbk. : alk. paper) 1. Climatic changes—Government policy. 2. Greenhouse gases—Environmental aspects. 3. Environmental policy—International cooperation. 4. Science and state. I. Title.
  QC981.8.C5S3558 2010
  363.738'7456—dc22   2009017103

Printed on recycled, acid-free paper
✹

Manufactured in the United States of America
10 9 8 7 6 5 4 3 2

Dedicated to a generation of our students who, over the years, have been the inspiration—and the guinea pigs—for the evolution of this book.

CONTENTS

In January 2006, Steve Schneider and Armin Rosencranz concluded that their earlier volume, *Climate Change Policy* (Island Press, 2002), was no longer current. In light of the many developments in climate change science and policy and the broad recognition that anthropogenic carbon loading of the atmosphere is having increasingly severe global consequences, a new and comprehensive book on the subject was needed. Schneider and Rosencranz invited their former student and recent PhD, Michael Mastrandrea, to join them in coediting such a volume.

The editors are indebted to a number of people who played critical roles in bringing this project to fruition. First and foremost is Kristin Kuntz-Duriseti, who had coauthored one of Schneider's chapters in the 2002 edition and who worked with authors to edit each chapter down to its essence for the final version. Kristin's work was so impressive that the three original editors asked her to join the project as a fourth editor.

Sixteen graduate students in Armin Rosencranz's climate change policy class at the University of Maryland's School of Public Policy—Sanjana Ahmad, Andre Aquino, Russ Conklin, Jonathan Dorn, Kate Durant, Dan Emerine, Jerry Hinkle, Christina Hodgson, Marisa London, Mandy Ma, Lisa McGoldrick, Julia Miller, Margaret Spearman, Rodolfo Tello, Brian Turnbaugh, and Elizabeth Vonhof—also carefully reviewed and evaluated earlier drafts of each chapter.

The editors also received countless hours of help in formatting and preparing the final chapters and artwork. Vikram and Sanjay Padval were the helpers on the formatting. The helpers on the artwork, who often needed to find original and high-resolution sources, were Lee Love-Anderegg and Sarah Jo Chadwick. Lee and Sarah spent countless hours binding all the loose strings together, and their work was essential to the completion of the book. Lee also played a critical role working with Kristin on final chapter edits.

Chapter authors are the heroes of this publication. They patiently and cooperatively revised and updated their material at least twice and often three times in order to make the book as up to date and as relevant as it could be.

It has been a pleasure to work again with Todd Baldwin at Island Press. With his help, we have sought to make this book complete, authoritative, and readable. We editors are responsible for any lacunae that the reader may find.

Stephen H. Schneider
Armin Rosencranz
Michael Mastrandrea
Kristin Kuntz-Duriseti

Introduction

John P. Holdren

The popular term *global warming* is a misnomer. It implies something uniform, gradual, mainly about temperature, and quite possibly benign. What is happening to global climate is none of those. It is occurring with uneven effects across geographic, economic, and social divisions. It is rapid compared with ordinary historic rates of climatic change, as well as rapid compared with the adjustment times of ecosystems and human society. It is affecting a wide array of critically important climatic phenomena besides temperature, including precipitation, humidity, soil moisture, atmospheric circulation patterns, storms, snow and ice cover, and ocean currents and upwellings. And its effects on human well-being are and undoubtedly will remain far more negative than positive. A more accurate, albeit more cumbersome, label for this phenomenon is *global climatic disruption*.

The disruption that the world is experiencing has been documented in thousands of scientific reports covering dozens of climatic phenomena in hundreds of locations around the globe. While natural processes affect climate and climate change on time scales both short and long, overwhelming scientific evidence supports the conclusion that the human influences on climate, growing steadily since the beginning of the Industrial Revolution, have become the dominant driver of the exceptionally rapid climatic change experienced during the last half century. The largest of these human influences is the atmospheric buildup of the heat-trapping gas carbon dioxide, which civilization emits in immense quantities through combustion of the coal, oil, and natural gas that fuel 80 percent of world energy use and through deforestation occurring largely in the tropics.

These fundamental conclusions about the problem of climate change are solid. They are based on:

- Basic and uncontroversial understandings of how energy interacts with the gases that make up the atmosphere and how the Earth's water and carbon cycles work.
- Temperature readings by global networks of thermometers spanning 125 years, as well as similarly long-running and wide-ranging measurements of other climate variables.

- The records of earlier climates preserved in tree rings, corals, ocean sediments, and layered ice cores.
- Complex and detailed computer models of the motions of atmosphere and oceans showing that the sum of human and natural influences as we understand them explains with high fidelity the pattern of observed changes in global climate that has been observed.

Subject to somewhat greater uncertainties (which are two-sided and thus not a rational basis for complacency) are projections of how global climatic disruption will evolve in the future. There is, after all, a range of possibilities for the future trajectories of population, economic activity, and technology—and thus of the emissions of carbon dioxide and other climate-altering substances emanating from the activities of civilization—and for any given emissions scenario there is a set of uncertainties, expanding as one moves further into the future, arising from imprecision in our predictions of how emissions will affect atmospheric concentrations, how concentrations will affect average temperatures, and how changes in average temperatures will affect the spatial and temporal patterns of hot and cold, wet and dry, and storm and calm that constitute the actual climate.

It is known with very little doubt, however, that the average surface temperature of the Earth has already gone up about 0.8 degree Celsius above its level in 1750 and would eventually go up about another 0.5 degree even if all the climate-altering gases and particles in the atmosphere could be "frozen" at today's concentrations (due to the long time lag for the ocean to reach equilibrium with the altered atmospheric energy transport those climate-altering substances cause). It is also

known with considerable confidence that, under a middle-of-the-road emissions trajectory going forward—not the highest plausible and also not the lowest—the global average temperature increase is likely to reach something like 2 degrees Celsius above the 1750 figure by 2050 and 3 to 4 degrees Celsius above it by 2100.

These increases may sound small to those not steeped in climate change science, but they are frightening to nearly all of those who are, for two reasons: (1) the changes in mid-continents (where agriculture is concentrated) are typically twice the global average change, and those in the far north are 3 to 4 times the global average (with huge impacts on tundra, permafrost, northern forests, and sea ice); and (2) small changes in global-average surface temperature correspond to large changes in the climatic *patterns* that strongly influence human affairs and the fate of ecosystems. The last time the Earth was 2 degrees Celsius warmer was 130,000 years ago, when the climate was markedly different from today's and sea level was 4 to 6 meters higher. The last time the Earth was 3 to 4 degrees Celsius warmer was 30 million years ago, when the climate was drastically different and sea level was 20 to 30 meters higher.

Climate change is already causing harm *now*. Major floods, droughts, heat waves, and wildfires have been on the rise all around the world, in patterns largely predicted by climate models for a world heating up as ours is. The evidence that the disruption is also driving an increase in the power of hurricanes and typhoons is becoming compelling. Coral reefs are being roasted by rising sea-surface temperatures and pickled by the increased ocean acidity that is caused by uptake of some of the excess carbon dioxide from the atmosphere. The World Health Organization estimated in

2002 that global climate change was responsible for 150,000 premature deaths worldwide already in the year 2000; the number would be much higher today.

Faced with the challenges posed by this human-driven disruption of global climate, civilization has only three options:

1. Mitigation, meaning measures to reduce the pace and magnitude of the changes in climate that will occur (e.g., emissions reductions, afforestation and reforestation, improved soil management, geo-engineering to create cooling effects to offset warming by heat-trapping gases).

2. Adaptation, meaning measures to minimize the harm done by the climate changes that do occur (e.g., developing heat- and drought-resistant crop strains, strengthening defenses against diseases favored by a warming world, building more dams to contain floods and dikes to cope with rising sea level).

3. Suffering the impacts that mitigation and adaptation don't avoid.

Society is already doing some of each: we are mitigating; we are adapting; and we are suffering. The open question is what the future mix will be.

Neither mitigation alone nor adaptation alone will do. No amount of effort at mitigation can stop climate change in its tracks; the disruption is certain to grow before it can be stabilized and perhaps rolled back. Adaptation, for its part, becomes more costly and less effective the larger the climatic changes to which we are trying to adapt. And there are some impacts for which no meaningful adaptation is possible. Many low-lying regions will not adapt to 2 or 3 meters of sea-level rise; they will be submerged. Nor will we adapt to the likely loss of the world's coral reefs and their rich biodiversity; we will just do without them, permanently impoverished by their absence. Clearly, minimizing suffering will require a strategy incorporating a great deal of both mitigation and adaptation.[1]

Specifying the degree of climate change that should be avoided has been a vexing issue. The 1992 UN Framework Convention on Climate Change, which under international law is the "law of the land" in all 191 countries that have ratified it (including the United States), called for "stabilization of greenhouse gas concentrations in the atmosphere at a level that would prevent dangerous anthropogenic interference with the climate system." The convention did not define what would constitute "dangerous anthropogenic interference," however, and the climate-science-and-policy literature has been increasingly populated ever since with articles proposing definitions of varying degrees of complexity and sophistication.

I find this question a sterile part of the debate. In my personal opinion the world is already experiencing "dangerous anthropogenic interference" by any ordinary understanding of the meaning of the word "dangerous." The question now, in my view, is whether we can avoid *catastrophic* anthropogenic interference with the climate system.

How much mitigation will be needed to avoid catastrophe? The probability of crossing a tipping point into potentially unmanageable degrees of climatic change rises rapidly for increases in global average surface temperature more than 2 to 2.5 degrees Celsius above the 1750 level.[2] The European Union reached agreement in 2002 that not exceeding a 2-degree-Celsius increase should be society's goal. Taking into account uncertainties about the exact sensitivity of global average surface temperature to greenhouse gas concentrations, one finds that achieving a 50 percent chance of not exceeding 2 degrees Celsius above the 1750 level requires stabilizing the atmospheric concentrations of climate-altering substances at the equivalent of 450 parts per million by volume of carbon dioxide. A 50 percent chance of not exceeding 2.5 degrees Celsius requires stabilizing at 500 parts per million by volume CO_2-equivalent.

As a number of the chapters in this book explain in greater detail, translating these concentration targets into corresponding emissions trajectories leads to the conclusion that global emissions of carbon dioxide from fossil fuel burning and deforestation must peak and begin to decline no later than 2015 to 2025 (depending on which target is chosen and on what the trajectories of the other climate-altering substances are). Achieving such a large deflection from "business as usual" will be a large challenge, however. The quantity of civilization's carbon dioxide emissions is immense—well over 30 billion tons per year from fossil fuel burning and tropical deforestation combined. Today, 75 to 80 percent of this is coming from burning the fossil fuels—coal, oil, and natural gas—that still supply 80 percent of the world's energy. The offending pollutant is emitted in volumes too large and from sources too diverse to be captured cheaply, and the fossil fuel–dominated global energy system is too big and too costly to be changed quickly. The CO_2 emissions coming from tropical forestation will also not be easy to reduce, because the forces driving this deforestation are deeply embedded in the economics of food, fuel, timber, trade, and development.

Although the challenge is large, there is a wide array of effective emission-reduction options; employing enough of them in parallel could give a reasonable chance of avoiding ca-

tastrophe. Many of these—particularly those focused on increasing the efficiency of energy end-use in buildings, transport, and manufacturing—can be undertaken at negative cost (that is, at a profit), because the value of the energy saved more than pays for them. Some others, such as measures to reduce soot emissions from inefficient engines, would be seen to be profitable propositions for society as a whole if their co-benefits in improved public health or other "public goods" were taken into account. Still others would at least be relatively inexpensive if their social benefits were subtracted from their costs—for example, reforestation, afforestation, and avoided deforestation (forest preservation).

It's very unlikely that emissions can be reduced as much as needed using only the profitable and inexpensive options, however. While one can imagine technological breakthroughs that might expand the scope for inexpensive reductions, the timing of the need means we cannot wait. Most important, we need to outfit the new coal-burning power plants scheduled to be built in the next twenty years to capture and sequester their carbon dioxide. Measures as costly as that will only be embraced if a high price is placed on greenhouse-gas emissions or if regulations require them. That's why government action is essential to move mitigation forward at an adequate pace.

There is every indication from the most careful studies that have been conducted to date that the cost of a portfolio of measures adequate to the task—comprising profitable, inexpensive, and costlier alternatives together—would probably not reduce world GDP in 2050 by more than a percent or two, nor reduce GDP in 2100 by more than twice that, which would represent one or two years' growth at the generally forecasted rates. In other words, the citizens of the world would need to wait until 2101 or 2102 to be as rich as they otherwise would have been in 2100. This seems to me to be a small price to pay for a large reduction in the probability of climatic catastrophe with a far bigger negative impact on future GDPs.

The most important conclusions about global climatic disruption—that it's real, that it's accelerating, that it's already doing significant harm, that human activities are responsible for most of it, that there is a growing danger of its becoming unmanageable, and that there is much that could be done to reduce the danger at affordable cost—have not been concocted by environmental extremists or enemies of capitalism. They are based on an immense edifice of painstaking studies published in the world's leading peer-reviewed scientific journals. They have been vetted and documented in excruciating detail by the largest, longest, costliest, most international, most interdisciplinary, and most thorough formal review of a scientific topic ever conducted—the now nearly twenty-year-long, multi-thousand-participant study by the Intergovernmental Panel on Climate Change organized under the auspices of the World Meteorological Organization and the UN Environment Programme.[3] They have been attested to by the leadership of the academies of science of Brazil, Canada, China, France, Germany, India, Italy, Japan, Russia, the United Kingdom, and the United States (in a joint statement issued in 2005); the American Geophysical Union; the American Association for the Advancement of Science; and the national meteorological offices of every country that has one. They are endorsed by all three winners of the only Nobel Prize ever awarded for atmospheric science.

The science of global climate disruption and its impacts, the technology of remedies, the economics of damage and evasive action, and the policy issues that must be resolved if the available insights from science, technology, and economics are to be integrated into an effective program of action have already been the topics of several long shelves' worth of books and book-length reports. Why another book now, and why the particular book that is before you?

This book's audience, in my view, is a broad one: students in interdisciplinary courses on climate change for upper-division undergraduate and graduate courses; graduate students and more senior researchers who work on one part or another of the science, technology, economics, or politics of the climate change problematique and seek a an authoritative but still accessible introduction to the parts they have not focused on in their own work; natural and social scientists who haven't worked on the climate problem, but want to survey the subject in a way that moves them efficiently up the learning curve; and journalists, policy makers, and members of the wider public who have the appetite and capacity for mastering the level of technical detail needed to understand—really understand—what global climate disruption is, where it is headed, what can be done, and how. No other book that I'm aware of offers this one's combination, for these purposes, of comprehensiveness, authoritativeness, currency, and readability.

To those who might ask whether the 2007 Fourth Assessment Report of the Intergovernmental Panel on Climate Change hasn't already met this need, I would answer, in part, that the three main volumes—the full reports of IPCC Working Groups I, II, and III—are too much (at 2,823 large-format pages of fine print), while the Summaries for Policymakers and even the Technical Summaries are too little.[4] The book before you is the "Goldilocks Solution"—the amount of intellectual sustenance on climate change that is "just right."

In addition, there is the matter of coverage of the most recent developments. The multi-layered review and consultative process followed by the IPCC dictated a December 2005 cutoff for science findings to be covered in an assessment issued in early to mid-2007. The climate change field is fast-moving—not only in its scientific but also in its technological and policy dimensions—and many important developments too recent to be treated by the 2007 IPCC documents are covered here.

One of the primary uses of this volume, in my view, will be to equip its most industrious readers with all they need to recognize and to rebut the misunderstandings and misrepresentations being propagated by climate change deniers, who continue to receive attention in the media disproportionate to their numbers, their qualifications, or the merit of their arguments. The attention and credence they receive are a menace, insofar as it delays the development of the political consensus that will be needed before society embraces remedies that are commensurate with the magnitude of the climate change challenge.

It is often said that there are three stages of denial in relation to issues at the science-society interface:

1. They tell you you're wrong, and they can prove it: "Climate isn't changing in unusual ways" or "Human activities are not the cause of climate change."

2. They tell you you're right, but it doesn't matter: "OK, the Earth's climate is changing, and humans are playing a role, but the damages will be small and there may even be some benefits."

3. They tell you it matters, but it's too late to do anything about it: "Yes, climate disruption is going to do some real damage, but it's too late, too difficult, or too costly to avoid that, so we'll just have to hunker down and suffer."

Individual deniers often move over time from stage 1 to 2 and from 2 to 3 as evidence becomes harder to ignore or refute. The very few deniers with any credentials in climate change science have virtually all shifted in the past few years from stage 1 to 2; jumps from 2 to 3 and even from 1 to 3 are becoming more frequent.

All three positions are deeply wrong, and the reasons why they are wrong are nowhere as clearly, comprehensively, and authoritatively laid out as in this book. The intellectual terrain it covers is vast and sometimes demanding, but it will repay the effort of all those able and willing to traverse it. May many do so.

Notes

1. The international Sigma Xi/UN Foundation Scientific Expert Group (SEG) on Climate Change and Sustainable Development, which provided its recommendations on mitigation and adaptation opportunities to the UN Commission on Sustainable Development and the Secretary General in February 2007, stated the dual mitigation-adaptation challenge concisely in the subtitle of its report: "Avoiding the Unmanageable and Managing the Unavoidable." Scientific Expert Group on Climate Change (SEG), 2007: Confronting Climate Change: Avoiding the Unmanageable and Managing the Unavoidable. Rosina M. Bierbaum, John P. Holdren, Michael C. MacCracken, Richard H. Moss, and Peter H. Raven, eds. Report prepared for the United Nations Commission on Sustainable Development. Sigma Xi, Research Triangle Park, NC, and the United Nations Foundation, Washington, DC, 144 pp.

2. Ibid.

3. See the Web site of the Intergovernmental Panel on Climate Change, www.ipcc.ch.

4. IPCC, 2007a: *Climate Change 2007: The Physical Science Basis. Contribution of Working Group I to the Fourth Assessment Report of the Intergovernmental Panel on Climate Change*, S. Solomon, D. Qin, M. Manning, M. Marquis, K. Averyt, M. M. B. Tignor, H. L. Miller, Jr., and Z. Chen, eds., Cambridge University Press, Cambridge, UK.

IPCC, 2007b: *Climate Change 2007: Impacts, Adaptation, and Vulnerability. Contribution of Working Group II to the Fourth Assessment Report of the Intergovernmental Panel on Climate Change*, M. L. Parry, O. F. Canziani, J. P. Palutikof, P. J. van der Linden, and C. E. Hanson, eds., Cambridge University Press, Cambridge, UK.

IPCC 2007c: *Climate Change 2007: Mitigation of Climate Change. Contribution of Working Group III to the Fourth Assessment Report of the Intergovernmental Panel on Climate Change*, B. Metz, O. R. Davidson, P. R. Bosch, R. Dave, and L. A. Meyer, eds., Cambridge University Press, Cambridge, UK.

Impacts of Climate Change

Chapter 1

Climate Change Science Overview

MICHAEL D. MASTRANDREA AND STEPHEN H. SCHNEIDER

Introduction

This chapter outlines the current state of scientific knowledge regarding the climate system and the effects of human activities on climate. Although uncertainty remains regarding knowledge about climate, the basic processes that cause climate change are scientifically well established, and human activities have been identified with very high confidence as the main driver of most observed climate-induced trends during the last several decades. Conclusions such as these are based on the vast preponderance of accumulated scientific evidence. To understand complex systems science like the study of climate change, it is essential to distinguish such conclusions from hypotheses that can be "falsified" by one or even several lines of argument that seem to contradict the mainstream consensus. This is how simple science used to be done—for example, testing whether the liquid in a tube is acidic or basic. One piece of litmus paper can falsify a wrong preliminary hypothesis. While it can take decades to reconcile incomplete elements of complex systems analysis, rarely will a few contrary results entirely overthrow a consensus built on decades of consistent lines of evidence.

Throughout this chapter, many research findings we refer to are taken from the multiply-peer-reviewed, government-approved Intergovernmental Panel on Climate Change (IPCC) Assessment Reports, which present the best approximation of a worldwide consensus on climate change science every five to six years. One important feature of IPCC reports is the quantified assessment of the likelihood of each major conclusion, and the explicit assignment of the authors' confidence in the underlying science to back up each conclusion. This practice clearly separates out aspects that are well established from those that are better described by competing explanations and from those best labeled as speculative. This contrasts markedly from most of the media and political debates in which well-established conclusions are often conflated with speculative ones, and public confusion results. Box 1.1 presents the likelihood and confidence definitions from the 2007 IPCC Fourth Assessment Report (AR4).[1]

Box 1.1

The IPCC defines the likelihood of an outcome or a result as: Virtually certain (greater than 99 percent probability of occurrence), extremely likely (greater than 95 percent), very likely (greater than 90 percent), likely (greater than 66 percent), more likely than not (greater than 50 percent), unlikely (less than 33 percent), very unlikely (less than 10 percent), and extremely unlikely (less than 5 percent).

The IPCC defines the level of confidence in the correctness of the science underlying a conclusion as: Very high confidence (at least a 9 out of 10 chance of being correct), high confidence (about an 8 out of 10 chance), medium confidence (about a 5 out of 10 chance), low confidence (about a 2 out of 10 chance), and very low confidence (less than a 1 out of 10 chance).

The Global Temperature Record

Modern temperature records date back to the mid-nineteenth century, when thermometers became accurate and widespread enough to allow scientists to calculate a meaningful global average temperature. These records show (figure 1.1) that the Earth's average surface temperature has increased by about 0.75 degree Celsius (around 1.4 degrees Fahrenheit) since the mid-nineteenth century (with an uncertainty of about a tenth of a degree Celsius).[2]

Year-to-year variation in temperature cannot override this long-term upward trend in global average temperature. Unfortunately, short-term variability in the temperature record is often inappropriately used to "refute" long-term climatic trends. Climate, however, refers to the state of atmospheric conditions over decades or longer, while weather refers to shorter-term variations in atmospheric conditions. Thus, the IPCC description of the warming trend of past century or so as "unequivocal" is indeed appropriate, and even decadal-scale exceptions do not disprove this long-term fact.

Looking back into history can tell us more about how the current anthropogenic (or human-caused) changes compare to naturally induced changes in the past, both in magnitude and in rate. Paleoclimatologists use proxy variables that vary with temperature to approximate temperature records that stretch back hundreds, thousands, and even millions of years (see figure 1.2). These proxies consider diverse factors such as tree rings, the extent of mountain glaciers, changes in coral reefs, and pollen in lake beds. Although there is considerable uncertainty in temperature, the averaged trend over the last 1,000 years is a gradual temperature decrease over the first 900 years, followed by a sharp upturn in the twentieth century (shown also in figure 1.1). The question is, Why?

In particular, there are three typical explanations of observed global mean surface air temperature trends: (1) natural internal variability, in which energy exchanges among atmosphere, oceans, ice sheets, and ecosystems cause random, unpredictable background noise; (2) natural forcings in the Earth's radiative energy input from volcanic dust veils or solar energy fluctuations; and (3) anthropogenic

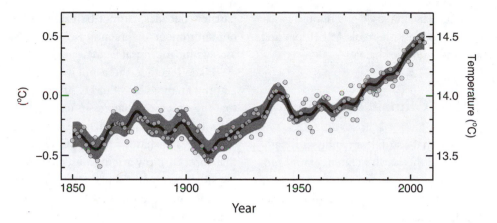

FIGURE 1.1. Observed global average temperature record (since 1850), shown relative to the average for 1961–1990 (vertical axis on the left), as well as in absolute terms (vertical axis on the right). Smoothed black line represents decadal average values, circles represent yearly values. Shading represents uncertainty in observations. Source: Intergovernmental Panel on Climate Change (IPCC), 2007(a), *Climate Change 2007: The Physical Science Basis. Contribution of Working Group I to the Fourth Assessment Report of the Intergovernmental Panel on Climate Change*, S. Solomon et al., eds., Cambridge University Press: Cambridge, United Kingdom.

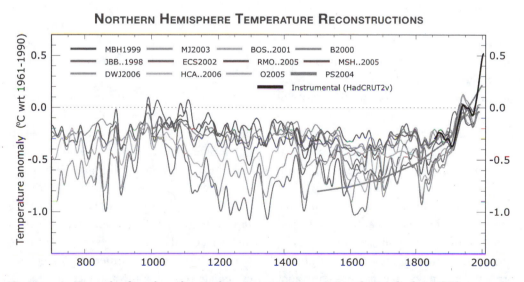

FIGURE 1.2. Records of northern hemisphere temperature variation during the last 1,300 years relative to the 1961–1990 average using multiple proxy records. Observed temperature record since 1850 shown in black. Source: Intergovernmental Panel on Climate Change (IPCC), 2007(a), *Climate Change 2007: The Physical Science Basis. Contribution of Working Group I to the Fourth Assessment Report of the Intergovernmental Panel on Climate Change*, S. Solomon et al., eds., Cambridge University Press: Cambridge, United Kingdom.

forcings, such as increased greenhouse gases, altered atmospheric aerosols (e.g., dust and smoke), and land-use changes.

A Natural Climate Variation?

Is it possible that natural variability and natural forcings of the Earth's climate could produce the temperature record of figures 1.1 and 1.2? Using a variety of methods to detect the human "fingerprint" on observed warming trends, scientists are finding overwhelming evidence that the answer to this question is "no" (see chapter 2).

Scientists can also use computer models of the climate system (see below) to investigate the contribution of natural and human factors to the observed warming. Figure 1.3 shows a comparison of the global average surface temperature record for the twentieth century (black line) with two sets of climate model simulations of this time period. The gray lines represent simulations that are driven only by estimates of purely natural forcings—solar variability and volcanic activity (see Solar Variability and Aerosols below). The range of simulations indicates an estimate of the degree of uncertainty in the model calculations. The estimated temperature variation due to natural forcing alone does not show an overall warming trend and is clearly a poor fit to the actual surface temperature record, especially in the second half of the century when temperatures made a significant upturn. The lighter lines represent simulations that also incorporate anthropogenic factors—emissions of greenhouse gases and aerosols. The fit between these simulations and the observed record is far better; they strongly suggest that the temperature changes observed in the twentieth century, particularly the rise of the

past few decades, cannot be explained without anthropogenic greenhouse gas emissions as a significant causal factor.

Taken together, these and many other fingerprint analyses provide very strong evidence that the observed changes in climate over at least the past several decades are anthropogenic.[3] This has led the IPCC to conclude that most of the warming observed over the last fifty years is attributable to human activities and, in addition, that the influences of anthropogenic climate change are now identifiable on warming ocean temperatures, changes in the life cycles of plants and animals (see chapter 3), atmospheric circulation patterns, and the increasing intensity of some extreme weather events.[4]

Keeping the Earth Warm

What ultimately determines climate and specifically the Earth's temperature? That question is at the heart of climate science and of the issues surrounding anthropogenic climate change.

About half of the light energy from the sun penetrates the atmosphere and is absorbed by the Earth's surface. The surface warms and re-emits some of the energy as infrared radiation. Certain naturally occurring gases and particles—greenhouse gases—absorb 80 to 90 percent of the infrared radiation emitted at the surface and radiate heat in all directions, both up to space and back down toward the surface, warming the surface further. This feedback cycle between the Earth's surface and the atmospheric greenhouse gases continues until the infrared radiation released to space is in balance with the sources of radiant energy.

Because the atmosphere functions, in a crude sense, like the heat-trapping glass of a

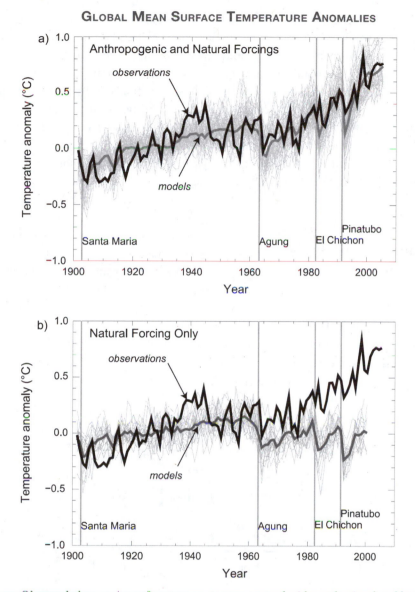

FIGURE 1.3. Observed changes in surface temperature compared with results simulated by climate models using natural and anthropogenic forcings. Decadal averages of observations are shown for the period 1906–2005 (black lines) relative to the corresponding average for 1901–1950. Gray lines depict model estimates; the ranges of estimates reflect model uncertainty. Gray lines use only natural forcings due to solar activity and volcanoes (dark gray line is multimodel average). Gray lines use both natural and anthropogenic forcings (dark gray line is multi-model average). Major volcanic eruptions are shown in both panels, corresponding to temporary cooling episodes. Source: Intergovernmental Panel on Climate Change (IPCC), 2007(a), *Climate Change 2007: The Physical Science Basis. Contribution of Working Group I to the Fourth Assessment Report of the Intergovernmental Panel on Climate Change*, S. Solomon et al., eds., Cambridge University Press: Cambridge, United Kingdom.

greenhouse, this heating process has earned the nickname "greenhouse effect." The natural greenhouse effect from gases and clouds effectively raises the Earth's surface temperature by 33 degrees Celsius (59 degrees Fahrenheit), which supports life as we know it on the Earth. However, increasing concentrations of atmospheric greenhouse gases due to human activities are intensifying the greenhouse effect and further increasing the Earth's temperature.

Greenhouse Gases Past and Present

Human activities add to the atmospheric concentrations of a number of naturally occurring greenhouse gases and introduce other potent greenhouse gases that are not naturally occurring. Increasing concentrations of the green-

house gas carbon dioxide (CO_2) due to human activities, primarily the burning of fossil fuels but also deforestation and other land-use changes, have contributed most to the intensification of the greenhouse effect. As shown in figure 1.4, before the Industrial Revolution, CO_2 concentrations were relatively stable for roughly 10,000 years, varying between 260 and 280 parts per million (ppm). In the last 150 years, atmospheric CO_2 concentrations have increased by more than 35 percent, from around 280 to around 380 ppm. The reality of this CO_2 increase is well documented and is well-established science.

Carbon dioxide entering the atmosphere does not just sit there. Huge quantities of carbon circulate between the atmosphere, the ocean, and land ecosystems. As the burning of fossil fuels and carbon dioxide emissions have increased, flows of carbon from the atmosphere into the ocean and into land ecosys-

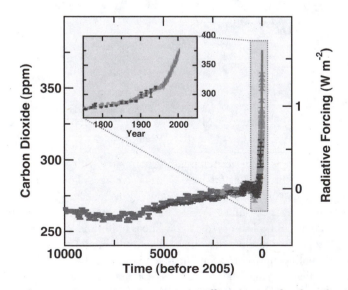

FIGURE 1.4. Atmospheric concentrations in parts per million (ppm) of carbon dioxide during the last 10,000 years and since 1750 (inset panel). Measurements from ice cores (different shades for different studies) and atmospheric samples. Corresponding radiative forcing is shown on the right side.

tems have also increased, but by a smaller amount. Currently, about half of the annual anthropogenic carbon dioxide emissions are taken up by ocean and land ecosystems. However, scientists have observed a decrease in the fraction of anthropogenic emissions absorbed by ocean and land ecosystems, and expect that fraction to continue to decrease as these "sinks" become saturated.[5]

How are scientists able to estimate the concentrations of these gases in the atmosphere for thousands of years in the past? Ice cores bored in Greenland and Antarctica provide estimates of both temperature and atmospheric greenhouse gases going back hundreds of thousands of years. So far, Antarctic ice cores

have yielded a continuous record of the past 740,000 years.[6] Variations in ice density associated with seasonal snowfall patterns provide a way to determine the age of specific points in some ice cores. By measuring the ratio of the hydrogen isotope deuterium (D) to hydrogen in the ice, scientists can calculate a proxy for the temperature at the time each layer of ice formed. By analyzing air bubbles trapped in this ancient ice, scientists can even measure the composition of the Earth's past atmosphere. The result of such an ice core analysis, shown in figure 1.5, gives dramatic evidence that temperature (measured by variations in deuterium; D) and greenhouse gas concentrations, particularly carbon dioxide,

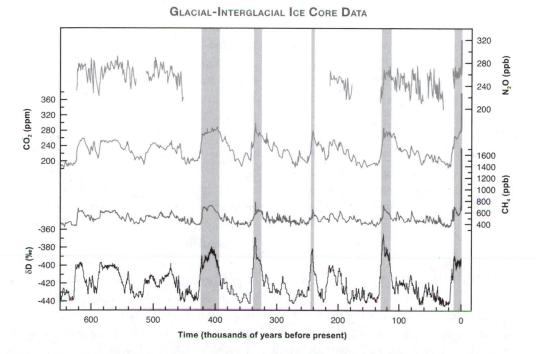

GLACIAL-INTERGLACIAL ICE CORE DATA

FIGURE 1.5. Variations of deuterium (δD; bottom line) in Antarctic ice core records, a proxy for local temperature, and the atmospheric concentrations of the greenhouse gases carbon dioxide (CO_2; line second from top), methane (CH_4; line second from bottom), nitrous oxide (N_2O; top lines) in air trapped within these ice cores and from recent atmospheric measurements. Data cover 650,000 years, and the shaded bands indicate previous interglacial warm periods.

are correlated over the long term. Although greenhouse gases are not the sole trigger for climate change historically—other factors like variations in the Earth's orbit are likely to initiate and end ice ages—greenhouse gases amplify processes that accelerate ice age formation and eventual deglaciation. The data support the mechanistic understanding of the role of greenhouse gases in climate changes and their ability to cause current and future climate changes as human activities increase atmospheric greenhouse gas concentrations.

The maximum CO_2 concentration in the ice core record of the past 650,000 years is less than 300 ppm. The present-day concentration of around 380 ppm is far above anything the Earth has seen, probably, for millions of years. Figure 1.5 shows the recent rise in CO_2 and other greenhouse gases relative to the rest of the ice core data. Clearly, the anthropogenic increase in CO_2 concentration is unprecedented in both its size and its rapidity over this time period. We have made truly dramatic changes in the Earth's atmosphere over the past century or so, and we are already observing impacts of climate change around the world that will continue to grow. To begin to predict the extent of these changes, we must examine *all* of the important influences human activities have on the climate system.

Greenhouse Gases and Radiative Forcing

Climatologists characterize the effect of a given atmospheric constituent by its radiative forcing, the rate at which it alters absorbed solar or outgoing infrared energy. Water vapor is the most important greenhouse gas, but is not directly influenced much by human activities—only indirectly as a feedback process

amplifying warming from the anthropogenic greenhouse gases. Carbon dioxide is the most important of the anthropogenic greenhouse gases, but other gases play a significant role, too. On a molecule-to-molecule basis, most other greenhouse gases are far more potent absorbers of infrared radiation than is CO_2, but they are released in much smaller quantities so their overall effect on climate is smaller. The second most prevalent anthropogenic greenhouse gas is methane. One methane molecule is roughly thirty times more effective at absorbing infrared than is one CO_2 molecule. Although CO_2 concentration increases tend to persist in the atmosphere for centuries or longer, methane typically disappears in decades, making its warming potential relative to that of CO_2 lower on longer timescales. Currently, the radiative forcing from anthropogenic methane is slightly less than one-third that of CO_2.

Other anthropogenic greenhouse gases include nitrous oxide and gases solely created through industrial processes, such as halocarbons used in refrigeration. Halocarbons include chlorofluorocarbons (CFCs), which are also the leading cause of stratospheric ozone depletion. Newer halocarbons do not cause severe ozone depletion but are still powerful greenhouse gases. They are hundreds to thousands of times more potent than carbon dioxide, molecule to molecule, and remain in the atmosphere for centuries to millennia, but appear in much lower concentrations than carbon dioxide and methane. Together, nitrous oxide and halocarbons account for approximately the same level of radiative forcing as methane. A number of other trace gases contribute a small amount of additional forcing. All the gases mentioned so far are well mixed, meaning that they last long enough to be distributed in roughly even concentrations throughout the troposphere, the lowest 10 kilometers of the atmosphere.

Finally, ozone (O_3), familiar because of the "ozone hole" and its depletion by anthropogenic CFCs, is also a greenhouse gas. Ozone in the troposphere near the surface is a potent component of smog, resulting largely from motor vehicle emissions. Tropospheric ozone contributes about one-fourth the radiative forcing of CO_2, although unlike the well-mixed gases, tropospheric ozone tends to be limited to industrialized regions, and it is of great concern for health effects as well as climatic influences.

The cooling of the stratosphere from added greenhouse gases has an effect on ozone, both by temperature-dependent atmospheric chemistry, which might slightly increase ozone levels in the tropical stratosphere, and by cooling of the polar stratosphere, which causes more high-altitude clouds that increase ozone destruction. Thus, there are many processes around the globe leading to climate and ozone changes arising from increasing the concentrations of greenhouse gases in the atmosphere above their natural levels.

Aerosols

Fuel combustion, and to a lesser extent agricultural and other industrial processes, produce emissions that create particulate matter. Coal-fired power plants burning high-sulfur coal, in particular, emit gases that become sulfate aerosols and reflect incoming solar energy, producing a cooling effect. Natural aerosols that produce a cooling effect are also created during volcanic eruptions and the evaporation of seawater, as well as from emissions of hydrocarbons in forested areas like the

Great Smoky Mountains—hazes that are largely from biological emissions. Conversely, diesel engines and some biomass burning produce black aerosols such as soot, which absorb the sun's energy and, depending on circumstances, can warm the climate.

Aerosol particles also affect radiative forcing indirectly. For example, they act as "seeds" for the condensation of water droplets to form clouds, affecting the color, size, and number of cloud droplets, and, in aggregate, likely offset some greenhouse warming. The IPCC estimates that the negative radiative forcing resulting directly from all anthropogenic aerosols (e.g., aerosol hazes) offsets about one-third of the positive forcing from greenhouse gases, with indirect effects (e.g., the change in cloud optical properties resulting from pollutant aerosols) offsetting, in aggregate, roughly another third.[7] However, there is considerable uncertainty regarding these figures (especially the indirect effects), which may be much larger or much smaller than these central estimates, although still likely to be a net negative forcing. Unfortunately, the uncertainty in aerosol radiative forcing complicates the assessment of "climate sensitivity": the amount the Earth's surface warms for a given increase in forcing—typically a doubling of CO_2 over pre-industrial levels. The climate sensitivity is an important parameter in projecting future climate change.

Solar Variability

Another important influence on the climate system not affected by human activities is the variation in the sun's energy output. Variations caused by the twenty-two-year sunspot cycle are typically estimated to amount to only about 0.1 percent of solar output and are too small and occur too rapidly to explain a significant climatic effect like the late-twentieth-century warming in figure 1.2. However, long-term solar variations, either from variability in the sun itself or from changes in the Earth's orbit and tilt, have substantially affected the Earth's climate over tens of thousands of years. Accurate, satellite-based measurements of solar output are available for only a few decades. To estimate past variations in solar activity, scientists use proxies such as the level of the isotope beryllium-10 in ice cores. Beryllium-10 is generated by cosmic rays entering the atmosphere, and its level in ice goes down when the sun is active and the "solar wind" of energetic electrons and protons repels more of these cosmic rays, and vice versa.

The IPCC estimates that current solar forcing is equivalent to about one-tenth of the forcing from CO_2, which contributes somewhat to observed global climate change but is far below what is needed to fully account for the warming of recent decades. There are many hypotheses suggesting that various solar effects have generated climate change, but none are considered likely explanations of the recent climate warming.[8]

Radiative Forcing: The Overall Effect

Figure 1.6 summarizes our current knowledge of radiative forcing caused by green-

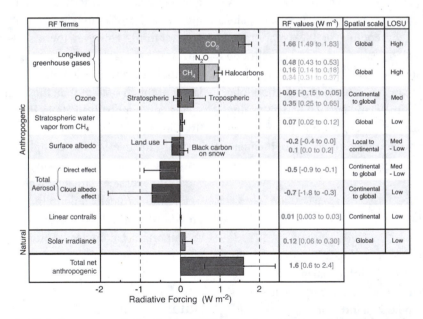

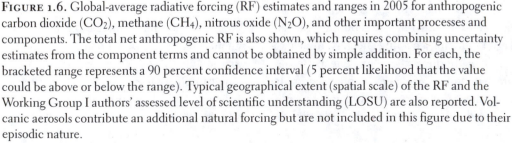

FIGURE 1.6. Global-average radiative forcing (RF) estimates and ranges in 2005 for anthropogenic carbon dioxide (CO_2), methane (CH_4), nitrous oxide (N_2O), and other important processes and components. The total net anthropogenic RF is also shown, which requires combining uncertainty estimates from the component terms and cannot be obtained by simple addition. For each, the bracketed range represents a 90 percent confidence interval (5 percent likelihood that the value could be above or below the range). Typical geographical extent (spatial scale) of the RF and the Working Group I authors' assessed level of scientific understanding (LOSU) are also reported. Volcanic aerosols contribute an additional natural forcing but are not included in this figure due to their episodic nature.

house gases, aerosols, land-use changes, solar variability, and other effects since the start of the industrial era.[9] The bottom bar presents an estimate of the total net current anthropogenic forcing. An important point to remember is that the individual forcings in the top panel have different levels of persistence and uncertainty. For example, different greenhouse gases remain in the atmosphere for different periods, as discussed previously. Therefore, the total net forcing is not a simple sum of the individual components. Comparing the top (CO_2) and bottom (total) forcing bars in figure 1.6, scientists estimate that total current forcing is roughly equal to the positive forcing from carbon dioxide.

Feedback Effects

Knowing the radiative forcing caused by changes in atmospheric constituents would be sufficient to project future climate, if there were no additional climatic effects beyond the direct change in energy balance. But a change in climate caused by simple forcing can have significant effects on atmospheric, geological, oceanographic, biological, chemical, and even social processes. These effects, in turn, can further change the climate. If that additional change is in the same direction as its initial cause, then the effect is called a positive or amplifying feedback. If that additional change is in the opposite direction, then it is a negative or dampening feedback. In reality, numerous feedback effects complicate the assessment of climate change. Here we list just a few feedback processes to give a sense of their variety and complexity.

Albedo is a planet's reflectance of solar radiation. The Earth's albedo is about 0.31, meaning that 31 percent of solar radiation is reflected back to space. A decrease in that number means that more radiation is absorbed. As the amount of radiation absorbed increases, global temperature also increases. One consequence of rising temperatures is the melting of snow and ice, which can already be observed in many parts of the world in the form of melting and receding mountain glaciers and decreasing snowpacks. Such melting eliminates a highly reflective surface and exposes the darker land or water beneath the ice. The result is a decreased albedo, increased solar energy absorption, and additional warming. This is a positive feedback.

Rising temperature also results in increased evaporation of water from the oceans into the atmosphere. Because water vapor is itself a greenhouse gas, this effect results in still more warming and is thus a positive feedback. Most assessments suggest that the overall effect of increased water vapor with global warming is a positive feedback that causes a temperature increase some 50 percent higher than would occur in the absence of this feedback mechanism.[10] But increased water vapor in the atmosphere can also mean more widespread cloudiness. More cloudy areas raise the Earth's albedo by reflecting more incoming solar radiation. This reflection results in less energy absorbed by the Earth-atmosphere system, a negative feedback if the increased cloud amount was caused by some positive forcing. On the other hand, more clouds mean greater absorption of outgoing infrared radiation from the Earth's surface. Furthermore, more evaporation or surface heating could mean increases in cloud top heights that would add to the greenhouse effect. Both of the latter processes are positive feedbacks. The net effect of increasing cloud amount depends on latitude and season, but averaged annually over the globe is often estimated to be a positive feedback.[11] However, uncertainty in the *net* cloud feedback—including changing cloud amount, top height, and microphysical properties like number, color, or size of droplets—makes it difficult to precisely estimate how sensitive the climate is to increasing greenhouse gas concentrations (see chapter 15).[12]

As mentioned above, huge amounts of carbon are continuously cycled among the atmosphere, ocean, land, and terrestrial biosphere as part of the global carbon cycle. In fact, a significant fraction of anthropogenic emissions are removed from the atmosphere by oceanic and terrestrial uptake. Increasing atmospheric greenhouse gas concentrations influence these processes in many ways.

For example, CO_2 dissolves in water. As CO_2 in the atmosphere increases, more CO_2 dissolves into surface waters, which is a negative feedback on CO_2 concentrations in the climate system. Some of this oceanic dissolved CO_2 is taken up by phytoplankton (tiny plants) and other organisms that are capable of photosynthesizing and thus converting it to organic material. Zooplankton (small marine animals) graze on the phytoplankton. When these phytoplankton and zooplankton die, their bodies sink, along with other organic matter, transporting the carbon to the deep ocean. Much of the carbon is redissolved along the way, but some reaches the ocean floor and is buried, becoming sediment. This small fraction becomes very significant to the carbon cycle over long timescales. Warmer water can hold less CO_2 than colder water, so as temperature increases, the uptake of atmospheric CO_2 will slow, which is a positive feedback. Scientists estimate that oceanic processes currently take up about one-fourth of CO_2 from fossil fuel burning, but this uptake may slow in the future as warming inhibits overturning of surface waters with the deep ocean, and as ocean acidification and increasing temperature reduce the rate of CO_2 uptake (see chapter 5).

In the terrestrial biosphere, increased atmospheric CO_2 stimulates plant growth, and plants in turn remove CO_2 from the atmosphere, which is a negative feedback. On the other hand, warmer soil temperatures stimulate microbial action that releases CO_2 from the decomposition of dead organic matter, which is a positive feedback. Scientists estimate that terrestrial processes currently take up about one-tenth of CO_2 emissions from fossil fuel burning, the so-called land sink. This represents a larger sink from plant growth partially offset by emissions from land-use change, such as deforestation. What will happen to this sink in the future is highly uncertain. Several studies simulating future climate indicate that this sink may become a source of additional emissions later this century even if deforestation decreases, primarily due to increased release of carbon from soils as temperatures warm beyond a degree or two Celsius.[13]

There are even social feedbacks. For example, rising temperature causes more people to install and use air conditioners. If the resulting increase in electrical consumption resulted in more fossil fuel-generated atmospheric CO_2, that would be another positive feedback. Increasing temperatures and climate impacts, combined with assessments of future risks, may encourage more stringent policies to reduce emissions, which will in turn reduce further intensification of those impacts, a negative social feedback also known as climate mitigation policy.

Accounting for all significant feedback effects entails not only identifying important feedback mechanisms, but also developing a quantitative understanding of how those mechanisms work. Such understanding often includes research at the boundaries of disciplines, including meteorology, atmospheric chemistry, oceanography, biology, and geology; social sciences such economics and sociology; and research on technological development.

Climate Models

Uncertainty in future greenhouse gas emissions and in scientific understanding of the response of the climate system to their influence makes projecting future climate change a complex task. The most sophisticated tools we have are global models of the climate system. Not only can they reproduce global temperature records, as shown in figure 1.3, but the best model results reproduce, although not completely, the detailed geographic patterns of temperature, precipitation, and other climatic variables seen on a regional scale, and can project changes in those patterns given scenarios for future greenhouse gas emissions.

A climate model is a set of mathematical statements describing physical, biological, and chemical processes that determine climate. What must go into a climate model depends on what one wants to learn from it. A few simple equations can give a reasonable range of estimates of the average global warming in response to specified greenhouse forcings. Our estimate above that the Earth's global average temperature in the absence of the greenhouse effect would be colder by about 33 degrees Celsius was based on a simple climate model. In that case, the Earth's surface is treated as a single point, with a simple height-varying atmosphere and no distinction between land and oceans. Simple models have the advantage that their predictions are easily understood on the basis of well-known physical laws. Furthermore, they produce re-

sults quickly and can, therefore, be used to test a wide range of assumptions by changing parameters of the model. More advanced are "multibox" models that treat land, ocean, and atmosphere as separate "boxes" and include flows of energy and matter between these boxes. More sophisticated multibox models may break the atmosphere and ocean into several layers or the Earth into several latitude zones.

Most sophisticated are the complex computer models known as general circulation models (GCMs). Such detailed models can only be run effectively on a limited number of supercomputers around the world. These divide the Earth's surface into a grid that can represent with reasonable accuracy the actual shape of the Earth's land masses and, to a lesser extent, mountains. The atmosphere above and ocean below each surface grid cell are further divided into layers, making the basic unit of the model a small, three-dimensional cell. Properties such as temperature, pressure, and humidity are averaged within each cell. Equations based in physics, chemistry, and biology regulate the various quantities within a cell, and other equations describe the transfer of energy and matter between adjacent cells. The newest models also include processes such as the cycling of carbon between the atmosphere, land, and ocean, the response of the Earth's vegetation to changing conditions and its feedbacks to the climate system, atmospheric chemistry, and the functioning of the cryosphere. Figure 1.7, panel A, displays the typical geographic resolution of the grid representing northern Europe at the time of each of the four IPCC assessment reports and the improvement in resolution (i.e., grid-box size) over this period. Panel B displays the progression in climate models since the 1970s in terms of the processes and com-

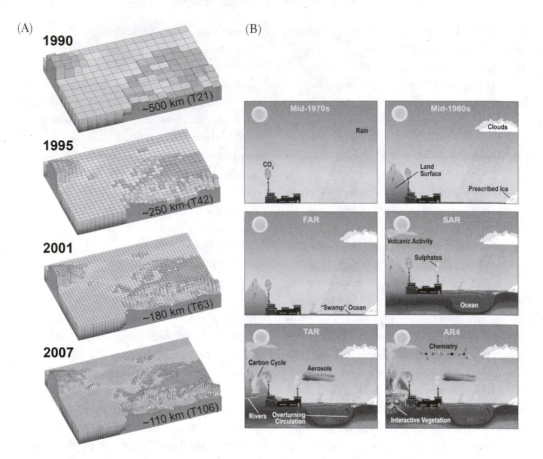

FIGURE 1.7. Panel A: Geographic resolution of GCMs at the time of each of the IPCC assessment reports. Vertical resolution in both atmosphere and ocean models is not shown, but has increased as well, beginning typically with a single-layer "slab" ocean and ten atmospheric layers in 1990 and progressing to about thirty levels in both atmosphere and ocean in 2007. Panel B: The complexity of climate models has increased during the last few decades. The series of pictures displays different features of the modeled world and when they were incorporated. Source: Intergovernmental Panel on Climate Change (IPCC), 2007(a), *Climate Change 2007: The Physical Science Basis. Contribution of Working Group I to the Fourth Assessment Report of the Intergovernmental Panel on Climate Change*, S. Solomon et al. (eds.), Cambridge University Press: Cambridge, United Kingdom.

ponents of the climate system that GCMs incorporate.[14]

Even with the rapid expansion of computational power, the best global climate models are currently limited to a geographic grid-box resolution of roughly 100 kilometers horizontally and 1 kilometer vertically. But climatically important phenomena occur on smaller scales, such as clouds or the substantial thermal differences between cities and surrounding areas. Because all physical, chemical, and biological properties are averaged over a single grid cell, it is impossible to represent these phenomena *explicitly* within a model. But they can be treated *implicitly* with what is called a parametric representation, or "parameterization." A parameterization connects small-scale processes to grid-box averages with semi-empirical rules designed to capture the major interactions between explicitly modeled grid-scale variables and sub-grid-scale processes. For example, a grid cell half covered by scattered clouds might be parameterized as a uniform blockage of somewhat less than half the incoming sunlight. Such an approximation manages not to ignore clouds altogether but doesn't quite handle them correctly. One can imagine that the effects of full sunlight penetrating to the ground in some parts of a grid box while other parts are in full shade might be different from those of a uniform light overcast, even with the same total energy reaching the ground averaged over the grid box.[15]

Model Validation

How can modelers be confident in their model results? How do they know that they have taken into account all climatologically significant processes and that they have satisfactorily parameterized processes whose scales are smaller than their models' grid cells? The answer lies in a variety of model validation techniques, most of which attempt to reproduce known climatic conditions in response to known forcings.

Major volcanic eruptions inject enough dust into the stratosphere to exert a global cooling influence that lasts several years. Such eruptions typically occur once a decade or so, and they constitute natural experiments that can be used to test climate models. The climatic effects of the largest recent major eruption, Mount Pinatubo in 1991, were forecast by a number of climate modeling groups to cool the planet by several tenths of a degree

Celsius for a few years. That is indeed what happened.

Seasonality provides another natural experiment for testing climate models. Winter predictably follows summer, averaging some 15 degrees Celsius colder than summer in the northern hemisphere and 5 degrees Celsius colder in the southern hemisphere (the southern hemisphere variation is smaller because a much larger portion of that hemisphere is water, with a high heat capacity that moderates seasonal temperature variations). Climate models do an excellent job of reproducing the timing and magnitude of the seasonal temperature variations, although the absolute temperatures themselves may not be completely accurate.

Still another way to gain confidence in a model's future climate projections is to model past climates. Starting in 1860 with known climatic conditions, for example, can the model reproduce a reasonable simulation of the temperatures observed during the twentieth century? The "experiments" of figure 1.3 discussed previously provide clear evidence that the answer is "mostly yes" and also help modelers understand what physical processes are significant in determining past climate trends.

Climate models certainly have room for improvement. For example, models are less accurate in representing climatic variations involving precipitation and other aspects of the hydrologic cycle. While temperature changes are driven by large-scale forcing such as greenhouse gas heat-trapping or continental-scale aerosol cooling, precipitation is influenced by complex local/regional processes like the nature of the land surface, proximity to topographical features (e.g., mountains), and temperature differences across the region. All of those interacting smaller-scale processes and drivers are more difficult to include accurately in models. Nevertheless, today's climate models can reproduce recognizable simulations of regional patterns of temperature, precipitation, and other climatic variables. These pattern-based comparisons of models and reality provide further confirmation of the models' broad-scale validity. No one model validation experiment alone is enough to give us high confidence in future climate projections. But considered together, results from the wide range of experiments probing the validity of climate models give considerable confidence that these models are treating the essential climate-determining processes with reasonable accuracy—certainly for temperature trends at continental scales, and with some skills for regional trends and/or precipitation changes in certain regions like high latitude continents and Mediterranean climates of the subtropics.[16] Furthermore, researchers have linked grid-box-scale changes in temperature with observed changes in the lifecycles of plants and animals during the last fifty years (see chapter 3).[17]

Conclusion

We have given a thumbnail sketch of the science of global climate change. The greenhouse effect and its intensification by human-induced emissions of greenhouse gases are well understood and solidly grounded in basic science. Likewise, observed warming is now unequivocal, and many impacts of that warming can already be observed around the world. Nevertheless, the future effects of climate change are characterized by deep uncertainty, compounded by the global scale of the problem and the fact that climate change is not just a scientific topic but also a matter of public and political debate. There are two general sources of uncertainty in projecting future climate change: what we do and how the natural climate system responds. Policy decisions can strongly influence the first source of uncertainty (future emissions), but will have little influence on the second source (climate response to emissions). We cannot know precisely what the severity of impacts will be for a specific trajectory for future emissions, but we can confidently say that the severity will be reduced if emissions are reduced. In very general terms, climate policy is about managing risk: assessing the potential impacts of climate change, judging how likely it is that various impacts will occur, and determining how our policy choices will affect those risks.

Notes

1. Intergovernmental Panel on Climate Change (IPCC), 2007(a), *Climate Change 2007: The Physical Science Basis. Contribution of Working Group I to the Fourth Assessment Report of the Intergovernmental Panel on Climate Change*, S. Solomon et al., eds., Cambridge University Press: Cambridge, United Kingdom, pp. 22–23.

2. Ibid.

3. Ibid.

4. Ibid. See also IPCC, 2007(b), *Climate Change 2007: Impacts, Adaptation, and Vulnerability. Contribution of Working Group II to the Fourth Assessment Report of the IPCC*, M. Parry et al., eds., Cambridge University Press: Cambridge, United Kingdom.

5. Denman, K. L. et al., 2007: Couplings between Changes in the Climate System and Biogeochemistry. In: *Climate Change 2007: The Physical Science Basis. Contribution of Working Group I to the Fourth Assessment Report of the Intergovernmental Panel on Climate Change*, S. Solomon et al., eds., Cambridge University Press, Cambridge, United Kingdom; P. Canadell et al., 2007: Contri-

butions to accelerating atmospheric CO_2 growth from economic activity, carbon intensity, and efficiency of natural sinks. *Proceedings of the National Academy of Sciences* 104, 18866–18870.

6. EPICA community members, 2004: Eight glacial cycles from an Antarctic Ice Core. *Nature* 429, 623–28.

7. IPCC, 2007(a) op. cit.

8. M. Lockwood and C. Fröhlich, 2008: Recent oppositely directed trends in solar climate forcings and the global mean surface air temperature. II. Different reconstructions of the total solar irradiance variation and dependence on response time scale. *Proceedings of the Royal Society* A 464, 1367–85; E. Bard and M. Frank, 2006: Climate change and solar variability: What's new under the sun? *Earth and Planetary Science Letters* 248, 1–14.

9. IPCC, 2007(a) op. cit.

10. Ibid.

11. B. J. Soden and I. M. Held, 2006: An assessment of climate feedbacks in coupled ocean atmosphere models. *Journal of Climate* 19, 3354–60.

12. S. H. Schneider, 1972: Cloudiness as a global climatic feedback mechanism: The effects on the radiation balance and surface temperature of variations in cloudiness. *Journal of the Atmospheric Sciences* 29, 1413–22.

13. P. M. Cox et al., 2000: Acceleration of global warming due to carbon-cycle feedbacks in a coupled climate model. *Nature* 408, 184–87.

14. For a detailed discussion of these models, see IPCC, 2007(a) op. cit.

15. S. H. Schneider and R. E. Dickinson, 1976: Parameterizations of fractional cloud amounts in climate models: The importance of modeling multiple reflections. *J. Appl. Meteorol* 15, 1050–56.

16. IPCC, 2007(a) op. cit.

17. T. I. Root, D. MacMynowski, M. D. Mastrandrea, and S. H. Schneider, 2005: Human-modified temperatures induce species changes: Joint attribution. *Proceedings of the National Academy of Sciences* 102, 7465–69.

Chapter 2

Progress in Detection and Attribution Research

B. D. Santer and T. M. L. Wigley

1. Introduction

In 1988, the Intergovernmental Panel on Climate Change (IPCC) was jointly established by the World Meteorological Organization and the United Nations Environment Programme. The goals of this panel were threefold: to assess available scientific information on climate change, to evaluate the environmental and societal impacts of climate change, and to formulate response strategies. The IPCC's first major scientific assessment, published in 1990, concluded that "unequivocal detection of the enhanced greenhouse effect from observations is not likely for a decade or more."[1]

Six years later, the IPCC's second scientific assessment reached a more definitive conclusion regarding human impacts on climate, and stated that "the balance of evidence suggests a discernible human influence on global climate."[2] This cautious sentence marked a paradigm shift in scientific understanding of the nature and causes of recent climate change. The shift arose for a variety of reasons. Chief among these was the realization that the cooling effects of anthropogenic

sulfate aerosols had partially obscured the warming signal arising from increasing atmospheric concentrations of greenhouse gases (GHGs).[3] A further major area of progress was the increasing use of so-called fingerprint studies, which involve detailed statistical comparisons of modeled and observed climate change patterns.[4,5,6] Fingerprinting relies on the fact that each climate forcing mechanism (e.g., changes in solar irradiance, volcanic dust, sulfate aerosols, or GHG concentrations) has a unique pattern of climate response (see figure 2.1). Fingerprint studies have greatly enhanced our ability to diagnose cause and effect relationships in the climate system.

The third IPCC assessment was published in 2001, and went one step further than its predecessor. It made an explicit statement about the magnitude of the human effect on climate, and concluded that "There is new and stronger evidence that most of the warming observed over the last 50 years is attributable to human activities."[7] This conclusion was based on improved estimates of natural climate variability, better reconstructions of temperature fluctuations during the last millen-

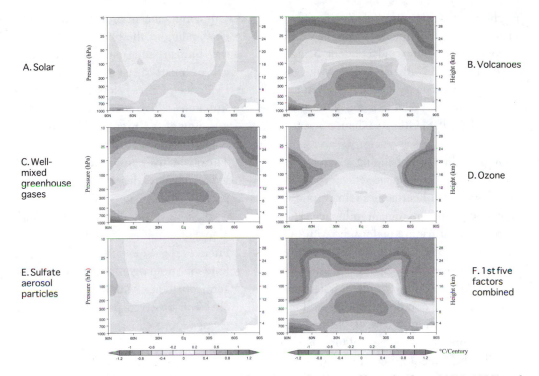

FIGURE 2.1. Zonally averaged temperature changes as a function of latitude (from 90°N–90°S) and height (from 1000 hPa to 10 hPa). Results are from single forcing experiments (A through E) with historical changes in five individual forcings, and from an experiment with simultaneous changes in all five forcings (F). All experiments were performed with the coupled atmosphere-ocean Parallel Climate Model (PCM).[8] Temperature changes are expressed as linear trends in degrees Celsius per century, and were calculated over the period from 1890 to 1999. All results are ensemble means (averages over four individual realizations).

nium, continued warming of the climate system, refinements in fingerprint methods, and the use of results from more (and improved) climate models, driven by more accurate and complete forcing estimates.

This gradual strengthening of scientific confidence in the reality of human influences on global climate continued in the IPCC AR4 report, which stated that "warming of the climate system is unequivocal," and "most of the observed increase in global average temperatures since the mid-twentieth century is *very likely* due to the observed increase in anthropogenic greenhouse gas concentrations"[9]

(meaning >90% probability that the statement is correct). The AR4 justified this increase in scientific confidence on the basis of "... longer and improved records, an expanded range of observations and improvements in the simulation of many aspects of climate and its variability."[10] In its contribution to the AR4, IPCC Working Group II concluded that anthropogenic warming has had a "discernible influence" not only on the physical climate system, but also on a wide range of biological systems.[11]

The fundamental conclusion that human activities have significantly altered not only the chemical composition of Earth's atmosphere, but also the climate system, has been corroborated by other independent bodies, such as the U.S. National Academy of Sciences[12], the Science Academies of eleven nations[13], and the first Synthesis and Assessment Product of the U.S. Climate Change Science Plan.[14]

Despite the overwhelming evidence of pronounced anthropogenic effects on climate, important uncertainties remain in our ability to quantify the human influence. The experiment that we are performing with the Earth's atmosphere lacks a suitable control: we do not have a convenient "undisturbed Earth," which would provide a reference against which we could measure the anthropogenic contribution to climate change. We must therefore rely on numerical models and paleoclimatic evidence [15] to estimate how the Earth's climate might have evolved in the absence of any human "forcing" (see figure 2.2). Such sources of information will always have significant uncertainties.

In the following, we provide a personal perspective on recent developments in the field of detection and attribution (D&A) research — that is, research directed toward detecting significant climate change, and attributing it to a specific cause or causes.[16,17,18,19]

2. Recent Progress in D&A Research

Physical Consistency and Robustness of D&A Results

The IPCC and National Academy findings that human activities are affecting global-scale climate are based on multiple lines of evidence:

- Our continually improving physical understanding of the climate system and the human and natural factors that cause climate to change.
- Evidence from paleoclimate reconstructions, which enables us to place the warming of the twentieth century in a longer-term context.[21,22]
- The qualitative consistency between observed changes in many different aspects of the climate system and model predictions of the changes that should be occurring in response to human influences.[23,24]
- Evidence from rigorous quantitative fingerprint studies, which compare modeled and observed patterns of climate change.

This Chapter focuses on fingerprint evidence. The underlying strategy in fingerprint studies is to search for a model-predicted pattern of climate change (the "fingerprint") in observational data. The fingerprint can be estimated in different ways, but is typically derived from a model experiment in which one or more human factors are varied according to the best-available estimates of their historical changes. Different statistical techniques are then applied to quantify the level of agreement between the fingerprint and observations and between the fingerprint and model estimates of climate noise. This enables researchers to make rigorous tests of competing hypotheses[25] regarding the possible causes of recent climate change.[26,27,28,29]

While early fingerprint work dealt almost exclusively with changes in near-surface or atmospheric temperature, more recent studies have applied fingerprint methods to a range of different variables, such as ocean heat

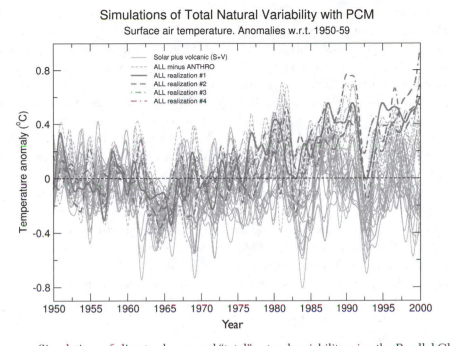

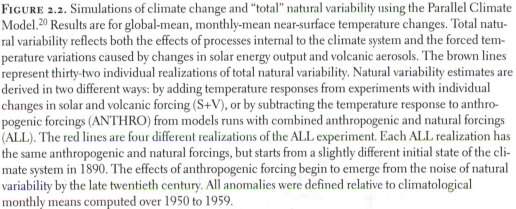

FIGURE 2.2. Simulations of climate change and "total" natural variability using the Parallel Climate Model.[20] Results are for global-mean, monthly-mean near-surface temperature changes. Total natural variability reflects both the effects of processes internal to the climate system and the forced temperature variations caused by changes in solar energy output and volcanic aerosols. The brown lines represent thirty-two individual realizations of total natural variability. Natural variability estimates are derived in two different ways: by adding temperature responses from experiments with individual changes in solar and volcanic forcing (S+V), or by subtracting the temperature response to anthropogenic forcings (ANTHRO) from models runs with combined anthropogenic and natural forcings (ALL). The red lines are four different realizations of the ALL experiment. Each ALL realization has the same anthropogenic and natural forcings, but starts from a slightly different initial state of the climate system in 1890. The effects of anthropogenic forcing begin to emerge from the noise of natural variability by the late twentieth century. All anomalies were defined relative to climatological monthly means computed over 1950 to 1959.

content[30,31], sea-level pressure[32], tropopause height[33], zonal-mean precipitation[34], and atmospheric moisture.[35] The general conclusion is that natural climate variability alone cannot explain the observed climate changes during the second half of the twentieth century. The best statistical explanation of the observed climate changes invariably involves a large human contribution. These results are robust to the processing choices made by different groups, and show a high level of physical consistency across different climate variables. For example, observed atmospheric water vapor increases[36] are physically consistent with increases in ocean heat content[37,38] and near-surface temperature.[39]

There are a number of popular misconceptions about fingerprint evidence. One misconception is that fingerprint studies consider global-mean temperatures only, and thus provide a very poor constraint on the relative contributions of human and natural factors to observed changes.[40] In fact, fingerprint studies rely on information about the detailed spatial structure (and often the combined space-time structure) of observed and simulated climate changes. Complex patterns provide much stronger constraints on the possible contributions of different factors to observed climate changes.[41,42,43]

Another misconception is that model-based estimates of natural internal climate variability ("climate noise") are accepted uncritically in fingerprint studies, and never tested against observations.[44] This is demonstrably untrue. Many fingerprint studies explicitly test whether model estimates of climate noise are realistic, at least on annual and decadal timescales where observational data are of sufficient length to obtain reliable estimates of observed noise.[45,46,47,48]

2.2. The MSU Debate: A Resolution?

For more than a decade, scientists critical of "fingerprint" studies have argued that tropospheric temperature measurements from satellites and weather balloons (radiosondes) show little or no warming of the troposphere over the last several decades, while climate models indicate that that the troposphere should have warmed markedly in response to increases in greenhouse gases (figure 2.1C). This apparent discrepancy between climate model estimates and observations has been used to cast doubt on the reality of a "discernible human influence" on the climate system.[49]

It is unquestionable that satellites have transformed our scientific understanding of the weather and climate of planet Earth. Since 1979, Microwave Sounding Units (MSUs) on polar-orbiting satellites have measured the microwave emissions of oxygen molecules in the atmosphere, which are proportional to atmospheric temperatures. Measurements of microwave emissions made at different frequencies can be used to obtain information about the temperatures of broad atmospheric layers. Most attention has focused on estimates of the temperatures of the lower stratosphere and mid- to upper troposphere (T_4 and T_2, respectively) as well as on a retrieval of lower tropospheric temperatures (T_{2LT}).[50]

The first attempts to obtain climate records from MSU data were made by scientists at the University of Alabama in Huntsville (UAH).[51,52,53] Until recently, the UAH group's analysis of the MSU data suggested that the tropical lower troposphere had cooled since 1979. Concerns regarding the reliability of the MSU-based tropospheric temperature trends were dismissed with the argument that weather balloons also suggested cooling of the tropical troposphere[54], and constituted a completely independent temperature monitoring system.[55,56]

Throughout most of the 1990s, only one group (the UAH group) was actively working on the development of temperature records from MSU data. In 1998, the Remote Sensing Systems (RSS) group in California identified a problem in the UAH data related to the progressive orbital decay and altitude loss over the lifetimes of individual satellites. This introduced a spurious cooling trend in the UAH data.[57] The RSS findings suggested that the lower troposphere had warmed over the satellite era.

The UAH group subsequently discovered two new corrections that approximately compensated for the cooling influence of orbital degradation. The first new correction was related to the effects of orbital drift on the sampling of Earth's diurnal temperature cycle. The second (the so-called instrument body effect) was due to variations in measured microwave emissions arising from changes in the temperature of the MSU instrument itself, caused by changes in the instrument's exposure to sunlight.[58]

Additional research cast doubt on the UAH results. Three separate groups found that the mid- to upper troposphere had warmed markedly during the satellite era[59,60,61,62,63,64], in contrast to the UAH results.[65,66] The UAH group, however, continued to claim close correspondence between their own MSU-based estimates of tropospheric temperature trends and trends derived from radiosondes.[67] This raised critical questions regarding the quality of radiosonde temperature measurements. Were weather balloons an unambiguous "gold standard"?

Recent research indicates that the answer to this question is "no." The temperature sensors carried by radiosondes have changed over time, as has the shielding that protects the sensors from direct solar heating. Solar heating of the sensors can affect the temperature measurements themselves. The introduction of progressively more effective shielding results in less solar heating, thus imparting a non-climatic cooling trend to the daytime measurements.

Sherwood et al.[68] discovered this effect by comparing the radiosonde-based temperature trends based on nighttime ascents (with no solar heating effects) and daytime launches. The former showed pronounced tropospheric warming, while the latter did not. These results were independently confirmed by Randel and Wu.[69] Accounting for the influence of solar heating yielded tropospheric temperature trends that were in better agreement with RSS estimates than with UAH results.[70]

Two papers shed further light on these issues. The first paper was by the RSS group, and described a new MSU retrieval of lower tropospheric temperatures.[71] RSS obtained substantially larger T_{2LT} trends than UAH.[72] Mears and Wentz[73] attributed most of these differences to an error in UAH's method of adjusting for drift in the time of day at which satellites sample the Earth's daily temperature cycle. This error was acknowledged by Christy and Spencer.[74] When the UAH group remedied this problem, however, their lower tropospheric trends increased by much smaller amounts than expected on the basis of the RSS analysis.[75]

The second paper addressed the physics that governs changes in atmospheric temperature profiles. It compared the relationship between surface and tropospheric temperature changes over a wide range of observational and climate model datasets.[76] The focus was on the deep tropics (20°N–20°S), where the UAH and RSS tropospheric temperature trends diverged most markedly. The intent was to investigate whether the simple physics that governs the vertical structure of the tropical atmosphere could be used to constrain the uncertainties in satellite-based trends.

This "simple physics" involves the release of latent heat when moist air rises due to convection and condenses to form clouds. Because of this heat release, tropical temperature changes averaged over large areas (and averaged over sufficient time to damp day-to-day "weather noise") are generally larger in the lower and mid-troposphere than at the surface. This "amplification" behavior is well-known from basic theory[77], observations[78], and climate model results.[79]

The UAH amplification results were puzzling. For month-to-month fluctuations in tropical temperatures, UAH T_{2LT} anomalies were 1.3 to 1.4 times larger than surface temperature anomalies, consistent with models, theory, and other observational datasets. But for decade-to-decade temperature changes, the UAH T_{2LT} trends were smaller than surface trends, implying that the troposphere *damped* surface warming. In contrast, model amplification results were consistent across all timescales considered, despite large differences in model structure, parameterizations, and forcings. The RSS data also showed similar amplification of surface warming on different timescales.

These results have at least two possible explanations.[80,81] The first is that the UAH data are reliable, and different physical mechanisms control the response of the tropical atmosphere to "fast" and "slow" surface temperature fluctuations. Such time-dependent changes in the physics seem unlikely given our present understanding, and mechanisms that might explain such changes have yet to be identified.

A second explanation is that significant inhomogeneities remain in the UAH tropospheric temperature records, leading to residual cooling biases in the UAH long-term trend estimates. This is both a simpler and more plausible explanation given the consistency of amplification results across models and timescales, our theoretical understanding of how the tropical atmosphere should respond to sustained surface heating[82], and the currently large uncertainties in observed tropospheric temperature trends.[83]

The extraordinary claim that the tropical troposphere had cooled since 1979 has not survived rigorous scrutiny. We have learned that uncertainties inherent in satellite estimates of tropospheric temperature change are far larger than originally believed, and now fully encompass the model results.[84] There is no longer a fundamental discrepancy between modeled and observed estimates of tropospheric temperature changes.[85]

2.3. Detecting Anthropogenic Effects at Sub-Global Scales

Because regional-scale climate changes will determine societal impacts, many recent D&A studies have shifted their focus from global to regional scales. One fundamental problem in regional D&A work is that climate noise typically becomes larger when averaged over increasingly finer scales.[86]

To illustrate this, figure 2.3 shows surface temperature changes in an "unforced" control run and in simulations of twentieth-century climate change. Results are averaged over the globe, the Northern Hemisphere, and the Western U.S. It is obvious that averaging over the entire globe produces the largest reduction in climate noise. Signal and noise are generally most easily separable in the global results, although in regions with large signals, the signal-to-noise ratio may actually increase as the spatial scale decreases.[87]

As attention shifts to smaller scales, it becomes more important to obtain reliable information about climate forcings. Some of these forcings are both uncertain and highly variable in space and time.[88,89] Examples include human-induced changes in land surface properties[90] or in the concentrations of carbon-containing aerosols.[91,92] Neglect or inaccurate specification of these forcings can hamper the identification of an anthropogenic fingerprint.

Despite such problems, and despite the

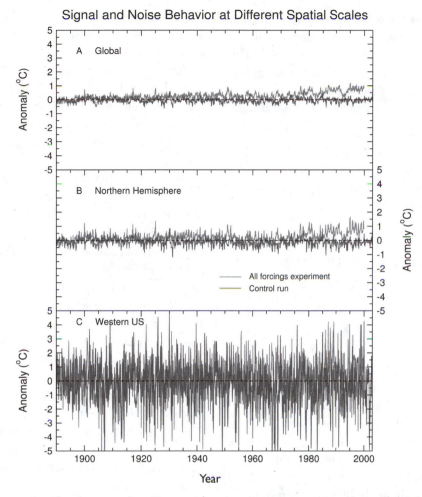

Signal and Noise Behavior at Different Spatial Scales

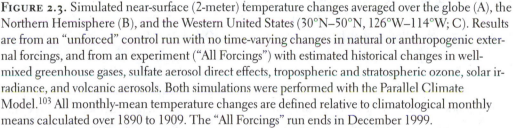

FIGURE 2.3. Simulated near-surface (2-meter) temperature changes averaged over the globe (A), the Northern Hemisphere (B), and the Western United States (30°N–50°N, 126°W–114°W; C). Results are from an "unforced" control run with no time-varying changes in natural or anthropogenic external forcings, and from an experiment ("All Forcings") with estimated historical changes in well-mixed greenhouse gases, sulfate aerosol direct effects, tropospheric and stratospheric ozone, solar irradiance, and volcanic aerosols. Both simulations were performed with the Parallel Climate Model.[103] All monthly-mean temperature changes are defined relative to climatological monthly means calculated over 1890 to 1909. The "All Forcings" run ends in December 1999.

signal-to-noise issues mentioned above, a number of authors have shown that the combined and individual signals of greenhouse gases and sulfate aerosols are now identifiable at continental and sub-continental scales in many different regions around the globe.[93,94,95,96,97,98]

Related work[99,100] suggests that an anthropogenic climate signal has already emerged from the background noise at even smaller spatial scales (at or below 500 km)[101], and may be contributing to regional changes in the distributions of plant and animal species.[102] This new body of regional D&A research provides evidence that we are on the verge of detecting human effects on climate at spatial scales and in factors that are of direct relevance to policy makers.

2.4. Assessing Risks of Changes in Extreme Events

Although we cannot confidently attribute any specific extreme event to human-induced climate change[104], we are capable of making informed scientific statements regarding the influence of human activities on the *likelihood* of extreme events.[105,106] This is an important distinction.

As noted previously, climate models can be used to perform the control experiment (no human effects on climate) that we cannot perform in the real world. Using the "unforced" climate variability from a multi-century control run, it is possible to determine how many times an extreme event of a given magnitude should have been observed in the absence of human interference. The probability of obtaining the same extreme event is then calculated in a perturbed climate—for example, in a model experiment with historical or future increases in anthropogenic forcings, or under some specified change in mean climate.[107] Comparison of the frequencies of extremes in the control and perturbed experiments allows one to make probabilistic statements about how human-induced climate change may have altered the likelihood of the extreme event.[108,109,110] This is sometimes referred to as an assessment of "fractional attributable risk."[111]

Recently, a "fractional attributable risk" study involving the European summer heat wave of 2003 concluded that "there is a greater than 90% chance that over half the risk of European summer temperatures exceeding a threshold of 1.6 K is attributable to human influence on climate."[112] A similar investigation into the causes of sea-surface temperature (SST) increases in Atlantic and Pacific hurricane formation regions found "an 84% chance that external forcing explains at least 67% of observed SST increases in the two tropical cyclogenesis regions."[113] The causes of SST changes in these hurricane breeding grounds are of considerable interest given scientific evidence of a link between rising SSTs and increases in hurricane intensity.[114,115]

These and related studies illustrate that the D&A community has moved beyond analysis of changes in the mean state of the climate, and now applies rigorous statistical methods to the problem of estimating how human activities may alter (or have altered) the probability of occurrence of extreme events. The demonstration of human culpability in changing these risks is likely to have significant implications for the debate on policy responses to climate change.

3. Conclusions

In evaluating how well a novel has been crafted, it is important to look at the internal consistency of the plot. Critical readers examine whether the individual storylines are neatly woven together, and whether the internal logic makes sense.

We can ask similar questions about the "story" contained in observational records of climate change. The evidence from numerous sources (paleoclimate data, rigorous fingerprint studies, and qualitative comparisons of modeled and observed climate changes) shows that the climate system is telling us an internally consistent story about the causes of recent climate change.

Over the last century, we have observed large and coherent changes in many different aspects of Earth's climate. The oceans and land surface have warmed.[116,117,118] Atmospheric moisture has increased.[119,120] Glaciers have retreated over most of the globe.[121,122,123] Sea level has risen.[124] Snow and sea-ice extent have decreased in the Northern Hemisphere.[125,126] The stratosphere has cooled[127], and there are now reliable indications that the troposphere has warmed.[128,129] The height of the tropopause has increased.[130] Individually, all of these changes are consistent with our scientific understanding of how the climate system should be responding to anthropogenic forcing. Collectively, this behavior is inconsistent with the changes that we would expect to occur due to natural variability alone.

There is now compelling scientific evidence that human activity has had a discernible influence on global climate. However, there are still significant uncertainties in our estimates of the size and geographical distribution of the climate changes projected to occur during the twenty-first century.[131] These uncertainties make it difficult for us to assess the magnitude of the mitigation and adaptation problem that faces us and our descendants. The dilemma that confronts us, as citizens and stewards of this planet, is how to act in the face of both hard scientific evidence that our actions are altering global climate and continuing uncertainty in the magnitude of the planetary warming that faces us.

References and Endnotes

1. Houghton, J. T., et al., 1990: *Climate Change. The IPCC Scientific Assessment*. Cambridge University Press, Cambridge, UK, page xxix.

2. Houghton, J. T., et al., 1996: *Climate Change 1995: The Science of Climate Change*. Cambridge University Press, Cambridge, UK, page 4.

3. Wigley, T. M. L., 1989: Possible climatic change due to SO_2-derived cloud condensation nuclei. *Nature*, 339, 365–367.

4. Hasselmann, K., 1979: On the signal-to-noise problem in atmospheric response studies. In: *Meteorology of Tropical Oceans* (Ed. D.B. Shaw). Royal Meteorological Society of London, London, UK, pp. 251–259.

5. Hasselmann, K., 1993: Optimal fingerprints for the detection of time dependent climate change. *Journal of Climate*, 6, 1957–1971.

6. North, G. R., K. Y. Kim, S. S. P Shen, and J. W. Hardin, 1995: Detection of forced climate signals. Part I: Filter theory. *Journal of Climate*, 8, 401–408.

7. Houghton, J. T., et al., 2001: *Climate Change 2001: The Scientific Basis*. Cambridge University Press, Cambridge, UK, page 4.

8. Washington, W. M. et al., 2000: Parallel Climate Model (PCM) control and transient simulations. *Climate Dynamics*, 16, 755–774.

9. IPCC, 2007: Summary for Policymakers. In: *Climate Change 2007: The Physical Science Basis*. Contribution of Working Group I to the Fourth Assessment Report of the Intergovernmental Panel on Climate Change [Solomon, S., D. Qin, M. Manning, Z. Chen, M. Marquis, K. B. Averyt, M. Tignor, and H. L. Miller (eds.)]. Cambridge University Press, Cambridge, United Kingdom and New York, NY, USA.

10. Ibid.

11. IPCC, 2007: Summary for Policymakers. In: *Climate Change 2007: Impacts, Adaptation and Vulnerability*. Contribution of Working Group II to the Fourth Assessment Report of the Intergovernmental Panel on Climate Change [Parry, M. et al. (eds.)]. Cambridge University Press, Cambridge, United Kingdom and New York, NY, USA.

12. NRC (National Research Council), 2001: *Climate Change Science: An Analysis of Some Key Questions*. Board on Atmospheric Sciences and Climate, National Academy Press, Washington D.C., 29 pp.

13. Prior to the Gleneagles G8 summit in July 2005, the Science Academies of eleven nations issued a joint statement on climate change

(www.nasonline.org/site). The statement affirmed the IPCC finding that "most of the warming observed over the last fifty years is attributable to human activities" (Houghton et al., 2001, op cit. 7). The signatories were from the Academia Brasiliera de Ciências, the Royal Society of Canada, the Chinese Academy of Sciences, the Academié des Sciences, France, the Deutsche Akademie der Naturforscher, the Indian National Science Academy, the Accademia dei Lincei, Italy, the Science Council of Japan, the Russian Academy of Sciences, the United Kingdom Royal Society, and the U.S. National Academy of Sciences.

14. Karl, T. R., S. J. Hassol, C. D. Miller, and W. L. Murray (eds.), 2006: *Temperature Trends in the Lower Atmosphere: Steps for Understanding and Reconciling Differences*. A Report by the U.S. Climate Change Science Program and the Subcommittee on Global Change Research. National Oceanic and Atmospheric Administration, National Climatic Data Center, Asheville, NC, USA, 164 pp.

15. Mann, M. E., and P. D. Jones, 2003: Global surface temperatures over the past two millennia. *Geophysical Research Letters*, 30, 1820, doi:10.1029/2003GL017814.

16. Mitchell, J. F. B. et al., 2001: Detection of climate change and attribution of causes. In: *Climate Change 2001: The Scientific Basis*. Contribution of Working Group I to the Third Assessment Report of the Intergovernmental Panel on Climate Change [Houghton, J.T. et al., (eds.)]. Cambridge University Press, Cambridge, United Kingdom and New York, NY, USA, pp. 695–738.

17. IDAG (International Detection and Attribution Group), 2005: Detecting and attributing external influences on the climate system: A review of recent advances. *Journal of Climate*, 18, 1291–1314.

18. Santer, B. D., J. E. Penner, and P. W. Thorne, 2006: How well can the observed vertical temperature changes be reconciled with our understanding of the causes of these changes? In: *Temperature Trends in the Lower Atmosphere: Steps for Understanding and Reconciling Differences*. A Report by the U.S. Climate Change Science Program and the Subcommittee on Global Change Research [Karl, T.R., S.J. Hassol, C.D. Miller, and W.L. Murray (eds.)]. National Oceanic and Atmospheric Administration, National Climatic Data Center, Asheville, NC, USA, pp. 89–108.

19. Hegerl, G. C., F. W. Zwiers, P. Braconnot, N. P. Gillett, Y. Luo, J. A. Marengo Orsini, J. E. Penner, and P. A. Stott, 2007: Understanding and Attributing Climate Change. In: *Climate Change 2007: The Physical Science Basis*. Contribution of Working Group I to the Fourth Assessment Report of the Intergovernmental Panel on Climate Change [Solomon, S., D. Qin, M. Manning, Z. Chen, M. Marquis, K. B. Averyt, M. Tignor, and H. L. Miller (eds.)]. Cambridge University Press, Cambridge, United Kingdom and New York, NY, USA, pp. 663–745.

20. Washington et al., 2000, op cit. 8.

21. National Research Council, 2006: *Surface Temperature Reconstructions for the Last 2,000 Years*. National Academies Press, Washington D.C., 196 pp.

22. A recent assessment of the U.S. National Academy of Sciences concluded that "It can be said with a high level of confidence that global mean surface temperature was higher during the last few decades of the twentieth century than during any comparable period during the preceding four centuries" (National Research Council, 2006, op cit. 21, page 3). The same study also found "it plausible that the Northern Hemisphere was warmer during the last few decades of the twentieth century than during any comparable period over the preceding millennium" (National Research Council, 2006, op cit. 21, pages 3–4).

23. IPCC, 2007, op cit. 9.

24. Examples include increases in surface and tropospheric temperature, increases in atmospheric water vapor and ocean heat content, sea-level rise, widespread retreat of glaciers, etc.

25. An example includes testing the null hypothesis that there has been no external forcing of the climate system against the alternative hypothesis that there has been significant external forcing. Currently, all such hypothesis tests rely on model-based estimates of "unforced" climate variability (also known as natural internal variability). This is the variability that arises solely from processes internal to the climate system, such as interactions between the atmosphere and ocean. The El Niño phenomenon is a well-known example of internal climate noise.

26. Mitchell, J. F. B. et al., 2001, op cit. 16.

27. IDAG, 2005, op cit. 17.

28. Santer, B. D., J. E. Penner, and P. W. Thorne, 2006, op cit. 18.

29. Hegerl, G. C., F. W. Zwiers, P. Braconnot, N. P. Gillett, Y. Luo, J. A. Marengo Orsini, J. E. Penner and P. A. Stott, 2007, op cit. 19.

30. Barnett, T.P. et al., 2005: Penetration of human-induced warming into the world's oceans. *Science*, 309, 284–287.

31. Pierce, D. W. et al., 2006: Anthropogenic warming of the oceans: Observations and model results. *Journal of Climate*, 19, 1873–1900.

32. Gillett, N. P., F. W. Zwiers, A. J. Weaver, and P. A. Stott, 2003: Detection of human influence on sea level pressure. *Nature*, 422, 292–294.

33. Santer, B. D. et al., 2003: Contributions of anthropogenic and natural forcing to recent tropopause height changes. *Science*, 301, 479–483.

34. Zhang, X. et al., 2007: Detection of human influence on 20th century precipitation trends. *Nature*, 448, 461–465.

35. Santer, B. D., et al., 2007: Identification of human-induced changes in atmospheric moisture content. *Proceedings of the National Academy of Sciences*, 104, 15248–15253.

36. Trenberth, K. E., J. Fasullo, and L. Smith, 2005: Trends and variability in column-integrated atmospheric water vapor. *Climate Dynamics*, 24, doi:10.1007/s00382-005-0017-4.

37. Levitus, S., J. I. Antonov, and T. P. Boyer, 2005: Warming of the world ocean, 1955–2003. *Geophysical Research Letters*, 32, L02604, doi:10.1029/2004GL021592.

38. Domingues, C.M., et al., 2008: Rapid upper-ocean warming helps explain multi-decadal sea-level rise, *Nature*, 453, 1090–1093.

39. Jones, P. D., M. New, D. E. Parker, S. Martin, and I. G. Rigor, 1999: Surface air temperature and its changes over the past 150 years. *Reviews of Geophysics*, 37, 173–199.

40. The argument here is that some anthropogenic "forcings" of climate (particularly the so-called indirect forcing caused by the effects of anthropogenic aerosols on cloud properties) are highly uncertain, so that many different combinations of these factors could yield the same global-mean changes. While this is a valid concern for global-mean temperature changes, it is highly unlikely that different combinations of forcing factors could produce the same complex spatio-temporal *patterns* of climate change (see figure 3.1 figure 2.1).

41. For example, some researchers have argued that most of the observed near-surface warming during the twentieth century is attributable to an overall increase in solar irradiance. The effect of such an increase would be to warm most of the atmosphere (from the Earth's surface through the stratosphere; see figure 2.1A). Such behavior is not seen in observations. While temperature measurements from satellites and radiosondes do show warming of the troposphere, they also indicate that the stratosphere has cooled over the past two to four decades (Karl, T. R., S. J. Hassol, C. D. Miller, and W. L. Murray (eds.), 2006, op cit. 14). Stratospheric cooling is fundamentally inconsistent with a "solar forcing only" hypothesis of observed climate change, but *is* consistent with simulations of the response to anthropogenic GHG increases and ozone decreases (figures 2.1C and D, respectively). The possibility of a large solar forcing effect has been further weakened by recent research indicating that changes in solar luminosity on multi-decadal timescales are likely to be significantly smaller than previously thought (Foukal, P., G. North, and T. M. L. Wigley, 2004, op cit. 42; Foukal, P., C. Fröhlich, H. Spruit, and T. M. L. Wigley, 2006, op cit. 43).

42. Foukal, P., G. North, and T. M. L. Wigley, 2004: A stellar view on solar variations and climate. *Science*, 306, 68–69.

43. Foukal, P., C. Fröhlich, H. Spruit, and T. M. L. Wigley, 2006: Physical mechanisms of solar luminosity variation, and its effect on climate. *Nature*, 443, 161–166.

44. In order to assess whether observed climate changes over the past century are truly unusual, we require information on the amplitude and structure of climate noise on timescales of a century or longer. Unfortunately, direct instrumental measurements are of insufficient length to provide such information. This means that D&A studies must rely on decadal- to century-timescale noise estimates from climate model control runs.

45. Allen, M. R., and S. F. B. Tett, 1999: Checking for model consistency in optimal fingerprinting. *Climate Dynamics*, 15, 419–434.

46. Thorne, P. W. et al., 2003: Probable causes of late twentieth century tropospheric temperature trends. *Climate Dynamics*, 21, 573–591.

47. Santer, B.D. et al., 2006: Causes of ocean surface temperature changes in Atlantic and Pacific tropical cyclogenesis regions. *Proceedings of the National Academy of Sciences*, 103, 13905–13910.

48. AchutaRao, K. M., M. Ishii, B. D. Santer, P. J. Gleckler, K. E. Taylor, T. P. Barnett, D. W. Pierce, R. J. Stouffer, and T. M. L. Wigley, 2007: Simulated and observed variability in ocean temperature and heat content. *Proceedings of the National Academy of Sciences*, 104, 10768–10773.

49. See, for example, an op-ed by James Schlesinger (former Secretary of Energy, Secretary of Defense, and Director of the CIA) in the January 22, 2004 edition of the Los Angeles Times. Mr. Schlesinger noted that ". . . the theory that increasing concentrations of greenhouse gases like carbon dioxide will lead to further warming is at least an oversimplification. It is inconsistent with the fact that satellite measurements over 35 years [sic] show no significant warming in the lower atmosphere, which is an essential part of the global-warming theory."

50. The designations T_4 and T_2 reflect the fact that the original MSU measurements employed MSU channels 4 and 2 (subsequently replaced by equivalent data from other channels in the latest Advanced Microwave Sounding Units). The bulk of the microwave emissions monitored by channel 4 is from roughly 14 to 29 km above Earth's surface (150 to 15 hPa). Channel 2 primarily samples emissions from the surface to 18 km (75 hPa). The T_{2LT} retrieval is constructed using the outer and inner "scan angles" of channel 2, and is a measure of temperatures from the surface to 8 km (350 hPa). For further details of the atmospheric layers sampled by MSU, see Karl, T. R., S. J. Hassol, C. D. Miller, and W. L. Murray (eds.), 2006, op cit. 14.

51. Spencer, R. W., and J. R. Christy, 1990: Precise monitoring of global temperature trends from satellites. *Science*, 247, 1558–1562.

52. Spencer, R. W., and J. R. Christy, 1992: Precision and radiosonde validation of satellite grid-point temperature anomalies, Part I, MSU channel 2. *Journal of Climate*, 5, 847–857.

53. Spencer, R. W., and J. R. Christy, 1992: Precision and radiosonde validation of satellite grid-point temperature anomalies, Part II, A tropospheric retrieval and trends during 1979–90. *Journal of Climate*, 5, 858–866.

54. Christy, J. R., R. W. Spencer, W. B. Norris, W. D. Braswell, and D. E. Parker, 2003: Error estimates of version 5.0 of MSU-AMSU bulk atmospheric temperatures. *Journal of Atmospheric and Oceanic Technology*, 20, 613–629.

55. The true degree of independence is uncertain. In their early work, the UAH group apparently relied on radiosonde-derived temperatures to choose between different possible "adjustment pathways" in the complex process of correcting MSU data for known inhomogeneities (Christy, J. R., R. W. Spencer, and E. S. Lobl, 1998, op cit. 56).

56. Christy, J. R., R. W. Spencer, and E. S. Lobl, 1998: Analysis of the merging procedure for the MSU daily temperature time series. *Journal of Climate*, 11, 2016–2041.

57. Wentz, F. J., and M. Schabel, 1998: Effects of orbital decay on satellite-derived lower-tropospheric temperature trends. *Nature*, 394, 661–664.

58. Christy, J .R., R. W. Spencer, and W. D. Braswell, 2000: MSU Tropospheric temperatures: Data set construction and radiosonde comparisons. *Journal of Atmospheric and Oceanic Technology*, 17, 1153–1170.

59. The three groups involved were at RSS (Mears, C. A., M. C. Schabel, and F. W. Wentz, 2003, op cit. 60), the University of Maryland (UMD) (Vinnikov, K. Y., and N. C. Grody, 2003, op cit. 61; Vinnikov, K. Y. et al., 2006; op cit. 62), and the University of Washington/NOAA (UW) (Fu, Q., C. M. Johanson, S. G. Warren, and D. J. Seidel, 2004, op cit. 63). The RSS and UMD groups independently reprocessed the raw T_2 data used by UAH, using different adjustment procedures to account for orbital drift and instrument body effects. In the RSS and UMD analyses, the mid- to upper troposphere warmed by 0.1 to 0.2°C/decade over the satellite era. These values were substantially larger than the satellite era T_2 trends estimated by UAH, which ranged from 0.02 to 0.04°C/decade. The UW group applied a statistical approach to quantify and remove the influence of stratospheric cooling on MSU T_2 data (Fu, Q., C. M. Johanson, S. G. Warren, and D. J. Seidel, 2004, op cit. 63; Fu, Q., and C. M. Johanson, 2005, op cit. 64). This enhanced the estimated warming of the mid- to upper troposphere in both the UAH and RSS datasets. The approach also highlighted the physically implausible behavior of the UAH group's tropical T_2 and T_{2LT} trends, which showed sustained warming and cooling (respectively) of adjacent layers of the troposphere (Fu, Q., and C.M. Johanson, 2005, op cit. 64).

60. Mears, C. A., M. C. Schabel, and F. W.

Wentz, 2003: A reanalysis of the MSU channel 2 tropospheric temperature record. *Journal of Climate*, 16, 3650–3664.

61. Vinnikov, K. Y., and N. C. Grody, 2003: Global warming trend of mean tropospheric temperature observed by satellites. *Science*, 302, 269–272.

62. Vinnikov, K. Y. et al., 2006: Temperature trends at the surface and the troposphere. *Journal of Geophysical Research*, doi:10.1029/2005jd006392.

63. Fu, Q., C. M. Johanson, S. G. Warren, and D. J. Seidel, 2004: Contribution of stratospheric cooling to satellite-inferred tropospheric temperature trends. *Nature*, 429, 55–58.

64. Fu, Q., and C. M. Johanson, 2005: Satellite-derived vertical dependence of tropical tropospheric temperature trends. *Geophysical Research Letters*, 32, L10703, doi:10.1029/2004GL022266.

65. Christy, J. R., R. W. Spencer, W. B. Norris, W. D. Braswell, and D. E. Parker, 2003, op cit. 54.

66. Christy, J. R., R. W. Spencer, and W. D. Braswell, 2000, op cit. 58.

67. Christy, J. R., R. W. Spencer, W. B. Norris, W. D. Braswell, and D. E. Parker, 2003, op cit. 54.

68. Sherwood, S. C., J. Lanzante, and C. Meyer, 2005: Radiosonde daytime biases and late 20th century warming. *Science*, 309, 1556–1559.

69. Randel, W. J., and F. Wu, 2006: Biases in stratospheric temperature trends derived from historical radiosonde data. *Journal of Climate*, 19, 2094–2104.

70. Sherwood, S. C., J. Lanzante, and C. Meyer, 2005, op cit. 68.

71. Mears, C. A., and F. W. Wentz, 2005: The effect of diurnal correction on satellite-derived lower tropospheric temperature. *Science*, 309, 1548–1551.

72. The RSS estimate of the global-mean T_{2LT} trend over 1979 to 2003 was $0.1°C/decade$ warmer than the UAH estimate (0.193 versus 0.087°C/decade, respectively). Differences were even larger in the tropics, where the lower troposphere cooled in the UAH data (by 0.015°C/decade), but warmed markedly in the RSS analysis (by 0.189°C/decade) (Mears, C. A., and F. W. Wentz, 2005, op cit. 71).

73. Mears, C. A., and F. W. Wentz, 2005, op cit. 71.

74. Christy, J. R., and R. W. Spencer, 2005:

Correcting temperature data sets. *Science*, 310, 972.

75. The UAH global-mean T_{2LT} trend over December 1978 to July 2005 increased by only 0.035°C/decade (to 0.123°C/decade). This change was stated to be "within our previously published error margin of ±0.05K/decade" (Christy, J. R., and R. W. Spencer, 2005, op cit. 74).

76. Santer, B. D. et al., 2005: Amplification of surface temperature trends and variability in the tropical atmosphere. *Science*, 309, 1551–1556.

77. Hess, S. L., 1959. *Introduction to Theoretical Meteorology*. Holt, Rinehart and Winston, New York, 362 pp.

78. Wentz, F. J., and M. Schabel, 2000: Precise climate monitoring using complementary satellite data sets. *Nature*, 403, 414–416.

79. Manabe, S., and R.J. Stouffer, 1980: Sensitivity of a global climate model to an increase of CO_2 concentration in the atmosphere. *Journal of Geophysical Research*, 85, 5529–5554.

80. Santer, B. D., J. E. Penner, and P. W. Thorne, 2006, op cit. 18.

81. Wigley, T.M.L. et al., 2006: Executive Summary. In: *Temperature Trends in the Lower Atmosphere: Steps for Understanding and Reconciling Differences*. A Report by the U.S. Climate Change Science Program and the Subcommittee on Global Change Research [Karl, T. R., S. J. Hassol, C. D. Miller, and W. L. Murray (eds.)]. National Oceanic and Atmospheric Administration, National Climatic Data Center, Asheville, NC, USA, pp. 1–14.

82. Prolonged surface warming should destabilize tropical temperature profiles, thus producing the conditions necessary for moist convection and readjustment of atmospheric temperatures to a moist adiabatic lapse rate (Hess, S. L., 1959, op cit. 77).

83. Karl, T. R., S. J. Hassol, C. D. Miller, and W. L. Murray (eds.), 2006, op cit. 14.

84. Santer, B. D., et al., 2008: Consistency of modeled and observed temperature trends in the tropical troposphere. *International Journal of Climatology*, 28, 1703–1722.

85. Karl, T. R., S. J. Hassol, C. D. Miller, and W. L. Murray (eds.), 2006, op cit. 14.

86. Wigley, T. M. L., and P. D. Jones, 1981: Detecting CO_2-induced climatic change. *Nature*, 292, 205–208.

87. Ibid.

88. Ramaswamy, V. et al., 2001: Radiative forcing of climate change. In: *Climate Change 2001: The Scientific Basis*. Contribution of Working Group I to the Third Assessment Report of the Intergovernmental Panel on Climate Change [Houghton, J. T. et al., (eds.)]. Cambridge University Press, Cambridge, United Kingdom and New York, NY, USA, pp. 349–416.

89. NRC (National Research Council), 2005: *Radiative Forcing of Climate Change: Expanding the Concept and Addressing Uncertainties*. Board on Atmospheric Sciences and Climate, National Academy Press, Washington D.C., 168 pp.

90. Feddema, J. et al., 2005: A comparison of a GCM response to historical anthropogenic land cover change and model sensitivity to uncertainty in present-day land cover representations. *Climate Dynamics*, 25, 581–609.

91. Penner, J. E. et al., 2001: Aerosols, their direct and indirect effects. In: *Climate Change 2001: The Scientific Basis*. Contribution of Working Group I to the Third Assessment Report of the Intergovernmental Panel on Climate Change [Houghton, J. T. et al. (eds.)]. Cambridge University Press, Cambridge, United Kingdom and New York, NY, USA, pp. 289–348.

92. Menon, S., J. Hansen, L. Nazarenko, and Y. F. Luo, 2002: Climate effects of black carbon aerosols in China and India. *Science*, 297, 2250–2253.

93. Stott, P. A., 2003: Attribution of regional-scale temperature changes to anthropogenic and natural causes. *Geophysical Research Letters*, 30, doi: 10.1029/2003GL017324.

94. Zwiers, F. W., and X. Zhang, 2003: Toward regional-scale climate change detection. *Journal of Climate*, 16, 793–797.

95. Karoly, D. J. et al., 2003: Detection of a human influence on North American climate. *Science*, 302, 1200–1203.

96. Min, S.-K., A. Hense, and W.-T. Kwon, 2005: Regional-scale climate change detection using a Bayesian detection method. *Geophysical Research Letters*, 32, L03706, doi:10.1029/2004GL021028.

97. Barnett, T. P. et al., 2008: Human-induced changes in the hydrology of the western U.S. *Science*, 319, 1080–1083.

98. Bonfils, C., P. B. Duffy, B. D. Santer, T. M. L. Wigley, D. B. Lobell, T. J. Phillips, and C. Doutriaux, 2008: Identification of external influences on temperatures in California. *Climatic Change*, 87, 43–55.

99. Karoly, D. J., and Q. Wu, 2005: Detection of regional surface temperature trends. *Journal of Climate*, 18, 4337–4343.

100. Knutson, T. R. et al., 2006: Assessment of twentieth-century regional surface temperature trends using the GFDL CM2 coupled models. *Journal of Climate*, 19, 1624–1651.

101. Knutson et al. (Knutson, T. R. et al., 2006, op cit. 100) state that their "regional results provide evidence for an emergent anthropogenic warming signal over many, if not most, regions of the globe."

102. Root, T. L., D. P. MacMynowski, M. D. Mastrandrea, and S. H. Schneider, 2005: Human-modified temperatures induce species changes: Joint attribution. *Proceedings of the National Academy of Sciences*, 102, 7465–7469.

103. Washington, W. M. et al., 2000, op cit. 8.

104. Allen, M.R., 2003: Liability for climate change. *Nature*, 421, 891–892.

105. Wigley, T. M. L., 1988: The effect of changing climate on the frequency of absolute extreme events. *Climate Monitor*, 17, 44–55.

106. Meehl, G. A., and C. Tebaldi, 2004: More intense, more frequent, and longer lasting heat waves in the 21st century. *Science*, 305, 994–997.

107. Stott, P. A., D. A. Stone, and M. R. Allen, 2004: Human contribution to the European heatwave of 2003. *Nature*, 423, 610–614.

108. Ibid.

109. Meehl, G. A., and C. Tebaldi, 2004, op cit. 106.

110. Tebaldi, C., K. Hayhoe, J. M. Arblaster, and G. A. Meehl, 2006: Going to the extremes: An intercomparison of model-simulated historical and future changes in extreme events. *Climatic Change*, 79, 185–211.

111. Stott, P. A., D. A. Stone, and M. R. Allen, 2004, op cit. 107.

112. Stott, P. A., D. A. Stone, and M. R. Allen, 2004, op cit. 107.

113. Santer, B. D. et al., 2006, op cit. 47.

114. Emanuel, K., 2005: Increasing destructiveness of tropical cyclones over the past 30 years. *Nature*, 436, 686–688.

115. Hoyos, C. D., P. A. Agudelo, P. J. Webster, and J. A. Curry, 2006: Deconvolution of the factors contributing to the increase in global hurricane intensity. *Science*, 312, 94–97.

116. Barnett, T. P. et al., 2005, op cit. 30.

117. Domingues, C. M., et al., 2008, op cit. 38.

118. Jones, P. D, and A. Moberg, 2003: Hemispheric and large scale surface air temperature variations: an extensive revision and an update to 2001. *Journal of Climate*, 16, 206–223.

119. Santer, B. D., et al., 2007, op cit. 35.

120. Trenberth, K. E., J. Fasullo, and L. Smith, 2005, op cit. 36.

121. Arendt, A. A. et al., 2002: Rapid wastage of Alaska glaciers and their contribution to rising sea level. *Science*, 297, 382–386.

122. Paul, F., A. Kaab, M. Maisch, T. Kellenberger, and W. Haeberli, 2004: Rapid disintegration of Alpine glaciers observed with satellite data. *Geophysical Research Letters*, 31, L21402,doi:10.1029/2004GL020816.

123. Meier, M. F., et al., 2007: Glaciers dominate eustatic sea-level rise in the 21st century. *Science*, 317, 1064–1067.

124. Cazenave, A., and R. S. Nerem, 2004: Present-day sea level change: Observations and causes. *Reviews of Geophysics*, 42, RG3001, doi:10.1029/2003RG000139.

125. Vinnikov, K. Y. et al., 1999: Global warming and Northern Hemisphere sea ice extent. *Science*, 286, 1934–1937.

126. Stroeve, J., et al., 2008: Arctic sea ice plummets in 2007. *EOS*, 89, 2, 13–14.

127. Ramaswamy, V. et al., 2006: Anthropogenic and natural influences in the evolution of lower stratospheric cooling. *Science*, 311, 1138–1141.

128. Karl, T. R., S. J. Hassol, C. D. Miller, and W. L. Murray (eds.), 2006, op cit. 14.

129. Trenberth K.E., et al., 2007: Observations: Surface and atmospheric climate change. In: *Climate Change 2007: The Physical Science Basis*. Contribution of Working Group I to the Fourth Assessment Report of the Intergovernmental Panel on Climate Change. [Solomon S., Qin D., Manning M., Chen Z., Marquis M., Averyt K. B., Tignor M., Miller H. L. (eds.)]. Cambridge University Press, Cambridge, United Kingdom and New York, NY, USA, pp. 235–336.

130. Santer, B. D. et al., 2003, op cit. 33.

131. IPCC, 2007, op cit. 9.

Chapter 3

Wild Species and Extinction

TERRY L. ROOT AND ELIZABETH S. GOLDSMITH

Climate has long been recognized as a primary driver of biotic systems.[1] It plays a central role in determining which types of species inhabit which parts of the world.[2] Between 1750 and 2007, the average global temperature increased by around 0.75 degree Celsius (around 1.3 degrees Fahrenheit).[3] Human activities have been linked to the rapid warming.[4] The rate of warming is expected to continue to escalate throughout the twenty-first century, increasing by a minimum of 1.1 degrees Celsius and potentially rising 6.4 degrees Celsius or more.[5] Even with a total increase of 1.1 degrees Celsius, many species will exhibit significant changes, making climatic considerations fundamental in a discussion of the status and trends of ecological conditions.

Until relatively recently, concerns about declining species densities focused primarily on habitat modification, overharvesting, invasive species, and other human-caused changes. Since the late 1990s, researchers have found that many species are also changing as a result of climate change: moving poleward, for example, and blooming earlier in the spring.[6] Meta-analyses of studies from

around the globe have found wild species exhibiting consistent responses to global warming.[7] Joint attribution research shows that the regional warming to which species respond is due in part to human activities.[8] Changes to species are already occurring, and temperatures are expected to escalate, pushing an increasing number of species toward extinction.[9] Coordinated policy interventions at many different scales are imperative. Research-informed, strategically comprehensive conservation programs are needed to stave off accelerating rates of extinctions.

What We Know

Plants and animals survived the less-rapid warming of the post–Ice Age transition by shifting their ranges poleward and up in elevation, similar to the range shifts we have witnessed more recently.[10] Also as contemporary species do, prehistoric species moved differentially, resulting in a reorganization of species in natural communities and ultimately developing into the species communities of today.[11] The past is of limited use, however, in understanding wild species' responses to unprecedented present changes. Rapidly increasing temperatures now confront species, as do many other human-induced challenges, such as obstacle courses of industrial parks, roads, and settlements blocking migration. The combined effects of rapid temperature increases and other anthropogenic changes collectively affect the flora and fauna of our planet more strongly—synergistically—than if each of these disturbances occurred alone.

Rapid Temperature Change and Species

Climate is a major factor influencing location and shape of geographical distributions of species.[12] We know that wild species are already showing significant changes in response to rapid warming, such as shifts in ranges and in the timing of spring activities. Effective adaptation policies must therefore take into account ecological responses to rapid climate changes, as well as to non-climatic forces such as land-use change. Unprecedented types and rates of ecological changes around the globe, however, challenge our ability to design adaptive-management plans that will ensure species survival. Such plans are necessary to help reduce the number of species extinctions, thereby helping to maintain ecosystem biodiversity.[13]

RANGE SHIFTS

Species have physiological temperature thresholds and must avoid habitats with temperatures beyond those thresholds. With rapid global warming, many species need to shift their ranges in order to inhabit areas with tolerable temperatures. Species in a wide variety of taxonomic groups and in numerous geographical areas have moved poleward and upward in elevation during the last century.[14] For example, spittlebugs (nymphs of the superfamily *Cercopoidea*) along the California coast, the Mountain pine beetle (*Dendroctonus ponderosae*) in Canada, and zooplankton—an essential part of almost all marine foodwebs—in the North Atlantic have all shifted measurably north, while the grey-headed flying-fox (*Pteropus poliocephalus*), a fruit bat in Australia, has expanded its southern range by 750 kilometers since the 1930s.[15] Species with range boundaries at the poleward edge of a continent or near the top of mountains, however, need human intervention if they are to avoid extinction, as natural dispersal is not possible without moving through habitat outside of their physiological thresholds.

PHENOLOGY

Common changes in the phenology—timing—of spring activities include earlier arrival of migrant birds, chorusing of frogs, and flowering of plants. Many recent studies reveal a consistent pattern of phenological change across the northern hemisphere in response to warming temperatures, and satellite data unequivocally confirm earlier greening in the northern hemisphere.[16] For all species showing springtime phenological changes, the average number of days changed over the last thirty years of the twentieth century was around 15.5 days, or about 5 days per decade.[17] The increase in global average temperatures over that time period was around 0.4 degree Celsius.[18] Given that the predicted global-average-temperature increase for the twenty-first century is between approximately 1.1 and 6.4 degrees Celsius, we might expect major phenological changes, which will highly likely cause major ecological changes.

Disruption of Natural Communities

Extension of species ranges into new habitats may significantly change the interactions in the communities into which the shifting species are inserting themselves. The invading species can cause significant disruptions, as seen in Canada's pine forests in western Canada. The number of trees killed by the mountain pine beetle, due to the combination of fire suppression with winter temperatures no longer cold enough to kill the beetle seasonally, is staggering. Within a decade, the volume of mature timber killed went from less than 5 million cubic meters in 1996 to more than 400 million cubic meters in 2005 (see figure 3.1). The impact of this beetle has affected not only the pines' survival, but also the welfare of the birds and other species throughout the area.[19]

The asynchronous changes of species within natural communities present addi-

tional challenges to species, disrupting biotic interactions based on timing. One casualty of a separating predator-prey relationship is the pied flycatcher, a migratory bird in Europe. Populations of pied flycatchers have already suffered a 15 percent decline in breeding success. They are not significantly changing their arrival time to their breeding grounds, but the caterpillars they rely upon to feed their young have shifted their timing to be significantly earlier. Hence, the flycatchers miss the peak-food-supply period, resulting in failed breeding attempts.[20]

Changes in Physical Structure and Genetics

Compared with other types of responsive change, fewer studies focus on shifts in morphology and genetics. Nevertheless, both are found to relate to temperature change. For example, labeled woodrats are effectively "paleothermometers" because changes in their body sizes are closely aligned with changes in temperature.[21] Genetic changes are also unexpectedly evident over this relatively short

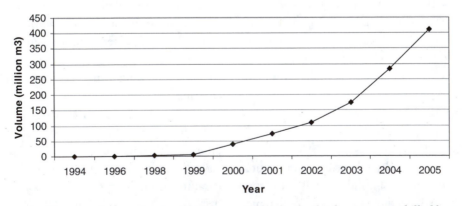

Figure 3.1. The cumulative volume of timber created by the death of mature trees killed by mountain pine beetle. Source: Nelson, H. (2007). Does a Crisis Matter? Forest Policy Responses to the Mountain Pine Beetle Epidemic in British Columbia. *Canadian Journal of Agricultural Economics* 55:459–470.

period of rapid global warming. For example, the genetically controlled photoperiod of the pitcher-plant mosquito (*Wyeomyia smithii*) has changed over the last thirty years.[22]

EXTINCTION

Highly disturbed ecosystems have lowered resistance to nonlinear, dynamic combinations of changes, especially those presented by extreme conditions, as predicted by global climate change scenarios.[23] For species in those systems, extinction is the ultimate irreversible outcome. Unless climate change and other disturbances, such as habitat loss, can be slowed, and unless we can enact well-designed adaptation policies and management plans, widespread extinctions are expected.[24] The warmer the planet gets, the more extinctions we can anticipate.

Non-Climatic Forces

The ability to cope with rapid climate change is severely compromised by the added stress of ongoing non-climate-related changes, such as habitat change, overharvesting, pollution, and exotic species; synergistic effects can make a combination of stressors larger than the sum of its parts. One way to manage conservation of species under climate change is to abate the affects of non-climate stressors.

HABITAT CHANGE

Urban and suburban development and agricultural production have divided the landscape into habitat patches of varying size. These divisions often disrupt interactions among organisms, block or detour migration pathways, and alter availability of resources.

Establishing or enhancing migration corridors to enable species movement is one management option for this problem, requiring great central coordination and presenting its own problems.[25]

OVERHARVESTING/OVEREXPLOITATION

Humans have historically used plants and animals, frequently unsustainably, to enhance their quality of life. Unsustainable biological-resource extraction is rooted in three main problems: (1) high demand for the resource creates a substantial harvesting profit; (2) high short-term return on investment encourages more people to harvest; and (3) development of a large-scale, cost-effective method of harvesting rapidly depletes the resource.[26] Besides depleting overharvested species, impacts can cascade onto other species in the same community.[27] In addition, habitat destruction due to extraction can have negative effects on all species in the area, which may result in irreversible impacts on species survival and ecosystem functions.

POLLUTION

Humans produce chemicals that often create injurious environments within which species must live. These include chemicals emitted into the environment both by design, such as fertilizers, and as by-products, such as carbon monoxide. All of these can adversely affect species and ecosystems.[28]

INTRODUCTION OF EXOTIC SPECIES

Humans introduce exotic species into unfamiliar habitats frequently, both intentionally and accidentally. For example, kudzu (*Pueraria montana*) was brought to the United

States from Japan to help stop erosion along highways, but has moved into forests and is smothering trees.[29] Zebra mussels (*Dreissena polymorpha*) were accidentally introduced in ballast water emptied into the Great Lakes, where they are causing massive declines in native species.[30] Introduced species can also carry novel diseases or parasites that can have devastating repercussions for native species.[31]

Synergistic Effects: Non-Climate and Climate Drivers

Currently, many plants and animals need to adapt to the rapidly changing climate, while simultaneously coping with other human-created stresses. The synergistically amplified impacts of climate change and other human-induced changes create a very different challenge in the twenty-first century than those faced by species during prehistoric warming.

These synergistic issues raise management problems of anticipating and responding to all global-change risk. For example, some propose to address the combined problem of temperature-motivated range shifts through patchy habitats by establishing interconnected nature reserves.[32] Constructing species-appropriate corridors or other management solutions will require resources, understanding, expertise, and political will, all of which may not be readily available.

Meta-Analyses and Joint Attribution

Recognizing how species tend to adapt to environmental change and understanding obstacles to that adaptation are key to understanding how wild species might respond to rapid climate change. How do we know,

though, that wild species are currently responding to climate change—and in particular to climate change attributed to human activity? Meta-analyses and assessment of joint attribution have demonstrated a strong signal of plant and animal responses to climate change during the last fifty to sixty years, a signal particularly important given the expected escalating rate of warming.

Meta-Analyses

Meta-analyses provide statistical methods to determine whether independent studies demonstrate consistent results. Various meta-analyses have found that significant impacts of recent climatic warming are discernible in the changes exhibited by plants and animals around the globe.[33] For example, more than 80 percent of species showing shifts in phenology, range, genetics, morphology, or other traits are shifting in the direction expected for a warming climate, based on species' known physiological constraints.[34] Using modeled temperatures as a more direct demonstration of the likely relationship between human-induced climate forcing and environmental systems change establishes direct associations, through a two-step process called "joint attribution."[35]

Human-Induced Warming and Changing Species: Joint Attribution

Joint attribution provides a method to assess the relationships among human emissions, regional warming, and species changes. The two attribution steps show that (1) human activities have a discernible impact on regional temperature change, and (2) plants and animals exhibit discernible impacts of regional temperature change.[36] Jointly, these imply a discernible effect of human-induced regional warming on plants and animals.[37]

Studies addressing spring phenological changes in flora and fauna around the northern hemisphere have been examined to test for a discernible human impact.[38] Associations between the timing of each species' spring event (e.g., flowers blooming) and *observed* local spring temperatures were quantified using linear regression. The frequency distribution of the number of species with similar regression coefficients (R^2) is plotted (see figure 3.2: dark histograms; N=83 species). Regressions were also calculated between the changes in spring events of species and

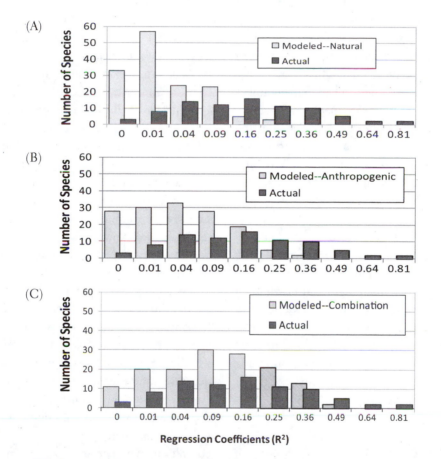

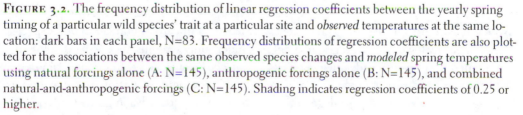

FIGURE 3.2. The frequency distribution of linear regression coefficients between the yearly spring timing of a particular wild species' trait at a particular site and *observed* temperatures at the same location: dark bars in each panel, N=83. Frequency distributions of regression coefficients are also plotted for the associations between the same observed species changes and *modeled* spring temperatures using natural forcings alone (A: N=145), anthropogenic forcings alone (B: N=145), and combined natural-and-anthropogenic forcings (C: N=145). Shading indicates regression coefficients of 0.25 or higher.

modeled local spring temperatures with only natural forcings (for example, solar activity and volcanic dust veils) included as drivers of temperature change (figure 3.2A, N=145). Similar analyses used only human-driven changes in the modeled temperatures (figure 3.2B, N=145), and combined natural-and-anthropogenic forcings in the model (figure 3.2C, N=145).

Modeled temperatures with combined (anthropogenic-and-natural) forcings agree most strongly with changes in species' spring events. Those temperatures explain at least 25 percent of the variability for thirty-six species, while human forcings explain that amount of variability for only seven species, and natural forcings only three species. Additionally, the frequency diagram of observed temperatures is most similar to that of the combined frequency plot, less similar to the anthropogenic plot, and least similar to the natural plot (figure 3.2).

Joint attribution indicates that human activities are likely (with a 66 to 90 percent probability) to be contributing to changes in regional surface temperatures and that human activities are likely to be contributing more to changing temperature than are natural forces.[39] Joint attribution thus also shows that human activities are likely to be contributing to climate change's ongoing impact of earlier spring warming on numerous species.

What We Can Do

Extinction is an irreversible consequence of a species' inability to adapt to changes in its environment. If warming exceeds 2 degrees Celsius above pre-industrial temperatures (the increase thus far is around 0.75 degree Celsius), around 20 percent of known species will likely be unable to adapt. If warming exceeds 4 degrees Celsius, this percentage of highly at-risk species increases to around 40 percent.[40] Given that between 1.7 and 1.8 million species have been identified thus far, approximately 340,000 and 680,000 species, respectively, could face extinction.[41] These numbers do not suffice as indicators of the extinctions' potential consequences for planetary ecosystems; they cannot, in fact, suffice as numerical indicators, as they cannot include any of the estimated 14 million unidentified species.[42] The numbers indicate that we may be at the brink of a mass-extinction event, precipitated by the actions of one species.

What will be the consequences of so many extinctions? Some may close off valuable resources for humans, such as biological sources of new medicines. Some could cause co-extinction of interdependent species, as when the loss of specific host plants caused Singaporean tropical butterfly species to die out.[43] Some extinctions may cause unexpected cascading effects, complicating the survival of many species. For example, extinction of a top predator in a community (for example, the coyote, *Canis latrans*) can cause mesopredator release: a density increase in middle predators (the fox and the house cat), which in turn can cause dramatic population decreases, and even extinctions, of their prey (various ground-loving birds).[44] Species interdependence within a community or ecosystem is complex, not linear, and often more complex than we can anticipate: extinction of one species frequently affects the population size of an uncertain number of other species, under some circumstances—particularly under ecosystem stress—causing or facilitating further extinctions. Extinctions can prompt more extinctions, leading, at worst, to self-perpetuating positive feedback.

The complexity and uncertainty of extinction have often proven so daunting that policy designers intending to maximize survival have not known where to begin. Although some uncertainty is inevitable in any scientific enterprise, species-preservation planning and actions call for a more complete and action-focused understanding of climate-plant-animal interactions. Major gaps in our knowledge result from at least a twenty-year mismatch between conservation priorities and conservation research, lacking studies on threatened systems such as desert and tundra as well as on threatened taxa such as amphibians.[45] Researchers currently propose various central criteria for prioritization: some have suggested redirecting conservation resources toward high-yield targets such as life cycle–enabling habitats or specific threats such as invasive species control.[46] Large-scale and coordinated approaches to conservation, on the whole, are still novel and often lack realistic incorporation of costs into assessment of conservation results; at the primary research level, costs themselves are still measured in widely varying ways.[47] Increased scope and integration of research on complex interactions among species is particularly and urgently needed to inform and guide highly productive conservation strategies.

Given the enormity of the task of recognizing and minimizing extinctions and their ecosystem-level consequences, humans lack sufficient funds, land, political will, policy-driven research, and skilled managers and scientists willing to focus on protecting species and ecosystems around the globe. Without a more comprehensive and cohesive plan, we are not and will not be conserving as much biodiversity as we could. Rik Leemans (see chapter 4) calls for a coordinated, international approach to conserving wild species.

Such a plan must involve controversial trade-offs and decisions. Unfortunately, resource limitations force us to prioritize, either implicitly or explicitly, as we coordinate efforts to protect species from extinction.

In this context of complex interactions and uncertainty, practical management policies require a type of informed simplification. We suggest applying the sorting technique of triage. Best known in human health crises, triage is demonstrably the best method to save the most that can be saved when resources are limited. Similarly, quickly and systematically determining conservation priorities when pressures are high and resources low can mitigate extinctions.[48] Though triage necessitates some losses, overall losses are higher without the system. We agree that a triage approach to crisis situations may help biodiversity at risk of extinction as it does patients in emergency medicine.

A working triage system requires sorting criteria, explicitly outlined in advance. Conservation triage might require categorizing criteria for species, populations, communities, or ecosystems; under some circumstances, developing plans suited to a particular region or biome may be most useful. One preliminary example of a species-level triage plan might sort extinction-risk species into three different groups: (1) those that will clearly become extinct without intervention, but can be saved relatively quickly and easily; (2) those needing minimal but measurable help; and (3) the unavoidable losses—those that could only be saved with intense, lengthy, and expensive effort, and even then are unlikely to survive. Species in the first category would be the first to receive intervention. Triage of species would mean consciously directing resources and management actions toward species that are in crisis yet have a chance of survival.

Management plans can be drafted based on such primary grouping criteria. Beyond cost of conservation efforts, potential for survival with intervention, and extinction risk, a triage plan might also need to consider exceptional cases, such as a species that is the only extant member of a family. Another species-level consideration is the capacity of the species' extinction—as far as current knowledge can recognize—to cause a cascading effect: is its extinction likely to cause other species' extinctions, or damage to an ecosystem? The same logic can apply to habitats and communities. Knowledge of such central ecosystem roles is currently very limited; we need increased research into complex relationships within ecosystems to inform policy decisions.

While designing categories for triage, we may well make mistakes: some that result from inevitable uncertainty, some from gaps in ecological knowledge that research has not yet filled, some for political reasons. The consequences of these mistakes, however, would likely be outweighed by the consequences of using resources solely for sporadic, uncoordinated efforts, as at present. At current extinction rates, there is still time to design triage plans for implementation over large areas—ideally, at the level of continents. We need increased funding for biodiversity research targeting complex or poorly understood aspects of actionable species, habitats, and ecosystems, and we need coordination between scientists and policy makers to develop widely and regionally acceptable prioritization criteria. We must act quickly and systematically to define triage plans, designing research priorities and conservation policies that address rather than avoid uncertainties and ongoing ecosystem responses. Coordinated efforts will save the most biodiversity possible, using the knowledge that we have to avoid unnecessary extinctions—as well as the consequences that we cannot yet and would rather not know.

Notes

1. See Andrewartha, H. G., and L. C. Birch (1954). *The Distribution and Abundance of Animals*. Chicago, IL: University of Chicago Press. For a more recent explanation, see Schneider, S. H., and R. Londer (1984). *Coevolution of Climate and Life*. Washington, DC: Sierra Club Books.

2. Root, T. L. (1988a). Energy constraints on avian distributions and abundances. *Ecology* 69(2): 330–39. Root, T. L. (1988b). Environmental factors associated with avian distributional limits. *Journal of Biogeography* 15(3): 489–505.

3. IPCC (2007a): *Climate Change 2007: The Physical Science Basis*. Contribution of Working Group I to the Fourth Assessment Report of the Intergovernmental Panel on Climate Change, S. Solomon, D. Qin, M. Manning, M. Marquis, K. Averyt, M. M. B. Tignor, H. L. Miller, Jr., and Z. Chen, eds., Cambridge University Press, Cambridge, UK, 996 pp.

4. See IPCC (1995a): *Climate Change 1995. The Science of Climate Change*. Contribution of Working Group I to the Second Assessment Report of the Intergovernmental Panel on Climate Change. J. T. Houghton, L. G. Meiro Filho, B. A. Callander, N. Harris, A. Kattenburg, and K. Maskell, eds., Cambridge University Press, Cambridge, UK. The work of Root and colleagues linking human activities to global warming is explained later in this chapter: Root, T. L., D. P MacMynowski, M. D. Mastrandrea, and S. H. Schneider (2005). Human-modified temperatures induce species changes: Joint attribution. *Proceedings of the National Academy of Sciences* 102: 7465–69.

5. IPCC (2007a) op. cit.

6. Peters, R., and T. Lovejoy (1992). *Global Warming and Biological Diversity*. New Haven, CT: Yale University Press, p. 386.

7. See, for example, these four pieces: Root, T. L., and S. H. Schneider (2002). Overview and implications for wildlife. In *Wildlife Responses to Climate Change: North American Case Studies*, T. L. Root and S. H. Schneider, eds., Washington, DC: Island Press, pp. 1–56. Root, T. L., J. T. Price, K. R. Hall, S. H. Schneider, C. Rosenzweig, and

J. A. Pounds (2003). Fingerprints of global warming on animals and plants. *Nature* 421: 57–60. Parmesan, C., and G. Yohe (2003). A globally coherent fingerprint of climate change impacts across natural systems. *Nature* 421: 37–42. Parmesan, C. (2006). Ecological and evolutionary responses to recent climate change. *Annual Review of Ecology, Evolution, and Systematics* 37: 637–69.

8. Root et al. (2005) op. cit.

9. Thomas, C., A. Cameron, et al. (2004). Extinction risk from climate change. *Nature* 427: 145–48. IPCC (2007b). *Climate Change 2007: Impacts, Adaptation, and Vulnerability*. Contribution of Working Group II to the Fourth Assessment Report of the Intergovernmental Panel on Climate Change. M. L. Parry, O. F. Canziani, J. P. Palutikof, P. J. van der Linden, and C. E. Hanson, eds., Cambridge University Press, Cambridge, UK, 976 pp.

10. Graham, R. W., and E. C. Grimm (1990). Effects of global climate change on the patterns of terrestrial biological communities. *Trends in Ecology and Evolution* 5(9): 289–92.

11. Overpeck, J. T., R. S. Webb, and T. Webb III (1992). Mapping eastern North American vegetation change over the past 18,000 years: No analogs and the future. *Geology* 20: 1071–74.

12. Root, T. L. (1988a, b) op. cit.

13. See Rik Leemans's chapter 4 in this volume.

14. See Parmesan (2006) op. cit., as well as other assessments: McCarty, J. P. (2001). Ecological consequences of recent climate change. *Conservation Biology* 15(2): 320–31. Walther, G.-R., E. Post, et al. (2002). Ecological responses to recent climate change. *Nature* 416: 389–95.

15. Karban, R., and S. Y. Strauss (2004). Physiological tolerance, climate change, and a northward range shift in the spittlebug, *Philaenus spumarius*. *Ecological Entomology* 29(2): 251–54. Carroll, A. L., S. W. Taylor, J. Regniere, and L. Safranyik. (2003). Effects of climate change on range expansion by the mountain pine beetle in British Columbia. In *Mountain Pine Beetle Symposium: Challenges and Solutions*, T. L. Shore, J. E. Brooks, and J. E. Stone, eds., Natural Resources Canada, Canadian Forest Service, Pacific Forestry Centre, Information Report BC-X-399, Victoria, BC, 298 pp. Beaugrand, G., P. C. Reid, F. Ibañez, J. A. Lindley, and M. Edwards (2002). Reorganization of North Atlantic marine copepod biodiversity

and climate. *Science* 296:1692–94. Spencer, H. J., C. Palmer, and K. Parryjones (1991). Movements of fruit-bats in eastern Australia, determined by using radio-tracking. *Wildlife Research* 18(4): 463–68.

16. Phenological change consistent with warming is observed in previously mentioned works: IPCC (2007b), Parmesan (2006), Root, Price, et al. (2003), Parmesan and Yohe (2003), and Walther et al. (2002). Also see IPCC (2001b): *Climate Change 2001: Impacts, Adaptation, and Vulnerability*. Contribution of Working Group II to the Third Assessment Report of the Intergovernmental Panel on Climate Change. James J. McCarthy, Osvaldo F. Canziani, Neil A. Leary, David J. Dokken, and Kasey S. White, eds., Cambridge University Press, UK. Myneni, R. B., C. D. Keeling, C. J. Tucker, G. Asrar, and R. R. Nemani (1997). Increased plant growth in the northern high latitudes from 1981 to 1991. *Nature* 386: 698–702.

17. Root et al. (2003) op. cit.

18. IPCC (2007a) op. cit.

19. See Nelson, H. (2007). Does a crisis matter? Forest policy responses to the mountain pine beetle epidemic in British Columbia. *Canadian Journal of Agricultural Economics* 55: 459–70.

20. Both, C., and M. E. Visser (2001). Adjustment to climate change is constrained by arrival date in a long-distance migrant bird. *Nature* 411: 296–98.

21. Smith, F. A., and J. L. Betancourt (1998). Response of bushy-tailed woodrats (*Neotoma cinerea*) to late Quaternary climatic change in the Colorado Plateau. *Quaternary Research* 50(1): 1–11. See also Parmesan, C., and H. Galbraith (2004). *Observed impacts of global climate change in the U.S.* Arlington, VA: Pew Center on Global Climate Change, 67 pp.

22. Bradshaw, W. E., and C. M. Holzapfel (2001). Genetic shift in photoperiodic response correlated with global warming. *Proceedings of the National Academy of Sciences of the United States of America* 98(25): 14509–11.

23. Schneider, S. H., and T. L. Root (1996). Ecological implications of climate change will include surprises. *Biodiversity and Conservation* 5: 1109–19. Easterling, D., G. Meehl, et al. (2000). Climate extremes: Observations, modeling, and impacts. *Science* 289: 2068–74. Meehl, G. A., F. Zwiers, J. Evans, T. Knutson, L. Mearns, and P.

Whetton (2000). Trends in extreme weather and climate events: Issues related to modeling extremes in projections of future climate change. *Bulletin of American Meteorological Society* 81(3): 427–36.

24. Sekercioglu, C. H., S. H. Schneider, J. P. Fay, and S. R. Loarie (2008). Climate change, elevational range shifts, and bird extinctions. *Conservation Biology* 22: 140–50.

25. McLachlan, J. S; J. J. Hellmann, and M. W. Schwartz (2007). A framework for debate of assisted migration in an era of climate change. *Conservation Biology* 12(2): 297–302.

26. Koh, L. P., and D. S. Wilcove (2007). Cashing in palm oil for conservation. *Nature* 448: 993–94. Robbins, W. D., M. Hisano, S. R. Connolly, and J. H. Choat (2006). Ongoing collapse of coral-reef shark populations. *Current Biology* 16(23): 2314–19. Serjeantson, D. (2001). The great auk and the gannet: A prehistoric perspective on the extinction of the great auk. *International Journal of Osteoarchaeology* 11(1–2): 43–55.

27. Waldram, M. S., W. J. Bond, and W. D. Stock (2008). Ecological engineering by a megagrazer: White rhino impacts on a South African savanna. *Ecosystems* 11(1): 101–12.

28. Hayes, T. B. (2005). Welcome to the revolution: Integrative biology and assessing the impact of endocrine disruptors on environmental and public health. *Integrative and Comparative Biology* 45(2): 321–29.

29. Webster, C. R., M. A. Jenkibs, and S. Jose (2006). Woody invaders and the challenges they pose to forest ecosystems in eastern United States. *Journal of Forestry* 104(7): 366–74.

30. Schloesser, D. W., J. L. Metcalfe-Smith, W. P. Kovalak, et al. (2006). Extirpation of freshwater mussels (*Bivalvia unionidae*) following the invasion of dreissenid mussels in an interconnecting river of the Laurentian Great Lakes. *American Midland Naturalist* 155(2): 307–20.

31. Bouwma, A. M., M. E. Ahrens, C. J. DeHeer, et al. (2006). Distribution and prevalence of Wolbachia in introduced populations of the fire ant *Solenopsis invicta*. *Insect Molecular Biology* 15(1): 89–93.

32. Sitzia, T. (2007). Hedgerows as corridors for woodland plants: A test on the Po Plain, northern Italy. *Plant Ecology* 188(2): 235–52.

33. For details, see again Root and Schneider (2002), Parmesan and Yohe (2003), and Root et al. (2003).

34. Root et al. (2003) op. cit.

35. Root et al. (2005) op. cit.

36. See previously mentioned IPCC reports (2007, 1995a), as well as others: IPCC (2001a): *Climate Change 2001: The Scientific Basis*. Contribution of Working Group I to the Third Assessment Report of the Intergovernmental Panel on Climate Change. J. T. Houghton, Y. Ding, D. J. Griggs, M. Noguer, P. J. van der Linden, X. Da, K. Maskell, and C. A. Johnson, eds., Cambridge University Press, UK. See also IPCC (1995b): *Climate Change 1995: Impacts, Adaptations and Mitigation of Climate Change: Scientific-Technical Analyses*. Contribution of Working Group II to the Second Assessment Report of the Intergovernmental Panel on Climate Change. R. Watson, M. C. Zinyowera, and R. Moss, eds., Cambridge University Press, Cambridge, UK. Also see IPCC (1995c): *Climate Change 1995: Economic and Social Dimensions of Climate Change*. Contribution of Working Group III to the Second Assessment Report of the Intergovernmental Panel on Climate Change. J. P. Bruce, H. Lee, and E. F. Haites, eds., Cambridge University Press, Cambridge, UK. See also IPCC (2001b) op. cit.

37. IPCC (2007b) op.cit.

38. Root et al. (2005) op. cit.

39. IPCC (2007b) op. cit.

40. Fischlin, A., G. F. Midgley, J. Price, R. Leemans, B. Gopal, C. Turley, M. D. A. Rounsevell, P. Dube, J. Tarazona, and A. A. Velichko (2007). Ecosystems, their properties, goods, and services. In *Climate Change 2007: Impacts, Adaptation and Vulnerability*. Contribution of Working Group II to the Fourth Assessment Report of the Intergovernmental Panel on Climate Change. M. L. Parry, O. F. Canziani, J. P. Palutikof, P. J. van der Linden, and C. E. Hanson, eds., Cambridge University Press, Cambridge, UK, 211–72. See also Thomas et al. (2004) and Rik Leemans's chapter 4 in this volume.

41. Lewinsohn, T. M., and P. I. Prado (2005). How many species are there in Brazil? *Conservation Biology* 19(3): 619–24.

42. Hammond, P., B. Aguirre-Hudson, M. Dadd, B. Groombridge, J. Hodges, M. Jenkins, M. H. Mengesha, and W. Stewart Grant (1995). The current magnitude of biodiversity. In *Global Biodiversity Assessment*, V. H. Heywood, ed., Cambridge University Press, Cambridge, UK, pp. 113–38.

43. Koh, L. P., N. S. Sodhi, and B. W. Brook (2004). Co-extinctions of tropical butterflies and their hostplants. *Biotropica* 36: 272.

44. Soulé, M. (1988) Reconstructed dynamics of rapid extinctions of chaparral-requiring birds in urban habitat islands. *Conservation Biology* 2(1): 75–92.

45. Lawler, J. J., J. E. Aukema, J. B. Grant, B. S. Halpern, P. Kareiva, C. R. Nelson, K. Ohleth, J. D. Olden, M. A. Schlaepfer, B. R. Silliman, and P. Zaradic (2006). Conservation science: A 20-year report card. *Frontiers in Ecology and the Environment* 4(9): 473–80.

46. Levin, P. S., and G. W. Stunz (2005). Habitat triage for exploited fishes: Can we iden-tify essential "Essential Fish Habitat?" *Estuarine Coastal and Shelf Science* 64(1): 70–78. Wilson, K. A., E. C. Underwood, S. A. Morrison, K. R. Klausmeyer, W. W. Murdoch, et al. (2007). Conserving biodiversity efficiently: What to do, where and when. *PLoS Biol* 5(9): e223.

47. Murdoch, W., S. Polasky, K. A. Wilson, H. P. Possingham, P. Kareiva, and R. Shaw (2007). Maximizing return on investment in conservation. *Biological Conservation* 139: 375–88.

48. Millar, C. I., N. L. Stephenson, and S. L. Stephens (2007). Climate change and forests of the future: Managing in the face of uncertainty. *Ecological Applications* 17(8): 2145–51.

Chapter 4

Ecosystems

RIK LEEMANS

Climate change is already altering ecosystems in such a way that we must ask ourselves about their capacity to adapt. While ecosystems are extremely resilient and typically only change rather slowly, there are growing signs from across the world that they can reach a critical point where they undergo "greenlash"—or a sudden flip into a new structure.

> Professor Jacqueline McGlade, Executive Director of the European Environment Agency in her foreword of "The European Environment—State and Outlook," 2005.

Introduction

IPCC's 2001 report already clearly stated that climate change is not just a future concern; climate change already has led to observed changes in ecosystems.[1] Some natural systems will be irreversibly damaged. Over the last decade empirical evidence from continents and oceans shows that many physical and biological systems are affected by regional climate change.[2] The latest IPCC report has concluded that natural ecosystems and human systems are responding and adapting to climate change; the vast majority of the responses are consistent with what would be expected from understanding ecosystem dynamics.[3] Discernable impacts on natural and human environments are emerging, although many are masked by adaptation and non-climatic drivers.

Recent literature regarding the impacts on plant species and ecosystems shows that the observed impacts of climate change during the last century are widespread, acute, and serious (see chapter 3). Beyond a 2-degree-Celsius warming, many species and ecosystems will not be able to adapt. In addition, a *rate* of warming exceeding 0.1 degree Celsius per decade will compromise, perhaps severely, the ability of ecosystems to adapt to climate change; the current rate is 0.3 degree Celsius per decade.

Sophisticated models and scenario analyses project the future effects of climate change on ecosystems. I will discuss some of these possible impacts on species and ecosystems. The analysis will show that climate change can jeopardize the biodiversity conservation goals currently set by national legislation and international conventions. Finally, I will discuss briefly the consequences for future research and action.

Future Changes: A Review of Impact Assessments

Future changes in climate and impacts are generally assessed by using models and scenarios. Currently, the most widely used scenarios are those based on the emission scenarios of IPCC, which include four different narratives on how society will evolve over the next century.[4] These narratives specify the type of energy carriers (fossil versus renewable), the level of international collaboration (regional blocks or globalization) and solidarity (equity versus personal wealth), and result in largely different emissions, concentrations, and climate change. The level of climate change is determined by general circulation models (GCM) that comprehensively simulate the dynamic atmospheric processes that determine climate conditions. The output of GCMs (together with atmospheric CO_2 concentrations) are used in different ecological models to determine impacts. Such a causal chain approach typically neglects linkages and feedbacks between processes (e.g., ecosystem processes control CO_2 uptake and release from the biosphere), although the more advanced models partially address this deficiency.[5]

Plants and Vegetation

Many physiological processes (e.g., photosynthesis and respiration) are influenced by temperature and moisture availability. Plant species generally increase productivity at higher temperatures until a species-specific temperature optimum, after which productivity rapidly declines. Respiration, on the other hand, increases exponentially with increasing temperatures. These relationships strongly define global and regional productivity and carbon sequestration patterns.[6]

Including algorithms to dynamically assign plant types on basis of climate, soil, and disturbances, dynamic global vegetation models (DGVMs) indicate that the world's ecosystems, which currently sequester approximately a quarter of the anthropogenic CO_2 emissions, could shift and become a CO_2 source.[7] These results are, however, strongly dependent on the model and climate-change scenario used.[8] The ATEAM project, for example, uses all the IPCC emission scenarios and four different climate scenarios and shows that in the second half of this century, Europe's ecosystems would shift from a sink to a source. Scholze and colleagues take an even more rigorous approach with eighteen different climate scenarios.[9] They indicate high risks of forest loss for Eurasia, eastern China, Canada, Central America, and Amazonia (see chapter 9), with forest extensions into the Arctic and semiarid savannas. This loss creates a considerable net carbon source from the biosphere. At global mean temperature increases of more than 3 degrees Celsius the global biosphere converts to a carbon source during the twenty-first century in approximately half of the scenarios. At around 2 degrees Celsius the fraction drops to 15 percent but remains about 5 to 10 percent at even lower temperatures.

Besides changes in carbon storage and fluxes, biome models studies show large shifts in vegetation for each degree of warming.[10] DGVMs calculate the magnitude of possible vegetation shifts but not the individual species response nor the mechanism of how species migrate. More recent DGVMs also include algorithms to deal with successive changes.[11] All these models show that shifts in vegetation increase with increases in global mean temperatures. With a 3-degree-Celsius warming, approximately 35 percent of all vegetation in the world will change character (forests to grassland, coniferous to deciduous, and so on), but the impact on each vegetation type will be different.[12] Alpine and tundra vegetation will, for example, be reduced in extent by almost 90 percent, while the extent of tropical forests will remain relatively stable.

Recent research shows that these impacts are more severe than previously imagined. One of the first species-oriented studies predicts that 15 to 37 percent of plant species could be "committed to extinction" locally by 2050.[13] Its conclusion, however, remains controversial because of the use of simple approaches that differed for different biomes and also strongly different scenarios. A more reliable study, using species-specific models and linked those to changes in European biodiversity, projects late-twenty-first-century distributions for 1,350 European plants species under seven climate change scenarios and applied IUCN's Red List criteria to these changes.[14] The study shows that many European plant species could become severely threatened regionally; more than half of the species studied could be vulnerable (i.e., less abundant) or threatened with extinction by 2080. Expected species loss and turnover proved to be highly variable across scenarios and across regions. Modeled species loss and arrival of new species were found to depend strongly on the degree of change in just two climate variables describing temperature and moisture conditions. Despite the coarse scale of the analysis, species from mountains could be seen to be disproportionably sensitive to climate change. The boreal region was projected to lose few species, while gaining many others from migration. The greatest changes are expected in the transition between the Mediterranean and Euro-Siberian regions. They concluded that risks of extinction for European plants may be large, even in moderate scenarios of climate change.

There are many indirect impacts of climate change. For example, increased drought could increase fire risk (see chapter 8).[15] Malcolm and colleagues have assessed the magnitude of this threat at the global scale.[16] They used major vegetation types (biomes) as proxies for natural habitats and, based on projected future biome distributions under changed climates, calculated changes in habitat areas and associated extinctions of endemic plant and vertebrate species in biodiversity hotspots. Like most recent studies, they looked at a series of scenarios and different factors including average migration rates. Projected percent extinctions ranged from less than 1 percent to 43 percent of endemic species (average 12 percent). Especially vulnerable hotspots were the Cape Region, Caribbean, Indo-Burma, Mediterranean Basin, southwest Australia, and the tropical Andes, where plant extinctions per hotspot sometimes exceeded 2,000 species. Under the assumption that projected habitat changes were attained in 100 years, estimated global warming–induced rates of species extinctions in tropical hotspots exceeded those due to deforestation. This rapidly emerging threat to biodiversity is also identified by the UN Convention on Bio-logical Diversity in its second Biodiversity Outlook.[17]

The Millennium Ecosystems Assessment also makes projections on the basis of both an analysis of current trends and by exploring scenarios of plausible futures that affect biodiversity loss.[18] In particular, the loss of species diversity and transformation of habitats is likely to continue for the foreseeable future.

This is largely due to inertia in ecological and human systems and to the fact that most of the direct drivers of biodiversity loss—habitat change, climate change, the introduction of invasive alien species, overexploitation, and nutrient loading—are projected to either remain constant or to increase in the near future. In their assessment, however, climate change was considered an important threat, already now for some vulnerable species, and gaining in importance rapidly compared to the other factors.

Mammals

Many climate impacts studies are carried out for plants; only a few studies examine climate impacts on mammals. Much emphasis has been laid on polar animals due to their iconic sensitivity to climate change. Warming leads to decreased ice and snow conditions, which in turn limits access to food in winter; this has negative consequences for their fitness and survival.[19] The vulnerability of mammals in temperate and other regions has been inadequately studied. One of the few studies on climate change impacts on mammals assessed the sensitivity of 277 mammals to macroclimatic variables in Africa, such as temperature and precipitation using generalized regression models.[20] Future projections of animal distributions are derived using different scenarios to estimate the spatial patterns of loss and gain in species richness that might ultimately result in the 141 national parks considered. Assuming species are unable to migrate, 10 to 15 percent of the species are projected to become critically endangered or extinct by the middle of this century and between 25 and 40 percent toward 2100. Allowing for unlimited species spread reduces the percentage

of endangered or extinct species to 10 to 20 percent by the end of this century. On balance, the national parks might ultimately realize a substantial shift in the mammalian species composition of a magnitude unprecedented in recent geological time. They conclude that the effects of climate change on wildlife communities are most noticeable as a fundamental change in community composition. Such a change will have a large impact on, for example, tourism.

Thuiller and colleagues did not consider the direct effects of CO_2 on the species composition of grasses.[21] Currently, the most palatable grasses, on which wildlife depends, consist of C4 grasses, well adapted to low atmospheric CO_2 concentrations and dry conditions. Increasing CO_2 concentrations will favor C3 grasses. These grasses take advantage of higher CO_2 concentrations by growing faster and improving their water and nutrient use efficiency. Unfortunately, such grasses are much less palatable.[22]

Birds

Birds, especially migratory birds, are better studied. In the future, global warming will affect birds indirectly through sea level rise, changes in fire regimes, vegetation changes, and land-use change. Climate change could eventually destroy or fundamentally change 35 percent of the world's existing terrestrial habitats. In the Arctic, where several hundred million migratory birds breed, warming of two to three times the global average is predicted to destroy more than 90 percent of some bird species' habitat at higher levels of warming. In Europe's Mediterranean coastal wetlands, critical habitat for migratory birds could be completely destroyed by the end of this century.[23] In North America, approximately 2 degrees Celsius of warming will reduce the world's most productive waterfowl habitat by half, also halving this zone's duck numbers.[24] Global warming of 3 to 4 degrees Celsius would eliminate 85 percent of all the world's remaining wetlands, critical habitat for migratory birds.[25]

The overall extinction risk to birds due to climate change is still being quantified (see table 4.1).[26] However, initial research suggests that more than a third of all European birds will become extinct if species are unable to shift to new ranges. Candidates for extinction in Europe include the red kite (expected to lose most of its habitat) and the Scottish crossbill (expected to lose all of its habitat). The situation is worse still in the Australia wet tropics bioregion, where climate change could force the extinction of almost three-quarters of bird species.

However, projections likely underestimate extinction rates of birds since they consider only a limited number of climate variables and have not factored in the devastating impact of climate extremes, which will exacerbate threats.[27] Furthermore, most research considers only the direct impacts of climate on shifting or contracting ranges, while neglecting important indirect and secondary effects of climate change.

Climate change will also wreak some of its most serious, but least predictable, impacts by shifting the timing of natural events and by shifting species' geographical distributions. This will rearrange plant and animal communities and ecosystems and disrupt relationships with predators, competitors, prey, and parasites. These changes are expected to alter the makeup and functions of most, if not all, the world's ecosystems.[28] Evidence suggests that many bird species will not be able to adapt.

Many species may be unable to shift with geographic changes because their habitat is already fragmented and disconnected from potentially suitable alternate areas, especially if they are located outside currently protected areas.[29] Physical barriers such as mountains, large bodies of water, and human development present further obstacles to migration. In addition, communities of birds and other interdependent plants and animals are unlikely to shift intact.

Illustrating the magnitude of this threat, 80 percent of the chicks of the yellow-eyed penguin (the world's rarest penguin) died from

TABLE 4.1

Predicted Bird Extinctions for Different Scenarios

Region	Current Number of Bird Species	Warming Scenario	Predicted Percentage of Bird Species Extinctions	
			With Dispersal	No Dispersal
Europe	526	>2°C	4–6%	13–38%
South Africa	951	1.8–2.0°C	28–32%	33–40%
Australian Wet Tropics	740	>2°C	49–72%	N/A
Mexico	1,060	1.8–2.0°C	3–4%	5–8%

Source: A. van Vliet, and R. Leemans. 2006. Rapid species' responses to changes in climate require stringent climate protection targets. In H. J. Schellnhuber, W. Cramer, N. Nakícénovic, T. Wigley, and G. Yohe, eds. *Avoiding Dangerous Climate Change.* Cambridge University Press, Cambridge, UK, pp. 135–43.

avian diphtheria in 2004 as virus-carrying pests proliferated in wetter springs and summers, a trend linked to climate change.[30] Shifts in prey species have also profoundly affected birds' breeding success. In the North Sea, climate change has been linked to extreme declines in sandeel (a small fish) populations in key areas to a hundredth or a thousandth of their former levels, which has caused complete breeding failure in some seabird colonies that rely on these fish for food.[31] Most analyses to date have not yet considered these secondary impacts, and future research that does is likely to greatly increase estimates of risk to birds of climate change. In fact, the expected combination of climate change and other human disturbances such as habitat loss has been termed an "extinction spasm" due to the potential to disrupt communities and wipe out entire populations.[32]

What Future Research Needs to Be Done?

Research on the impact of extreme events, development of appropriate indicators of biodiversity, long-term monitoring of species distribution, and conservation efforts are all urgently needed to assess the impacts of climate change on ecosystems.

Van Vliet and Leemans conclude that by ignoring extreme weather events, traditional impacts assessments likely underestimate impacts on ecosystems, especially at lower levels of climate change.[33] Extreme weather has changed more profoundly than average weather. Ecosystems have responded more to extreme events than to changes in average climate conditions, which is a primary indicator in most climate scenarios used in model forecasts. This could explain the rapid appearance

of ecological responses throughout the world. Unfortunately, most climate scenarios do not routinely provide information on extreme events, although some have begun to address this void.[34] Integration of these effects into impacts studies in order to determine the real vulnerability of ecosystems to climate change is urgently needed.

Another limitation of impact studies is the lack of proper indicators. The Millennium Ecosystem Assessment concludes that there is not a single indicator to unambiguously measure the impacts of climate change on biodiversity.[35] One needs a number of indicators, including those on trends in extent of selected biomes, ecosystems and habitats, abundance and distribution of selected species, and incidence of human-induced ecosystem failure that can serve to derive trends where more detailed data are not available. Some of those aggregate indicators have been proposed.[36] Because small, fragmented ecosystems are more affected by changes in climate than large contiguous ecosystems with a more balanced micro-climate, trends in connectivity/fragmentation of ecosystems provide an indicator of the vulnerability of ecosystems to climate change.

Most of the convincing observed trends in Europe and in the Netherlands were based on long-term monitoring of vegetation, a group of species, or single species.[37] All these monitoring networks were not established to determine the impacts of climate change but for other scientific interests or environmental monitoring. However, their importance to the study of the impacts of climate change has become apparent. This makes such rare studies very valuable, convincing, and of much broader interest than just their original purpose. Many other short-term studies often remain anecdotal and consequently are much less convincing. Careful analysis of other long-term ecological research networks will probably also reveal additional and compelling evidence.

Finally, conservation strategies are focused on reducing the impacts of overexploitation, habitat destruction, air pollution, and eutrofication. One of the approaches is to develop protected areas. This strategy has been successful during the last decades. Just over 10 percent of all land is legally protected, and marine reserves are now being developed. However, protected areas are not the suitable way of dealing with climate change when the current distribution of a protected species does not comply with the future distribution. Species have to be able to shift. This means that connecting corridors between protected areas must be developed. This signifies that the 90 percent of the land that is not protected should be evaluated in terms of its capacity to allow species and ecosystems to cope with climate change. Ecological research is needed to understand what effective networks are. Research on institutions is required to assess the societal consequences. In other words: integrated, multidisciplinary research is urgently needed.

What Policies Have to Be Implemented?

Species have no insight into the future and only respond to changes in their environment. Only one species, *Homo sapiens*, is an exception and can predict, plan, manage, and learn. Adaptation measures targeted at all other species should therefore be focused on not only maintaining, but also trying to enhance resilience of these species and ecosystems. In 2001, world leaders at the World Sustainability Summit agreed to reduce the decline of biodiversity by 2010. Progress toward this 2010 target is challenging and depends on protecting the critical habitats, populations of species, and genetic diversity that contribute to resilience and facilitate adaptation in the face of climate change.[38] There is a large need for international conventions, national governments, and others to address this threat, including the United Nations Framework Convention on Climate Change. At the same time, activities aimed at the conservation and sustainable use of biodiversity also require a full account of climate change. This review clearly shows that many species, ecosystems, and regions are vulnerable to climate change. There is a need to develop and implement adaptation measures in all thematic programs of biodiversity conservation.

To limit the risks for impact of climate change on ecosystems, two approaches have to be taken. First, climate change has to be mitigated by reducing greenhouse gas emissions. A recent analysis shows that this is economically and technologically feasible.[39] Second, the resilience of species and ecosystems has to be increased, especially by reducing and removing the other stresses on ecosystems and enhancing conservation efforts.[40] However, conservation organizations and ecosystem managers are currently poorly prepared for climate change impacts, especially if based solely on protecting static areas systems and traditional forestry, agricultural, and fishery practices. Responses now remain limited to generalized, no-regrets strategies, while a much more proactive attitude is needed. Finally, the cascade of uncertainties from climate change projections (especially in rainfall patterns) through the limited knowledge of species to ecosystem responses remains significant barriers to developing coherent and detailed regional policy planning responses.

Conclusion

The observed impacts already show that ecosystems will be altered everywhere. Many places will experience future local and regional extinctions; habitats, especially in the polar regions, will disappear. Probably the most vulnerable areas will be those regions with many endemic species, such as mediterranean regions (including the South African Fynbos) and mountainous areas. Forests will also potentially be affected when droughts reduce their resilience and increase fire frequencies. These changes entail grave consequences for the effectiveness of mitigation strategies because they probably release carbon into the atmosphere.

This chapter shows that the magnitude and rate of climate change pose a major threat. Human-induced climate change will cause rates of change to species, ecosystems, and biodiversity that are historically unprecedented. This will exceed the ability of many plant and animal species to migrate or adapt and will lead to irreversible impacts.

Although some species and ecosystems will profit, most will be adversely affected by climate change, which will accelerate the decline of biodiversity. This phenomenon has been observed in the past when biodiversity declined during periods with rapid climate change, such as the Younger Dryas of 12,000 years ago.

The threats of climate change pose large challenges for conservation, especially since effective efforts to protect habitats and create ecological networks require international cooperation and concerted action. Developing successful conservation strategies must include support for developing countries. Only a global climate policy in conjunction with a conservation plan will reduce the threat of a major extinction event.

Notes

1. McCarthy, J. J., O. F. Canziani, N. Leary, D. J. Dokken, and K. S. White, eds. 2001. *Climate Change 2001: Impacts, adaptation and vulnerability*. Cambridge University Press, Cambridge.

2. See, e.g., Parmesan, C., and G. Yohe. 2003. A globally coherent fingerprint of climate change impacts across natural systems. *Nature* 421:37–42; Root, T. L., J. T. Price, K. R. Hall, S. H. Schneider, C. Rosenzweigh, and J. A. Pounds. 2003. Fingerprints of global warming on wild animals and plants. *Nature* 421:57–60; Root, T. L., D. P. MacMynowski, M. D. Mastrandrea, and S. H. Schneider. 2005. Human-modified temperatures induce species changes: Joint attribution. *Proceedings of the National Academy of Science* 102:7465–69; Both, C., S. Bouwhuis, C. M. Lessells, and M. E. Visser. 2006. Climate change and population declines in a long-distance migratory bird. *Nature* 441:81–83; Rosenzweig, C., Karoly, D., Vicarelli, M., Neofotis, P., Wu, Q., Casassa, G., Menzel, A., Root, T. L., Estrella, N., Seguin, B., Tryjanowski, P., Liu, C., Rawlins, S., Imeson, A., 2008. Attributing physical and biological impacts to anthropogenic climate change. *Nature* 453:353–57.

3. Parry, M. L., O. F. Canziani, J. P. Palutikof, C. E. Hanson, and P. J. Van der Linden, eds. 2007. *Climate Change 2007: Impacts, adaptation and vulnerability*. Contribution of Working Group II to the Fourth Assessment Report of the Intergovernmental Panel on Climate Change. Cambridge University Press, Cambridge.

4. Nakícenovíc, N., J. Alcamo, G. Davis, B. de Vries, J. Fenhann, S. Gaffin, K. Gregory, A. Grübler, T. Y. Jung, T. Kram, E. Emilio la Rovere, L. Michaelis, S. Mori, T. Morita, W. Pepper, H. Pitcher, L. Price, K. Riahi, A. Roehrl, H.-H. Rogner, A. Sankovski, M. E. Schlesinger, P. R. Shukla, S. Smith, R. J. Swart, S. van Rooyen, N. Victor, and Z. Dadi. 2000. Special report on emissions scenarios. Cambridge University Press, Cambridge.

5. Bouwman, A. M., T. Kram, and K. Klein Goldewijk, eds. 2006. *Integrated modelling of global environmental change.* Netherlands Environmental Assessment Agency, Bilthoven.

6. Valentini, R. 2000. Respiration as the main determinant of carbon balance in European forests. *Nature* 404:861; King, A. W., C. A. Gunderson, W. M. Post, D. J. Weston, and S. D. Wullschleger. 2006. Plant respiration in a warmer world. *Science* 312:536–37; Reich, P. B., M. G. Tjoelker, J.-L. Machado, and J. Oleksyn. 2006. Universal scaling of respiratory metabolism, size and nitrogen in plants. *Nature* 439:457–61.

7. Steffen, W. L., W. Cramer, M. Plochl, and H. Bugmann. 1996. Global vegetation models: Incorporating transient changes to structure and composition. *Journal of Vegetation Science* 7:321–28; Thonicke, K., S. Venevsky, S. Sitch, and W. Cramer. 2001. The role of fire disturbance for global vegetation dynamics: Coupling fire into a Dynamic Global Vegetation Model. *Global Ecology and Biogeography* 10:661–77; Betts, R. A., P. M. Cox, S. E. Lee, and F. I. Woodward. 1997. Contrasting physiological and structural vegetation feedbacks in climate change simulations. *Nature* 387:796–99; Woodward, F. I., M. R. Lomas, and R. A. Betts. 1998. Vegetation-climate feedbacks in a greenhouse world. *Philosophical Transactions of the Royal Society London, Series B* 353:29–38; Cramer, W., A. Bondeau, F. I. Woodward, I. C. Prentice, R. A. Betts, V. Brovkin, P. M. Cox, V. Fisher, J. A. Foley, A. D. Friend, C. Kucharik, M. R. Lomas, N. Ramankutty, S. Sitch, B. Smith, A. White, and C. Young Molling. 2001. Global response of terrestrial ecosystem structure and function to CO_2 and climate change: Results from six dynamic global vegetation models. *Global Change Biology* 7:357–73.

8. Parry et al. 2007, op. cit.

9. Schröter, D., W. Cramer, R. Leemans, et al. 2005. Ecosystem service supply and vulnerability to global change in Europe. *Science* 310:1333–37; Scholze, M., W. Knorr, N. W. Arnell, and I. C. Prentice. 2006. A climate-change risk analysis for world ecosystems. *PNAS* 103:13116–20.

10. Woodward, F. I. 1987. *Climate and Plant Distribution.* Cambridge University Press, Cambridge; Neilson, R. P., and R. J. Drapek. 1998. Potentially complex biosphere responses to transient global warming. *Global Change Biology* 4:505–21; Leemans, R., and B. Eickhout. 2004. Another reason for concern: Regional and global impacts on ecosystems for different levels of climate change. *Global Environmental Change* 14:219–28; Rickebusch, S., Thuiller, W., Hickler, T., Araújo, M. B., Sykes, M. T., Schweiger, O., and Lafourcade, B., 2008. Incorporating the effects of changes in vegetation functioning and CO_2 on water availability in plant habitat models. *Biology Letters* 23:556–59. Doi: 10.1098/rsbl.2008.0105.

11. Cramer et al. 2001, op. cit.

12. Cf. Leemans and Eickhout, 2004, op. cit.

13. Thomas, C. D., A. Cameron, R. E. Green, M. Bakkenes, L. J. Beaumont, Y. C. Collingham, B. F. N. Erasmus, M. F. de Siqueira, A. Grainger, L. Hannah, L. Hughes, B. Huntley, A. S. van Jaarsveld, G. F. Midgley, L. Miles, M. A. Ortega Huerta, A. T. Peterson, O. L. Phillips, and S. E. Williams. 2004. Extinction risk from climate change. *Nature* 427:145–48.

14. Thuiller, W., S. Lavorel, M. B. Araújo, M. T. Sykes, and I. C. Prentice. 2005. Climate change threats to plant diversity in Europe. *Proceedings National Academy of Sciences* 102:8245–50.

15. Schröter et al. 2005, op. cit.

16. Malcolm, J. R., C. Liu, R. P. Neilson, L. Hansen, and L. Hannah. 2006. Global warming and extinctions of endemic species from biodiversity hotspots. *Conservation Biology* 20:538–48.

17. Secretariat of the Convention on Biological Diversity. 2006. Global Biodiversity Outlook 2. Convention on Biological Diversity, Montreal.

18. Carpenter, S., P. Pingali, E. Bennett, and M. Zurek, eds. 2005. *Ecosystems and Human Well-Being: Scenarios.* Island Press, Washington, DC; Reid, W. V., H. A. Mooney, A. Cropper, D. Capistrano, S. R. Carpenter, K. Chopra, P. Dasgupta, et al. 2005. *Millennium Ecosystem Assessment Synthesis Report.* Island Press, Washington, DC.

19. Weladji, R. B., and O. Holand. 2003. Global climate change and reindeer: Effects of winter weather on the autumn weight and growth of calves. *Oecologia* 136:317–23.

20. Thuiller, W., O. Broennimann, G. Hughes, J. R. M. Alkemade, G. F. Midgley, and F. Corsi. 2006. Vulnerability of African mammals to anthropogenic climate change under conservative land transformation assumption. *Global Change Biology* 12:424–36.

21. Ibid.

22. Walker, B. 2001. Tropical savanna. In F. S. Chapin III, O. E. Sala, and E. Huber-Sannwald, eds. *Global Biodiversity in a Changing Environ-*

ment: Scenarios for the 21st Century. Springer Verlag, New York, 139–56.

23. Gitay, H., A. Suárez, R. T. Watson, et al. 2002. Climate change and biodiversity. IPCC Technical Paper V, Intergovernmental Panel on Climate Change, Geneva.

24. Sorenson, L. G., R. Goldberg, T. L. Root, and M. G. Anderson. 1998. Potential effects of global warming on waterfowl populations breeding in the Northern Great Plains. *Climatic Change* 40:343–69.

25. Secretariat of the Convention on Biological Diversity 2006, op. cit.

26. Thomas et al. 2004, op. cit.

27. Leemans, R., and A. van Vliet. 2004. Extreme weather: Does nature keep up? Observed responses of species and ecosystems to changes in climate and extreme weather events: Many more reasons for concern. Report Wageningen University and WWF Climate Change Campaign, Wageningen; Parmesan, C., T. L. Root, and M. R. Willig. 2000. Impacts of extreme weather and climate on terrestrial biota. *Bulletin of the American Meteorological Society* 81:443–50; van Vliet, A., and R. Leemans. 2006. Rapid species' responses to changes in climate require stringent climate protection targets. In H. J. Schellnhuber, W. Cramer, N. Nakíc énovic, T. Wigley, and G. Yohe, eds., *Avoiding Dangerous Climate Change*. Cambridge University Press, Cambridge, 135–43.

28. Lovejoy, T. E., and L. Hannah, eds. 2005. *Climate Change and Biodiversity*. Yale University Press, New Haven.

29. Lemoine, N., and K. Bohning-Gaese. 2003. Potential impact of global climate change on species richness of long-distance migrants. *Conservation Biology* 17:577–86.

30. Refer to www.yellow-eyedpenguin.org.nz for further information and research.

31. Robinson , R. A., J. A. Learmonth, A. M. Hutson, C. D. Macleod, T. H. Sparks, D. I. Leech, G. J. Pierce , M. M. Rehfisch, and H. Q. P. Crick. 2005. Climate change and migratory species. BTO Research Report 414, British Trust for Ornithology, The Nunnery, Thetford, Norfolk.

32. Root et al. 2003 and Root et al. 2005, op. cit.

33. van Vliet and Leemans 2006, op. cit.

34. See, for example, Giorgi, F. 2006. Climate change hot-spots. *Geophysical Research Letters* 33. Doi:10.1029/2006GL025734.

35. Carpenter et al. 2005 and Reid et al. 2005, op. cit.

36. de Groot, R., J. van der Perk, A. Chiesura, and A. van Vliet. 2003. Importance and threat as determining factors for criticality of natural capital. *Ecological Economics* 44:187–204; Scholes, R. J., and R. Biggs. 2005. A biodiversity intactness index. *Nature* 434:45–49.

37. See, for example, Tamis, W. 2005. Changes in the flora of the Netherlands in the 20th century. *Gorteria* Supplement 6:1–233; Kuchlein, J. H., and W. N. Ellis. 1997. Climate-induced changes in the microlepidopter fauna of the Netherlands and the implications for nature conservation. *Journal of Insect Conservation* 1:73–80. See also Both, C., and M. E. Visser. 2001. Adjustment to climate change is constrained by arrival date in a long-distance migrant bird. *Nature* 411:296–98; Both et al. 2006, op. cit.

38. Balmford, A., L. Bennun, B. ten Brink, et al. 2005. The Convention on Biological Diversity's 2010 Target. *Science* 307:212–13.

39. Grubb, M., C. Carraro, and J. Schellnhuber. 2006. Technological change for atmospheric stabilization: Introductory overview to the Innovation Modeling Comparison Project. *Energy Journal* 27:1–16; Stern, N., Peters, S., Bakhshi, V., et al., 2006. *The Economics of Climate Change*. Cambridge University Press, Cambridge.

40. Noble, I., J. K. Parikh, R. Watson, R. Howarth, R. J. T. Klein, A. Abdelkasder, and T. Forsyth. 2005. Climate change. In K. Chopra, R. Leemans, P. Kumar, and H. Simons, eds. *Ecosystems and Human Well-Being. Policy Responses. Findings of the Responses Working Group*. Island Press, Washington, DC, 373–400; Fischlin, A., G. F. Midgley, J. Price, R. Leemans, B. Gopal, C. Turley, M. D. A. Rounsevell, P. Dube, J. Tarazona, and A. A. Velichko. 2007. Ecosystems, their properties, goods and services. In M. L. Parry, O. F. Canziani, J. P. Palutikof, C. E. Hanson, and P. J. Van der Linden, eds. *Climate Change 2007: Impacts, Adaptation and Vulnerability*. Contribution of Working Group II to the Fourth Assessment Report of the Intergovernmental Panel on Climate Change. Cambridge University Press, Cambridge, 211–72.

Chapter 5

Marine Ecosystems

CAROL TURLEY

Introduction

Marine ecosystems play a major role in the working of the Earth's life-support system. Comprising approximately two thirds of the Earth's surface, marine ecosystems carry out about 50 percent of global primary production and support extensive biodiversity.[1] Oceans play an important role in the transfer of heat around the planet and in determining weather systems and climate at sea and on land. In addition, they play a major role in the cycling and storage of the Earth's biogeochemically active substances. For example, the oceans provide the second largest reservoir of carbon (other than that in rocks), which is around twenty and fifty times greater than those in the terrestrial biosphere or in the atmosphere, respectively. Marine ecosystems also provide protein sources and livelihoods for millions of people through fisheries, aquaculture, transport, tourism, and recreation.

Increases in atmospheric carbon will affect marine ecosystems through warming sea temperatures and alterations in the chemical composition of our oceans.

Ocean Warming

Marine ecosystems will be affected by climate change through ocean warming, although the precise impacts are still uncertain. Long-term time-series studies during the last forty to sixty years have already revealed that major changes have occurred in planktonic and benthic community composition and productivity.[2] Also, the reduced survival of young cod in the North Sea may be due to the increase in the sea surface temperature of around 1 degree Celsius. Increases in sea surface temperature are also predicted to increase the depth of the warmer surface layer (thermal stratification).[3] Increased thermal stratification will act as a greater barrier to the upwelling of nutrients required for primary production from the deeper, colder nutrient-rich waters into the sunlit upper waters where primary production occurs; and it is predicted that this will lead to an expansion of the less productive waters in the mid-ocean, which are sometimes called "ocean deserts" (see figure 5.1). Models also predict that reduced upwelling will result in around a 5 per-

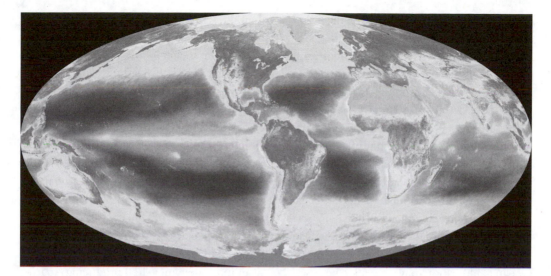

FIGURE 5.1. Satellite image showing the primary productivity of land and sea. The lighter the ocean color, the higher is the productivity, whereas the darker the color, the less the productivity. Notice the central oceanic gyres (dark areas) are low in productivity—these are predicted to expand with future ocean warming. The highly productive waters off the west coasts of South America and Africa are caused by upwelling of nutrient-rich waters that are used to fuel the high productivity in these regions, which is also predicted to decline.

cent decrease in global oceanic primary production with a doubling of CO_2.

Both the melting of the Greenland and Antarctic ice sheets and the expansion of seawater when warmed will increase sea level.[4] Thermohaline circulation, large ocean currents that carry heat around the ocean, warm areas that would otherwise be cooler. For example, the Atlantic Heat Conveyor (the Gulf Stream plays a part in this) helps to maintain mild temperatures in northwestern Europe and its seas. Some observations suggest that it is losing its strength, but more data are needed to distinguish this trend from natural variability. Scientists predict that there will also be an increase in wave heights and their frequency. Sea level rise and increased wave height will impact coastal ecosystems and human inhabitants, and protection or adaptation, particu-

larly in low-lying areas, will be very costly. Another effect, loss of marginal sea ice in polar regions, is already occurring, with a 42 percent loss already in the northern hemisphere and a 17 percent loss in the southern hemisphere. However, in the last couple of years loss may be even greater. Marginal sea ice habitats are highly productive and support the food web in the surrounding waters, so their reduction may have profound impacts on Arctic Ocean productivity and biodiversity with a potential loss of charismatic species such as polar bears.

Ocean Acidification

The same man-made CO_2 that we observe to be the major greenhouse gas causing climate change is also altering the chemical balance of the oceans. This—"the other CO_2 problem"—has received little attention until quite recently. This newly recognized issue emerged with the groundbreaking symposiums The Ocean in a High-CO_2 World in 2004 and 2008, the publication of several seminal papers in the high-profile journals *Nature* and *Science* between 2000 and 2004, the publication of a report by the Royal Society in 2005, and a summary paper on ocean acidification presented to climate change scientists and policy makers at the Avoiding Dangerous Climate Change symposium in 2005.[5] Ocean acidification may turn out to be as serious as the more familiar greenhouse warming effect.[6]

The world's oceans currently absorb on average about 1 metric ton of the CO_2 produced by each person every year. It is estimated that the surface waters of the oceans have taken up more than 500 billion tons of CO_2 (500 Gt CO_2), around 27 to 34 percent of all that generated by fossil-fuel burning, cement manufacturing, and land-use change since 1800.

By absorbing CO_2, the oceans have buffered the effects of atmospheric climate change—but at a cost. CO_2 reacts with seawater to form a weak acid (carbonic acid) that increase seawater acidity (expressed as a reduction in pH). A measurement of the number of hydrogen ions (H+), the pH level in surface ocean waters has already declined by about 0.1 unit since preindustrial times, which means that the number of H+ ions has increased by 30 percent. Potentially, ocean pH could fall further—by as much as 0.4 unit by the year 2100 and 0.7 unit by 2300. Ocean pH will take 10,000 years to return to close to preindustrial levels.

Predicted increases in ocean pH that we are likely to see this century are far greater than the naturally occurring annual variation in pH that organisms currently experience. Further, such changes have certainly not occurred for at least the last 420,000 years and probably not for the past several tens of millions of years. Marine organisms have therefore had a fairly constant pH environment in which to evolve and so have had no need to develop adaptations to rapid changes. About 50 to 55 million years ago, at the boundary between the Paleocene and Eocene geological epochs, the ocean pH did decline to the levels we can expect to see at around 2300, and this resulted in the extinction of many bottom-dwelling, calcifying (shell-producing) marine organisms even though it took thousands of years for the pH to fall. The cause then is thought to have been the release of massive quantities of methane from marine sediments. The methane was broken down to CO_2 and reacted with seawater—just like the CO_2 we are currently putting into the atmosphere through burning fossil fuels. It is worth noting that the current decline in ocean pH will happen far more rapidly, over only a few centuries, compared to the decline experienced 55 million years ago. Thus, it is not surprising that scientists are concerned with the level of decline in ocean pH as well as the speed at which it will happen.

Ocean acidification is now a serious scientific concern for the majority of international marine scientific research associations and organizations. As the scientific research on the impacts of ocean acidification is just emerging or is being planned, there will undoubtedly be impacts and adaptations discovered that have not been addressed here. Predicting the state of future marine ecosystems and understanding the feedbacks to the functioning of the Earth's life-support system will be one of the biggest challenges for marine scientists in future decades. A greater, but achievable, challenge will be for every one of us to reduce our carbon emissions.

Effects of Changing Ocean Temperatures and Chemical Composition

Changes to CO_2 in seawater also affect the rest of the carbonate cycle, with increased CO_2 resulting in a decrease in the amount of carbonate ions in seawater (see figure 5.2). Carbonate ions are used by calcifying organisms to make calcium carbonate shells, skele-

tons, and liths (small platelets).[7] Currently, most surface waters of the world's oceans are supersaturated with carbonate ions. However, the lower the concentration of carbonate ions, the harder it will be for calcifying organisms to make their shells or skeletons. In waters that are undersaturated in carbonate ions, the shells of organisms will dissolve. Recent studies predicting future carbonate ion concentration, at double the preindustrial atmospheric CO_2 concentration (i.e., 560 parts per million), show that aragonite, used by corals to make their hard skeletal reefs, will become so low in tropical waters that coral calcification will be reduced by about 30 percent. This

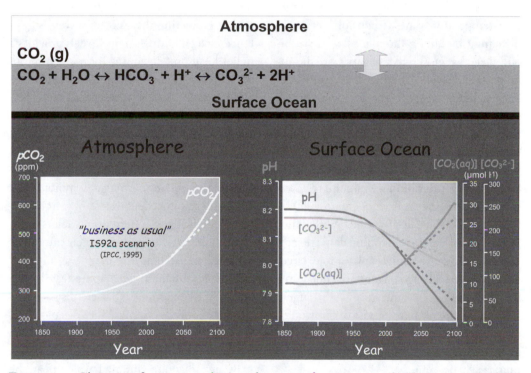

FIGURE 5.2. Changes in future atmospheric and seawater chemistry using the "business as usual" scenario of fossil-fuel consumption. The oceans are becoming more acidic (falling pH), and carbonate ions (CO_3^{2-}) are decreasing as the oceans are taking up more CO_2 (CO_2 [aq]) from the increasing concentration of CO_2 found in the atmosphere (pCO_2). Source: Wolf-Gladrow et al. (1999). Direct effects of CO_2 on growth and isotopic composition of marine plankton. *Tellus Series B-Chemical and Physical Meterology* 51B, 461.

means that by around 2050 reef formation would be less than reef erosion, making reefs unsustainable.

In polar and sub-polar waters aragonite is predicted to become marginal or undersaturated by 2100 with some areas achieving this shell-dissolving state within decades. Affected organisms include shelled pteropods, also known as sea butterflies, which can form an important part of the food web. These organisms may shift in their geographic distribution toward low latitudes where aragonite concentrations are higher. However, we have little understanding of how they will react to warmer temperatures or whether they will be able to exist there.

In addition to loss of sea ice, rising sea surface temperatures are responsible for coral bleaching by killing the microscopic symbiotic algae that grow in partnership with the animal coral polyps. Together, they contribute to the building of productive warm-water coral ecosystems in often-unproductive waters. These microscopic algae contain pigmented chloroplasts; when they die the coral loses its color and looks bleached. As a result, the animal partner in this symbiotic relationship loses the carbon fixed by photosynthesis by its partner algae and can die. While other factors such as increased fishing damage, pollution, and disease may also play a role, large-scale coral bleaching and mortality due to rising sea surface temperatures are largely blamed for the 30 percent loss of the world's corals in the last twenty years. In some regions, such as the Caribbean, loss may be as much as 80 percent.

There is hope that the corals may eventually adapt or migrate to cooler waters, as they did under climatic excursions during past glacial to interglacial transitions. However, studies have shown that while corals could possibly adapt by incorporating different symbiotic algae with a higher temperature tolerance, the increasing seawater temperature would be too warm even for these thermotolerant algae by 2100. Another adaptation may be to migrate toward the higher latitudes where water is cooler. However, the decrease in aragonite saturation already realized in high-latitude waters makes it unlikely that corals will flourish there. Our current understanding suggests the "double trouble" of both ocean warming and acidification means that corals could become rare on tropical and subtropical reefs by 2050.

Coral reef ecosystems harbor a huge number of species and are the most diverse marine habitats. They are also important socioeconomically through tourism, fishing, and their role in protecting shores from waves. Subsistence food gathering on these reefs provides a major local protein source for many millions of people, especially in developing nations and Pacific Island countries. Many more people are supported through tourism and fisheries that bring in billions of dollars each year.[8]

Unlike their warm-water relatives, cold-water corals do not have algal symbionts and can thus grow in deeper waters away from sunlight. Often living hundreds of meters deep, cold-water corals remained largely unnoticed until recently. Their importance as a habitat and their substantial geographic distribution are only just emerging. Now scientists are concerned about the vulnerability of cold-water corals to the rising of the aragonite saturation horizon. This horizon is the dividing line between aragonite-saturated water near the ocean's surface and aragonite-undersaturated water below. This horizon is currently hundreds to thousands of meters deep in the northern Atlantic, but as the surface oceans take up more and more CO_2 this horizon will move upward, toward the sea surface. The horizon may even surface in high latitudes by 2100 so that high-latitude surface waters will be aragonite undersaturated. It is therefore unlikely that a large proportion of cold-water corals growing in deep water at higher latitudes will survive beyond the end of the twenty-first century. However, not all cold-water corals exist in deep waters, and as long as they are still well above the aragonite saturation horizon, they may survive.[9]

Another important group of microscopic, single-celled plants called coccolithophores are also under threat. These phytoplankton may be tiny, but their annual blooms cover about 1.4 million square kilometers of the ocean and are so extensive that you can see them from space (see figure 5.3). They are currently thought to be the largest producers of calcite on the planet. When they die their calcium carbonate platelets, which are known as liths (see figure 5.4), rain down to the ocean floor and are buried over time. This locks away carbon in the sediment, which in time can form vast structures such as the White Cliffs of Dover. When the blooms of coccolithophores die, the liths can be incorporated into aggregates of dying or dead organic material. Such incorporations may act as ballasts, making the aggregates sink faster to the deep sea bed. This process transfers carbon to the deep sea bed before it has time to be recycled and respired back to CO_2 in the surface of the

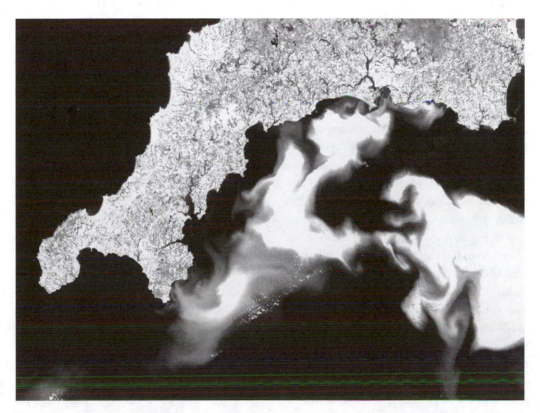

FIGURE 5.3. A coccolithophore bloom off the southwestern coast of England captured by satellite. The calcium carbonate liths produced by these microscopic algae make the sea look milky. Courtesy of Peter Miller (PML).

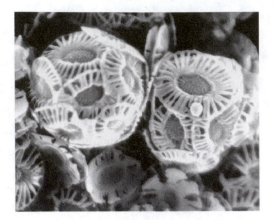

FIGURE 5.4. Scanning electron microscope image of the microscopic marine algal coccolithophore, *Emiliania huxleyi*, showing the liths made from calcium carbonate. Courtesy of Peter Miller (PML).

ocean and exchange with the atmosphere. This "biological pump" helps to control the exchange of carbon between the oceans, atmosphere, and sediment. A number of scientists have shown that the ability of some species of coccolithophores to form calcite (calcium carbonate) liths is impaired when grown at the CO_2 concentrations expected by the end of the century, so much so that the liths are deformed. The concern is that if coccolithophore blooms or their degree of lith calcification decreases this may reduce the extent of the biological pump and the removal of carbon to the sediments through ballasting, which could result in large changes in the Earth's carbon cycle. Coccolithophores also play a role in climate feedbacks through their production of dimethyl sulphide (DMS) and reflection of incoming solar radiation by the calcite liths (a process similar to the increased albedo effect of snow). DMS is a significant source of cloud condensation nuclei so a reduction in coccolithophore liths could result in fewer clouds. Reduced albedo and clouds

would result in global warming, but we currently do not know how severe the effect on our climate would be.

The study of the impact of ocean acidification on these organisms is still in its infancy. Scientists are currently using seawater mesocosms (large-volume natural seawater enclosures) dosed with future CO_2 concentrations to study phytoplankton growth and physiology and the production of important climate change gases such as DMS.[10] Other mesocosms are being used to examine the impact of a high CO_2 ocean on the biodiversity and biogeochemistry of animals that live on the seabed and within the sediments. Some of these animals, which burrow and plough (bioturbators) through the sediments, play a key role in maintaining the biodiversity and important chemical feedback processes of the overlying seawater. One such process is the breakdown of the organic matter that has sunk to the seabed after being produced in the sunlit upper ocean through photosynthesis by microscopic algae, the phytoplankton. Microbes in the sediment, with the help of the bioturbating sediment animals, recycle nutrients to the seawater by breaking down this organic matter. These nutrients are mixed by currents, tides, and wave action into the upper sunlit layer of the ocean and this sustains the next bloom of phytoplankton growth. These animals are the focus of other studies because there may be domino effects if some of them are lost due to their inability to cope with decreasing pH and carbonate ions for shell-making. Such species include starfish, sea urchins, and shellfish, many of which are also important food for fish and humans.

The chemical changes that occur through the increased oceanic uptake of CO_2 may also directly affect fish and shellfish by affecting the success of fertilization and development of their eggs, and indirectly through impacting the planktonic and benthic organisms on which they rely for food. The consequences for marine food webs are not yet well understood, but could be very serious.

Conclusion

Unless we urgently introduce effective ways to reduce CO_2 emissions, organisms and ecosystems are going to have to deal with a number of major rapid global changes simultaneously, namely rising sea temperatures, loss of ice, and ocean acidification. It is likely that these changes will have significant negative effects on marine ecosystems, with cascading global consequences.

Notes

1. Primary production is the production of organic compounds from atmospheric or aquatic CO_2, principally through photosynthesis. All life on Earth is directly or indirectly dependent on primary production.

2. Planktic organisms are plants and animals that live by drifting around in the seawater. Benthic organisms are those that live on or in the seabed sediments.

3. Upwelling is the upward movement of cold, nutrient-rich deep water toward the ocean surface, replacing the warmer, usually nutrient-depleted surface waters.

4. IPCC (2007). *Climate Change 2007: The Physical Science Basis.* The Intergovernmental Panel on Climate Change Contribution from Working Group I to the Fourth Assessment Report. Summary for Policymakers, 1–21.

5. The Intergovernmental Oceanographic Commission (IOC) of UNESCO and Scientific Committee of Oceanic Research (SCOR) of the International Council of Scientific Unions (ICSU) cohosted a symposium on the potential environmental consequences of using the deep ocean for intentional storage of CO_2 and to address for the first time the consequences of higher atmospheric CO_2 on the oceans, their chemistry, and the organisms and ecosystems within them. The symposium The Ocean in a High-CO_2 World was held in Paris in May 2004 (http://ioc.unesco.org/iocweb/co2panel/HighOceanCO2.htm). Key scientific papers from the symposium were published in a special issue of the *Journal of Geophysical Research* (2005, vol. 110). The seminal papers include Riebesell, U., et al. (2000). Reduced calcification of marine plankton in response to increased atmospheric CO_2. *Nature* 407:364–67; Caldeira, K., and M. E. Wickett (2003). Anthropogenic carbon and ocean pH. *Nature* 425:365; Sabine, C. L., et al. (2004). The oceanic sink for anthropogenic CO_2. *Science* 305:367–71; Feely, R. A., et al. (2004). Impact of anthropogenic CO_2 on the $CaCO_3$ system in the ocean. *Science* 305:362–66; and Orr, J. C., et al. (2005). Anthropogenic ocean acidification over the twenty-first century and its impact on calcifying organisms. *Nature* 437:681–86. Royal Society (2005). Ocean acidification due to increasing atmospheric carbon dioxide. Policy document 12/05 Royal Society: London, The Clyvedon Press, Ltd., Cardiff, UK, 68 pp. www.royalsoc.ac.uk/displaypagedoc.asp?id=13539. Turley, C., et al. (2006). Reviewing the impact of increased atmospheric CO_2 on oceanic pH and the marine ecosystem. In Schellnhuber, H. J., W. Cramer, N. Nakicenovic, T. Wigley, and G. Yohe, eds., *Avoiding Dangerous Climate Change.* Cambridge University Press, Cambridge, UK, pp. 65–70. www.stabilisation2005.com/55_Jerry_Blackford.pdf.

6. Royal Society (2005), op. cit.

7. Kleypas, J. A., R. A. Feely, V. J. Fabry, C. Langdon, C. L. Sabine, and L. L. Robbins (2006). Impacts of ocean acidification on coral reefs and other marine calcifiers: A guide for future research. Report of a workshop held April 18–20, 2005, St. Petersburg, FL, sponsored by NSF, NOAA, and the US Geological Survey. www.ucar.edu/communications/Final_acidification.pdf.

8. Royal Society (2005), op. cit.

9. C. M. Turley, J. M. Roberts, and J. M. Guinotte (2007). Corals in deep-water: Will the unseen hand of ocean acidification destroy cold-water ecosystems? *Coral Reefs* 26:445–48. DOI 10.1007/s00338-007-0247-5.

10. Royal Society (2005), op. cit. IOC (2005), op. cit. Refer also to papers cited in note 5.

Chapter 6

Water

PETER H. GLEICK

Introduction

Water resources are an integral part of the global hydrologic cycle, and, hence, the global climate cycle. Precipitation originates as evaporation from land and the oceans and feeds natural ecosystems and human water-management systems. Soil moisture is used by plants, which return more moisture to the atmosphere. Water that does not evaporate or transpire seeps into aquifers or runs off to form streams and rivers, which are used for hydropower, irrigation, and urban use. Mountain snow stored in winter provides water for rivers and deltas in the spring and summer. Storms bring extra moisture, while protracted periods of low rainfall cause drought. Scientists anticipate that climate change will entail significant and potentially severe consequences for the hydrologic cycle, with attendant repercussions for dependent human systems.

During the last century, humans have built a vast and complex infrastructure to provide clean water for drinking and for industry, dispose of wastes, facilitate transportation, generate electricity, irrigate crops, and reduce the risks of floods and droughts. Largely invisible, this extensive infrastructure has mitigated the effects of the natural variability in our climate. Although many uncertainties remain regarding the effects of climate change on the hydrologic cycle, we can no longer take for granted the ability of this infrastructure to continue to insulate us from our dependence on climate. Complex impacts affecting every sector of society—water resources especially—now seem unavoidable.[1]

There are limitations in the ability of climate models to incorporate and reproduce important aspects of the hydrologic cycle. Many hydrologic processes such as the formation and distribution of clouds and precipitation occur on a spatial scale smaller than most climate models are able to resolve. Regional data on water availability and use are often poor. Tools for quantifying many impacts are imperfect, at best.

At the same time, not everything is uncertain. We know with a high degree of confidence that climate change will cause rising sea levels, rising temperatures, enhanced evaporation, and accelerating snowmelt.[2] We have learned important things about the vul-

74

nerability and sensitivity of water systems and management rules to climate change. Progress has been made in exploring the strengths and weaknesses of technologies and policies that might help us cope with adverse impacts on water systems and take advantage of possible beneficial effects.

Prudent planning requires that strong climate and water research programs be maintained, that decisions about future water planning and management be flexible, and that the risks and benefits of climate change be incorporated into all long-term water planning. Rigid, expensive, and irreversible actions in climate-sensitive areas can increase vulnerability and long-term costs. Water managers and policymakers must start considering climate change as a factor in all decisions about water investments and the operation of existing facilities and systems.[3] Accordingly, two of the most important coping strategies must be to try to understand what the consequences of climate change will be for water resources and to then begin planning for and adapting to those changes.

Climate and Water Impacts: Some Key Concerns

The current state-of-the-science suggests that plausible climate changes raise a wide range of concerns that should be addressed by water managers and planners, climatologists, hydrologists, policymakers, and the public.

Impacts on Natural Hydrology

A wide variety of impacts are anticipated; I describe here only a few of the complex set of anticipated consequences. Climate changes have the potential to alter water quality significantly by changing temperatures, flows, runoff rates and timing, and the ability of watersheds to assimilate wastes and pollutants. Changes in precipitation can lead to both positive and negative impacts on water quality. Increased water flows could increase erosion and lead to greater sediment buildup, while decreased water flows could concentrate pollutants or reduce dissolved oxygen levels. The net effect on water quality depends on the interaction of climate change, land-use and agricultural practices, regulatory measures, and technical advances. The ability of existing systems and operating rules to manage these changes has not been adequately assessed.

Research indicates with very high confidence that watersheds with a substantial snowpack in winter, such as the Sierra Nevada in California and the Rocky Mountains, will experience major changes in the timing and intensity of runoff as average temperatures rise. Although detailed predictions for specific regions are difficult, in general, reductions in spring and summer runoff, increases in winter runoff, and earlier peak runoff are all common responses to rising temperatures; the implications are increased winter flooding and summer dryness.

One of the earliest reports of the Intergovernmental Panel on Climate Change concluded that

> the flood-related consequences of climate change may be as serious and widely distributed as the adverse impacts of droughts. . . . There is more evidence now that flooding is likely to become a larger problem in many temperate regions, requiring adaptations not only to droughts and chronic water shortages, but also to floods and associated

damages, raising concerns about dam and levee failure.[4]

Actual changes in flood risks will depend not only on climate, but on land-use designs, flood control infrastructure development and operation, and insurance policies.

Relative sea-level rise will adversely affect groundwater aquifers and freshwater coastal ecosystems through an increase in the intrusion of salt water into coastal aquifers. Shallow island aquifers (such as those found in Pacific Island nations, Hawaii, Cape Cod, and along the southeastern seaboard of the United States) together with coastal aquifers supporting large amounts of water for human use (such as those in Long Island, New York, and central coastal California) are at greatest risk.[5] Rising sea levels will also increase flood damages associated with current storm frequency and intensity. Synergistic effects with changes in storm frequency and intensity must also be evaluated. Other impacts of sea-level rise are likely to include changes in salinity distribution in estuaries, altered coastal circulation patterns, destruction of transportation infrastructure in low-lying areas, and increased pressure on coastal levee systems. A new assessment of the impacts of sea-level rise for California has identified dozens of wastewater treatment plants that are vulnerable to inundation from future increases, and other water-related infrastructure may also be vulnerable.[6]

Lakes are known to be sensitive to a wide array of changes in climate. Even small changes in climate can produce large changes in lake levels and salinity. As air temperatures increase, fewer lakes and streams in high-latitude areas will freeze to the bottom and the number of ice-free days will increase, leading to increases in nutrient cycling and productivity. Other effects include higher thermal stress for cold-water fish, improved habitat for warm-water fish, increased biological productivity, and lower dissolved oxygen.

Depending on the nature of the change, the systems affected, and the nature and scope of intentional interventions by humans, climate change will affect freshwater ecosystems through changes in vegetation patterns, possible extinction of endemic fish species already close to their thermal limits, declining area of wetlands with reductions in waterfowl populations, general decline in stream health, and major habitat loss.[7] Researchers express concern for the limited ability of natural ecosystems to adapt to or cope with climate changes that occur over a short time frame. This limited ability to adapt to rapid changes may lead to irreversible impacts, such as extinctions. While some research has been done on these issues, far more is needed. Nonlinear or threshold events are likely to occur, but are difficult to project. Examples include a fall in lake level that cuts off outflows or separates a lake into two separate parts, an increase in flood intensity that passes specific damage thresholds, and surpassing water-quality limits.

Impacts on Water Systems

There is a growing body of literature about how different climate changes may affect the infrastructure and complex systems built to manage water resources. Research has been conducted on potential impacts over a wide range of water-system characteristics, including reservoir operations, hydroelectric generation, navigation, and other concerns. At the same time, significant knowledge gaps remain and far more research is needed. Regional impacts will further depend upon the economic, institutional, and structural conditions in any region. Dynamic management strategies can be effective in mitigating the adverse impacts of climate change, but such policies need to be implemented before these changes occur to maximize their effectiveness.[8] Priorities and directions for future work should come from water managers and planners as well as from the more traditional academic and scientific research community.

Little work has been done on the impacts of climate change for specific groundwater basins, or for general groundwater recharge characteristics or water quality. Some studies suggest that regional groundwater storage volumes can be very sensitive to even modest changes in available recharge.[9] Although large changes in the reliability of water yields from reservoirs could result from small changes in inflows due to climate change, long-term demand growth can have a greater impact than climate changes. Uncertainties in projecting future water demands complicate evaluating the relative effects of these two forces.

Variability in climate already causes fluctuations in hydroelectric generation. Changes in the timing of hydroelectric generation can affect the value of the energy produced. More vulnerable regions include California (where 20 percent of electricity comes from hydroelectric plants), China, and parts of South America.

Climate change will have a wide range of consequences for coastal regions. In general, low-lying coastal regions such as Bangladesh, the Nile Delta, the Netherlands, and barrier islands are vulnerable to sea-level rise and changes in storm intensity and possibly storm frequency. Coastal wetlands are vulnerable to sea-level rise and changes in freshwater inflows from rivers. Ecosystem health will be affected by changes in the quality and quantity of freshwater runoff into coastal wetlands, higher water temperatures, extreme runoff rates or altered timing, and the ability of watersheds to assimilate wastes and pollutants. The net effect on coastal systems depends not only on climate change but also on a wide range of other human actions, including construction and operation of dams that trap sediments and nutrients, water withdrawal rates and volumes, disposal of wastes, extensive land development, and more.

A change in flood risks is one of the potential effects of climate change with the greatest implications for human well-being. Few studies have looked explicitly at the implications of climate change for flood frequency, in large part because of the lack of detailed regional precipitation information from climate models and because of the substantial influence of both human settlement patterns and water-management choices on overall flood risk.

Climate change can lead to dramatic long-term changes in forest health and distribution, depending in part on how precipitation and runoff patterns will change.[10] In turn, changes in forest conditions will have important feedback effects on runoff, soil erosion, soil salinization, groundwater quality, and more. These effects have not been adequately assessed.

In addition to the direct effects of climate change on water systems, climate change will entail significant indirect effects, via water systems, on societal functioning. Climate change will play a role in power production from conventional fossil fuel and nuclear power plants by raising cooling water temperatures and reducing plant efficiency (medium confidence). In some circumstances, higher water temperatures will constrain plant operations. Waterborne shipping and navigation are sensitive to changes in flows, water depth, ice formation, and storm frequency and intensity. Warming would increase the potential length of the shipping season in northern regions that typically freeze in winter. Decreases in river flows could reduce the periods when navigation is possible, increase transportation costs, or increase the conflicts over water allocated for other purposes. Research in specific watersheds has shown that the water in some major river basins (Colorado, Columbia, Yellow River, Nile, and others) is so heavily subscribed, with such complicated overlapping management layers, that the ability of local water users to adapt to changes in climate may be constrained.[11] Dynamic management strategies can be effective in mitigating the adverse impacts of climate change, but such policies need to be implemented before such changes occur to maximize their effectiveness.[12]

All of the physical and ecological impacts of climate change will entail social and economic costs and benefits. On top of the uncertainties in evaluating both climate change and potential impacts, evaluating the economic implications of water-related impacts is fraught with additional difficulties, and few efforts to quantify them have been made. Ultimately, however, comprehensive efforts to evaluate costs will be necessary in order to assist policymakers and the public in understanding the implications of both taking and not taking actions to reduce or adapt to the impacts of climate change.

The socioeconomic impacts of climate change vary depending on specific climate projections and on the methods and assumptions adopted by the researchers; thus policymakers should have less confidence in specific quantitative estimates. Even so, some general conclusions may guide assessment: (1) even small climate changes have the potential to affect the hydrologic cycle; (2) the economic consequences for changes in the hydrologic cycle are possibly quite large (for example, the United States could face costs up to 0.5 percent of its gross domestic product); and (3) opportunities exist to adapt to changing hydrologic conditions. The cost of adaptive measures is sensitive to water allocation mechanisms and management institutions.

Even given the uncertainties, research indicates that the possible economic impacts of reductions in water availability could be very large and that some water systems are highly sensitive to climate.[13] Under some climate scenarios, the additional costs imposed by climate changes are considerably larger than the additional costs imposed by future population growth, industrial changes, and changing new demands for water to grow food.

Regional Water/Climate Hotspots and Cross-Cutting Issues

Coastal Zones: Water resources in these areas are especially vulnerable to two likely impacts: sea-level rise and changes in storm frequency and intensity. Examples: Bangladesh; the Nile delta; Netherlands; barrier islands; regions with extensive development and building; island nations with limited or shallow groundwater resources.

Coastal Wetlands/Marshes: These areas are vulnerable to sea-level rise and changes in freshwater inflows from rivers. Examples: the Everglades; Louisiana/Gulf mangroves and marshes; Sacramento/San Joaquin delta; many other regional wetlands.

Snowmelt and Glacier-Dominated Water Supplies: All regions where snowfall and snowmelt play an important part in water supply will face significant impacts on timing and magnitude of runoff. Examples: California; the Rocky Mountains; the European Alps; the Himalayas (with a focus on glacier-melt systems).

Regions with High Dependence on Hydroelectricity: There is great uncertainty about changes in regional runoff, which affects hydropower generation, but some regions with high dependence will be more vulnerable to changes in water flows. Among the examples of possible problems: California (where 20 percent of electricity comes from hydroelectric plants); China; parts of South America.

Semiarid Regions: There is growing evidence from global climate models (GCMs) that dry areas will get drier, while wet areas will get wetter.

Food Production: There is growing concern about future water availability for food production on both irrigated and rainfed lands. Recent studies of agriculture suggest that overall production of food is tightly coupled to climate and water availability. The indirect effects of climate change on hydrology and water have not been incorporated into agroclimatic models, particularly effects of changes on pests, soil conditions, disease vectors, and socioeconomic factors. Even less work has evaluated the impacts of changes in climate variability for agriculture. Integrating these and other links between water and food should remain a high priority for researchers. Other concerns, such as impacts on ecosystems, water quality, and human health, are addressed in other chapters.

Climate change is just one of a number of factors influencing hydrological systems and water resources. Population growth, changes in land use, restructuring of the industrial sector, and demands for ecosystem protection and restoration are all occurring simultaneously. Current laws and policies affecting water use, management, and development are often contradictory, inefficient, or unresponsive to changing conditions. There are also growing connections among a wide range of intranational and international water policy initiatives and the risks of climate changes for shared watersheds. Current international agreements do not include provisions that explicitly address the risks of climate change for water quality, availability, and distribution. In the absence of explicit efforts to address these issues, the societal costs of water problems are likely to rise as competition for water grows and supply and demand conditions change.

Conclusion

There are many opportunities to reduce the risks of climate variability and change for water resources. Water managers have a long history of adapting to changes in supply and demand. Past efforts have focused on minimizing the risks of natural variability and maximizing system reliability. Tools for reducing these risks have traditionally included supply-side options such as new dams, reservoirs, and pipelines; and more recently, demand-management options, such as improving efficiency, modifying demand, altering water-use processes, and changing land-use patterns in floodplains. These management practices can be used to address the impacts of climate change as well. However, there are important differences between past experience dealing with natural climate variability and future needs to deal with climate change:

- Climate changes are likely to produce—in some places and at some times—hydrologic conditions and extremes of a *different nature* than current systems were designed to manage.
- Climate changes may produce similar kinds of variability but *outside of the range* for which current infrastructure was designed and built.
- Relying solely on traditional methods assumes that sufficient time and information will be available before the onset of large or irreversible climate impacts to permit managers to respond appropriately. This approach assumes that no special efforts or plans are required to protect against surprises or uncertainties.

Recommendations

- Improve long-term flood and drought planning.
- Develop technology options for better managing both supply and demand.
- Develop new economic tools for better managing both supply and demand.
- Maintain strong national climate and water monitoring and research programs.
- Reevaluate legal, technical, and economic approaches for managing water resources in the light of potential climate changes.

- Avoid expensive and irreversible actions in climate-sensitive areas.
- Improve planning under climate uncertainty.
- Systematically reexamine engineering designs, operating rules, contingency plans, and water allocation policies under a wider range of climate conditions and extremes than have been used traditionally. For example, the standard engineering practice of designing for the worst case in the historical observational record may no longer be adequate.
- Facilitate cooperation between water agencies and leading scientific organizations about the state-of-the-art thinking on climate change and impacts on water resources.

Notes

1. U.S. National Assessment Water Sector Report (Gleick, P. H., et al.). 2000. *Water: Potential consequences of climate variability and change for the water resources of the United States.* Report of the water sector assessment team of the national assessment of the potential consequences of climate variability and change. U.S. Geological Survey and Pacific Institute. Oakland, California.

2. See the various IPCC reports, including the 2007 Assessment, for details, especially Working Group II assessments, www.ipcc.ch.

3. As recommended early on, for example, in American Water Works Association. 1997. Climate change and water resources: Committee Report of the Public Advisory Forum. *Journal of the American Water Works Association*, Vol. 89, No. 11, pp. 107–10.

4. Intergovernmental Panel on Climate Change. 1995. *Impacts, Adaptations and Mitigation of Climate Change*: Scientific-Technical Analyses Contribution of Working Group II to the Second Assessment of the Intergovernmental Panel on Climate Change. R. T. Watson, M. C. Zinyowera, and R. H.Moss, eds. Cambridge University Press, Cambridge, UK.

5. Burns, W. C. G. 2002. Pacific Island developing country resources and climate change. In P. H. Gleick, ed., *The World's Water 2002–2003*, Island Press, Washington, D.C., pp. 113–31.

6 Heberger, M., H. Cooley, P. Herrera, P. H. Gleick, and E. Moore. 2009. The impacts of sea-level rise on the California coast. Pacific Institute, Oakland, California.

7. See, for example, P. S. Murdoch, J. S. Baron, and T. L. Miller. 2000. Potential effects of climate change on surface water quality in North America. *Journal of the American Water Resources Association*, Vol. 36, pp. 347–66; also V. Burkett and J. Kusler. 2000. Climate change: Potential impacts and interactions in wetlands of the United States. *Journal of the American Water Resources Association*, Vol. 36, pp. 313–20.

8. Yao, H., and A. Georgakakos. 2001. Assessment of Folsom Lake response to historical and potential future climate scenarios; 2. Reservoir management. *Journal of Hydrology*, Vol. 249 (1–4), pp. 176–96.

9. See, for example, K. Sandstrom. 1995. Modeling the effects of rainfall variability on groundwater recharge in semiarid Tanzania. *Nordic Hydrology*, Vol. 26, pp. 313–20; and T. R. Green, et al. 1997. Simulated impacts of climate change on groundwater recharge in the subtropics of Queensland, Australia. In *Subsurface Hydrological Responses to Land Cover and Land Use Changes*. Kluwer Academic Publishers, Norwell, Massachusetts, pp. 187–204.

10. McNulty, S. G., and J. D. Aber. 2001. Assessment on forest ecosystems: An introduction. *BioScience* September 2001, Vol. 51, No. 9, pp. 720–22.

11. Arnell, N. W. 2006. Climate change and water resources: A global perspective. In H. J. Schellnhuber, W. Cramer, N. Nakicenovic, T. Wigley, and G. Yohe, *Avoiding Dangerous Climate Change*. Cambridge University Press, Cambridge, UK, pp. 167–75.

12. Yao and Georgakakos, 2001. Op. cit.

13. Strzepek, K. M., and D. N. Yates. 2000. Responses and thresholds of the Egyptian economy to climate change impacts on the water resources of the Nile River. *Climatic Change*, Vol. 46, No. 3, pp. 339–56.

Chapter 7

Hurricanes

JUDITH A. CURRY AND PETER J. WEBSTER

This chapter examines the observations that describe variations and trends in hurricane activity, and summarizes the key points in the scientific debate surrounding the detection of increased hurricane activity and the attribution of this increase to greenhouse warming. Some speculations are given as to what the future might hold for hurricane activity in the North Atlantic, including a discussion of the uncertainties in projections of future hurricane activity. Finally, some perspective is given on our vulnerability to damage from hurricanes and policy responses. (See "Editors' Note, July 2009" at the end of this chapter for more recent information.)

The Data: Detection of Increased Hurricane Activity

Do the data indicate that global warming is causing increased hurricane activity? Our ability to detect an increase in hurricane activity from the historical database and to attribute any increase to greenhouse warming is plagued by uncertainties. The most reliable data on tropical cyclones (which includes tropical storms and hurricanes) are in the North Atlantic. The HURDAT data prepared by the National Hurricane Center go back to 1851.[1] Prior to 1944, only surface-based data were available (e.g., on landfalling storms and ship observations), and it is likely that some storms were missed. Since 1944, aircraft reconnaissance flights have been made in nearly all of the North Atlantic tropical cyclones, improving the estimates of hurricane intensity (wind speed). Since 1970, satellite observations have made observing and monitoring tropical cyclones more accurate.

Figure 7.1 shows the time series in the North Atlantic of the numbers of named storms (tropical cyclones; minimum wind speed 39 miles per hour), hurricanes (wind speeds exceeding 74 miles per hour), and category 4 and 5 hurricanes (wind speeds exceeding 131 miles per hour). To highlight the decadal and longer-term variability, the data have been smoothed (eleven-year running mean) to eliminate the year-to-year variability. A nominal seventy-year cycle is evident from peaks around 1880 and 1950 and minima around 1915 and 1985. However, the most striking aspect of the time series is the overall

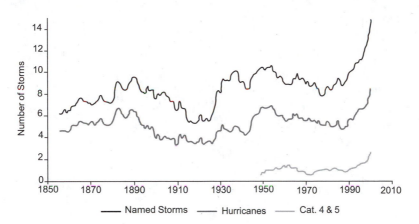

FIGURE 7.1. Number of total named storms, hurricanes, and category 4 and 5 storms since 1851, filtered by an eleven-year running mean. Data are obtained from www.aoml.noaa.gov/hrd/hurdat/. Figure courtesy of J. Belanger.

TABLE 7.1

Comparison of North Atlantic hurricane statistics for the periods 1945 to 1955 and 1995 to 2005 (data from www.aoml.noaa.gov/hrd/hurdat/).

	1945–1955	1995–2005
Named storms	115	165
Hurricanes	74	112
Category 4 +5	19	28

Source: Curry, J. A., P. J. Webster, and G. J. Holland, 2006: Mixing politics and science in testing the hypothesis that greenhouse warming is causing a global increase in hurricane intensity. *Bull. Amer. Meteorol. Soc.*, 87(8), 1025–37.

increasing trend since 1970 and the high level of activity since 1995.

Table 7.1 compares the statistics for the period 1995 to 2005 with the previous period of peak activity, 1945 to 1955. It is seen that the level of hurricane activity in the current period is 50 percent higher than the earlier period in every category—named storms, hurricanes, and category 4 and 5 storms—than the period around 1950. It is clear that the current period is not analogous to the 1950s and 1960s, since we are just entering the active phase.

While the data since 1944 are generally agreed to be reliable, what about the quality of data earlier in the record? Figure 7.2 shows the time series of total named storms and the average sea surface temperature (SST) in the main development region of the North Atlantic. Comparison of the two time series shows coherent variations of the number of storms and the SST for periods greater than twenty years. In particular, the period between 1910 and 1920 with low storm activity is associated with anomalously cool sea surface temperatures. The coherence between the total number of tropical storms and the sea surface temperature on multidecadal time scales lends credence to the tropical storm data in the earlier part of the period, although the storm intensity in the earlier part of the record is arguably much less accurate.

The most reliable dataset in the early part of the record is the count of the number of storms that have made U.S. landfall (figure 7.3). Strong evidence of the seventy-year cycle

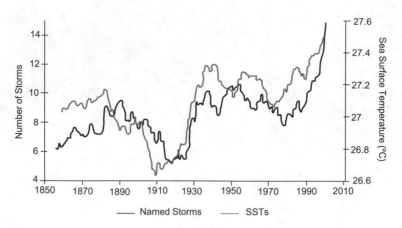

FIGURE 7.2. Number of total named storms in the North Atlantic and the average sea surface temperature in the main development region, filtered by an eleven-year running mean. Data are obtained from www.aoml.noaa.gov/hrd/hurdat/. Figure courtesy of J. Belanger.

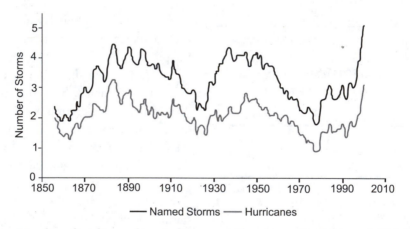

FIGURE 7.3. Number of total named tropical storms and hurricanes that have made U.S. landfall, filtered by an eleven-year running mean. Data are obtained from www.aoml.noaa.gov/hrd/hurdat/. Figure courtesy of J. Belanger.

is seen in these plots. Again, the activity during the past decade, particularly in terms of the total number of landfalling tropical storms, has surpassed slightly the previous peak period in the 1930s through the 1950s.

While the data provide strong support for elevated hurricane activity in the North Atlantic that is significantly beyond what has been seen in the historical record, is there evidence of elevated hurricane activity in the other oceanic regions where hurricanes form? Webster and colleagues examined the global hurricane activity since 1970 (the advent of reliable satellite data).[2] The most striking finding from this study is that while the total number of hurricanes has not increased globally, the number and percentage of category 4 and 5 hurricanes has nearly doubled since 1970 (see figure 7.4).

Scientists are debating the quality of the

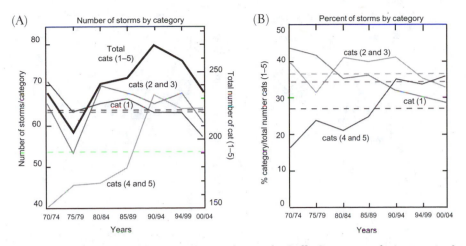

FIGURE 7.4. Intensity of global hurricanes according to the Saffir-Simpson scale (categories 1 to 5), in five-year periods. (A) The total number of storms and (B) the percent of the total number of hurricanes in each category class. Source: Webster, P. J., J. A. Curry, J. Liu, and G. J. Holland, 2006: Response to comment on "Changes in tropical cyclone number, durations, and intensity in a warming environment." *Science*, 311.

data upon which these analyses are based, particularly the data on the Indian Ocean, the Western North Pacific Ocean, and the South Pacific Ocean. There is an obvious need for an improved climate data record for global hurricane characteristics. Efforts are underway to reprocess the satellite data, although it will be a considerable challenge to assemble the data prior to 1977. A consistent method of determining surface wind speed combined with careful assessment of the satellite data integrity and sampling errors are essential elements of a reanalysis.

Cause and Attribution of the Increased Hurricane Activity

The increase in global hurricane intensity since 1970 and the increase in the number of named storms in the North Atlantic since 1995 have been associated directly with a global increase in tropical sea surface temperature.[3] Figure 7.5 shows the variation of tropical sea surface temperature (SST) in each of the ocean regions where tropical cyclone storms form. In each of these regions, the sea surface temperature has increased by approximately 0.5 degree Celsius (or 1 degree Fahrenheit) since 1970. The causal link between SST and hurricane intensity was established more than fifty years ago, when it was observed that tropical cyclones do not form unless the underlying SST exceeds 26.5 degrees Celsius and that warm sea surface temperatures are needed to supply the energy to support development of hurricane winds. The role of SST in determining hurricane intensity is generally understood and is supported by case studies of individual storms and by the theory of potential intensity. By contrast, no trend is seen in wind shear (see figure 7.6). Wind shear is the change of wind speed and direction with height in the atmosphere; small wind shear is conducive to tropical cyclone formation. While wind shear is an important

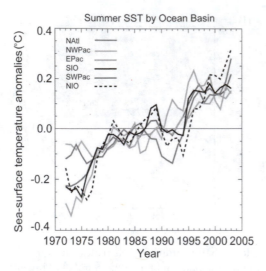

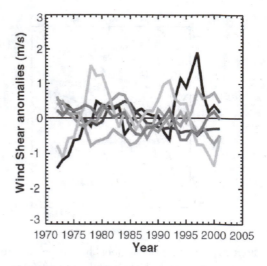

FIGURE 7.5. Evolution of the sea surface temperature anomalies relative to the 1970 to 2004 period for the North Atlantic, Western Pacific, East Pacific, South Indian Ocean, Southwest Pacific, and North Indian Ocean basins. Source: Curry, J. A., P. J. Webster, and G. J. Holland, 2006: Mixing politics and science in testing the hypothesis that greenhouse warming is causing a global increase in hurricane intensity. *Bull. Amer. Meteorol. Soc.*, 87(8), 1025–37.

FIGURE 7.6. Evolution of the wind shear anomalies relative to the 1970 to 2004 period for the North Atlantic, Western Pacific, East Pacific, South Indian Ocean, Southwest Pacific and North Indian Ocean Basins. Hoyos, C. D., P. A. Agudelo, P. J. Webster, and J.A. Curry, 2006: Devolution of the factors contributing to the increase in global hurricane intensity. *Science*, 312, 94–97.

determinant of the intensity of individual storms and even in the population of storms in an individual season, there is no trend in wind shear that can explain the observed increase in hurricane intensity since 1970.

A number of natural internal oscillations of the atmosphere/ocean system have a large impact on SST (e.g., El Niño, North Atlantic Oscillation). However, decadal-scale oscillations tend to be specific to each ocean basin and are often anti-correlated from one basin to another. The data show that the tropical SST increase is global in nature and occurs consistently in each of the ocean basins (see figure 7.5). This tropical warming is consistent with a similar increase in global surface temperatures and suggests a global forcing mechanism for the tropical SST increase.

The North Atlantic hurricanes deserve special discussion in light of the repeated assertions from the U.S. National Hurricane Center that the recent elevated hurricane activity is associated with natural variability, particularly the Atlantic Multidecadal Oscillation (AMO). Figures 7.1 through 7.4 suggest that natural modes of multidecadal variability, notably the AMO (around a seventy-year cycle), do have an influence on North Atlantic hurricane activity. However, recent examination of the data by Mann and Emanuel and Trenberth and Shea suggest that the impact of the AMO on tropical sea surface temperature and hurricane activity has been overestimated.[4] Variability in natural phenomena does not account completely for the observational record. Observations of globally averaged temperature are tracked by the *combination* of natural and anthropogenic forcings. Thus, analyses that rely solely on SST to identify the AMO may have convoluted the effects of the natural and anthropogenic factors in the identification of the phase and amplitude of the AMO signal. In addition, the AMO is expected to intensify during the current phase until 2020, which will exacerbate the effects of global warming. The strength of the tropical cyclone activity during the period of 1995 to 2005 suggests that the AMO alone cannot explain the elevated tropical cyclone activity observed in the North Atlantic during the last decade.

What can we conclude from the above analysis regarding the global increase in hurricane intensity and the increase in the North Atlantic of the total number of tropical cyclones? The arguments for natural variability to explain the increase are refuted by the known range of natural variability in the existing database and the absence of a convincing mechanism for natural variability that can explain the global increase in both oceanic temperatures and the frequency of intense hurricanes. The best available evidence supports the hypothesis that greenhouse warming is contributing to an increase in hurricane activity, although the evidence is by no means conclusive at this point. The primary issue is the magnitude of the observed increase in hurricane intensity, especially given concerns about the quality of the data.

Projections of Future Hurricane Activity

While both groups of scientists (those that support the natural variability explanation and those that support the global warming contribution) agree that hurricane activity in the North Atlantic will remain elevated for some years, the implications for future projections of hurricane activity are quite different. Based upon the hypothesis of natural variability being the cause of the high hurricane activity in the North Atlantic since 1995, there have been several predictions of a forthcoming downturn in hurricane activity: Goldenberg and colleagues imply a downturn in ten to forty years, while Gray anticipates a downturn in three to eight years associated with a global cooling.[5]

If our hypothesis is correct that greenhouse warming is causing both an increase in hurricane intensity globally and an increase in the number of storms in the North Atlantic, what does this imply for future hurricane activity as sea surface temperatures continue to rise and the oceanic warm pool continues to expand, especially in the spring and fall? We consider a range of projections from two different approaches: climate model sensitivity to increasing greenhouse gases and simplistic projections based upon the historical data record. A projection is made for average conditions in the year 2025 (such that high-frequency fluctuations from short-term oscillations such as El Niño are ignored), corresponding to an increase in tropical SST of 1 degree Fahrenheit that is attributable to greenhouse warming.

The observations of Webster and colleagues extrapolate to a 6 percent increase in maximum wind speeds for a 1 degree Fahrenheit SST increase.[6] By contrast, high-resolution climate model simulations have found a 2 percent increase in intensity when scaled for a 1 degree Fahrenheit SST increase, which is a factor three times smaller than that determined from the observations.[7] Oouchi also found that the number of North Atlantic tropical cyclones increased by 30 percent for a 2.5 degree Celsius increase in SST, which interpolates to an increase of one tropical cyclone per 1 degree Fahrenheit increase in SST.[8] By contrast, based upon the historical data record in the North Atlantic, an increase of 1 degree Fahrenheit in tropical SST implies an additional five tropical storms per season (inferred from figure 7.2), which is a factor of five greater than the number inferred from climate model simulations.

Projections of future hurricane variability must include both natural variability and greenhouse warming. Estimates of the magnitude of the impact of the Atlantic Multidecadal Oscillation (AMO) on the total number of tropical cyclones per year range from 0 (no effect) to 4 to 6 (the AMO explains the entire magnitude of the trough to peak variability in figure 7.1). Assuming that the AMO continues with a seventy-year periodicity, the peak of the next cycle is expected in 2020 (seventy years after the previous 1950 peak), so 2025 is very near the peak of the AMO cycle. Proponents of the natural variability explanation refer to active and quiet phases rather than actual cyclic behavior; their analysis states that we are currently in an active phase that will last ten to forty years, and there is no implicit assumption that the level of activity in 2025 will be higher than the activity of the past decade.

Based upon these assumptions, consider the following simple statistical model. The average number for the past decade of total North Atlantic tropical cyclones is 14.4. For simplicity, we assume that the effects of greenhouse warming and the AMO are separable and additive. Table 7.2 compares the simple statistical projections including both greenhouse warming (AGW) and AMO, AGW only, and AMO only. The range of projections given in table 7.2 provides some broad constraints on the conceivable elevation of North

Atlantic tropical cyclone activity in coming decades. The combination of AGW plus AMO would result in the greatest elevation in the number of named storms and an unprecedented level of tropical cyclone activity. The different assumptions range from an increase of 0 to 6.5 named storms per year. In terms of the intensity of the storms, figure 7.1 suggests that the distribution of the storm intensity is changing with warming, whereby the increase is in the number of tropical storms and in the number of category 4 and 5 storms, rather than in the weaker hurricanes. Consideration of U.S. landfalling hurricanes (see figure 7.3) suggests a continued increase over the next two decades in the ascending mode of the AMO, which when combined with AGW may result in an unprecedented number of U.S. landfalling hurricanes. Once the AMO begins its descending mode around 2020, continued warming makes it doubtful that we will ever again see the low levels of hurricane activity of the 1980s, and we can expect a leveling-off rather than significant decrease in activity until the next ascending phase of the AMO.

The projections in table 7.2 show a broad range, and any projections of future hurricane activity must be viewed as highly uncertain at this time, and it may take a decade before it is clear which of these models has the better predictive capability. Nevertheless, the range of projections given in table 7.2 does provide some broad constraints on the conceivable elevation of North Atlantic tropical cyclone activity in coming decades.

Politics and Policy Responses

Hurricane-induced economic losses have increased steadily in the United States during the last fifty years, with estimated total losses

TABLE 7.2

Projections for the average total number of North Atlantic tropical cyclones (named storms) for 2025. AGW refers to anthropogenic greenhouse warming; AMO refers to Atlantic Multidecadal Oscillation.

	AGW + AMO	AGW only	AMO only
Average last decade:	14.4	14.4	14.4
Global warming increases SST 1°F:	+1 to +5	+1 to +5	0
Continued increase of AMO:	+1.5	0	0
Total	16.9–20.9	15.4–19.4	14.4

averaging $35.8 billion per year during the last five years.[9] The 2005 season was exceptionally destructive, with damage loss from Hurricane Katrina exceeding $100 billion. During 2004 and 2005, nearly 2,000 lost lives were attributed to landfalling hurricanes. Damage from landfalling hurricanes is associated with high winds, storm surges, rainfall, and tornadoes. While coastal regions are the most vulnerable, the most intense precipitation and tornadic activity may occur inland and far from the actual center of the storm.

The devastation associated with Hurricane Katrina has served as a focusing event for the political debate on global warming. Although a single storm or a single season cannot be attributed to global warming, the connection between Hurricane Katrina and global warming was made in the public consciousness by the media attention given to the publication of the Emanuel's and Webster and colleagues' papers linking an increase in hurricane intensity and global warming.[10] Furthermore, projected rises in sea level will compound the impact of increased hurricane activity through increased storm surge vulnerability. The prospect of increased hurricane activity with global warming has substantially raised public awareness and concern about the potential adverse impacts of global warming.

There is certainly a risk of elevated hurricane activity with increased global warming, although the magnitude of this risk is presently uncertain. The uncertainties in the hurricane data are sufficient that hurricanes cannot be used as any kind of "smoking gun" for global warming; rather, the risk of elevated hurricane activity arguably represents the most devastating short-term impact of global warming, at least for the United States.

Researchers on both sides of the hurricane and global warming debate agree that the risk from hurricanes will remain high over the next few decades. To place the U.S. vulnerability in perspective, 50 percent of the U.S. population lives within 50 miles of a coastline and the physical infrastructure along the Gulf and Atlantic coasts represents an investment of more than $3 trillion; during the next several decades this investment is expected to double.[11] Globally, Bangladesh is particularly vulnerable to the combination of increased hurricane activity and sea-level rise; several hundred million people live in the southern part of the country where the elevation is only a few feet above sea level, and three tropical cyclones during the twentieth century each killed more than 100,000 people. The combination of the coastal demographics with the increased hurricane activity will continue to escalate the socioeconomic impact of hurricanes.

Any conceivable policy for reducing CO_2 emissions or sequestering CO_2 is unlikely to have a noticeable impact on sea surface temperatures and hurricane characteristics during the next few decades; mitigation strategies only have the potential to impact the longer-term effects of global warming.[12] Therefore, policy responses must include adaptive measures. The effects of increasing concentrations of population and wealth in vulnerable coastal regions must be confronted.[13] Rapidly escalating hurricane damage in recent decades owes much to government policies that serve to subsidize risk and hence promote risky behavior. Decreasing our vulnerability to damage from hurricanes will require a comprehensive evaluation of coastal engineering, building construction practices, insurance, land use, emergency management, and disaster relief policies in vulnerable regions. Political will at the local, national, and international levels is needed to develop the appropriate policy and technological options that are practically feasible, cost effective, and politically viable. Adaptation strategies will vary regionally, based upon the local geographic risks and nature of the economic dependence on coastal development and activities. The vulnerability of the developing world to increased hurricane activity and sea-level rise raises not only the obvious humanitarian and economic issues, but potential regional instabilities associated with mass migrations raise serious national security issues.

Editors' Note, July 2009

More recently, the correlation between SST and the Power Dissipation Index (PDI), combining storm frequency, intensity and duration,[14] has been challenged. Qualifying earlier statements, Kerry Emanuel acknowledges uncertainty and variability in the relationship between global warming and tropical cyclone activity across regions and models. In fact, the most recent projections indicate a decrease in the number of tropical storms, as a result of an increase in the threshold SST for atmospheric convective heating in a warming world, although the intensity of these storms may increase.[15] Even more critical of the SST and PDI correlation, Kyle Swanson argues that the PDI is sensitive to deviations from mean temperatures,[16] which has dramatically different implications for storm forecasts in a changing climate.

Questioning the reliability and accuracy of the record, Gabriel Vecchi and Thomas Knutson argue that early data points likely underestimate the number of storms. Adjusting for low counts early in the record supports a small, nominally positive trend in storm frequency since 1878, but this trend is not statistically significant.[17] Vecchi and Brian Soden do not rule out that global warming contributes to increased tropical storm activity, but they suggest that other factors, possibly in addition to global warming, are likely to have been substantial contributors. While an increase is SST may lead to an increase in the number and intensity of storms, warming of the troposphere and an increase in wind shear are linked to decreased frequency and intensity.[18] Which factors are more likely to prevail under climate change is difficult to determine.

Although it is probably premature to conclude definitively that greenhouse warming has already had a discernable impact on hurricane activity,[19] the generally emerging view seems to be that global warming may cause some increase in hurricane intensity that will develop slowly over time and likely will lead to a few more Category 4 and Category 5 storms. Given the major importance of tropical storm intensity to the vulnerability of coastal infrastructure and any flood-prone areas, it is certain that this topic will be the subject of intense scientific and political debate over the next decade.

Notes

1. Curry, J. A., P. J. Webster, and G. J. Holland, 2006: Mixing politics and science in testing the hypothesis that greenhouse warming is causing a global increase in hurricane intensity. *Bull. Amer. Meteorol. Soc.*, 87(8), 1025–37.

2. Webster, P. J., G. J. Holland, J. A. Curry, and H.-R. Chang, 2005: Changes in tropical cyclone number, duration, and intensity in a warming environment. *Science*, 309, 1844–46.

3. Emanuel, K., 2005: Increasing destructiveness of tropical cyclones over the past 30 years. *Nature*, 436, 686–88.

4. Mann, M. E., and K. A. Emanuel, 2006: Atlantic hurricane trends linked to climate change. *EOS*, 87, 233–44. Trenberth K. E., and D. J. Shea, 2006: Atlantic hurricanes and natural variability in 2005. *Geophys. Res. Lett.*, 33, L12704, doi:10.1029/2006GL026894.

5. Goldenberg, S. B., C. W. Landsea, A. M. Mestas-Nuñez, and W. M. Gray, 2001: The recent increase in Atlantic hurricane activity: Causes and implications. *Science*, 293, 474–79. Gray, W. M., 2006: Global warming and hurricanes. 27th American Meteorological Society Conference on Tropical Meteorology, Monterey, CA., paper 4C.1 http://ams.confex.com/ams/pdfpapers/107533.pdf.

6. Webster et al., 2005 op. cit.

7. Knutson, T. R., and R. E. Tuleya, 2004: Impact of CO_2-induced warming on simulated hurricane intensity and precipitation: Sensitivity to the choice of climate model and convective parameterization. *J. Clim.*, 17, 3477–95.

8. Oouchi, K., J. Yoshimura, H. Yoshimura, R. Mizuta, S. Kusunoki, A. Noda, 2006: Tropical cyclone climatology in a global-warming climate as simulated in a 20 km-mesh global atmospheric model: Frequency and wind intensity analyses. *J. Meteorol. Soc. Japan*, 84(2): 259.

9. Pielke, Jr., R. A., C. Landsea, M. Mayfield, J. Laver, and R. Pasch, 2005. Hurricanes and global warming. *Bull. Amer. Meteorol. Soc.*, 86, 1571–75.

10. Emanuel, 2005 op. cit. Webster et al., 2005 op. cit.

11. Mann and Emanuel, 2006 op. cit.

12. Trenberth and Shea, 2006 op. cit.

13. Meehl et al., 2004 op. cit.

14. Emanuel, 2005 op. cit. Emanuel, K. 2007. Environmental factors affecting tropical cyclone power dissipation. *Journal of Climate*, 20, 5497–509.

15. Emanuel, K. A. 2008. The hurricane-climate connections. *Bull. Amer. Meteorol. Soc.*, 89, ES10–20. Emanuel, K. A., R. Sundararajan, and J. Williams, 2008. Hurricanes and global warming results from downscaling IPCC AR4 simulations. *Bull. Amer. Meteorol. Soc.*, 89, 347–67. Knutson, T. R., J. J. Sirutis, S. T. Garner, I. M. Held, and R. E. Tuleya, 2007. Simulation of the recent multidecadal increase of Atlantic hurricane activity using an 18-km-grid regional model. *Bull. Amer. Meteorol. Soc.*, 88, 1549–65. Knutson, T. R., J. J. Sirutis, S. T., Garner, G. A Vecchi, and I. M. Held, 2008. Simulated reduction in Atlantic hurricane frequency under twenty-first century warming conditions, *Nature Geoscience*, doi:10.1038/ngeo 202.

16. Swanson, K. L., 2008: Nonlocality of Atlantic tropical cyclone intensities. *Geochemistry, Geophysics, Geosystems*, 9, Q04V01, doi:10.1029/2007GC001844.

17. Vecchi, G. A., and T. R. Knutson, 2008. On estimates of historical North Atlantic tropical cyclone activity. *Journal of Climate*, 21, 3580–600.

18. Vecchi, G. A., and B. J. Soden, 2007. Effect of remote sea surface temperature change on tropical cyclone potential intensity. *Nature*, 450, 1066–70, doi:10.1038/nature06423.

19. Knutson, T. R., 2008. Global warming and current research. Online at www.gfdl.noaa.gov/global-warming-and-hurricanes.

Chapter 8

Wildfires

ANTHONY L. WESTERLING

Introduction

There is a perception that large and severe wildfires have increased in many parts of the world in recent decades, and these increases are often attributed to anthropogenic climate change. Actually documenting increases in large wildfire frequency in many parts of the world is, however, very difficult. This is because, on the one hand, the incidence of wildfire naturally varies greatly on interannual to decadal timescales, necessitating a long record in order to detect significant trends in wildfire activity. On the other hand, long records that document wildfire activity are often not readily available.

Where fire histories are available, they tend to be affected by several problems. Older records are less comprehensive than recent records, meaning fires can appear to be increasing merely because of improved reporting. Changes in strategies and resources for managing wildfires over time may also affect apparent trends in wildfire as the comprehensiveness and effectiveness of suppression varies. Changes in population and land use can have immediate and dramatic effects on the number and sources of ignitions and on the availability and flammability of fuels. Over the long term, fire management and land uses that suppress surface fires can lead to changes in the density and structure of the vegetation that fuels wildfires, changing the likelihood of a large or severe fire occurring. Consequently, while there are good reasons to expect that climate change will lead to changes in wildfire, detecting a climate change signal in historical wildfire records is usually not possible.

While comprehensive long-term (i.e., century-scale) wildfire histories are often unavailable, we can still use the available records to infer the impact of climate change scenarios on wildfire. Where they can be compiled, accurate, comprehensive short-term documentary wildfire histories for recent decades analyzed in conjunction with climate and vegetation data provide insight into how wildfire responds to variations in climate.

Similarly, reconstructions of past wildfire from fire scars preserved in trees and from charcoal records from sedimentary cores, combined with reconstructions of past climate from tree rings, ice cores, corals, and other natural recorders, can also give us insights into how wildfire responds to climate variability. Reconstructions can span long time periods, but usually can't give a clear indication of climate-related long-term trends through recent decades for a variety of reasons, including the effects of land-use changes and fire suppression. However, from relationships observed between climate, vegetation, and wildfire in recent documentary records and fire reconstructions, we can infer how wildfire in diverse locations will respond to a warmer climate.

Climate—primarily temperature and precipitation—influences the occurrence of large wildfires through its effects on the availability and flammability of fuels. Climatic averages and variability over long (seasonal to decadal) time scales influence the type, amount, and structure of the live and dead vegetation that composes the fuel available to burn in a given location.[1] Climatic averages and variability over short (seasonal to interannual) time scales determine the flammability of these fuels.[2]

The relative importance of climatic influences on fuel availability versus flammability can vary greatly by ecosystem and wildfire regime type.[3] Fuel availability effects are most important in arid, sparsely vegetated ecosystems, while flammability effects are most important in moist, densely vegetated ecosystems. Consequently, climate scenarios' changes in precipitation can have very different implications than do changes in temperature in terms of the characteristics and spatial location of wildfire regime responses.

While climate change models generally agree that temperatures will increase over time, changes in precipitation tend to be much more uncertain. Therefore, in ecosystems where wildfire risks are strongly affected by precipitation, it is uncertain how these wildfire regimes may change. However, in ecosystems where wildfire risks are strongly driven by temperature, climate change is likely to lead to substantial increases in wildfires. Also, as climate change alters the potential spatial distribution of vegetation types, vegetation and associated wildfire regimes will be transformed synergistically.

While policies to mitigate climate change can help limit changes in wildfire regimes, some level of additional warming is going to occur regardless, requiring adaptation. Fire suppression, fuels management, and development policies (zoning and building codes) are the primary means by which wildfire risks are managed, in descending order of prominence. As means to adapting to a warmer climate, however, their priority may need to be reversed. Development policies and fuels management show more promise than intensified fire suppression for reducing some of the economic impacts of increased wildfire risks.

The first section of this chapter uses fire histories and climate records for the western United States to demonstrate some important climate-vegetation-wildfire interactions. In the second section we extrapolate these relations to some important ecosystems outside this region. Section three discusses the impacts of climate change on these climate-vegetation-wildfire interactions, and section four concludes with a summary of policy implications.

Climate-Vegetation-Wildfire Interactions in the Western United States

The type of vegetation that can grow in a given place is governed by moisture availability, which is a function of both precipitation (via its effect on the supply of water) and temperature (via its effect on evaporative demand for water).[4] As a result, the spatial distribution of vegetation types is strongly correlated with long-term average precipitation and temperature.

We consider four coarse vegetation types found in the western United States: forests, woodlands, shrublands, and grasslands.[5] Instead of representing the location of vegetation types spatially—by longitude and latitude (as in figure 8.1)—we can instead plot them climatologically: by their long-term average

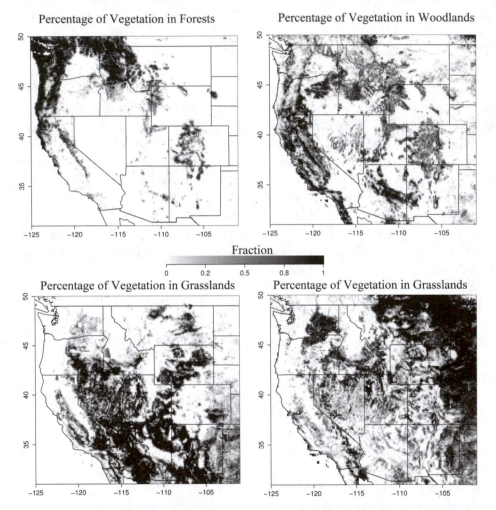

FIGURE 8.1. Fraction of vegetated area in the western United States covered with each of four vegetation types on a 1/8-degree grid, using University of Maryland vegetation classifications with fractional vegetation adjustment, from the North American Land Data Assimilation System. Upper left: forest (evergreen, deciduous, and mixed-cover types). Upper right: woodland (woodland and wooded grass/shrubland cover types). Lower left: shrubland (closed and open shrubland cover types). Lower right: grassland (grassland cover type). Source: K. E. Mitchell et al., *J. Geophys. Res* 109, D07S90, 2004.

annual precipitation and long-term average summer temperature. Forests are concentrated in parts of the western United States with the highest average annual precipitation and lowest average summer temperatures. Conversely, the highest fractions of shrubland cover are concentrated in locations with the least precipitation and warmest temperatures. In between these extremes there is a gradient from forest to woodland to grassland to shrubland cover types, with evident trade-offs between temperature and precipitation. This is because higher temperatures produce higher evaporation, reducing the moisture available to plants.[6] Vegetation types that require large amounts of water are consequently sensitive to temperature as well as precipitation. Woodlands, for example, can occur at higher tem-

peratures where precipitation is plentiful, but in regions where precipitation is more moderate, woodlands tend to be more plentiful in locations with lower summer temperatures (see figure 8.2).

The response of wildfire regimes in each location to interannual variability in climate varies in a manner consistent with a set of commonly employed hypotheses about the relative importance of fuel availability versus fuel flammability in diverse vegetation types. These hypotheses can be summarized as follows: (1) fuel availability becomes more of a limiting factor for wildfire activity as average moisture availability and biomass decrease; (2) fuel flammability becomes more of a limiting factor as average moisture availability and biomass increase.[7] These hypotheses make

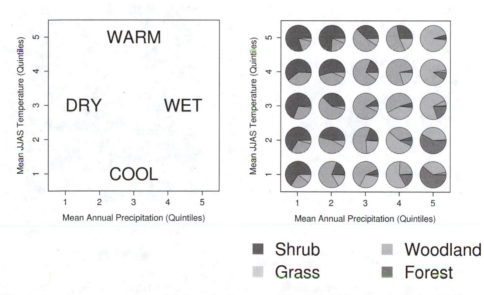

FIGURE 8.2. Average vegetation fractions for forest, woodland, shrubland, and grassland. Vegetation classifications are plotted for all the 1/8-degree grid cells in each quintile of annual precipitation and summer temperature for the contiguous western United States. Forests and woodlands are concentrated in cooler and/or wetter locations, while grasslands and shrublands tend to be in drier, warmer climates. Left: Quintiles of summer temperature (y-axis) and annual precipitation (x-axis) for the western United States. Right: Each pie chart shows the fractional vegetation coverage for lands corresponding to a pair of temperature and precipitation quintiles.

intuitive sense: the moist conditions that foster high biomass on average also tend to reduce fuel flammability, while the dry conditions that foster low average biomass imply high flammability in most years.

For our western U.S. example, we show an index of drought conditions ("moisture deficit") for large wildfires (i.e., fires greater than 200 hectares in burned area) during the season they burned and one year prior (see figure 8.3), again plotting them by their long-term average annual precipitation and long-term average summer temperature.[8] Average moisture deficits at the time of discovery for large wildfires are driest for cool, wet locations with mostly forest and woodland vegetation types (figure 8.3, left panel). That is, the locations with the highest average moisture availability

and biomass have the most fires when conditions are much drier than normal. This is consistent with the hypothesis that fuel flammability is the most important factor determining interannual variability in fire risks in these locations.

Moisture deficits were somewhat wetter than normal at the time of fire discovery for the warmest, driest locations that were mostly covered with shrubland vegetation (figure 8.3, left panel). This indicates that fires in the hottest, driest locations tended to occur in relatively wet years. Wet winter conditions in these locations foster the growth of grasses and forbs that quickly cure out in the very hot summer dry season typical of these locations, providing a load of fine fuels that can foster the ignition and spread of large wildfires.[9] This

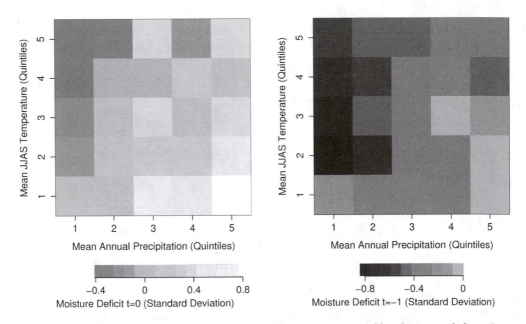

FIGURE 8.3. Deviations from average drought conditions for the year of fire discovery (left) and one year prior (right) for more than 8,000 western U.S. wildfires. Drought conditions are represented by normalized cumulative water-year moisture deficits, averaged for fires in locations whose long-term average annual precipitation and long-term average summer temperature correspond to quintiles in western U.S. precipitation and summer temperature. Normalized moisture deficits are shown in standard deviations from the mean.

pattern is consistent with the hypothesis that the availability of fine fuels is the limiting factor for fire risks in arid locations with less biomass.[10]

Moisture deficits a year prior to fire occurrences (figure 8.3, right panel) indicate wetter than normal conditions for a large part of the western United States, particularly for those areas with lower average annual precipitation that are primarily shrub- and grassland. Again, the tendency for fires to follow wet years is consistent with the idea that climate-driven variability in the availability of fuels drives the risk of large wildfires in more arid, lower biomass regions.[11]

To summarize the discussion thus far, the climatic controls (temperature and precipitation) on vegetation type largely determine the biomass loading in a given location, as well as the sensitivity of vegetation in that location to interannual variability in the available moisture. These in turn shape the response of the wildfire regime in each location to interannual variability in the moisture available for the growth and wetting of fuels. Cooler, wetter areas have greater biomass, and wildfires there tend to occur in dry years. Warmer, drier areas tend to have less biomass, and wildfires there tend to occur after one or more wet years.[12]

From a global perspective, an important consequence of this variability in wildfire regime response to climate is that wildfire is much more sensitive to variability in temperature in some locations than in others. In the western United States, cool, wet, forested locations tend to be at higher elevations and latitudes (figure 8.4) where snow can play an important role in determining summer moisture availability.[13] Above-average spring and summer temperatures in these forests have a dramatic impact on wildfire, with a highly

nonlinear increase in the number of large wildfires above a certain temperature threshold. A recent study by Westerling and colleagues concluded that this increase is due to earlier spring snowmelt and a longer summer dry season in warm years.[14] They found that years with early arrival of spring account for most of the forest wildfires in the western United States (56 percent of forest wildfires and 72 percent of area burned, as opposed to 11 percent of wildfires and 4 percent of area burned occurring in years with a late spring). This effect of the timing of spring was particularly pronounced in the higher-latitude (greater than 42 degrees north) mid-elevation (1,680 to 2,590 meters) forests of the Rocky Mountains, which accounted for 60 percent of the increase in forest wildfires in the western United States.[15] Higher-elevation forests in the same region have been buffered against these effects by available moisture, while lower elevations have a longer summer dry season on average and are consequently less sensitive to changes in the timing of spring.

The number of large wildfires in western U.S. grass- and shrublands is not significantly correlated with average spring and summer temperatures. In the western United States these types of vegetation tend to occur at lower elevations and latitudes, and consequently do not have as much snowfall, nor snow on the ground for as long, as do the forests of the Northern Rockies or forests at higher elevations in the Sierra Nevada or Colorado Rocky Mountains. The incremental effect of warmer temperatures on the duration and intensity of summer drought is less pronounced in areas with little or no snow on the ground for most of the year, and wildfire in these vegetation types appears to be limited more by fuel availability than by flammability. Given the importance of fuel availability, the

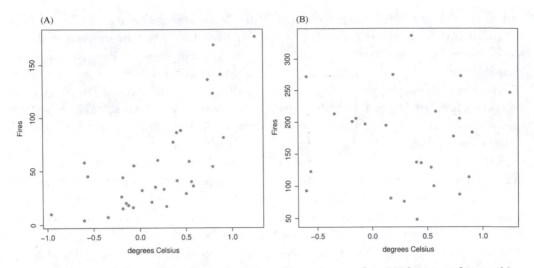

FIGURE 8.4 [LEFT]. Scatterplot of annual number of large (greater than 200 hectares) forest wildfires versus average spring and summer temperature for the western United States. Forest Service, Park Service, and Bureau of Indian Affairs management units reporting 1972 to 2004. Fires reported as igniting in forested areas only. [RIGHT]. Scatterplot of annual number of large (greater than 200 hectares) nonforest wildfires versus average spring and summer temperature for the western United States. Forest Service, Park Service, and Bureau of Indian Affairs and Bureau of Land Management management units reporting 1980 to 2004. Fires reported as igniting in nonforested areas only.

moisture available during the growing season is an important consideration, but it is probably less affected by spring and summer temperatures than by variability in precipitation.

Climate, Vegetation, and Wildfire Interactions Outside the Western United States

While the examples presented above are derived for the western United States, the climate-vegetation-wildfire linkages they describe are not unique to the region. Climatic influences on wildfire risks everywhere involve trade-offs between fuel availability and fuel flammability. The characteristics of a locale's vegetation and climate determine the relative importance of availability versus flammability at different time scales. Here we briefly discuss some additional examples from the literature on wildfire in boreal forests and tropical forests.

Wildfire in the boreal forests of Canada is associated with warm temperature anomalies and dry conditions caused by persistent high-pressure systems in the upper atmosphere.[16] Not unlike the Northern Rockies, the Canadian boreal forests are another example of a flammability-limited, high-biomass fire regime, where drought and warm temperatures increase the risk of large wildfires. However, antecedent moisture does not have a significant impact on fuel availability on interannual and shorter time scales. Impacts of earlier spring snowmelt on wildfire in the Canadian boreal forests have not been documented.

Goldammer and Price note that temperature is "not necessarily a critical factor influencing forest and savanna fire regimes in the tropical environment, which is characterized by high daytime temperatures anyway."[17] Recently, tropical Southeast Asia, Mexico, and the Amazon Basin have experienced very active fire years with large forest fires. These have been associated with El Niño conditions in the Pacific Ocean, which brought reduced precipitation in all three regions.[18] Like the boreal forest and the mountain forests of the western United States, fuel flammability is an important factor for fire risks in these tropical forests. Unlike in the higher latitudes, however, temperature anomalies are not as important. While reduced precipitation can play a role in promoting fire risks in boreal, tropical, and western U.S. mountain forests, its effects are probably most immediate in tropical forests due to their higher temperatures.

Implications for Climate Change

The direct effects of anthropogenic climate change on wildfire are likely to vary considerably according to current vegetation types and whether fire activity is currently more limited by fuel availability or flammability. In the long run, climate change is likely to lead to changes in the spatial distribution of vegetation types, implying that transitions to different fire regimes will occur in locations with substantial changes in vegetation.

Climate change will result in higher temperatures and more frequent and intense drought. In forests where wildfire is very sensitive to variations in temperature, the result is likely to be an increase in the frequency of very active fire seasons and an increase in the number of large wildfires. There have been substantial increases documented in the frequency of large wildfires in the boreal forests of Canada, Alaska, and Siberia, as well as in mid-elevation forests of the Northern Rocky Mountains of the western United States.[19] These increases have been associated with warmer temperatures in those areas in recent years.

Changes in precipitation, combined with increased temperatures, may have very different effects on wildfire in different ecosystems. In tropical forests for example, decreased precipitation will likely result in increased wildfire.[20] Conversely, higher temperatures and decreased precipitation could result in decreased wildfire activity in some dry, fuel-limited wildfire regimes, as the reduced moisture available to support the growth of fine fuels leads to less biomass and less continuous fuel coverage.[21] In both cases, any increases in precipitation might be counterbalanced to some extent by increased evaporative demand from higher temperatures.

The overall direction and spatial pattern of changes in precipitation under diverse climate change scenarios varies considerably across both future greenhouse gas emissions scenarios and global climate models.[22] In ecosystems where climatic influences on fire risks are dominated by precipitation effects, this implies greater uncertainty about climate change impacts on wildfire in those locations.[23]

Policy Implications

Climate scenarios (even those with rapid reductions in global greenhouse gas emissions) project increases in temperature substantially greater than those observed in recent decades, which have been associated with substantial increases in wildfire activity in some ecosystems.[24] Strategies for adapting to a warmer world will therefore need to consider the impacts of climate change on wildfire.

Currently, the primary strategies for managing wildfire risks fall into three general categories: fire suppression, fire prevention, and development policies. Suppression involves the active extinguishing of wildfires. Prevention measures seek to reduce the number of large fires and their economic and ecological impacts, primarily through vegetation management (e.g., mechanical thinning, managed fires, cleared buffers) and ignition reduction (e.g., burn controls, park closures, warnings, and educational campaigns). Development strategies include measures designed to reduce the impact of wildfires on structures, and of structures on the ability to manage wildfires safely and effectively. Measures include zoning ordinances to reduce the spread of development in fire-prone wild areas and regulations to enhance the ability of structures to resist fire (e.g., fireproof materials, thermal barriers, cleared perimeters, fire-resistant landscaping).

Developed countries devote considerable resources to suppressing wildfires, and the technologies employed have increased in sophistication during the last century. However, fire suppression technologies are still not effective under climatic conditions that foster the rapid spread of wildfires. Absent revolutionary technological developments, it is unlikely that additional investment in fire suppression will significantly reduce future fire risks in a warmer world. Furthermore, the ecological consequences of this kind of intervention might turn out to have their own undesirable consequences. Reducing fire activity in the short run may increase risks in the long term by contributing to the buildup of

fuels in otherwise fuel-limited wildfire regimes. This has already become a major problem in ponderosa pine forests in the Sierra Nevada and the southwestern United States due to fire suppression and land uses (such as grazing livestock).[25] Conversely, if fires could be effectively suppressed, this might be a desirable course of action in naturally dense forest ecosystems where very long return times between fires was previously the norm, if the result of climate change is that these forests would not regenerate post-fire and a substantial portion of the carbon stored in them would be released into the atmosphere.

Among prevention strategies, fuel management is likely to continue to be an important tool for building buffers around communities at risk from wildfire. It may also reduce the severity of wildfires in locations where forests have accumulated biomass due to fire suppression and land use. However, thinning forests that are naturally densely vegetated may not actually reduce wildfire risks. In the Amazon Basin, for example, thinning and clearing forests makes the remaining vegetation drier, increasing the risk of wildfire and of forest conversion.[26]

Development policies could make a substantial difference in the economic impact of wildfire in a warmer world by reducing the capital losses associated with catastrophic wildfires. By reducing the need to actively protect structures during a wildfire, these measures could also free up suppression resources that could be better employed protecting resources with cultural and natural conservation values. All of these measures (suppression, prevention, development) have been emphasized to varying degrees around the world. In places like the western United States, where there is a substantial and rapidly growing wildland-urban interface in fireprone areas, development strategies hold out the greatest promise to reduce the economic impact of wildfires in a changed climate. However, they have only limited applicability to preserving ecosystem and resource values.

Conclusion

The effects of climatic change on wildfire will depend on how past and present climates have combined with human actions to shape extant ecosystems. Climate controls the spatial distribution of vegetation, and the interaction of that vegetation and climate variability largely determines the availability and flammability of the live and dead vegetation that fuels wildfires. In moist forest ecosystems where snow plays an important role in the hydrologic cycle and fuel flammability is the limiting factor in determining fire risks, anthropogenic increases in temperature will lead to more fire activity.

In dry ecosystems where fire risks are limited by fuel availability, warmer temperatures may not increase fire activity significantly. Warmer temperatures and greater evaporation in some places could actually reduce fire risks over time if the result is reduced growth of grasses and other surface vegetation that provide the continuous fuel cover necessary for large fires to spread. The effect of climate change on precipitation is also a major source of uncertainty for fuel-limited wildfire regimes. However, in some places these are the same ecosystems where fire suppression and land uses that reduce fire activity in the short run have led to increased fuel loads today as formerly open woodlands have become dense forests. For the immediate future, this increases the risk of large, difficult-to-control fires with ecologically severe impacts.

Thus, the combined long-term impact of diverse human activities has been to increase the risks of large wildfires in many places in ways that cannot be easily reversed. Even if prompt action is taken now to reduce future emissions of greenhouse gases, the legacy of increased atmospheric concentrations of these gases means that the risk of large fires will remain high and will continue to increase in many forests. Consequently, societies will need to adapt.

Given that the time frame for climate change extends decades and centuries into the future, it is feasible to consider development strategies that reduce societies' economic vulnerability to wildfire. By putting fewer structures in places where fire risks are high and increasing, and by taking measures to increase the ability of structures to resist fire, capital losses due to wildfire can be reduced.

Notes

1. Stephenson, N. L. Actual evapotranspiration and deficit: Biologically meaningful correlates of vegetation distribution across spatial scales. *J. Biogeog* 25, 1998: 855–70.

2. Westerling, A. L., T. J. Brown, A. Gershunov, D. R. Cayan, and M. D. Dettinger. Climate and wildfire in the western United States. *Bulletin of the American Meteorological Society* 84(5), 2003: 595–604.

3. Ibid.

4. Stephenson, 1998 op. cit.

5. We use the Land Data Assimilation System (LDAS) 1/8-degree gridded vegetation layer using the University of Maryland vegetation classification scheme with fractional vegetation adjustment ("UMDvf") (see K. E. Mitchell et al., *J. Geophys. Res* 109, D07S90, 2004). Four coarse vegetation types were derived from the UMDvf classifications: forest (evergreen, deciduous, and mixed-cover types), woodland (woodland and wooded grass/shrubland cover types), shrubland (closed and

open shrubland cover types), and grassland (grass-land cover type). Figure 8.1 shows the fraction of vegetated area in each 1/8-degree grid cell comprised of these vegetation types.

6. Stephenson, 1998 op. cit.

7. See, for example, Swetnam, T. W., and J. L. Betancourt. Mesoscale disturbance and ecological response to decadal climatic variability in the American Southwest. *Journal of Climate* 11, 1998: 3128–47. Veblen, T. T., T. Kitzberger, and J. Donnegan. Climatic and human influences on fire regimes in ponderosa pine forests in the Colorado Front Range. *Ecological Applications* 10, 2000: 1178–95. Westerling et al. 2003, op. cit. Westerling, A. L., H. G. Hidalgo, D. R. Cayan, and T. W. Swetnam. Warming and earlier spring increases western U.S. forest wildfire activity. *Science* 313, 2006: 940–43.

8. The index of drought conditions used here is defined as the average, normalized, cumulative water-year moisture deficit. Moisture deficit is the difference between the moisture that could evaporate from soils and vegetation based on the observed temperature (potential evapotranspiration) and the actual evaporation constrained by the available moisture (actual evapotranspiration). It is a more reliable indicator of drought stress in plants than many other hydrologic measures and neatly incorporates the trade-offs between temperature and precipitation that determine the moisture available to plants (see Stephenson, 1998 op. cit.).

We use cumulative water-year (October through September) moisture deficits here to represent the cumulative drought stress on vegetation during the fire season, which is usually in the summer in this region. For much of the western United States, most precipitation occurs in fall, winter, and spring, rather than the summer, so the water year begins in October to capture the effect of variations in the years' water supply. By normalizing the moisture deficit, we get a measure of the deviation from average conditions that can be compared across locations.

9. Osmond, C. B., L. F. Pitelka, and G. M. Hidy, Eds. Plant biology of the basin and range: Ecological studies. *Springer-Verlag* 80, 1990.

10. A complicating factor is that in locations with an active summer monsoons, fires are ignited by dry lightning in the early summer and extinguished by monsoon rains later in the season. Greater fire activity can be associated with more lightning strikes, which in turn may be associated with subsequent precipitation. In this case, moist conditions could also indicate fires associated with an active summer monsoon rather than with just a wet winter immediately preceding the fire season.

11. In this case, associations between lightning ignitions and precipitation do not play a role because of the long lag time (one year) between the above-average moisture and the occurrence of a large fire.

12. Note that the regions with the wettest conditions on average the year of the fire (upper left corner of figure 8.3, left panel) do not show as strong a wet signal a year prior to the fire as do regions that are somewhat cooler (upper left versus middle left of figure 8.3, right panel). This may have to do with the effects of summer temperature on the time it takes for vegetation to dry out sufficiently to burn. Fine fuels grown in very warm locations may be more likely to burn the same year, while in cooler locations there may be more of a trade-off between moisture effects on availability and flammability, resulting in longer lags on average between the growth, drying out, and burning of vegetation.

Notice also that the average change in moisture a year prior to fires in forest areas is negligible (figure 8.3, right panel), implying that moisture effects on fuel production are probably not important on interannual time scales in these locations. The greater biomass available in these forest areas (where canopy cover is greater than 60 percent) means the incremental effect on fuel load of one years' growth is negligible.

13. Sheffield, J., G. Goteti, F. H. Wen, and E. F. Wood. *J. Geophys. Res.* 109, D24108, 2004.

14. Westerling et al., 2006 op. cit.

15. "Northern" from a U.S.-centric viewpoint, of course. The Canadian Rockies lie farther to the north.

16. See Girardin, M. P., Y. Bergeron, J. C. Tardif, S. Gauthier, M. D. Flannigan, and M. Mudelsee. A 229 year dendroclimatic-inferred record of forest fire activity for the Boreal Shield of Canada. *International Journal of Wildland Fire* 15, 2007: 375–88 for a summary citing Newark 1975; see also Johnson, E.A., and D. R. Wowchuck. Wildfires in the southern Canadian Rockies and their relationship to mid-tropospheric anomalies. *Canadian Journal of Forest Research* 23, 1993: 1213–22. Bessie, W. C., and E. A. Johnson. The

relative importance of fuels and weather on fire behavior in subalpine forests in the southern Canadian Rockies. *Ecology* 26, 1995: 747–62. Skinner, W. R., B. J. Stocks, D. L. Martell, B. Bonsal, and A. Shabbar. The association between circulation anomalies in the mid-troposphere and the area burned by wildland fire in Canada. *Theoretical and Applied Climatology* 63, 1999: 89–105. Skinner, W. R., M. D. Flannigan, B. J. Stocks, D. L. Martell, B. W. Wotton, J. B. Todd, J. A. Mason, K. A. Logan, and E. M. Bosch. A 500-hPa synoptic wildland fire climatology for large Canadian forest fires, 1959–1996. *Theoretical and Applied Climatology* 71, 2002: 157–69. Flannigan, M. D., and J. B. Harrington. A study of the relation of meteorological variables to monthly provincial area burned by wildfire in Canada (1953–1980). *Journal of Applied Meteorology* 27, 1988: 441–52.

17. Goldammer, J. G., and C. Price. Potential impacts of climate change on fire regimes in the tropics based on MAGICC and a GISS gcm-derived lightning model. *Climatic Change* 39, 1988: 273–96.

18. Román-Cuesta, R. M., M. Gracia, and J. Retana. Environmental and human factors influencing fire trends in Enso and non-Enso years in tropical Mexico. *Ecological Applications* 13(4), 2002: 1177–92. See also Schimel, D., and D. Baker. The wildfire factor. *Nature* 420, 2002: 29–30. Nepstad, D., et al. Amazon drought and its implications for forest flammability and tree growth: A basin-wide analysis. *Global Change Biology* 10(5), 2004: 704–17.

19. Gillett, N. P., A. J. Weaver, F. W. Zwiers, and M. D. Flannigan. Detecting the effect of climate change on Canadian forest fires. *Geophysical Research Letters* 31, 2004. Soja, Amber J., et al. Climate-induced boreal forest change: Predictions versus current observations. *Global and Planetary Change*, 56(3–4), April 2007: 274–96. Westerling et al., 2006 op. cit.

20. Goldammer and Price, 1998 op. cit.

21. Westerling, A. L., and B. P. Bryant. Climate change and wildfire in California. *Climatic Change* 87, 2007: 231–49.

22. Dettinger, M. D. A component-resampling approach for estimating probability distributions from small forecast ensembles. *Climatic Change* 76(1–2), 2006: 149–68.

23. Westerling and Bryant, 2007 op. cit.

24. Intergovernmental Panel on Climate Change (IPCC). *Climate Change 2007: The Physical Science Basis*. Cambridge, UK: Cambridge University Press, 2007. Gillett et al. 2004, Westerling et al., 2006, and Soja et al., 2007 op. cit.

25. Allen, C. D., et al. Ecological restoration of southwestern ponderosa pine ecosystems: A broad perspective. *Ecol. Appl.* 12, 2002: 1418–33.

26. Laurance, W. F., et al. Ecosystem decay of Amazonian forest fragments: A 22-year investigation. *Conservation Biology* 16(3), 2002: 605–18. Cochrane, M. A. Fire science for rainforests. *Nature* 421, 2003: 913–19.

Chapter 9

Tropical Forests of Amazonia

PHILIP M. FEARNSIDE

Introduction

Tropical forests are vulnerable to climate change, and large areas of these forests will not survive under business-as-usual scenarios. Projected climate changes could threaten the biodiversity of these forests and the traditional peoples and others who depend upon the forests for their livelihoods. Climate change also threatens global economic interests and the environmental services supplied by the forests to locations both near to and far from the forests themselves. Greenhouse-gas emissions provoked by forest dieoff are part of a potential positive feedback relationship leading to more warming and more dieoff. The Amazon forest is a focus of concern both because of the particularly severe impacts of climate changes predicted for this area and because the vast extent of this forest gives it a significant role in either intensifying or mitigating future climate change. If all of Brazil's Amazon forest were replaced with the landscape implied by land-use trends in the region's deforested areas today, the net committed emission that would be released totals 68.7 Gt CO_2-equivalent C, or gigatons (billion metric tons) of carbon in the form of carbon dioxide, with non-CO_2

gases converted to CO_2 equivalents using the global warming potentials adopted under the Kyoto Protocol.[1] The equivalent value for all vegetation types in the tropics as a whole would be 225 Gt CO_2-equivalent C— an astronomical amount were it to enter the atmosphere.

This chapter discusses modeled scenarios for climate change impacts in Amazonia and explains climate-forest interactions that magnify the threat to forest survival. These include synergisms with deforestation and loss of evapotranspiration and feedbacks with El Niño, forest fires, and the release of soil carbon. The chapter concludes with an explanation of the role of Amazonian rainforests in defining "dangerous" climate change. Unfortunately, the climate has already become dangerous for Amazonian forest.

Climate-Forest Interaction in Amazonia

Scenarios

Modeled scenarios for future climate in tropical forest areas vary widely, which can easily be misleading from a policy perspective for three reasons. First, the ghost of resolved uncertainties can continue to haunt not only popular discourse, but also scientific discussion of the topic for years or decades.[2] Second, there is a strong tendency to fall victim to the "Goldilocks fallacy"—when presented with a range of projections, one naturally assumes that the one in the middle will be "just right," even though the best result may well be at either the high or the low end of a range of available estimates.[3] Third, the existence of uncertainty commonly provokes the response of "let's wait and see what the experts decide," when this uncertainty should instead lead to even more vigorous action based on the precautionary principle.[4]

The case of predicted climate changes and their impacts on Amazonian forest is a highly relevant example of the danger of applying a "Goldilocks" approach to interpreting modeled scenarios. In 2000, the Hadley Center model of the UK Meteorological Office (UKMO) was the first model to forecast a catastrophic dieoff of Amazonian forest by the year 2080 under a business-as-usual scenario due to the inclusion of important feedback effects previously ignored (figure 9.1).[5] Other global climate models, which lacked these feedbacks, did not indicate any such catastrophe.[6] By 2005, most of the global climate models had been revised to include feedbacks that had previously been restricted to the

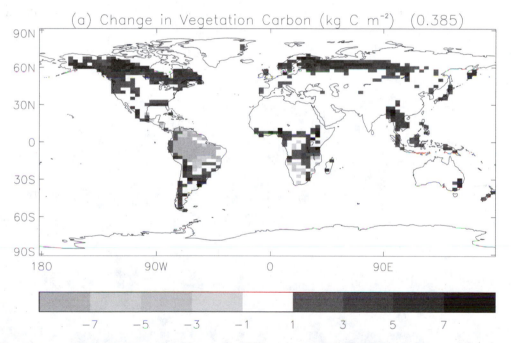

Figure 9.1. Change in vegetation biomass by 2080 projected with the Hadley center model under median climate sensitivity. Subsequent improvements of the model also show collapse of the Amazon forest on the same time scale. Source: Cox, P. M., R. A. Betts, C. D. Jones, S. A. Spall, and I. J. Totterdell. 2001. Modelling vegetation and carbon cycle interactive elements of the climate system. Hadley Centre Technical Note 23, UK Meteorological Office, Hadley Center, Exeter, UK. 28 pp.

Hadley model, with the result that five out of seven models show permanent "El Niño-like conditions" in the Pacific, with reduced rainfall and increased temperature in Amazonia. The Hadley model indicates Amazonia experiencing temperature increases ranging from 8.7 to 14 degrees Celsius by the end of the century depending on the climate sensitivity (figure 9.2).[7]

This calculation assumed the equilibrium concentration of CO_2 double the preindustrial level, a mark that should be reached around 2070 if there is no mitigation of the greenhouse effect. Under a business-as-usual emissions scenario, projected increases by 2100 are approximately 40 percent higher than the corresponding value for climate sensitivity (i.e., 3.5 degrees Celsius as a "most likely" value in 2100 versus 2.5 degrees Celsius for climate sensitivity).

An analysis of indicators of past climatic changes has recently reduced the estimates

for the probability of the true value of climate sensitivity being at the extreme high end of the range of possible values, the point that corresponds to a 95 percent margin of safety decreasing from 9.7 to 6.2 degrees Celsius (figure 9.3).[8] Proportionally, the 14-degree-Celsius increase in Amazonia in approximately 2070 under high climate sensitivity would fall to an increase of 8.7 degrees Celsius, which would still be a catastrophe that threatens both the forest and the human population in the area.

El Niño, the phenomenon where a warming of surface water in the tropical Pacific Ocean triggers changes in air currents that alter rainfall at many places around the world, causes droughts in some locations such as Amazonia and floods in others, such as southern Brazil. A causal connection between climate change and El Niño phenomenon has major policy implications because El Niño weather patterns have unambiguous and dev-

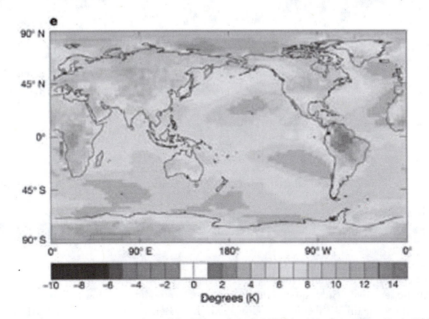

FIGURE 9.2. Temperature map. Adapted by Hergerl et al. 2006, op. cit. from Kerr, R. A. 2006. Latest forecast: Stand by for a warmer, but not scorching, world. *Science* 312: 351.

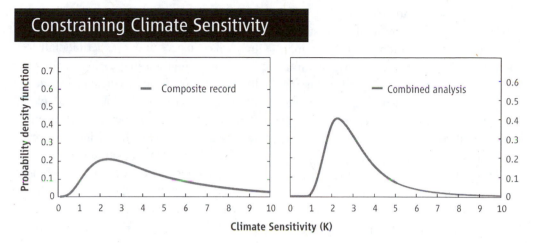

FIGURE 9.3. Revision of climate sensitivity, leaving the most probable value unchanged but reducing the probability of very low or very high sensitivities: Climate sensitivity is the increase in equilibrium global mean temperature above preindustrial levels when the preindustrial concentration of atmospheric CO_2 is doubled (i.e., in approximately 2070). Source: Hergerl et al. 2006, from Kerr, R. A., 2006. Latest forecast: Stand by for a warmer, but not scorching, world. *Science* 312: 351.

astating global consequences that are known already. Of twenty-one models, Hadley Center's Had3CM model provides the best representation of the link between sea-surface temperature in the Pacific and Amazonian droughts.[9] The connection between water temperatures in the Pacific and Amazonian droughts is known from direct observations and does not rely on model results. Even the more modest climate changes in Amazonia indicated by models with no El Niño connection would be sufficient to cause a large part of the Amazon forest to be replaced by savanna within the current century.[10]

Sea-surface temperatures in the Atlantic Ocean, which are also affected by global warming, have a significant effect on droughts in the southern and western parts of Amazonia, as occurred in 2005.[11] The Hadley model indicates a dramatic increase in this kind of drought beginning almost immediately assuming business-as-usual emissions, with the annual probability of such droughts increasing from 5 percent in 2005 to 50 percent in 2025 and 90 percent in 2060.[12]

Synergisms and Feedbacks

Flammability of Amazonian forest is expected to increase under numerous climatic scenarios, which will have significant feedback effects on climate change as carbon is released into the atmosphere and carbon sinks are removed.[13] The logical result of reducing rainfall and increasing temperature is to dry out the litter on the forest floor that serves as fuel for forest fires. Tree mortality increases the amount of litter available to burn, forming a positive feedback loop with fire occurrence.[14] Forest flammability is further increased through logging, which greatly increases the risk of fire by opening canopy and by the logging operations that kill many trees in addition to those that are harvested.[15] In addition, loss of forest both through deforestation and through dieback from climate change would lead to reduced evapotranspiration in the region, thereby cutting off part of the supply of water vapor needed to maintain large amounts of rainfall in the region—forming another positive feedback relationship leading to forest degradation and loss.[16]

Current El Niño conditions already result in wide areas of the region becoming susceptible to fire.[17] Forest fires have become a major threat to forests both in Amazonia and in Southeast Asia. These forests are not adapted to fire, and the thin bark of the trees makes them more susceptible to mortality when fires do occur than is the case for trees such as those in savannas or coniferous forests. In Amazonia, fire entering surrounding forest from burning in agricultural clearings or in cattle pastures was practically unknown to most Amazonian residents prior to the 1982/1983 El Niño event. Nevertheless, droughts caused by severe El Niños in the past resulted in forest burning as in the "big smoke" of 1926 and in four "mega–El Niño" events over the last 2,000 years when forest fires left charcoal in the soil.[18] But the 1982/1983 El Niño was a change, with substantial areas burning both in Amazonia and in Indonesia.[19] Fires are favored both by the greater frequency of El Niño and by the greatly increased presence of ignition sources from the spread of human agriculture and ranching in these areas.

Carbon in the biomass of standing Amazonian forests is released to the atmosphere during El Niño events.[20] These forests can subsequently reabsorb the carbon during La Niña and "normal" years, but the observed shift toward more frequent El Niños, together with the prediction of a permanent El Niño after the middle of the current century, suggest that carbon stocks will be steadily drawn down in the remaining forest. Forest degradation takes place under experimentally induced dry conditions in the Amazonian forest that mimic conditions predicted by models.[21] In these experimental plots, where plastic sheeting intercepts 60 percent of the moisture, large trees are the first to die, thus greatly increasing the release of carbon.[22] The same occurs at forest edges, where microclimatic conditions are hotter and drier than in the interior of a continuous forest.[23]

Unfortunately, fire risk is virtually never included in forest-management plans. Logging is rapidly spreading to formerly inaccessible areas of the forest. Outside of fully protected parks and reserves, management for timber is expected to be the use to which large areas of forest will be put. Fire risk will increase in the large areas subject to illegal logging, in legally managed areas on private land, and in new areas of public land to be opened for forest management in accord with a law enacted in January 2006 allowing forty-year concessions in up to 13 million hectares of "public forests."

The future role of soil carbon under climate change is a worldwide concern.[24] An early model indicating the possibility of substantial loss of soil carbon in Amazonia was developed by Townsend and coworkers.[25] The Hadley Center model predictions are much more severe—by 2080, approximately two-thirds of the soil carbon is lost.[26]

These soil carbon stocks represent a veritable time bomb. As the soil releases its carbon store, global temperatures will rise even more and trigger a "runaway greenhouse effect" that could escape from human control. The magnitude of this feedback effect is staggering since the total of soil carbon emissions potentially exceeds the combined fossil fuel and de-

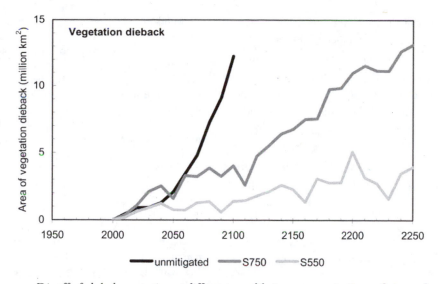

FIGURE 9.4. Dieoff of global vegetation at different equilibrium concentrations of atmospheric CO_2 using the Hadley Center model. The global vegetation dieoff is dominated by mortality in Amazonia (figure 9.1). The atmospheric CO_2 concentration passed the 380 ppmv mark in 2006 and is increasing at 2.6 ppmv/year. Source: Arnell, N. W., M. G. R. Cannell, M. Hulme, R. S. Kovats, J. F. B. Mitchell, R. J. Nichols, M. L. Parry, M. T. J. Livermore, and A. White. 2002. The consequences of CO_2 stabilisation for the impacts of climate change. *Climatic Change* 53(4): 413–46.

forestation emissions from human activity. Unlike emissions from fossil fuels and defor- estation, we humans do not have the option of solving the problem by altering our actions. The need for intensified research to quantify soil emissions under different climatic scenar- ios is urgent.

Conclusion

The consequences of climate change for the Amazon are globally significant due to the im- portant feedback effects of the Amazon forest. Loss of forest in Amazonia through defores- tation and increased fire risk exacerbates the trends driving climate change, especially changes in El Niño effects, while the green- house gases released from loss of forest bio- mass and soil carbon augments carbon forc-

ings. In addition to its substantial contribution to the impacts of climate-change around the world, loss of Amazonian forest destroys other environmental services such as maintenance of biodiversity and water cycling and deprives the region's traditional inhabitants of their livelihoods. Restricting emissions to keep at- mospheric CO_2 concentrations from rising much above their current levels would avert this disaster.[27] Stabilizing atmospheric CO_2 concentration at 750 parts per million by vol- ume (ppmv) would stave off the demise of Amazonian forest by approximately 100 years, while limiting the concentration to 550 ppmv would postpone the disaster for more than 200 years (figure 9.4). However, more recent re- sults indicate that greenhouse-gas concentra- tions would have to be stabilized at much lower levels, a maximum of 400 ppmv of CO2-equivalent carbon (including trace-gas effects) being needed to achieve 80% assurance of keeping the average global temperature within 2°C above the pre-industrial level.[28] Limiting the rise in average global temperature to 2 degrees Celsius would be necessary to avoid substantial forest degradation in Amazonia and consequent carbon releases.[29] A global average temperature rise of 2 degrees Celsius is close to the amount of temperature increase that has been set in motion by emissions that have already occurred.[30]

Notes

The Conselho Nacional de Desenvolvimento Científico e Tecnológico (CNPq: Proc. 470765/01-1, 306031/2004-3, 557152/2005-4, 420199/2005-5) and the Instituto Nacional de Pesquisas da Amazônia (INPA: PPI 1-1005, PRJ05.57) provided financial support. I thank the editors, Stanford student volunteer "Group 2" (Andre, Christine, Julia, and Meg), and an anonymous reviewer for helpful suggestions.

1. This estimate is updated from: Fearnside, P. M. 2000. Global warming and tropical land-use change: Greenhouse gas emissions from biomass burning, decomposition, and soils in forest conversion, shifting cultivation, and secondary vegetation. *Climatic Change* 46(1–2): 115–58; based on wood-density adjustments of: Nogueira, E. M., B. W. Nelson, and P. M. Fearnside. 2005. Wood density in dense forest in central Amazonia, Brazil. *Forest Ecology and Management* 208(1–3): 261–86; Nogueira, E. M., P. M. Fearnside, B. W. Nelson, and M. B. França. 2007. Wood density in forests of Brazil's "arc of deforestation": Implications for biomass and flux of carbon from land-use change in Amazonia. *Forest Ecology and Management* 248(3): 119–35; and adjustments for hollow trees and form factor based on Nogueira, E. M., B. W. Nelson, and P. M. Fearnside. 2006. Volume and biomass of trees in central Amazonia: Influence of irregularly shaped and hollow trunks. *Forest Ecology and Management* 227(1–2): 14–21; and Nogueira, E. M., P. M. Fearnside, B. W. Nelson, R. I. Barbosa, and E. W. H. Keizer. 2008. Estimates of forest biomass in the Brazilian Amazon: New allometric equations and adjustments to biomass from wood-volume inventories. *Forest Ecology and Management* 256(11): 1853–57; this includes unchanged values for uptake by the replacement landscape (4.9 GtC) and for soil carbon loss to a 1-meter depth (4.9 GtC); proportionate adjustment applied to the all-tropics value derived in Fearnside 2000, op. cit.

2. Fearnside, P. M. 2000. Effects of land use and forest management on the carbon cycle in the Brazilian Amazon. *Journal of Sustainable Forestry* 12(1–2): 79–97.

3. Fearnside, P. M., and W. F. Laurance. 2004. Tropical deforestation and greenhouse gas emissions. *Ecological Applications* 14(4): 982–86.

4. Schneider, S. H. 2004. Abrupt non-linear climate change, irreversibility, and surprise. *Global Environmental Change* 14: 245–58.

5. Cox, P. M., R. A. Betts, C. D. Jones, S. A. Spall, and I. J. Totterdell. 2000. Acceleration of global warming due to carbon-cycle feedbacks in a coupled climate model. *Nature* 408: 184–87.

6. See review by Nobre, C. A. 2001. Mudanças climáticas globais: Possíveis impactos nos ecossistemas do País. *Parecerias Estratégicas* 12: 239–58.

7. Stainforth, D. A., T. Aina, C. Christensen, M. Collins, N. Faull, D. J. Frame, J. A. Kettleborough, S. Knight, A. Martin, J. M. Murphy, C. Piani, D. Sexton, L. A. Smith, R. A. Spicer, A. J. Thorpe, and M. R. Allen. 2005. Uncertainty in predictions of the climate response to rising levels of greenhouse gases. *Nature* 433: 403–6.

8. Hegerl, G. C., T. J. Crowley, W. T. Hyde, and D. J. Frame. 2006. Climate sensitivity constrained by temperature reconstructions over the past seven centuries. *Nature* 440: 1029–32.

9. Cox, P. M., R. A. Betts, M. Collins, P. Harris, C. Huntingford, and C. D. Jones. 2004. Amazonian dieback under climate-carbon cycle projections for the 21st century. *Theoretical and Applied Climatology* 78: 137–56.

10. Oyama, M. D., and C. A. Nobre. 2003. A new climate-vegetation equilibrium state for Tropical South America. *Geophysical Research Letters* 30(23): 2199–2203; Salazar, L. F., C. A. Nobre, and M. D. Oyama. 2007. Climate change consequences on the biome distribution in tropical South America. *Geophysical Research Letters* 34: L09708, doi:10.1029/2007GL029695.

11. Trenberth, K. E., and D. J. Shea. 2006. Atlantic hurricanes and natural variability in 2005. *Geophysical Research Letters* 33, L12704, doi:10.1029/2006GL026894. Marengo, J. A., C. A.

Nobre, J. Tomasella, M. D. Oyama, G. Sampaio de Oliveira, R. de Oliveira, H. Camargo, L. M. Alves, and I. F. Brown. 2008. The drought of Amazonia in 2005. *Journal of Climate* 21: 495–516. Evan, A. T., D. J. Vimont, A. K. Heidinger, J. P. Kossin, and R. Bennartz. 2009. The role of aerosols in the evolution of tropical North Atlantic ocean temperature anomalies. *Science* 324: 778–81.

12. Cox, P. M., P. P. Harris, C. Huntingford, R. A. Betts, M. Collins, C. D. Jones, T. E. Jupp, J. A. Marengo, and C. A. Nobre. 2008. Increasing risk of Amazonian drought due to decreasing aerosol pollution. *Nature* 453: 212–15.

13. Cardoso, M., G. C. Hurtt, B. Moore III, C. A. Nobre, and E. M. Prins. 2003. Projecting future fire activity in Amazonia. *Global Change Biology* 9(5): 656–69.

14. Cochrane, M. A. 2003. Fire science for rainforests. *Nature* 421: 913–19; Cochrane, M. A., A. Alencar, M. D. Schulze, C. M. Souza Jr., D. C. Nepstad, P. Lefebvre, and E. A. Davidson. 1999. Positive feedbacks in the fire dynamic of closed canopy tropical forests. *Science* 284: 1832–35.

15. Cochrane, 2003 op. cit.; Cochrane et al., 1999 op. cit; and Nepstad, D., G. Carvalho, A. C. Barros, A. Alencar, J. P. Capobianco, J. Bishop, P. Moutinho, P. Lefebvre, U. L. Silva Jr., and E. Prins. 2001. Road paving, fire regime feedbacks, and the future of Amazon forests. *Forest Ecology and Management* 154: 395–407.

16. Fearnside, P. M. 1995. Potential impacts of climatic change on natural forests and forestry in Brazilian Amazonia. *Forest Ecology and Management* 78(1–3): 51–70.

17. Alencar, A. C., L. A, Solórzano, and D. C. Nepstad. 2004. Modeling forest understory fires in an eastern Amazonian landscape. *Ecological Applications* 14(4): S139–S149; Nepstad, D. C., A. Alencar, C. Nobre, E. Lima, P. Lefebvre, P. Schlesinger, C. Potter, P. Moutinho, E. Mendoza, M. Cochrane, and V. Brooks. 1999. Large-scale impoverishment of Amazonian forests by logging and fire. *Nature* 398: 505–8; Nepstad, D. C., P Lefebvre, U. L Silva Jr., J. Tomasella, P. Schlesinger, L. Solorzano, P. Moutinho, D. Ray, and J. G. Benito. 2004. Amazon drought and its implications for forest flammability and tree growth: A basin-wide analysis. *Global Change Biology* 10(5): 704–12.

18. Sternberg, H. O. 1968. Man and environmental change in South America, pp. 413–45 in E. J. Fittkau, T. S. Elias, H. Klinge, G. H. Schwabe, and H. Sioli, eds., *Biogeography and Ecology in South America*, vol. I. D. W. Junk & Co., The Hague, Netherlands. Meggers, B. J. 1994. Archeological evidence for the impact of mega-Niño events on Amazonia during the past two millenia. *Climatic Change* 28(1–2): 321–38.

19. Malingreau, J. P., G. Stephens, and L. Fellows. 1985. Remote sensing of forest fires: Kalimantan and North Borneo in 1982–1983. *Ambio* 14(6): 314–21.

20. Rice, A. H., E. H. Pyle, S. R. Saleska, L. Hutyra, M. Palace, M. Keller, P. B. de Camargo, K. Portilho, D. F. Marques, and S. C. Wofsy. 2004. Carbon balance and vegetation dynamics in an old-growth Amazonian forest. *Ecological Applications* 14(4) Supplement: S55–S71; Tian, H., J. M. Mellilo, D. W. Kicklighter, A. D. McGuire, J. V. K. Helfrich III, B. Moore III, and C. Vörömarty. 1998. Effect of interanual climate variability on carbon storage in Amazonian ecosystems. *Nature* 396: 664–67.

21. Nepstad, D. C., P. Moutinho, M. B. Dias-Filho, E. Davidson, G. Cardinot, D. Markewitz, R. Figueiredo, N. Vianna, J. Chambers, D. Ray, J. B. Gueireros, P. Lefebvre, L. Sternberg, M. Moreira, L. Barros, F. Y. Ishida, I. Tohver, E. Belk, K. Kalif, and K. Schwalbe. 2002. The effects of partial rainfall exclusion on canopy processes, aboveground production and biogeochemistry of an Amazon forest. *Journal of Geophysical Research* 107(D20): 1–18.

22. Nepstad, D. C., I. M. Tohver, D. Ray, P. Moutinho, and G. Cardinot. 2007. Mortality of large trees and lianas following experimental drought in an Amazon forest. *Ecology* 88(9): 2259–69.

23. Nascimento, H. E. M., and W. F. Laurance. 2004. Biomass dynamics in Amazonian forest fragments. *Ecological Applications* 14(4) Supplement: S127–S138; Laurance, W. F., S. G. Laurance, L. V. Ferreira, J. M. Rankin-de-Merona, C. Gascon, and T. E. Lovejoy. 1997. Biomass collapse in Amazonian forest fragments. *Science* 278: 1117–18.

24. Davidson, E. A., and I. A. Janssens. 2006. Temperature sensitivity of soil carbon decomposition and feedbacks to climate change. *Nature* 440: 165–73.

25. Townsend, A. R., P. M. Vitousek, and E. A. Holland. 1992. Tropical soils could dominate the short-term carbon cycle feedbacks to increase global temperatures. *Climatic Change* 22: 293–303.

26. Cox, P. M., R. A. Betts, C. D. Jones, S. A. Spall, and I. J. Totterdell. 2000. Acceleration of global warming due to carbon-cycle feedbacks in a coupled climate model. *Nature* 408: 184–87; Cox et al., 2004 op. cit.

27. Arnell, N. W., M. G. R. Cannell, M. Hulme, R. S. Kovats, J. F. B. Mitchell, R. J. Nichols, M. L. Parry, M. T. J. Livermore, and A. White. 2002. The consequences of CO_2 stabilisation for the impacts of climate change. *Climatic Change* 53: 413–46.

28. Hare, B., and M. Meinshausen. 2006. How much warming are we committed to and how much can be avoided? *Climatic Change* 75: 111–49; Meinshausen, M., N. Meinshausen, W. Hare, S. C. B. Raper, K. Frieler, R. Knutti, D. J. Frame, and M. R. Allen. 2009. Greenhouse-gas emission targets for limiting global warming to 2°C. *Nature* 458: 1158–62.

29. Huntingford, C., P. O. Harris, N. Gedney, P. M. Cox, R. A. Betts, J. A. Marengo, and J. H. C. Gash. 2004. Using a GCM analogue model to investigate the potential for Amazonian forest dieback. *Theoretical and Applied Climatology* 78: 177–85.

30. Op cit., note 28.

Chapter 10

Global Crop Production and Food Security

David B. Lobell

Introduction[1]

Despite great advances in technology and widespread use of irrigation, global crop production and food prices today remain quite sensitive to variations in weather. Any changes in temperature, rainfall, cloudiness, or other weather variables caused by climate change may therefore affect the future cost and availability of food. This chapter provides a brief review of climate change impacts on agriculture, with a focus on global-scale impacts for the major food crops (e.g., rice, wheat, maize) that provide the bulk of human calories and protein.

This chapter discusses four main points. First, the impacts of a CO_2 doubling on crop production will most likely be fairly small at the global scale, but significant in tropical regions, while larger CO_2 increases would likely result in significant global losses. Second, the many sources of uncertainty involved in modeling future impacts of climate change are discussed, some of which have been addressed in global assessments and others which have been largely ignored. The overall uncertainty

regarding global impacts, and thus the risk of severe impacts from climate change, remains poorly quantified. Third, the large uncertainties associated with agricultural impact analyses have important implications for policymakers, scientists, and institutions involved with agricultural development. Fourth, climate change is just one of several ongoing trends that will shape the future of food security, and investments in mitigation or adaptation therefore should consider these other dynamics.

Uncertainty and Sensitivity Analysis

The critical issue of uncertainty underlies all discussions of climate change. In the case of assessing agricultural impacts, climate model outputs are typically fed into crop models whose outputs are then often fed into economic trade models. Each of these models contains literally hundreds of parameters and assumptions that can affect conclusions. This "cascade" of uncertainties (see chapter 15) results in a wide range of possible future outcomes for global agricultural impacts, and even larger ranges for regional impacts.

Unfortunately, thorough uncertainty analyses are rarely conducted in agricultural impact assessments. This reflects the difficulty of explicitly analyzing the hundreds of assumptions embedded within models. Model intercomparisons, such as those performed with climate models, provide useful estimates of model uncertainty, but such exercises have not been attempted for global agricultural impact studies.

Equally important, if not more so, is understanding which model parameters and assumptions contribute most to model uncertainty through sensitivity analysis. Although most assessments conduct some form of sensitivity analysis, analyses often vary parameters one at a time, which ignores interactions between parameters.[2]

Expected Impacts

Analyses of global agricultural impacts typically combine climate model projections, crop growth models, and global trade models to forecast impacts of increasing atmospheric CO_2.[3] In general, the main conclusion is that the most likely impacts of a doubling of CO_2 will be moderate at the global scale (see table 10.1), but significant in tropical regions; larger CO_2 increases will likely result in significant global losses. Global results mask great disparities between developed and developing countries. Developed countries are expected to experience gains in all but the warmest climate scenarios, while developing countries fair far worse due to their warmer current climates and limited human adaptive capacity. This disparity will exacerbate existing food in-

security for people living in the tropics, affecting up to 300 million additional people as a result of doubled CO_2.

Sources of Uncertainty

On what assumptions are these assessments based, and what are the magnitude and sources of uncertainty? Here I provide a subjective review with a distinction between processes that are currently modeled explicitly or implicitly in global assessments and those that have been largely omitted to date.

Modeled Processes

CLIMATE CHANGES OVER AGRICULTURAL REGIONS

Thanks to the availability of output from many different general circulation models (GCMs), most impact assessments have been able to consider how conclusions vary as a function of climate model and emission scenario. In general, the choice of climate model has perhaps the largest single impact on projections of global impacts (table 10.1). Even greater differences between climate models are observed at regional scales, where precipitation changes are especially difficult to predict.[4]

A comparison of different climate models, while a useful measure of climate uncertainty, does not address potential issues that are common to all models, such as coarse grid resolution, simple representation of the land surface that ignores factors such as irrigation, and inability to model abrupt climate changes.[5] Several studies have shown a significant dependence of impact estimates on climate model resolution, although the overall

TABLE 10.1

The projected impacts of doubled CO_2 on global cereal production (percent change) for different climate models, adaptation levels, and with and without CO_2 fertilization. See text for details on adaptation levels. Climate models: GISS=Goddard Institute for Space Studies (4.2, 11); GFDL=Geophysical Fluid Dynamics Laboratory (4.0, 8); UKMO=United Kingdom Meteorological Office (5.2, 15). Numbers in parenthesis are global average change in temperature (in degrees Celsius) and precipitation (percent) for each model. Source: C. Rosenzweig and M. L. Parry, Potential impact of climate change on world food supply. Nature 367, no. 6459 (1994).

Scenario	GISS	GFDL	UKMO
Climate change only	−11	−12	−20
With CO_2 fertilization	−1	−3	−8
With CO_2 and adaptation Level 1	0	−2	−6
With CO_2 and adaptation Level 2	1	0	−2

uncertainty in global scale assessments due to climate model resolution remains unclear.[6]

CO_2 FERTILIZATION EFFECT

Although it is widely acknowledged that increased levels of atmospheric CO_2 are qualitatively beneficial to plant photosynthesis and crop yields, accurately quantifying the fertilization effect has proven difficult. Many early studies evaluated the effects of elevated CO_2 levels in chamber experiments that excluded known interactions of CO_2 with temperature, water, nutrients, and other factors (figure 10.1).[7] More recent studies have used free-air (chamberless) field CO_2 enrichment (FACE) experiments.[8] Although these experiments nearly always confirm the positive effect of CO_2, some have argued that they show a substantially lower magnitude of yield response than chamber experiments.[9] The uncertain nature of CO_2 effects led Wolfe and Erickson to suggest a conservative policy approach that assumes no CO_2 fertilizer effect.[10] Several recent assessments include projections both with and without a modeled yield increase

from CO_2. In a study by Parry and colleagues, the number of people at risk of hunger from climate change was found to range from 20 to 30 million with CO_2 fertilization, but around 50 to 600 million without.[11] This high sensitivity to CO_2 effects is also evident in table 10.1.

YIELD RESPONSE TO CHANGES IN MEAN GROWING SEASON CLIMATE

The effect of changes in mean growing season temperature and precipitation on crop yields has been widely studied both experimentally and by analyzing historical records. Overall, most crops in most parts of the world exhibit reduced yields in warmer years. Exceptions are high latitude regions such as Canada, the Northern Great Plains, Northern China, and Russia, where warm years allow for a longer growing season. For example, 13.2 percent of the land surface is currently considered too cold for successful crop cultivation, and much of this area is anticipated to be brought into production under warming scenarios.[12]

Warm temperatures reduce yields in most regions by altering the length of each phase of

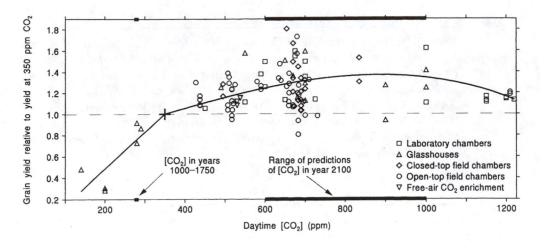

FIGURE 10.1. The experimental response of wheat to elevated CO_2 in fifty studies reviewed by Amthor. Elevated CO_2 up to approximately 900 ppm generally raised yields, but the quantitative effect was quite variable. Only treatments with ample water and nutrients are shown. Responses tend to be lower when N is limiting and slightly higher in drought conditions. Source: J. S. Amthor, Effects of atmospheric CO_2 concentration on wheat yield: Review of results from experiments using various approaches to control CO_2 concentration. *Field Crops Research* 73, no. 1 (2001).

crop development, including critical yield-determining stages such as grain filling, affecting dormancy or vernalization requirements such as for winter wheat, and increasing respiration rates and water stress.[13] A differential effect of day versus night temperature is exhibited by some observations and crop models (see figure 10.2), but the ability of models to capture these differences remains poorly tested.[14]

The sensitivity of model projections to uncertainties in crop responses to climate has been considered less frequently than climate change or CO_2 uncertainties. For example, while climate models may predict a 2-degree-Celsius warming over a crop field with a high degree of certainty, the crop response to the 2-degree-Celsius rise in temperature may be uncertain and vary considerably across models. Mearns and colleagues found significant

differences between two commonly used crop models—CERES and EPIC—for corn and wheat studies in the central Great Plains that were comparable to differences obtained when varying climate model resolutions (see table 10.2).[15] Aggarwal and Mal compared the ORYZA1N and CERES rice models in India and found differences that were nearly as large as those due to an optimistic versus pessimistic climate change scenario.[16] Lobell and colleagues estimated yield impacts of climate change on California crops using empirical yield models and found that crop model uncertainties were significant but smaller than those from climate models.[17] Thus, it appears that while climate change uncertainty is important, uncertainties in the processes that govern crop responses to climate are nearly as large but relatively underappreciated.

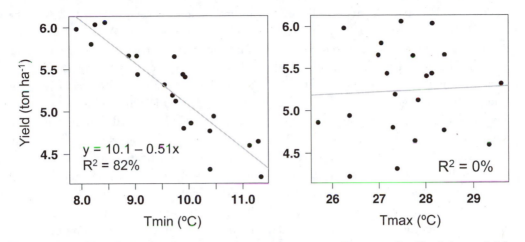

FIGURE 10.2. The relationship between average growing season (January to April) minimum (night-time) and maximum (daytime) temperatures and irrigated wheat yields in the Yaqui Valley of Mexico, 1982 to 2002. Source: D. Lobell, and I. Ortiz-Monasterio, Impacts of day versus night temperatures on spring wheat yields: A comparison of empirical and Ceres Model predictions in three locations. *Agronomy Journal* 99, no. 2 (2007).

TABLE 10.2

Average simulated yield changes (percent) for corn and wheat in the U.S. central Great Plains, using two climate model resolutions and two crop models. CO_2 effects were not included. For each crop, the range of results using different climate model resolutions and different crop models was roughly equal. The two uncertainties appear to interact strongly; e.g., for corn the crop models had a large effect with RegCM but not CSIRO. Source: L. O. Mearns, et al., Comparative responses of EPIC and CERES crop models to high and low spatial resolution climate change scenarios. Journal of Geophysical Research-Atmospheres 104, no. D6 (1999).

Crop	Global Climate Model (CSIRO)		Regional Climate Model (RegCM)	
	EPIC crop model	CERES crop model	EPIC crop model	CERES crop model
Corn	−10.9	−12.7	2.3	−15.2
Wheat	−3.3	6.4	−6.4	−8.7

YIELD RESPONSE TO CHANGES IN CLIMATE VARIABILITY

Some climate models project increased variability in temperature and rainfall as a result of greenhouse gas emissions.[18] Increased climate variability is generally believed to have negative impacts on cropping systems, al-though the magnitude of this impact is uncertain. One main reason is that crop yields in most regions exhibit a nonlinear response to temperature, with highest yields observed when temperatures are close to average and yield declines observed both in extremely warm or extremely cool years. Increased climate variability therefore increases the frequency of low-yielding years, even if the average climate state remains unchanged.[19]

Another important but less direct consequence of climate variability is that farmers cannot match management practices to the climate of an individual year, but must manage based on a range of possible outcomes. In other words, variability leads to uncertainty, which leads to suboptimal practices. One way to measure this is the yield gains possible with perfect forecasts of climate (i.e., a clairvoyant farmer).[20] Another measure is that yields tend to be roughly 50 percent more variable at the scale of an individual field than at the county scale, which indicates that the relative performance of different management practices depends on climate.[21] With increased variability, yield losses attributable to the uncertainty faced by farmers can be expected to increase, although the magnitude of this effect is not well known.

ADAPTATION

As climate change occurs, it is generally expected that farmers and other institutions will respond and, in so doing, reduce the costs or increase the benefits of the imposed changes. As with the factors above, this qualitative consensus gives way to great disagreement and uncertainty regarding the quantitative impact that adaptation can have. Reilly summarized the debate by stating: "The general conclusion that adaptation (to the extent it is economically justified) makes sense is tautological. But the value of the empirical work for identifying particular adaptation options is negligible or nonexistent."[22]

There are two scales of adaptation that are widely modeled: changes in practices at the farm level (agronomic adaptation) and shifts in production between different parts of the world (trade adaptations). Rosenzweig and Parry considered three scenarios of agronomic adaptation: (1) no adaptation; (2) Level 1 adaptation, where tactical decisions such as planting date and cultivar choice were optimized; and (3) Level 2 adaptation, which included more costly adaptations such as development of new irrigation infrastructure and new crop varieties.[23] These adaptations result in significant reductions of food prices and hunger (see table 10.1), but are generally more effective in developed than developing countries and thus exacerbate the disparity between temperate and tropical responses to warming. Similar studies that evaluate impacts of prescribed adaptation measures have found that adaptation can greatly reduce impacts.[24]

Because models consider only a limited number of possible actions, some have suggested that modeling studies may underestimate the true ability of farmers to adapt to change.[25] At the same time, however, models likely overestimate the impact of any single adaptation measure because they often assume that a farmer adjusts precisely to optimize yields or profits given the new climate. The true response of farmers therefore almost certainly lies between the assumptions of no adaptation and perfect adaptation.[26] In addition, natural climate variability may mask more gradual trends in climate and therefore slow the response of farmers to change or even cause adaptation to proceed in the wrong direction.[27]

Another important factor when assessing the benefits of adaptations is that most studies simply evaluate the benefits of adaptation measures without consideration of their costs.[28] For example, a model that projects zero net change in food prices should not be interpreted to mean that climate change will have no costs if the model result relied on potentially expensive adaptations such as new irrigation systems and new cultivar development.

Trade adaptations have been modeled in most global assessments using general equilibrium trade models, which include many of the major economic .sectors (including agriculture) in some twenty regions of the world and endogenously calculate commodity prices to satisfy demand. These models are needed given that much of food is traded across national boundaries, and thus food prices and availability can be affected as much or more by what happens in another country as at home. For instance, modeled trade flows from temperate to tropical countries generally increase in response to the differential effects of climate change. Unfortunately, the assumptions and uncertainties embedded in economic models are rarely made explicit.

4.2 Un-Modeled Processes

In addition to the uncertainties associated with processes that are modeled in most global assessments, there are many other factors that recent research suggests may be important.

- *Extreme events*. Many climate extremes, such as heat waves, heavy precipitation events, hurricanes, and so on, are expected to increase in climate change scenarios. Crop models generally do not capture the effects of rare but potentially harmful events. Rosenzweig and colleagues estimated that adding an effect of excess precipitation on crop growth (through effects on soil aeration) re-

sulted in $1.5 billion of simulated losses due to such events in the United States in current climate, with potentially doubled losses under climate change.[29] Porter and Semenov argue that crop injuries from extreme temperatures will become increasingly important with climate change as critical thresholds are more frequently exceeded.[30] In general, understanding of climate extreme effects on crop growth is limited.

- *Ozone effects on yields*. Fossil fuel burning and increased temperatures will likely result in significant increases in tropospheric ozone levels, which are known to be harmful to crops and already cause significant production losses.[31] Recent FACE studies have documented soybean yield losses of roughly 25 percent for around a 25 percent increase in ozone concentration, similar to the expected increase in ozone by 2050.[32]

- *Pest and disease damage*. Current estimates of the global impact of weeds, animal pests, fungal and bacterial pathogens, and viruses total roughly 30 to 40 percent of current production for each of the major food crops.[33] Even relatively minor increases due to climate change could thus cause billions of dollars in losses. While numerous approaches have been used to estimate pest and pathogen response to climate, global impact assessments have generally ignored these factors.[34]

- *Sea-level rise*. Increased sea levels will likely lead to increased risk of floods that could potentially devastate crops in coastal areas. Salinity levels in irrigation water may also be affected by sea-level rise. Again, these are poorly understood

aspects that appear more likely to reduce crop production than improve it.
Rice appears to be the most vulnerable
of the major crops to sea-level rise, with
roughly 5 percent of global rice area at
or below sea level.

- *Grain quality effects*. While most studies
  have focused on the quantity of production, a major concern to farmers and industry is grain quality. Increased CO_2
  has resulted in lower grain protein for
  wheat in experiments, while high temperature was found to reduce dough
  strength.[35]

- *Impacts on non-major crops*. Global
  assessments generally focus on major
  commodities such as rice, wheat, maize,
  soybean, and cotton. It is often assumed
  that impacts on other crops will be similar, but in reality these assumptions may
  not hold. For example, perennial crops
  such as fruit and nut trees remain in the
  ground for many years and thus adaptations may occur more slowly in these
  systems.[36]

Policy and Research Implications

The large uncertainties associated with agricultural impact projections have at least three
important implications.

First, climate change is unlikely to have
strong positive effects on global food production and security, but the possibility of severely negative effects cannot be dismissed.
This is especially true when considering that
most of the processes omitted from model assessments are more likely to reduce yields under climate change scenarios than improve
them. Policy makers should be cognizant not
only of the expected or most likely effects of
climate change on agriculture, but the real

(yet still poorly defined) possibility of major
impacts on food security, even in developed
countries.

Next, better quantification of uncertainties is an important and substantial challenge
to the research community. Efforts to reduce
uncertainties should also be a high priority.
Better constraints on climate changes at regional scales and yield responses to CO_2, for
instance, would greatly improve confidence
in projections. New opportunities to test assessment models with harvest and climate
data from recent years should also receive increased attention. For example, studies in India, China, Australia, and Argentina document negative impacts of recent warming
trends on crop growing season length and
yields.[37] Moreover, a recent study based on
global datasets estimated that warming trends
from 1981 to 2002 resulted in annual global
production losses for the major cereal crops
worth $5 billion.[38] Policy makers should be
aware, however, that these are substantial scientific challenges and waiting to make decisions until uncertainties are reduced may be a
very risky strategy.

Finally, adaptation is clearly an important
strategy for reducing climate change impacts,
but should not be viewed as cost-free or risk-
free panacea. The success of efforts by farmers
and other institutions to adapt to climate
change represents a key source of uncertainty.
At the same time, many studies indicate that
adaptation can only partially offset losses, especially in extreme climate change scenarios

Climate Change in Context

When focusing on the need for climate
change mitigation or adaptation to enhance
food security, it is important to recognize that
climate change is just one of several rapid

changes that will shape the future of food security. Others include human population dynamics, changes in living standards and consumption patterns, development of new crop varieties and other agricultural technologies, and possible degradation of water and soil resources. For example, a study of future food supply reported that uncertainties in climate change impacts were less important than uncertainties in management and technology improvements or future fertilizer use, but roughly equal to land degradation and pests and diseases.[39]

The amount and type of investments made to adapt to climate change should therefore depend on consideration of the trade-offs relative to other investment options. Fortunately, many of the changes that reduce vulnerability to climate change, such as development of heat- and drought-resistant crop varieties, use of seasonal climate forecasts, and improvement of rural infrastructure, are the same changes needed to enhance production in the absence of climate change. Climate change therefore does not necessarily change the types of investments that are needed in agriculture as much as it increases their potential returns to society.

Unfortunately, it must be noted that investments in the institutions most directly involved with agricultural development, particularly in developing countries, have slowed dramatically as of this writing since around 1990.[40] Annual budgets for the International Rice Research Institute and the International Maize and Wheat Improvement Center, the leading institutes in developing world research on these crops with impressive records of investment returns, are each approximately just $35 million.[41] Increases in private agricultural research and development investments have compensated for public investment losses in developed countries, where roughly half of investments are private, but more than 95 percent of investment in developing countries comes from public funds.[42] Given that food security concerns are concentrated in developing countries and that the substantial annual rates of return (around 80 percent) on agricultural investments typically take ten or more years to be realized, the risk that climate change will have serious impacts on food security is likely increasing as a result of these investment trends.[43]

Conclusion

Climate change clearly poses some risk to global crop production and food security, but the precise nature and magnitude of that risk remain unclear. Broadly speaking, possible global effects range from negligible, with increased production in temperate regions compensating for production losses in developing countries, to severely negative. While previous research has focused largely on the most likely outcomes, future work should attempt to better quantify the chance of extreme impacts characterized by major increases in food prices and hunger, and associated effects on the environment and national security. These less likely but higher consequence outcomes are especially relevant to policy decisions, such as how much to invest in agricultural research and infrastructure. Many opportunities exist to reduce the vulnerability of food production to climate change, but the necessary adaptations will require substantial time, money, and expertise.

Notes

1. This chapter was written in May 2006.
2. A. Saltelli, K. Chan, and E. M. Scott, eds., *Sensitivity Analysis* (New York: Wiley, 2000).

3. R. M. Adams, et al., Global climate change and United States agriculture. *Nature* 345, no. 6272 (1990); C. Rosenzweig, and M. L. Parry, Potential impact of climate-change on world food supply. *Nature* 367, no. 6459 (1994); C. Rosenzweig, et al., *Climate Change and World Food Supply* (Oxford: Univ. of Oxford, 1993).

4. U. Cubasch, G. A. Meehl, et al., Projections of future climate change, in *Intergovernmental Panel on Climate Change Working Group 1, Climate Change 2001: The Scientific Basis* (IPCC Working Group 1, 2001); F. Giorgi, B. Hewitson, et. al., Regional climate information — Evaluation and projections, in *Intergovernmental Panel on Climate Change Working Group 1, Climate Change 2001: The Scientific Basis* (IPCC Working Group 1, 2001).

5. D. B. Lobell, et al., Potential bias of model projected greenhouse warming in irrigated regions. *Geophysical Research Letters* 33 (2006); D. B. Lobell, G. Bala, and P. B. Duffy, Biogeophysical impacts of cropland management changes on climate. *Geophysical Research Letters* 33, no. 6 (2006).

6. L. O. Mearns, et al., Comparison of climate change scenarios generated from regional climate model experiments and statistical downscaling. *Journal of Geophysical Research-Atmospheres* 104, no. D6 (1999); L. O. Mearns, et al., Comparison of agricultural impacts of climate change calculated from high and low resolution climate change scenarios: Part I. The uncertainty due to spatial scale. *Climatic Change* 51, no. 2 (2001); W. E. Easterling, et al., Comparison of agricultural impacts of climate change calculated from high and low resolution climate change scenarios: Part II. Accounting for adaptation and CO_2 direct effects. *Climatic Change* 51, no. 2 (2001); E. A. Tsvetsinskaya, et al., The effect of spatial scale of climatic change scenarios on simulated maize, winter wheat, and rice production in the Southeastern United States. *Climatic Change* 60, no. 1–2 (2003).

7. B. A. Kimball, Carbon-dioxide and agricultural yield — An assemblage and analysis of 430 prior observations. *Agronomy Journal* 75, no. 5 (1983); J. D. Cure, and B. Acock, Crop responses to carbon-dioxide doubling — A literature survey. *Agricultural and Forest Meteorology* 38, no. 1–3 (1986).

8. E. A. Ainsworth and S. P. Long, What have we learned from 15 years of free-air CO_2 enrichment (face)? A meta-analytic review of the responses of photosynthesis, canopy. *New Phytologist* 165, no. 2 (2005).

9. Stephen P. Long, et al., Global food insecurity: Treatment of major food crops with elevated carbon dioxide or ozone under large-scale fully open-air conditions suggests recent models may have overestimated future yields. *Philosophical Transactions: Biological Sciences* 360, no. 1463 (2005); Stephen P. Long, et al., Food for thought: Lower-than-expected crop yield stimulation with rising CO_2 concentrations. *Science* 312, no. 5782 (2006).

10. D. W. Wolfe and J. D. Erickson, Carbon dioxide effects on plants: Uncertainties and implications for modeling crop response to climate change, in *Agricultural Dimensions of Global Climate Change*, ed. H. M. Kaiser and T. E. Drennen (Delray Beach, FL: St. Lucie Press, 1993).

11. M. L. Parry, et al., Effects of climate change on global food production under SRES emissions and socio-economic scenarios. *Global Environmental Change* 14, no. 1 (2004).

12. G. Fischer, et al., Socio-economic and climate change impacts on agriculture: An integrated assessment, 1990–2080. *Philosophical Transactions: Biological Sciences* 360, no. 1463 (2005).

13. M. P. Russelle, et al., Growth analysis based on degree days. *Crop Science* 24 (1984).

14. D. B. Lobell and J. I. Ortiz-Monasterio, Impacts of day versus night temperatures on spring wheat yields: A comparison of empirical and CERES Model predictions in three locations. *Agronomy Journal* 99, no. 2 (2007).

15. Mearns et al., 1999 op. cit.

16. P. K. Aggarwal and R. K. Mall, Climate change and rice yields in diverse agro-environments of India. II. Effect of uncertainties in scenarios and crop models on impact assessment. *Climatic Change* 52, no. 3 (2002).

17. David B. Lobell, et al., Impacts of future climate change on California perennial crop yields: Model projections with climate and crop uncertainties. *Agricultural and Forest Meteorology* 141, no. 2–4 (2006).

18. J. Räisänen, CO_2-induced changes in interannual temperature and precipitation variability in 19 cmip2 experiments. *Journal of Climate* 15, no. 17 (2002).

19. L. O. Mearns, C. Rosenzweig, and R. Goldberg, The effect of changes in daily and inter-

annual climatic variability on CERES-wheat: A sensitivity study. *Climatic Change* 32, 257–92 (1996).

20. R. W. Katz and A. H. Murphy, eds., *Economic Value of Weather and Climate Forecasts* (New York: Cambridge University Press, 1997).

21. D. B. Lobell, J. I. Ortiz-Monasterio, and W. P. Falcon, Yield uncertainty at the field scale evaluated with multi-year satellite data. *Agricultural Systems* 92, no. 1–3 (2007).

22. J. Reilly, What does climate change mean for agriculture in developing countries? A comment on Mendelsohn and Dinar. *The World Bank Research Observer* 14, no. 2 (1999).

23. Rosenzweig and Parry, 1994 op. cit.

24. R. S. J. Tol, S. Fankhauser, and J. B. Smith, The scope for adaptation to climate change: What can we learn from the impact literature? *Global Environmental Change* 8, no. 2 (1998).

25. IPCC, *Climate Change 2001: Impacts, Adaptation and Vulnerability* (Cambridge, UK: Cambridge University Press, 2001).

26. J. Reilly, 1999 op. cit.

27. S. H. Schneider, W. E. Easterling, and L. O. Mearns, Adaptation: Sensitivity to natural variability, agent assumptions, and dynamic climate changes. *Climatic Change* 45, no. 1 (2000).

28. Tol, Fankhauser, and Smith, 1998 op. cit.

29. C. Rosenzweig, et al., Increased crop damage in the U.S. from excess precipitation under climate change. *Global Environmental Change-Human and Policy Dimensions* 12, no. 3 (2002).

30. J. R. Porter and M. A. Semenov, Crop responses to climatic variation. *Philosophical Transactions of the Royal Society B-Biological Sciences* 360, no. 1463 (2005).

31. X. Wang and D. L. Mauzerall, Characterizing distributions of surface ozone and its impact on grain production in China, Japan, and South Korea: 1990 and 2020. *Atmospheric Environment* 38, no. 26 (2004); W. L. Chameides, et al., Growth of continental-scale cetro-agro-plexes, regional ozone pollution, and world food production. *Science (Washington)* 264, no. 5155 (1994); W. L. Chameides, et al., Is ozone pollution affecting crop yields in China? *Geophysical Research Letters* 26, no. 7 (1999).

32. Long, et al., 2005 op. cit.

33. E. C. Oerke, Crop losses to pests. *Journal of Agricultural Science* doi:10.1017/S0021859605005708 (2005).

34. S. M. Coakley, H. Scherm, and S. Chakraborty, Climate change and plant disease management. *Annual Review of Phytopathology* 37 (1999); C. Rosenzweig, et al., Climate change and extreme weather events: Implications for food production, plant diseases, and pests. *Global Change and Human Health* 2, no. 2 (2001).

35. IPCC, 2001 op. cit.

36. Lobell, et al., 2006 op. cit.

37. H. Pathak, et al., Trends of climatic potential and on-farm yields of rice and wheat in the Indo-Gangetic plains. *Field Crops Research* 80, no. 3 (2003); F. Tao, et al., Climate changes and trends in phenology and yields of field crops in China, 1981–2000. *Agriculture and Forestry Meteorology* 138, no. 1–4 (2006); V. O. Sadras and J. P. Monzon, Modelled wheat phenology captures rising temperature trends: Shortened time to flowering and maturity in Australia and Argentina. *Field Crops Research* 99, no. 2–3 (2006).

38. D. B. Lobell and C. B. Field, Global scale climate-crop yield relationships and the impacts of recent warming. *Environmental Research Letters* 2, no. 1 (2007).

39. B. R. Döös and R. Shaw, Can we predict the future food production? A sensitivity analysis. *Global Environmental Change-Human and Policy Dimensions* 9, no. 4 (1999).

40. P. G. Pardey and N. K. Beintema, *Slow Magic: Agricultural R&D a Century after Mendel* (Washington, DC: International Food Policy Research Institute, 2001).

41. Consultative Group on International Agricultural Research, *CGIAR Annual Report 2000: The Challenge of Climate Change: Poor Farmers at Risk* (2001); available from www.worldbank.org/html/cgiar/publications/annreps/cgar00/cgar2000.html.

42. Pardey and Beintema, 2001 op. cit.

43. J. M. Alston, et al., *A Meta Analysis of Rates of Return to Agricultural R&D: Ex Pede Herculem?* (Washington, DC: International Food Policy Research Institute, 2000).

Chapter 11

Human Health

Kristie L. Ebi

Weather and climate influence the distribution and incidence of a variety of health outcomes, including those resulting from extreme weather events, poor air quality, infectious diseases, and malnutrition. Climate change has the potential to affect any health outcome that is climate-sensitive, with the projected negative impacts largest for lower-income populations living predominantly within tropical/subtropical countries.[1]

The causal chain from climate change to changing patterns of health determinants and outcomes is complex and includes factors such as wealth, distribution of income, status of the public health infrastructure, provision of medical care, and access to adequate nutrition, safe water, and sanitation.[2] Therefore, the severity of future impacts will be determined by changes in climate as well as by concurrent changes in nonclimatic factors, and by the adaptation measures implemented to reduce negative impacts.

This chapter summarizes the potential climate change–related health impacts of extreme weather events, infectious diseases, and air pollutants, followed by a summary of aggregate assessments projecting future health burdens attributable to climate change. Approaches to reduce impacts are briefly discussed. Although the chapter focuses on low-income countries that are expected to suffer a larger burden of health impacts, high-income countries also are expected to experience increased risks.

Extreme Weather Events

Extreme weather events including heat waves, floods, windstorms, and droughts affect millions of people and cause billions of dollars of damage annually.[3] In 2003, in Europe, Canada, and the United States, floods and storms resulted in 101 people dead or missing and caused $9.73 billion in insured damages.[4] More than 35,000 excess deaths were attributed to the extended heat wave in Europe the same year.[5] The health impacts of extreme events in developing countries are substantially larger. A growing body of scientific research projects that climate change will increase the frequency and intensity of extreme weather events, suggesting that the associated health impacts also are likely to increase.[6]

The impacts of an extreme event, including loss of life and livelihood, infrastructure damage, population displacement, and economic disruption, are determined by the physical characteristics of the event, attributes of the location affected, and interactions of these with human actions and social, economic, institutional, and other systems. The adverse health consequences of flooding and windstorms often are complex and far-reaching, and include the physical health effects experienced during the event or cleanup process, or from effects brought about by damage to infrastructure, including population displacement. The physical effects largely manifest themselves within the weeks or months following the event, and may be direct (such as injuries) or indirect (such as water and food shortages and increased rates of vector-borne and other diseases).[7] Extreme weather events are also associated with mental health effects, such as post-traumatic stress disorder, resulting from the experience of the event or from the recovery process. These psychological effects tend to be much longer-lasting and may be worse than the direct physical effects.[8]

Heat waves affect human health through heat stress, heatstroke, and death as well as exacerbating underlying conditions that can lead to an increase in mortality from all causes of death.[9] Older adults, children, city-dwellers, the poor, and people taking certain medications are at the highest risk during a heat wave. The numbers of heat-related deaths are projected to increase with climate change.[10] Adaptive responses, including behavioral, physiological, and technological factors, may reduce the projected negative impacts.

The effects of drought on health include malnutrition, infectious diseases, and respiratory diseases.[11] Malnutrition, in addition to causing serious health consequences, increases the risk of dying from an infectious disease. The loss of livelihoods due to drought is a major trigger for population movements, which may cause additional adverse health burdens.

Projections suggest that, under a range of climate scenarios, the world will have sufficient food to feed everyone up to the end of the twenty-first century, assuming that people in low-income countries, where climate change impacts are predominantly negative, will have access to food produced in high-income countries.[12]

Although data are limited, malnutrition associated with drought and flooding may be one of the most important consequences of climate change due to the large number of people that may be affected. For example, one study projected that climate change would increase the percentage of the Malian population at risk of hunger from 34 percent to anywhere from 64 to 72 percent by the 2050s, although this may be reduced by implementation of a range of adaptive strategies, including migration of cropping patterns, development of heat-resistant cultivars, cropland expansion, adoption of improved cultivars, and changes in trade.[13]

Malaria and Other Infectious Diseases

Climate is a primary determinant of whether a particular location has environmental conditions suitable for the transmission of several infectious diseases. A change in temperature may hinder or enhance vector and parasite development and survival, thus lengthening or shortening the season during which vectors and parasites survive. Small changes in temperature or precipitation may cause previously inhospitable altitudes or ecosystems to become conducive to disease transmission, or cause currently hospitable conditions to become inhospitable. For example, some models suggest that both season of transmission and the geographic range of malaria may expand in Africa.[14] Although important, climate is not the only driver of infectious diseases. Non-climatic factors include drug and pesticide resistance, deterioration of health care and public health infrastructure (including vector control efforts), demographic changes, and land-use changes.

In addition, several food- and waterborne diseases are climate-sensitive, suggesting that climate change may affect their incidence and distribution. For example, studies report an approximately linear association between temperature and common forms of foodborne diseases such as salmonellosis.[15]

Air Pollutants, Including Ozone

Climate change may increase concentrations of selected air pollutants, particularly ozone, in some regions, and could decrease concentration of other pollutants, such as particulate matter. Air pollution concentrations are the result of interactions among local weather patterns, atmospheric circulation features, wind, topography, and other factors. Establishing the scale (local, regional, global) and direction of change (improvements or deterioration) of air quality is challenging.[16]

There is extensive literature documenting the adverse health impacts of exposure to elevated concentrations of air pollution, especially particulates with aerodynamic diameters under 10 and 2.5 micrometers, ozone, sulphur dioxide, nitrogen dioxide, carbon monoxide, and lead.[17] In 2000, 800,000 deaths from respiratory problems, lung disease, and cancer were attributed to urban air pollution, with the largest burden in low-income countries in the Western Pacific region and Southeast Asia.[18] In addition, there were 1.6 million deaths attributed to indoor air pollution caused by burning biomass fuels such as wood and dung.

More is known about the potential impacts of climate change on ground-level ozone than on other air pollutants. Changes in concentrations of ground-level ozone driven by scenarios of future emissions and/or weather patterns have been projected for Europe and North America.[19] Increases in ozone concentrations will likely increase respiratory problems in susceptible individuals. Based on projections of county-level pollutant concentrations, summer ozone-related mortality was projected to increase by 4 percent in the New York area by the 2050s based on climatic changes alone.[20] Despite the heavier pollution burdens, no studies have been conducted for cities in low- or middle-income countries.

Global Assessments of the Health Impacts of Climate Change

The most comprehensive evaluation of the health burden due to climate change used a comparative risk assessment approach to project total health burdens from climate change in 2000 and 2030, and to project how much of this burden might be avoided by stabilizing greenhouse gas (GHG) emissions.[21] The health outcomes included (diarrhea, malaria, malnutrition, heat-related mortality, and injury from floods and landslides) were chosen based on sensitivity to climate variations, likely future importance, and availability of quantitative global models (or feasibility of constructing them) for analysis. The projected relative risks attributable to climate change in 2030 vary by health outcome and region and are largely negative, with the majority of the projected health burden due to increases in diarrheal disease and malnutrition, primarily in low-income populations already experiencing a large burden of disease.

These results are consistent with a review that concluded health risks are likely to increase with increasing global mean surface temperature, particularly in low-latitude countries.[22] Actual health burdens depend on assumptions of population growth, future baseline disease incidence, and the extent of adaptation.

The relative direction, magnitude, and certainty of climate change–related health impacts, as summarized by the Human Health chapter of the Intergovernmental Panel on Climate Change Fourth Assessment Report, are shown in figure 11.1.

Particularly Vulnerable Populations and Regions

Vulnerability to climate change will vary according to population characteristics; local climatic effects; human, institutional, social, and economic capacity; distribution of income; provision of medical care; and access to adequate nutrition, safe water, and sanitation. In general, the most vulnerable include slum dwellers and homeless people in large urban areas, particularly in low-income countries, those living in water-stressed regions,

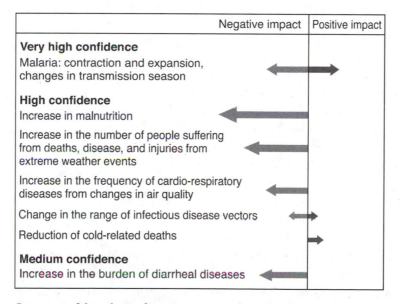

FIGURE 11.1. Summary of the relative direction, magnitude, and certainty of climate change–related health impacts. Source: Confalonieri U., B. Menne, R. Akhtar, K. L. Ebi, M. Hauengue, R. S. Kovats, B. Revich, and A. Woodward, 2007: Human health, in *Climate Change 2007: Impacts, Adaptation and Vulnerability. Contribution of Working Group II to the Fourth Assessment Report of the Intergovernmental Panel on Climate Change.* Parry, M. L., O. F. Canziani, J. P. Palutikof, P. J. van der Linden, and C. E. Hansson, eds. Cambridge University Press, Cambridge, U.K., pp. 391–431.

settlements in coastal and low-lying areas, and populations highly dependent on natural resources. However, as shown during Hurricane Katrina (2005) and the 2003 heat event in Europe, developed countries may not be prepared to cope with the projected increase in the frequency and intensity of extreme weather events.

Adaptation

Climate change will make more difficult the control of climate-sensitive health determinants and outcomes. Therefore, health policies need to explicitly incorporate climate-related risks in order to maintain current levels of control.[23] In many cases, the primary response will be to enhance current health risk management activities. Most health determinants and outcomes that are projected to increase with climate change are problems today. In some cases, programs will need to be implemented in new regions; in others, climate change may reduce current infectious disease burdens. The degree to which programs and measures will need to be augmented to address the additional pressures due to climate change will depend on factors such as the current burden of climate-sensitive health outcomes; the effectiveness of current interventions; projections of where, when, and how the health burden may change with changes in climate and climate variability; the feasibility of implementing additional cost-effective interventions; other stressors that might increase or decrease resilience to impacts; and the social, economic, and political context within which interventions are implemented.[24] Although there are uncertainties about future climate change, failure to invest in adaptation may leave com-

munities and nations poorly prepared and increase the probability of severe adverse consequences.[25] Examples of adaptation measures range from developing and deploying early warning systems and emergency response plans that specifically incorporate projections of climate change–related health risks to establishing surveillance programs in regions where projections suggest disease vectors may change their geographic range. Adaptation policies and measures need to consider how to effectively and efficiently reduce climate-related health risks in the context of sustainable development, considering projected demographic, economic, institutional, technologic, and other changes.

Because fossil fuel combustion is a source of urban air pollutants and greenhouse gases, policies to reduce GHG emissions may have health benefits in the near and long term. There are potential synergies in reducing GHG and improving population health via sustainable transport systems that make more use of public transport, walking, and cycling, especially in rapidly developing countries such as China and India.[26] For other energy sources, health impact assessments should be conducted to identify possible solutions to unintended consequences.

Conclusion

Climate change is already causing injuries, disease, and death from a range of climate-sensitive health outcomes. Health impacts are projected to increase over the coming decades, particularly in lower-income populations and regions that currently suffer from high burdens of climate-sensitive health outcomes. The degree to which impacts will be experienced will depend not only on the rate and degree of climate change, but also on the effectiveness and timeliness of adaptation and mitigation policies, strategies, and measures. However, even effective and timely actions will not prevent all health impacts.

Notes

1. McMichael, A., A. Githeko, R. Akhtar, R. Carcavallo, D. Gubler, A. Haines, R. S. Kovats, P. Martens, J. Patz, and A. Sasaki, 2001: Human population health, in *Climate Change 2001: Impacts, adaptations and vulnerability. Contribution of Working Group II to the Third Assessment Report of the Intergovernmental Panel on Climate Change*, edited by J. J. McCarthy, O. F. Canziani, N. Leary, D. J. Dokken, and K. S. White (New York: Cambridge University Press), 453–85.

2. Woodward, A., S. Hales, and P. Weinstein, 1998: Climate change and human health in the Asia Pacific region: Who will be the most vulnerable? *Climate Research* 11:31–37.

3. Windstorms include hurricanes and cyclones.

4. Swiss Reinsurance Co., 2004: Natural catastrophes and man-made disasters in 2003: Many fatalities, comparatively moderate insured losses, *Sigma* 1/2004.

5. Kosatsky, T., 2005: The 2003 European heat waves, *Euro Surveill* 10:148–49.

6. Easterling, D. R., J. L. Evans, P. Y. Groisman, T. R. Karl, K. E. Kunkel, and P. Ambenje, 2000: Observed variability and trends in extreme climate events, *Bulletin of the American Meteorological Society* 81:417–25; Meehl, G., and C. Tebaldi, 2004: More intense, more frequent and longer lasting heat waves in the 21st century, *Nature* 305:994–97.

7. Ahern, M. J., R. S. Kovats, P. Wilkinson, R. Few, and F. Matthies, 2005: Global health impacts of floods: Epidemiological evidence, *Epidemiol Rev* 27:36–45; Hajat S., K. Ebi, S. Kovats, B. Menne, S. Edwards, and A. Haines, 2003: The human health consequences of flooding in Europe and the implications for public health: A review of the evidence, *Applied Environmental Science and Public Health* 1:13–21.

8. Ibid.

9. Kilbourne, E. M., 1997: Heat waves and hot environments, in *The Public Health Consequences of Disasters*, edited by E. K. Noji (New York: Oxford University Press), 245–69; Kovats, R. S., and C. Koppe, 2005: Heat waves: Past and future impacts, in *Integration of Public Health with Adaptation to Climate Change: Lessons Learned and New Directions*, edited by K. L. Ebi, J. B. Smith, and I. Burton (London: Taylor & Francis), 136–60.

10. Keatinge, W. R., G. C. Donaldson, R. S. Kovats, and A. McMichael, 2002: Heat and cold related mortality morbidity and climate change, in *Health Effects of Climate Change in the UK* (London: Department of Health); Dessai, S., 2003: Heat stress and mortality in Lisbon, Part II: An assessment of the potential impacts of climate change, *Int J Biometeorol* 48:37–44; McMichael, A., R. Woodruff, P. Whetton, K. Hennessy, N. Nicholls, S. Hales, A. Woodward, and T. Kjellstrom, 2003: *Human Health and Climate Change in Oceania: Risk Assessment 2002* (Canberra: Commonwealth of Australia); Hayhoe, K., D. Cayan, C. B. Field, P. C. Frumhoff, E. P. Maurer, N. L. Miller, S. C. Moser, S. H. Schneider, K. Nicholas Cahill, E. E. Cleland, L. Dale, R. Drapek, R. M. Hanemann, L. S. Kalkstein, J. Lenihan, C. K. Lunch, R. P. Neilson, S. C. Sheridan, and J. H. Verville, 2004: Emission pathways, climate change, and impacts on California, *PNAS* 101:12422–27.

11. Menne, B., and R. Bertollini, 2000: The health impacts of desertification and drought, *Down to Earth* 14:4–6.

12. Parry, M. L., C. Rosenzweig, A. Iglesias, M. Livermore, and G. Fischer, 2004: Effects of climate change on global food production under SRES emissions and socio-economic scenarios, *Global Environmental Change* 14:53–67.

13. Butt, T., B. McCarl, J. Angerer, P. Dyke, and J. Stuth, 2005: The economic and flood security implications of climate change in Mali, *Climatic Change* 68:355–78.

14. Ebi, K. L., J. Hartman, J. K. McConnell, N. Chan, and J. Weyant, 2005: Climate suitability for stable malaria transmission in Zimbabwe under different climate change scenarios, *Climatic Change* 73:375–93; Tanser, F. C., B. Sharp, and D. le Sueur, 2003: Potential effect of climate change on malaria transmission in Africa, *Lancet* 362:1792–98; Thomas, C. J., G. Davies, and C. E. Dunn, 2004: Mixed picture for changes in stable malaria distribution with future climate in Africa,

Trends Parasitol 20:216–20;Van Lieshout, M., R. S. Kovats, M. T. J. Livermore, and P. Martens, 2004: Climate change and malaria: Analysis of the SRES climate and socio-economic scenarios, *Global Environmental Change* 14:87–99.

15. For example, D'Souza, R. M., N. G. Becker, G. Hall, and K. B. Moodie, 2004: Does ambient temperature affect foodborne disease? *Epidemiology* 15:86–92; Kovats, R. S., S. J. Edwards, S. Hajat, B. G. Armstrong, K. L. Ebi, and B. Menne, 2004: The effect of temperature on food poisoning: A time-series analysis of salmonellosis in ten European countries. *Epidemiol Infect* 132:443–53; Fleury, M., D. Charron, J. Holt, O. Allen, and A. Maarouf, 2006: A time series analysis of the relationship of ambient temperature and common bacterial enteric infections in Canadian provinces, *Int J Biometeorol* 50:385–91.

16. Bernard, S. M., and K. L. Ebi, 2001: Comments on the process and product of the Health Impacts Assessment component of the United States National Assessment of the potential consequences of climate variability and change, *Environ Health Perspect* 109 (Suppl 2):177–84.

17. The aerodynamic diameter of a particle determines the depth to which it will be inhaled into the lungs and, therefore, the degree of damage that may be caused to various parts of the lung.

18. World Health Organization, 2002: *World Health Report 2002: Reducing Risks, Promoting Healthy Life* (Geneva: World Health Organization).

19. Stevenson, D. S., C. E. Johnson, W. J. Collins, R. G. Derwent, and J. M. Edwards, 2000: Future estimates of tropospheric ozone radiative forcing and methane turnover: The impact of climate change, *Geophys Res Lett* 27:2073–76; Derwent, R. G., W. J. Collins, C. E. Johnson, and D. S. Stevenson, 2001: Transient behaviour of tropospheric ozone precursors in a global 3-D CTM and their indirect greenhouse effects, *Climatic Change* 49:463–87; Johnson, C. S., D. S. Stevenson, W. Collins, and R. Derwent, 2001: Role of climate feedback on methane and ozone studied with a coupled ocean-atmosphere-chemistry model, *Geo-*phys Res Lett* 28:1723–26; Taha, H., 2001: *Potential Impacts of Climate Change on Troposphereic Ozone in California: A Preliminary Episodic Modeling Assessment of the Los Angeles Basin and the Sacramento Valley* (Berkeley, CA: Lawrence Berkeley National Laboratories); Hogrefe, C., J. Biswas, B. Lynn, K. Civerolo, J. Y. Ku, J. Rosenthal, C. Rosenzweig, R. Goldberg, and P. L. Kinney, 2004: Simulating regional-scale ozone climatology over the eastern United States: Model evaluation results, *Atmos Environ* 38:2627. Future emissions are, of course, uncertain and depend on assumptions of population growth, economic development, and energy use.

20. Knowlton, K., J. E. Rosenthal, C. Hogrefe, B. Lynn, S. Gaffin, R. Goldberg, C. Rosenzweig, K. Civerolo, J. Y. Ku, and P. L. Kinney, 2004: Assessing ozone-related health impacts under a changing climate, *Environ Health Perspect* 112:1557–63.

21. McMichael, A. J., D. Campbell-Lendrum, S. Kovats, S. Edwards, P. Wilkinson, T. Wilson, R. Nicholls, S. Hales, F. Tanser, D. LeSueur, M. Schlesinger, and N. Andronova, 2004: Global climate change, in *Comparative Quantification of Health Risks: Global and Regional Burden of Disease due to Selected Major Risk Factors*, edited by M. Ezzati, A. Lopez, A. Rodgers, and C. Murray (World Health Organization, Geneva), 1543–1649.

22. Hitz S., and J. Smith, 2004: Estimating global impacts from climate change. *Global Environmental Change* 14:201–18.

23. Ebi, K. L., J. Smith, I. Burton, and J. Scheraga, 2006: Some lessons learned from public health on the process of adaptation, *Mitigation and Adaptation Strategies for Global Change* 11:601–20.

24. Ibid.

25. Haines, A., R. S. Kovats, D. Campbell-Lendrum, and C. Corvalan, 2006: Climate change and human health: Impacts, vulnerability, and public health, *Lancet* 367:1–9.

26. Ibid, page 9.

Chapter 12

Unique and Valued Places

W. NEIL ADGER, JON BARNETT, AND HEIDI ELLEMOR

Introduction

In this chapter, we explore the issue of values at risk from climate change. We argue that localized material and symbolic values have hitherto remained undervalued in standard political and economic calculus of climate change policy and science. The scientific discourse around undesirable and irreversible impacts of climate change has focused on the idea of dangerous anthropogenic interference in the *climate* system. However, the proposition that there are identifiable levels of danger appears to be illusory as deciding what is dangerous necessarily involves making value judgments. We evaluate the potential cultural and social impacts of climate change through considering the risks climate change poses to valuable and unique places, namely the Arctic region and atolls in the Pacific Islands. Projected climate changes may result in the loss of many of their unique natural and cultural components. We then explore two major challenges these cultural and social impacts present to climate change negotiations and policy: the problem of managing the risk of irreversible loss, and the problem of reconciling nonmarket valuations and noninstrumental components of place with the economic metrics used in decision making about climate change. We suggest that the possibility of losing unique places requires decision making based on the principles of human rights and justice.

Places at Risk

Because places are unique and valued by people, managing them often entails balancing conservation and development. The effects of climate change may cause disputes in response to changes in the environmental conditions and associated meanings of place due to climate change. David Harvey, a geographer and anthropologist at City University of New York, in his classic text *Justice, Nature, and the Geography of Difference*, argues that a consideration of environmental issues is almost impossible without confronting the idea of place, writing that "some of the fiercest movements of opposition to the political economy of capitalistic place construction are waged over the issue of the preservation or upsetting of valued environmental qualities in particular places."[1]

In the last thirty years, growing environmental consciousness coupled with an increasingly pervasive media has heightened awareness of the importance of iconic places. Television and newspapers carry stories and pictures of places at risk that need to be saved in some way. Whether or not these places are saved depends on a number of factors including the ability of residents and others to demonstrate the universal qualities of the place such that a broader community agrees that it is worth saving. Indeed, icons such as the Great Barrier Reef in Australia and polar bears in the Arctic are used to great effect to represent the risks of climate change to places and species that are valued well beyond their proximate human populations.[2] In the politics of conservation, who stands to benefit from preservation is crucial for determining whether or not a place or species is actually protected and conserved.

For example, in the case of climate change, unique places may harbor benefits with global value and application. Many subsistence and indigenous societies retain traditional ecological knowledge of their environments that allow them to monitor, observe, and manage environmental change. Such knowledge is manifest not only through knowledge of land, animals, and plants by people in their particular environments but also through knowledge about governing resource management systems.[3]

Traditional ecological knowledge is complementary to modern resource management and often offers additional insights for conventional science and environmental monitoring. Thus, traditional ecological knowledge is an important resource for guiding adaptation to climate change. There are numerous examples of how this knowledge is used to manage water resources, as well as predict the weather and the onset of periodic climatic events such as El Niño.[4]

Pacific Atolls

Some places are in the frontline of impacts and vulnerability to climate change. Foremost among these are the approximately 260 atoll islands of the South Pacific.[5] Four Pacific Island territories—Kiribati, the Marshall Islands, Tokelau, and Tuvalu—are comprised entirely of atolls. There is general agreement among a wide range of scientists that these are among the most vulnerable of all *countries* to climate change.[6]

Atolls are considered to be the most vulnerable of islands to climate change. Their high ratio of coastline to land area, average elevation of only 2 meters above sea level, poor soils, limited water resources, fragile artisanal fisheries and migratory pelagic fish stocks, and sensitivity to coral bleaching (coral provides the building material for the elevated *motu* on which people live) contribute to their vulnerability. Their adaptation options are limited by their small land area and high population densities, limited economic resources, economic marginalization due to isolation, and generally low levels of human resource development.

While there is little doubt that atoll peoples are prima facie vulnerable to climate change, there may well be substantial intrinsic resilience as well. If it is acceptable to generalize across all atolls in this way, some general features of atoll life suggest resilience. First, there is a relatively high degree of reciprocity among people, communities, and neighboring islands, especially in nonurban areas.[7] This facilitates the kinds of exchanges of materials and information that assist in coping with surprises. Second, atoll communities have a long history of exposure to short-term environmental perturbations and have various strategies that enable learning and adjustment.[8] Third, most atoll societies possess a high degree of traditional ecological knowledge, and hence there are opportunities for building on traditional resource management institutions.[9]

Recently, there has been an array of initiatives to implement sustainable resource management techniques that build on traditional knowledge and institutions in atoll societies. Many of these concern fishing. For example, in Aitutaki (Cook Islands), there has been successful implementation of an Individual Transferable Quota (ITQ) scheme on trochus harvesting.[10] This was based on an initial reserve area established by the Island Council, which later implemented the ITQ system with the assistance of the Cook Islands Ministry of Marine Resources. This system of management, which blends modern and traditional techniques, has produced a resilient fishery.[11] Also in the Cook Islands, a lapsed traditional system of customary prohibition — the *Ra'ui* — has been re-implemented by traditional chiefs with the assistance of the World Wildlife Fund for Nature and the Ministry of Marine Resources. This system, which is also based on customary practices, has resulted in higher densities of coral and fish inside the restricted areas.[12] In Kiribati and Niue alike, island councils have restricted the use of certain fishing techniques, restricted the access of outsiders to fishing grounds, and placed prohibitions on fishing in certain areas.[13] These and many other practices across the atoll islands suggest that while climate change poses grave risks to these critical coastal environments, atoll cultures have some capacity to respond. This capacity is the product of interactions across international, national, and local institutions using both traditional and contemporary forms of knowledge.

Atolls are unique places: they contain unique biophysical systems and species and they sustain unique material cultures, social orders, diets, stories, languages, habits, and skills; they are the homes of peoples; and they are at risk from climate change. That they are small in every way and distant from the world's centers of political, economic, and cultural power does not make them any less unique or valuable, and it does not mean the problem of the risk of loss is any less morally important.

Arctic

The Arctic is also at the front line of the impacts of climate change because of relatively high observed and projected increases in temperature compared to lower latitudes. The potential impacts of climate change on Arctic ecosystems are among the most significant anywhere on the planet. The Arctic Climate Impact Assessment (ACIA) finds a clear warming trend for the Arctic, particularly in the winter, at rates twice the global average over the last fifty years.[14] This warming trend has now been attributed partly to human influence on the climate in work by Nathan Gillett and colleagues.[15] In 2007 this warming led to a record loss of summer sea ice and projections of ice-free summers within the next two decades. There is likely to be significant future warming in Polar Regions, regardless of whatever global preventative action is taken. Tundra areas are projected to shift and contract to their lowest recorded area, and permafrost thawing is already disrupting infra-structure in Canada, Alaska, and Russia.

In the Arctic, as in the South Pacific, tradi-tional cultures have integrated modern tech-nologies and aspects of western culture into lifestyles and resource management. There is a rich heritage of cultural adaptations to deal with environmental and social change. Docu-mentary evidence presented in the ACIA sug-gests that switching between sedentary and nomadic ways of life has been the key to re-silience in the past.[16] Local knowledge of the environment in these regions is the key to sur-vival. Corroborating observations by local res-idents and researchers include:[17]

- Increased weather variability and more extreme weather conditions observed by Nunavut in Canada both in Hudson Bay and in Yukon in western Canada and Alaska
- Longer winters and cooling trends in the James Bay and Hudson Bay areas in Canada; warmer autumn and early win-ters observed by Saami in northern Scandanavia
- Increased sea ice during the spring in the Hudson Bay, but a decline in sea ice and more unpredictability in all Arctic regions
- Numerous related changes in availabil-ity and numbers of whales, walrus, seals, and caribou throughout the region

Recent ethnographic and human ecology research documenting changes in climate and resources demonstrate the dilemmas and responses of indigenous Inuvialuit societies in northwestern Canada. This research shows that through the 1990s communities have been forced to adapt by adjusting hunting practices in the short term. Recognizing the

threat that such changes pose, communities have been forced to develop comanagement institutions to enhance their capacity for learning and self-organization. Experienced Inuit hunters adapt to changes in ice and wild-life conditions by drawing on their traditional knowledge to alter the timing and location of harvesting.

The ACIA documents a number of suc-cessful adaptations to changing conditions by using new technologies and switching eco-nomic activities. However, they also argue that there are absolute limits to adaptations by traditional societies in these regions. There will be winners and losers in these changes, and some traditional lifestyles of hunting, eco-system management systems, and valued and unique places and settlements may be lost in such adaptations.[18]

Climate change has already altered pat-terns of resource use and will continue to do so. Some of these changes can and do disrupt traditional institutions and ecological knowl-edge. These changes are apparent in frontline regions such as the Pacific Islands and in Artic communities. Hence, climate change is likely to impose irreversible change to unique and valued places and their cultures and environ-ments. These impacts pose challenges for defining dangerous climate change, for policy analysis of climate change as a global public good, and for justice and equity.

Challenges of Places at Risk to Policy

Existing climate policy does not account for the risks to places and their cultures and envi-ronments. There are two major and related problems: the problem of managing the risk of irreversible loss and the problem of reconciling the nonmarket valuation with the economic metrics used in decision making about climate change.

Irreversible Loss

How to value irreplaceable losses continues to spark considerable debate. Some researchers adjust the discount rate or add sustainability constraints to standard economic analyses. Others argue that irreplaceable losses cannot be handled meaningfully with standard cost-benefit analysis and should be protected at all costs, implying an infinite price or a zero-discount rate. Furthermore, many irreplaceable losses cannot be reasonably priced using a monetary metric. Stephen Schneider of Stanford University and colleagues have proposed five irreducible metrics for assessing climate change:[19] market costs, human lives lost, distributional effects, quality of life changes, and refugees displaced. These metrics could be used to assess losses across the human and natural systems affected by climate change, but may not easily be incorporated into standard cost-benefit analyses. Despite the difficulties in valuing natural and cultural capital, the political and ethical ramifications of climate change compel us to include these factors in the intertemporal decision-making framework.

One of the basic principles of almost every legal system is compensation of loss, whether intentional or inadvertent. This is the basis of the polluter pays principle sustaining many environmental laws including the UNFCCC. Despite the moral imperative for the countries with the highest cumulative emissions of greenhouse gases to recompense the countries most adversely affected by climate change, it will be a challenge to implement such transfers against the desire for nations to protect their self-interest and maintain their national sovereignty. Indeed the discourse of compensation may delay action for consensus and cooperative partnerships to promote adaptation in the most vulnerable countries.

Decision Making around Intangible Losses

Even if it were possible to establish enforceable mechanisms to compensate climate losses, some losses are simply not amenable to compensation. For example: it is possible to price the replacement cost of a damaged house, but not the loss of home; it is possible to price the replacement cost a destroyed museum, but not the loss of the heritage; it is possible to price the cost of relocating island populations, but not the loss of culture and community.[20] If financial transfers cannot adequately compensate losses, then decision makers must look beyond cost-benefit analysis of the distribution of climate damages and consider how to formulate a just climate policy that protects inalienable rights.

Justice Perspectives

Given the issues of incommensurability and irreversibility for some climate damages, it seems apparent that cultures can never be satisfactorily compensated for the loss of their physical bases. Thus, a more deontological conception of fairness as a set of rights and rules is in some ways more intuitive than a utilitarian approach based on a fair distribution that is enshrined in traditional cost-benefit analyses. Climate change justice perhaps can be best framed by the precautionary principle of avoiding harm and irreversible change, but rights have to be argued for through governance structures, rather than simply asserted.

The international system, built on the sovereign rights of states, frequently struggles to reconcile sovereign rights with human rights, and at present the rights of states are largely upheld, but at the frequent expense of the rights of particular groups including indigenous peoples. For example, the indigenous people of northern Canada are already experiencing the impacts of climate change. These impacts are, in effect, breaching the right of individuals, groups, and nations to an environment safe from anthropogenic harm. The Inuit Circumpolar Conference is seeking to invoke the 1948 American Declaration on the Rights and Duties of Man in the United States to make this case. They claim that native peoples of North America and Eurasia are bearing the brunt of climate change. The case being brought by the indigenous communities raises fundamental issues of the rights of cultural groups, particularly the oppressed and marginalized, compared to those of nations.[21] In effect these cases demonstrate that climate change is a fundamentally unjust burden, an externality from past and present polluters that use the global atmosphere as an open access resource. The aspects of danger highlighted in this chapter include the serious risk climate change poses to places and their cultures, and even to the sovereignty of some of these countries.

There has already been discussion of the possibility that sea-level rise will make it impossible for human populations to remain on specific islands. There would be enormous economic, cultural, and human costs if large populations were to abandon their long-established home territories and move to new places. In the present international order,

each country is granted considerable autonomy in controlling its borders and in setting policies on immigration. It would be a major turnaround, unprecedented as far as we know, if countries began to encourage all their citizens to emigrate.[22] If islanders were free to migrate, rates of international migration from island countries threatened with climate change may pass a critical threshold that constitutes danger for a society. This may arise through increasing dependency on remittances or aid rather than domestic production for income, or through the adverse effects on culture arising from migration.

The recent example of New Zealand's Pacific Access Category of migrants from Tuvalu in response to concerns about climate change is instructive. The scheme allows for up to seventy-five people from Tuvalu to migrate each year, but since it began in July 2002 less than half the number of places have been filled, suggesting that even in Tuvalu, where there is widespread concern about climate change, people are not eager to leave their homeland. This points to the need for policies and measures that enable people to adapt to climate change in ways that allow them to continue to lead the kinds of lives they value in the places they call home, rather than to simply foster migration. As the climate change officer for Kiribati said in 2000, "I think of emigration as being the stage where you know you're losing the battle. We're nowhere near that."[23] Part of the right to avoid harm includes, we argue, the right to some autonomy and ability to affect one's own destiny and resilience in choosing appropriate adaptation strategies.

Conclusion

We have argued that the nonmaterial impacts of climate change, such as those associated with place, are undervalued in the present geopolitical calculus of response and nonresponse to climate change. While the risks of climate change are particularly acute in Pacific atolls and the Arctic, the dangers of climate change are no less real for many other places around the world. The rights of non-state places, the differences in climate risks they face, and the cultural ramifications of these are largely ignored within international debate and negotiation.

There are limits to adaptation options. For example, there are limits to what money and engineering skills can do in atolls that do not have available land for retreat from sea-level rise, or in the Arctic where ice and snow cannot be remade. Adaptive strategies such as migration may be intolerable if they imply loss of traditions, knowledge, social orders, identities, and material cultures.

There is a need for more geographically and culturally nuanced risk appraisals that allow policy makers to recognize the diverse array of climate risks to places and cultures as well as countries and economies. There is a need for decision making based on a politics of principle that takes seriously these alternative risk appraisals, rather than the shallow politics of national interest. This is, in effect, a call for both a new precautionary science and new institutions for decision making at the global scale that seek to promote social learning for just solutions to the consequences of climate change on diverse places.

Notes

1. Harvey, D. 1996. *Justice, Nature, and the Geography of Difference*. Blackwell: Oxford, p. 303.

2. Slocum, R. 2004. Polar bears and energy-efficient lightbulbs: Strategies to bring climate change home. *Environment and Planning D: Society and Space* 22, 413–38.

3. Berkes, F. 1999. *Sacred Ecology: Traditional Ecological Knowledge and Resource Management*. Taylor and Francis: Philadelphia.

4. See, for example, Riedlinger, D., and Berkes, F. 2001. Contributions of traditional knowledge to understanding climate change in the Canadian Arctic. *Polar Record* 37, 315–28; Strauss, S., and Orlove, B. 2003: *Weather, Climate, Culture*. Oxford: Berg; Lefale, P. 2003. Seasons in Samoa. *Water & Atmosphere* 11, 10–11.

5. Atolls are rings of coral reefs that enclose a lagoon. Around the rim of the reef there are islets called *motu* with a mean height above sea-level of approximately 2 meters.

6. Pernetta, J. 1990. Projected climate change and sea level rise: A relative impact rating for the countries of the Pacific basin, in *Implications of Expected Climate Changes in the South Pacific Region: An Overview*, ed. J. Pernetta and P. Hughes, UNEP Regional Seas Report and Studies No. 128, UNEP, Nairobi, pp. 14–24; Hoegh-Guldberg, O., Hoegh-Guldberg, H., Stout, D., Cesar, H., and Timmerman, A. 2000, *Pacific in Peril: Biological, Economic and Social Impacts of Climate Change on Pacific Coral Reefs*. Greenpeace: Amsterdam; Barnett, J., and Adger, W. N. 2003. Climate dangers and atoll countries. *Climatic Change* 61, 321–37.

7. Connell, J. 1993. Climatic change: A new security challenge for the atoll states of the South Pacific. *The Journal of Commonwealth and Comparative Politics* 31(2), 173–92.

8. See Hooper's account of cyclone impacts on Tokelau, for example: Hooper, A. 1990. Tokelau, in *Climate Change: Impacts on New Zealand*, New Zealand Ministry for Environment, Auckland, 210–14.

9. Overton, J. 1999. A future in the past? Seeking sustainable agriculture, in *Strategies for Sustainable Development: Experiences from the Pacific*, ed. J. Overton and R. Scheyvens. Sydney: UNSW Press and Zed Books, pp. 227–40.

10. After determination of the Total Allowable Catch for a year, each household is allocated a quota of the catch, which they can trade. Excess catches above the quotas cannot be sold, reducing pressure to overexploit the resource.

11. Adams, T. 1998. The interface between traditional and modern methods of fishery management in the Pacific Islands. *Ocean and Coastal Management* 40, 127–42.

12. Hoffman, T. 2002. The reimplementation

of the Ra'ui: Coral reef management in Rarotonga, Cooks Islands. *Coastal Management* 30, 401–18.

13. Thomas, F. 2001. Remodelling marine tenure on the atolls: A case study from Western Kiribati, Micronesia. *Human Ecology* 29, 399–423; observations from Niue are from the author's (Barnett's) fieldnotes.

14. Correll, R., ed. 2005. *Arctic Climate Impacts Assessment*. Cambridge: Cambridge University Press.

15. Gillett, N. P., Stone, D. A., Stott, P. A., Nozawa, T., Karpechko, A. Y., Hegerl, G. C., Wehner, M. F., and Jones, P. D. 2008. Attribution of polar warming to human influence. *Nature Geoscience* 1, 750–54.

16. Nuttall, M. 2005. Hunting, herding, fishing and gathering: Indigenous peoples and renewable resource use in the Arctic, in *Arctic Climate Impacts Assessment*. Cambridge: Cambridge University Press.

17. Huntington, H. P., S. Fox, I. Krupnik, and F. Berkes 2005. The changing Arctic: Indigenous perspectives, in *Arctic Climate Impacts Assessment*. Cambridge: Cambridge University Press.

18. Nuttall, 2005 op. cit.

19. Schneider, S. H., Kuntz-Duriseti, K., and Azar, C. 2000. Costing nonlinearities, surprises, and irreversible events.

20. This is a real example: Cyclone Heta in 2004 destroyed Niue's national museum and its contents. The building will be replaced, but the collection of artifacts has been lost and cannot be replaced and is a significant loss to Niueans' cultural heritage.

21. Kymlicka, W., and Norman, W., eds. 2000. *Citizenship in Diverse Societies*. Oxford: Oxford University Press.

22. Patel, S. S. 2006. Climate science: A sinking feeling. *Nature* 440, 734–36.

23. Cited in Pearce, F. 2000. Turning back the tide. *New Scientist* 165(2225), 44–47.

Policy Analysis

Chapter 13

Assessing Economic Impacts

STÉPHANE HALLEGATTE AND PHILIPPE AMBROSI

Assessments of the economic impacts of climate change span a broad spectrum, from low-positive benefits to the global economy with small increases in globally averaged temperature to catastrophic losses, especially to particularly vulnerable populations and regions. As a consequence, pessimists, predicting dire outcome for climate change, recommend accepting significant short-term costs in terms of reduced economic growth to mitigate climate change, while optimists, believing in human ingenuity in adapting to climate change, oppose aggressive and costly mitigation strategies. In the lack of a consensus on this question, alternative approaches, other than global cost-benefit analysis, need to be developed to design and assess climate change policies.

This chapter aims to summarize the current knowledge on this issue and highlight the research directions that may provide needed understanding in the future. To do so, we first identify the sources of uncertainty on future impacts and their socioeconomic consequences, or damages. Next, we review the literature on the assessment of climate change damages, distinguishing between global assessments and sectoral analyses. Finally, we draw summary conclusions.

Why Is It So Difficult to Assess Climate Change Impacts and Their Consequences on Welfare?

First, because the economy is a highly complex system, there is uncertainty in modelling the various pathways from the physical impacts of climate change (sea-level rise, fluctuations in crop yields, and so on) to their economic (monetary) and other welfare (non-monetary) consequences. We possess only very fragmented, fragile, and debatable information on the possible socioeconomic consequences of climate change impacts. Despite the fact that increasing information is available on the general impacts of climate change, less is known about how the impacts vary by region and by sector for a wide range of climate change scenarios that may have a greater bearing on economic consequences. There is typically strong evidence of thresholds beyond which we lose our ability to cope with changes, but determining the exact position of these thresholds is often a challenge. Also, we have a limited understanding of the synergies between local impacts and their propagation among different affected sectors and regions (e.g., through trade flows or climate change refugees). Moreover, societies and economic systems are dynamic and will change dramatically during this century, which makes it even more difficult to assess how they will be affected by climate change. For example, rapid economic development can increase exposure to climate impacts through a concentration of population along coastal zones. On the other hand, it could also reduce vulnerability through increased diversification in the economy, or through a strengthening of the institutional and technical capacity to cope with natural disasters. Future impacts of climate change will, therefore, depend heavily on somewhat unpredictable development patterns.

Second, assessing climate change damages is tricky because a continuous adaptation process will presumably take place as climate is altered. Theoretically, ideal adaptation could result in avoiding some damages on societies (even taking advantage of climate change), but ill-designed adaptation could also potentially amplify those damages, depending on our ability to accurately detect climate change, predict its direction, and implement pertinent adaptation strategies. In addition, development patterns will play a great role in determining which adaptation options are feasible by setting out the technical, financial, and institutional context for their implementation. The fact that many meteorological damages could be avoided today through relatively cheap preventative measures that have not yet been implemented (e.g., urban planning in floodprone areas) shows that barriers to adaptation can be for-

midable and are not necessarily reflective of cost considerations only but may also depend on weak institutions, lack of awareness, lack of skills, and so forth. Moreover, in a context of a rapidly changing climate, it may become increasingly more difficult to perceive, predict, and adapt to climate variability.

Third, while damage assessments draw on undeniably fragile information, as explained above, they are made more controversial by the huge variety of value judgments that must be made to assess threats or opportunities across sectors or across regions variously affected by climate change. Such assessments require the creation of a common set of global ethical values, on which no agreement has been reached so far. Furthermore, assessments at the national or international level averages away the differentiated impacts of climate change, or implicitly assumes compensation among the "winners" and "losers" of climate change. For example, agricultural production is projected to decline in the tropics and subtropics, and to expand in northern latitudes, so that the global net economic cost as measured may be negligible, but will certainly produce benefits for northern regions at the expense of losses to equatorial regions. Similar controversies arise when considering the welfare of future generations. Even if climate damages, averaged over one century or more, are limited, the effect on a particular generation could be significant, especially for future generations, who are expected to bear the brunt of climate damages.

In spite of these difficulties, from the very disparate set of regional and sectoral impact studies available in the literature economists have proposed a summary of the different facets of climate change consequences, assessed using a monetary metric. These pioneering studies, which are presented in the next section, are landmarks in the history of integrated model assessment. They have helped formulate damage functions linking temperature rise to economic losses and tried to establish the orders of magnitude of regional and sectoral vulnerabilities.

Global Assessments Using the Enumerative Approach

Early analyses of potential climate change damages assumed: (1) climate change affects the current economy only, with no considerations of how the future economy may differ; (2) damages are evaluated independently in each sector and each region, and then summed up to obtain an estimation of the global climate change costs, neglecting all complex economic interactions; and (3) only the impacts on markets are taken into account, because they are quantifiable and easier to valuate than nonmarket losses (e.g., biodiversity). Such studies are often limited to the United States; for those studies with a global scope (often extrapolated from the U.S. situation), economic damages range between 1.3 and 1.9 percent of the Gross World Production (GWP), for a warming of 2.5 degrees Celsius.[1]

A second wave of studies, published in the late 1990s, addressed several major shortcomings of these first studies, namely the absence of information about countries other than the United States, the omission of impacts that cannot be valued through a market price, the lack of information on the sensitivity of damages to the magnitude of temperature change and to adaptation measures, and the need to introduce equity considerations. For instance, Tol produced world estimates of climate change damages using three different methods to aggregate the various gains and losses

from a 1-degree-Celsius global mean temperature rise: (1) the simple summation of all gains and losses, also referred to as "one dollar, one vote," which gives more importance to impacts in richer regions; (2) the summation of all gains and losses, but evaluated using world prices instead of local prices; and (3) the summation of welfare losses, weighted by population, which is often referred to as "one man, one vote."[2] His results ranged from global *benefits* amounting to 2.3 percent of GWP (using the net sum of gains and losses), to global *losses* amounting to 2.7 percent of GWP (using population weights). The large spread of these results shows that climate change damage estimates must be produced region by region, and that aggregated values can be dangerously misleading.

Nordhaus and Boyer took into account nonmarket impacts through estimates of *willingness to pay*, which is the maximum amount of money one is ready to pay to achieve a given goal, such as protecting a particular animal species from extinction.[3] For a global mean temperature rise of 2.5 degrees Celsius, they found global negative impacts of climate change, amounting to 1.5 percent of GWP with a simple summation, and to 1.9 percent with a population-weighted summation.

Mendelsohn and colleagues looked at the climate change impacts for several levels of warming, but considered only a few economic sectors (agriculture, energy, and so on). They also assumed that optimal adaptation measures are implemented.[4] This study estimated climate change impacts as negligible, lying between –0.09 percent and 0.3 percent of GWP, depending on the intensity of climate change.

All these studies shared the same shortcomings. In particular, they neglect all interrelationships in the economic system. For instance, they did not consider additional impacts on the agricultural sector that stem from pressure on access to water (competition with other users). They did not account for dynamic processes, in which a reduction in investment (because the resources used for adaptation expenditures cannot be invested elsewhere) could slow down economic growth. Also, several important types of impacts were not included in these analyses, such as the effects of natural disasters.[5]

Further studies, often called *meta-analyses*, were carried out based on published assessments. These meta-analyses aimed at summarizing the findings of previous studies and assessing the uncertainty from the spread between model results. They focused on marginal damage, which is the amount of damage caused by the emission of one additional ton of carbon dioxide, or, equivalently, the amount of damage avoided through a one-ton reduction in carbon dioxide emissions.

The results of these meta-analyses are, unsurprisingly, contradictory and highly controversial: Clarkson and Deyes evaluated the marginal damage of climate change at $140 for one ton of emitted carbon (tC; one ton of carbon is equivalent to 3.7 tons of CO_2). Pearce evaluated damages between $8/tC and $60/tC; and Tol stated that damages are unlikely to exceed $50/tC.[6] The large team involved in Downing and colleagues also tried to encompass nonmarket impacts, socially contingent values, and climate surprises, and to produce consensus bounds for marginal damages.[7] While they could not reach any consensus on an upper bound for damages, they propose a lower bound at £35/tC. The main sources of uncertainty for all the studies mentioned here were: (1) climate sensitivity, that is, how much warming is caused by one additional ton of carbon; (2) the discounting

scheme, that is, how impacts occurring at different points in time are weighted against each other; (3) equity weighting among different populations; and (4) the adaptive capacity of future economies.

More recently, new estimates of aggregated climate change damages were produced by Mendelsohn and Williams, and Nordhaus, using statistical analyses of geographical, climatic, and economic data.[8] These authors use statistical relationships among climate variables (mean temperature and annual precipitation) and economic variables (farm values, energy expenditures, and others for Mendelsohn and Williams; local GDP for Nordhaus). Mendelsohn and Williams, working at the national scale, found that for all climate scenarios and all adaptation assumptions, there is a quasi-neutral impact of climate change, with total losses always lower than 0.25 percent of GWP, although significant consequences were found for developing countries. Nordhaus worked in an innovative manner on a 1-degree by 1-degree grid. His study suggests higher damages than past studies. It predicts GWP losses between 1 percent (output weights) and 3 percent (population weights) for a doubling of CO_2 concentration.

The latest attempt to assess the global socioeconomic costs of climate change was carried out for the *Stern Review on the Economics of Climate Change*.[9] Drawing on published monetary estimates of sectoral and regional damages, this exercise above all emphasized the risks of underestimating damages by neglecting the large uncertainties that surround the climate and impact sciences and illustrated the importance of equity considerations. The *Stern Review*, indeed, assumed approximately the same level of damages as Nordhaus *for a given level of warming*. But, instead of considering only one warming level, the Stern review was based on a "risk-management" approach and considered a probability distribution of possible changes (including high climate sensitivity and abrupt climate shifts), so that all potential future scenarios are evaluated but their outcome is weighted by the probability that they will occur. Unsurprisingly, the inclusion of the most pessimistic climate scenarios in the analysis raised the expected value of the damages. Moreover, the *Stern Review* considered a particularly long time horizon (up to year 2200) and a (much-discussed) low discount rate (1.4 percent), which also explained why its estimate of the net present value of climate damages was very high, with an average reduction in global per capita consumption of at least 5 percent now and forever compared to the baseline scenario. The authors then showed that including the so-called nonmarket impacts might double the toll of climate change, from a 5 percent to an 11 percent average loss in global per capita consumption, and that using "equity-weighting"—to reflect the higher burden on the developing countries—tended to increase global cost of climate change by 25 percent.

Sectoral Studies of Climate Change Economic Impacts

Recent studies have focused on regions or sectors and explicitly consider the cost burdens of uncertainty, adaptation, and indirect (or higher-order) social costs of climate change.

Among them, Schlenker and colleagues improved the work of Mendelsohn by taking into account the specific role of irrigation in determining climate impacts on the agriculture sector in the United States[10] Fischer and colleagues carried out extensive analyses of climate change consequences on global agriculture based on regional assessments.[11] They predicted no disruption of overall global food production, but indicated growing regional imbalances, with possible impacts on food security and large economic consequences at the national scale for developing countries. For instance, Africa was predicted to suffer from losses in the agriculture sector representing between 2 and 9 percent of GDP in 2080, increasing risks to food security with imports rising by as much as 25 percent. These results were consistent with findings by Mendelsohn and Williams as well as Bosello and Zhang.[12] Regarding another important sector for developing countries, Hamilton and colleagues investigated the influence of climate change on tourism and found potentially significant impacts. However, this influence was weaker than other drivers, like population and economic growth.[13] Berrittella and colleagues reported similar results and stressed that as tourism has been excluded from former economic impact studies of climate change, regional economic impacts might have been underestimated by more than 20 percent.[14]

Examining extreme events and natural disasters, Hallegatte and colleagues suggested that, if extreme event frequency and intensity were modified by climate change, short-term financial and technical constraints on reconstruction might increase significantly the long-term costs of natural disasters.[15] This effect has the potential to exacerbate poverty traps for the poorest countries, unable to cope with repeated disasters. Neumann and colleagues, as well as Nicholls and Tol, investigated the consequences of sea-level rise, taking into account possible adaptation measures, and found lower impact estimates than previous studies.[16] These results, compared with studies that disregard adaptation, showed how efficient adaptation actions could reduce climate change impacts. They highlighted the importance of predicting, or detecting in due time, climate change signals, thus making it possible to implement measures that may take time to become fully effective (e.g., dam or barrier construction, or population relocation).

There are still large uncertainties for future climate change, especially at the local scale, and natural variability will make it difficult to detect climate change signals in the coming decades. Therefore, it may not be possible to anticipate and implement optimal adaptation strategies. This has been illustrated by Hallegatte and colleagues, who focused on the adaptation of urbanism to higher temperatures in European cities.[17] They found that climate uncertainty prevented any low-cost adaptation and that adaptation expenditures for this sector could amount to several percent of regional GDP. Comparable issues facing adaptation have been stressed for other types of impacts, like sea-level rise, by Barnett.[18]

Most of the recent economic impact studies have added the indirect social costs of climate change to direct costs traditionally considered (e.g., the indirect effects for industries of energy-sector impacts). The majority of these studies used a static general equilibrium framework, where the economy-wide and country-wide propagations of a climate change–induced shock on a specific sector are scrutinized at a given point in time (sea-level rise—Bosello et al.; agriculture—Bosello and Zhang; tourism—Hamilton et al., Berrittella et al.; or human health—Bosello et al.).[19] Kemfert assessed synergies between impacts in several sectors, their resulting effects, and their implications for growth.[20] Starting from Tol's assessment of direct impacts in various sectors, her results indicated a global loss

equivalent to 1.8 percent of GWP for a warming of 0.25 degree Celsius by 2050; a very large increase compared to the direct impacts her work is based on (less than 1 percent). This analysis demonstrates the potential importance of impact propagations and dynamic mechanisms.[21]

Such findings echoed conclusions from Fankhauser and Tol, based on an analytical one-sector one-region growth model, indicating that "higher order" costs of climate change may be of the same magnitude as direct costs in the medium to long term and may also result in a slowdown of growth.[22] Also noteworthy is the assessment by Link and Tol, which is the only attempt to value catastrophic climate change scenarios by exploring the possible economic impacts of a shutdown of the thermohaline circulation.[23]

On the whole, this brief overview of the direct economic damage assessments and indirect socioeconomic assessments of climate change demonstrates the field is still maturing, as indirect social costs are just starting to be valued and more attention is being paid to uncertainty in adaptation. In particular, some of the new studies highlight damage overestimations by previous assessments, while others highlight damage underestimations, indicating that cost uncertainty is likely to be larger than suggested by the range of previous estimates.

Conclusion

Most estimates of the global economic impacts of climate change average out regional, sectoral, socioeconomic, and generational differences in climate change impacts to conclude that global climate change will produce small benefits or losses of a few percent of

GWP. All studies agree that strong losses are to be expected in developing tropical countries, while some of them predict benefits in developed high-latitude nations. In its latest assessment report, the Second Working Group to the IPCC, on impacts, adaptation, and vulnerability, states that "it is very likely that all regions will experience either declines in net benefits or increases in net costs for increases in temperature greater than about 2 to 3 degrees Celsius" and that "global mean losses could be 1 to 5 percent of GWP for 4 degrees Celsius of warming."[24] These estimates are characterized by a huge amount of uncertainty. This uncertainty arises from: (1) value judgments (the aggregation bias, the discounting scheme, and the valuation of nonmarket impacts); (2) the complexity of the economic system (the role of sector interactions, of adaptation strategies, and of market imperfections); and (3) uncertainty in future development pathways that influence both exposure and vulnerability to climate change.

Analyses on single sectors and single regions, however, have highlighted important processes that may cause major damage and are not yet accounted for by these aggregated studies. Examples of such processes are extreme events and natural catastrophes, vulnerability thresholds, inadequate adaptation strategies, migration, and international security concerns. More research is needed to get a better understanding of these mechanisms to introduce them into aggregate impact assessments. Other areas of potential improvement in our knowledge of impacts relates to the understanding of impacts on ecosystems, assessing their contribution to welfare, and, above all, estimating the risks induced by the irreversible decline or disappearance of certain ecosystems.

Interestingly, even in the presence of persisting uncertainties, the balance of evidence seems to suggest there are significant risks to further postponing global action to mitigate climate change. Further, there are reasons to think that aggregated assessments are poor estimates of future damages. Cost-benefit analysis may not, therefore, be the best tool to design climate policies, and alternative approaches, based on multiple indicators, should be investigated. In this context, Schneider and colleagues, among others, suggested the use of five "numeraires," as the best compromise between accuracy (with which the benefits of climate policies can be described) and relevance (of the metric to capture impacts of climate policy on welfare) in the description of climate policy benefits: (1) a monetary assessment of impacts on economic activities and human settlements; (2) the number of lives at risk and health risks; (3) a quality of life index, including psychological dimensions such as having to migrate, the loss of landscapes with their cultural value, and so on; (4) a measure of the risks to ecosystems; and, finally, (5) an indicator of the distribution of risks among different populations.[25] Using such a set of indicators would allow decision makers to assess the impact of climate policy on human welfare in a more comprehensive fashion and to incorporate the latest findings from the (physical) impacts of climate change, without being paralyzed by the controversies raised by the monetary valuation of impacts.

Notes

The views expressed in this paper are the sole responsibility of the author. They do not necessarily reflect the views of the World Bank, its executive directors, or the countries they represent.

1. Cline, W., (1992), *The Economics of Global Warming*, Institute for International Economics, Washington: D.C.; Fankhauser, S., (1995), *Valuing Climate Change: the Economics of Greenhouse*, Earthscan, London: UK; Nordhaus, W., (1991), To slow or not to slow: The economics of the greenhouse effect, *Econ. J.*, 101, 920–37; Titus, J. G., (1992), The cost of climate change to the United States, in *Global Climate Change: Implications, Challenges, and Mitigation Measures*, ed. S. K. Majumdar, L. S. Kalkstein, B. Yarnal, E. W. Miller, and L. M. Rosenfeld (dir.), Pennsylvania Academy of Science, Easton: PA; Tol, R. S. J., (1995), The damage costs of climate change: Toward more comprehensive calculations, *Environmental and Resource Economics*, 5, 353–74.

2. Tol, R. S. J., (2002), New estimates of the damage costs of climate change, Part I: Benchmark estimates, *Environmental and Resource Economics*, 21 (1), 47–73; Tol, R. S. J., (2002), New estimates of the damage costs of climate change, Part II: Dynamic estimates, *Environmental and Resource Economics*, 21 (2), 135–60.

3. Nordhaus, W., and Boyer, R., (2000), *Warming the World: Economics Models of Climate Change*, MIT Press, Cambridge: MA.

4. Mendelsohn, R., Morrison, W., Schlesinger, M., and Andronova, N., (2000), Country-specific market impacts of climate change, *Climatic Change*, 45, 553–569.

5. Munich-Re, (2006), *Topics: Annual Review: Natural Catastrophes 2005*, Munich Reinsurance Group, Geoscience Research Group, Munich: Germany.

6. Clarkson, R., and Deyes, K., (2002), *Estimating the Social Cost of Carbon Emissions*, GES Working Paper 140, HM Treasury, London: UK, available at www.hm-treasury.gov.uk/Documents/ Taxation_Work_and_Welfare/Taxation_and_the_ Environment/tax_env_GESWP140.cfm; Pearce, D. W., (2003), The social cost of carbon and its policy implications, *Oxford Review of Economic Policy*, 19, 362–84; Tol, R. S. J., (2005), The marginal damage costs of carbon dioxide emissions: An assessment of the uncertainties, *Energy Policy*, 33 (16), 2064–74.

7. Downing, T. E., Anthoff, D., Butterfield, B., Ceronsky, M., Grubb, M., Guo, J., Hepburn, C., Hope, C., Hunt, A., Li, A., Markandya, A., Moss, S., Nyong, A., Tol, R. S. J., and P. Watkiss, (2005), *Scoping Uncertainty in the Social Cost of Carbon*. DEFRA, London: UK, available at www.defra .gov.uk/environment/climatechange/carbon-cost/ sei-scc/index.htm.

8. Mendelsohn, R., and Williams, L., (2004), Comparing forecasts of the global impacts of climate change, *Mitigation and Adaptation Strategies for Global Change*, 9 (4), 315–33; Nordhaus, W., (2006), Geography and macroeconomics: New data and new findings, *Proceedings of the National Academy of Sciences*, May 2006.

9. See http://www.hm-treasury.gov.uk/inde pendent_reviews/stern_review_economics_climate _change/stern_review_report.cfm.

10. Schlenker, W., Hanemann, W. M., and Fisher, A. C. (2005), Will U.S. agriculture really benefit from global warming? Accounting for irrigation in the hedonic approach, *American Economic Review*, 95 (1), 395–406; Schlenker, W., Hanemann, W. M., and Fisher, A. C., (2006), The impact of global warming on U.S. agriculture: An econometric analysis of optimal growing conditions, *Review of Economics and Statistics*, 88 (1), 113–25; Schlenker, W., Hanemann, W. M., and Fisher, A. C., (2007), Water availability, degree days, and the potential impact of climate change on irrigated agriculture in California, *Climate Change* 81 (1), 19–38.

11. Fischer, G., Shah, M., and Van Velthuizen, H., (2002), *Climate Change and Agricultural Vulnerability*, IIASA, Laxenburg: Austria.

12. Mendelsohn, R., and Williams, L., (2004), Comparing forecasts of the global impacts of climate change, *Mitigation and Adaptation Strategies for Global Change*, 9 (4), 315–33; Bosello, F., and Zhang, J., (2005), *Assessing Climate Change Impacts: Agriculture*, Nota di Lavoro 94.2005, FEEM, Milan: Italy.

13. Hamilton, J. M., Maddison, D. J., and R. S. J. Tol, (2005), Climate change and international tourism: A simulation study, *Global Environmental Change*, 15 (3), 253–66.

14. Berrittella, M., Bigano, A., Roson, R., and R. S. J. Tol, (2006), A general equilibrium analysis of climate change impacts on tourism, *Tourism Management*, 27, 913–24.

15. Hallegatte, S., Hourcade, J.-C., and Dumas, P., (2006), Why economic dynamics matter in the assessment of climate change damages: Illustration extreme events, *Ecological Economics*, 62 (2), 330–40.

16. Neumann, J. E., Yohe, G., Nicholls, R.,

and Manion, M., (2000), *Sea-level Rise and Global Climate Change: A Review of Impacts to U.S. Coasts*, The Pew Center on Global Climate Change, Washington, D.C.; Nicholls, R. J., and Tol, R. S. J., (2006), Impacts and responses to sea-level rise: A global analysis of the SRES scenarios over the 21st century, *Philosophical Transaction of the Royal Society*, London, UK.

17. Hallegatte, S., Hourcade, J.-C., and Ambrosi, P., (2006), Using climate analogues for assessing climate change economic impacts in urban areas, *Climatic Change*, 82 (1–2), 47–60.

18. Barnett, J., (2001), Adapting to climate change in Pacific Island countries: The problem of uncertainty, *World Development*, 29 (6), 977–93.

19. Bosello, F., Lazzarin, M., Roson, R., and Tol, R. S. J., (2004), *Economy-Wide Estimates of the Implications of Climate Change: Sea-Level Rise, Research Unit Sustainability and Global Change*, Centre for Marine and Climate Research, Hamburg University, Hamburg, Germany; Bosello, F., and Zhang, J., 2005 op. cit; Hamilton, Maddison, and Tol, 2005 op cit.; Berrittella, M., Bigano, A., Roson, R., and Tol, R. S. J., 2006 op.

cit.; Bosello, F., Roson, R., and Tol, R. S .J., (2006), Economy-wide estimates of the implications of climate change: Human health, *Ecological Economics*, 58, 579–91.

20. Kemfert, C., (2002), An integrated assessment model of economy-energy-climate: The model Wiagem, *Integrated Assessment*, 3 (4), 281–98.

21. Tol, R. S. J., 2002 op. cit.

22. Fankhauser, S., and Tol, R. S .J., (2005), On climate change and economic growth, *Resource and Energy Economics*, 27, 1–17.

23. Link, P. M., and Tol, R. S .J., (2004), Possible economic impacts of a shutdown of the thermohaline circulation: An application of FUND, *Portuguese Economic Journal*, 3, 99–114.

24. *Climate Change 2007: Impacts, Adaptation and Vulnerability*, Working Group II Contribution to the Intergovernmental Panel on Climate Change, Fourth Assessment Report. Summary for Policymakers available at www.ipcc.ch.

25. Schneider, S. H., Duriseti, K. K., and Azar, C., (2000), Costing nonlinearities, surprises and irreversible events, *Pacific and Asian Journal of Energy*, 10 (1), 81–91.

Chapter 14

Integrated Assessment Modeling

HANS-MARTIN FÜSSEL AND MICHAEL D. MASTRANDREA

Introduction

Integrated assessment (IA) of climate change combines scientific knowledge from the natural and social sciences to provide relevant information for the management of global climate change and its impacts. Its application to global climate change is dominated by formal computer models (IA models) that integrate simplified representations of relevant subsystems (e.g., demography, energy, economy, carbon cycle, climate, agriculture, natural ecosystems, coastal zones, human health).

Global climate change presents unprecedented challenges for scientific analysis and its management by society.[1] The most important challenges are: (1) the global scale of the problem, which requires cooperation by many stakeholders with diverse interests; (2) the long inertia of relevant subsystems, which requires analysis and management over very long time scales; (3) the regional as well as social disparity between those mainly responsible for climate change and those most vulnerable to it, which brings up difficult equity issues; (4) the pervasiveness of causes and ef-

fects, which prohibits a simple technological fix; (5) the complexity of the involved systems, which results in large scientific uncertainties, in particular regarding regional and local impacts; and (6) the possibility of catastrophic and/or irreversible effects, which makes policy preferences very sensitive to (subjective) risk attitudes of policy makers.

Since even the most sophisticated IA models cannot include all relevant information for any policy decision, the application of IA models for policy analysis requires a critical examination of their limitations in a particular decision context. Otherwise, IA of climate change faces the risk of being exploited to support narrowly defined political positions.

A comprehensive treatment of IA is beyond the scope of this chapter. Instead, we highlight selected aspects of this field and provide references for a more in-depth discussion.[2] The second section presents various climate policy decision problems to which IA has been applied and introduces the main approaches to IA modeling. The third section discusses challenges as well as recent developments in the field. The fourth section presents two applications of IA models.

Framing Integrated Assessments
Questions for Climate Policy Analysis

The various stakeholders involved in addressing the climate problem have a wide range of informational needs that can only be addressed by integrating multiple aspects of the climate problem. Some of the policy questions to which IA has been applied include:

- What are the expected impacts of business-as-usual emission scenarios on different sectors and regions?
- What are the costs and benefits of deferring emissions control until we know more about the science of climate change and have developed cheaper low-carbon energy sources?
- What are the trade-offs between mitigation, sequestration, and adaptation policies?
- What are the relative merits of different policy options for achieving a given environmental target?
- How important are the impacts of potential large-scale climate instabilities for climate policy decisions?

The answers to most of these questions are contingent on assumptions regarding (1) key scientific uncertainties (e.g., climate sensitivity for a doubling of the atmospheric CO_2 concentration); (2) socioeconomic development (e.g., population growth, pace of technological change, level of adaptation); and (3) subjective value judgments (e.g., time discount rates, regional equity weights). It is gen-

erally argued that the most important findings of the IA enterprise are not the specific results of a particular model analysis, but broader insights into the structure of the problem that help frame the political debate and provide direction for future research. Such insights are typically gained from more focused questions such as:

- What levels of emissions control over time are consistent with limiting impacts from climate change and costs of controlling emissions to predefined levels?
- What are the relative importance of scientific uncertainties, beliefs about socioeconomic trends, the acceptability of specific policies, and alternative ethical judgments for the evaluation of different climate policies?
- What cost savings can potentially be achieved by allowing flexibility in when, where, and how to reduce greenhouse gas emissions?
- What are the main differences in results between policy analyses that consider several heterogeneous decision makers and analyses that assume a single "representative" decision maker?

Classification of IA Models

The wide variety of IA models reflects the diversity of information needs of climate policy makers and the range of scientific approaches to the problem. Important topics for model design are the relationship between scientific analysis and subjective value judgments, the relative importance of economics and moral philosophy, and the tension between detail/

realism and simplicity/transparency. The following dimensions are commonly included in categorizations of IA models.

Policy Evaluation versus Policy Optimization versus Policy Guidance

A fundamental categorization for IA models is between policy evaluation models, which are designed to project the consequences of a particular climate policy, and policy optimization models, which are designed to determine the best climate policy based on a specified policy goal or objective function. Policy guidance models combine aspects of these two categories.

- Policy optimization models (e.g., DICE/RICE, FUND, MERGE) may seem superior since they claim to identify the "best" policy.[3] However, their complexity is severely limited by the numerical algorithms used to solve optimization problems. Scientific simplifications include the representation of the climate system and climate impacts by very few equations, the neglect of key uncertainties, and the assumption of a single "global" decision maker. Furthermore, wide-ranging normative assumptions are necessary to aggregate all consequences of alternative policies in a common metric for social welfare. Typical assumptions include that social well-being can be fully described in terms of economic wealth, that equity considerations are irrelevant, and that intergenerational discounting should be based on observed market interest rates.[4]

  The main applications of optimizing models are as follows: in cost-benefit analyses, the preferences of climate policy makers are assumed to be completely represented by the social welfare function. In cost-effectiveness analyses, the optimization is subject to an environmental constraint (e.g., maximum amount of global mean temperature increase). In robustness analyses, the choice of policies is limited to a few discrete alternatives, and the optimal policy strategy is determined across a range of uncertain scientific parameters and normative assumptions.

- Policy evaluation models (e.g., AIM, IMAGE) evaluate the effects of specific policies on various social, economic, and environmental parameters.[5] Since these models are not subject to the computational constraints of optimizing models, they can include a much higher level of process and regional detail and provide more detailed information on the consequences of alternative policies. As Kolstad states, "at some point policymakers will want to know if an increment of GDP loss is the result of fewer recreational amenities in Southern California [. . .] or more deaths from Monsoon-driven flooding in Asia."[6]

- Policy guidance models (e.g., ICLIPS) determine all policies that are compatible with a set of normatively specified constraints ("guardrails").[7] Their ability to consider multiple independent criteria in evaluating the acceptability of a given policy strategy does not require the heroic normative assumptions necessary for formulating an aggregated welfare function in policy optimization models. However, because the algorithms applied by policy guidance models are similar to those of optimization models, they also require a highly simplified representation of dynamic system components.

DETERMINISTIC VERSUS PROBABILISTIC VERSUS ADAPTIVE

- Deterministic analyses apply best-guess values for all model parameters. The effect of alternative parameter choices can be determined through sensitivity and importance analyses.[8]
- Probabilistic (or stochastic) analyses specify probability distributions for some or all uncertain model parameters and inputs. The nature of probabilistic analyses differs according to the modeling approach: policy evaluation models determine probability distributions for output variables; policy optimization models typically determine the policy that maximizes an expected welfare function; and policy guidance models determine the set of policy strategies that is compatible with probabilistic guardrails.
- Adaptive (or hedging) analyses denote probabilistic applications of optimizing models that allow for future learning about key scientific or policy uncertainties. More specifically, they aim to determine the optimal near-term strategy under given assumptions about the resolution of these uncertainties in the future.

Challenges and Recent Developments

IA of climate change faces several challenges, and the associated limitations need to be considered in a policy context. Some of the main challenges include the following.

Uncertainty in Climate Projections

Despite significant advances in recent years, considerable uncertainty exists in projecting the global and regional climate effects of a particular emissions scenario. In general, climate projections are more reliable for temperature than for precipitation and other highly variable climate parameters, for changes in average values than for changes in extreme events, and for large regions than for small regions.

A related but separate issue is the limited capacity of climate models to simulate potential large-scale climate instabilities (also known as "large-scale climate singularities," "abrupt climate change," or "tipping points of the climate system"). However, it is not currently possible to predict under which conditions these events occur, and even a probabilistic treatment requires many subjective assumptions (see discussion later in this chapter).

Uncertainty about Non-Climatic Developments

Impacts of climate change, in particular on social systems, are strongly influenced by concurrent non-climatic developments, including adaptation. The social impacts of sea-level rise, for instance, depend on future migration patterns, on the economic resources of future generations, and on their willingness to invest them into coastal protection. Similarly, increases in climate-related malaria risk would be largely irrelevant if an effective, safe, and affordable vaccine should become globally available. None of these developments can be predicted with high certainty.

Aggregation and Valuation

Climate policy analysis requires the consideration and aggregation of information about diverse impact categories, regions, and societies over a long time horizon, whereby relevant information is available with varying degrees of accuracy and reliability. Aggregation of this information cannot be done on a purely scientific basis. It depends on normative choices such as the relative importance of current and future generations and of different social groups within a generation, the valuation of nonmarket impacts on human health and ecosystems, and the risk attitude.

Policy-optimizing models perform the aggregation by assigning monetary values to all relevant pieces of information. While most normative choices are included in the model, the analyst can explore the implications of alternative choices by varying relevant model parameters (e.g., time discount factors, equity weights). Policy-evaluating models, in contrast, leave most controversial choices to the decision maker, who is provided with a maximum amount of information. However, some selection or aggregation by the analyst is still necessary when model outputs are compiled for presentation. Policy guidance models limit the number of normative choices in the model by leaving the selection of "guardrails" to the decision maker.

Treatment of Data-Scarce Regions and Sectors

IA of global climate change faces the challenge of combining information from regions and sectors with different availability of relevant data and models. For instance, the dam-age functions representing monetary climate impacts in optimizing IA models have generally been extrapolated from a few (or just one) industrialized countries to the rest of the world. Similarly, the impacts of climate change on extreme weather events and associated social disasters have generally been neglected from the analysis due to the scarcity of reliable data. The exclusion of some kinds of (adverse) climate impacts is particularly problematic in optimizing analyses, because this practice causes a systematic bias in the determination of optimal climate policy strategies.

Diversity of Policy Options

Different policy options are available to manage the risks from anthropogenic climate change. The fundamental distinction is between mitigation of climate change and adaptation to climate change. Some classifications consider carbon sequestration, research and development, capacity building, and/or compensation for climate impacts as separate options. A comprehensive and efficient climate policy comprises different policy options and considers the links between climate change and other societal issues such as sustainable development, disaster reduction, and public health promotion. Few analyses consider all policy options together due to large differences in their characteristic spatial and temporal scales and in the relevant actors. Most IA models focus on the mitigation of global climate change, which is more amenable to global analysis than the highly context-specific analysis of regional adaptation policies.

Considerable progress has been achieved recently in the following aspects of IA.

Comprehensive Uncertainty Analysis

IA of climate change faces multiple uncertainties due to natural variability, lack of scientific knowledge, social choice (reflexive uncertainty), and value diversity. The natural variability in the climate system can in principle be characterized by frequentist probabilities. Uncertainties related to lack of knowledge (e.g., about the "true" value of climate sensitivity), in contrast, have to be represented by Bayesian probabilities that always contain substantial elements of subjectivity. Another approach that has recently been developed involves generalized models of probability such as interval probabilities, possibility theory, and belief functions.[9] Whether and under what conditions future social choice can be described in probabilistic terms has been the subject of considerable scientific debate.[10] Value diversity cannot be meaningfully characterized probabilistically. It is often addressed through sensitivity analysis, in which different value systems are explicitly represented and contrasted.

Probabilistic analysis is increasingly applied for IA, often in a risk-management framework.[11] Even though not all aspects of uncertainty can be represented probabilistically, these analyses have provided a wealth of valuable information about the relationship between climate policies and the risks from anthropogenic climate change (see the section on Illustrative Applications for an example).

Consideration of Large-Scale Climate Instabilities

The consideration of large-scale climate instabilities (such as a potential breakdown of the thermohaline ocean circulation, the melting of the Greenland ice sheet, or the disintegration of the West Antarctic ice sheet) presents many challenges for IA:[12] it is highly uncertain under which conditions they would occur and what their biophysical and social effects would be, and the economic valuation of these effects requires controversial subjective assumptions. As a result of these difficulties, IA of climate change has long excluded climate instabilities from the analysis and instead focused on best-guess scenarios of gradual climate change (dubbed "the same, only warmer"). Recognizing the limitations of this approach, many recent IA studies have attempted to consider climate instabilities.[13] While the numerical results of these studies differ significantly, there is wide agreement that the risk from climate instabilities is a very significant element of climate policy analysis. The consideration of climate instabilities in policy optimization models substantially increases the optimal level of emissions control.

Multi-Gas Assessments

Even though anthropogenic climate change is caused by the emission of several GHGs and aerosol precursors, most early IA models focused on policies to mitigate emissions of energy-related CO_2 emissions. Recent policy analyses that have considered the importance of different GHGs have found that a multi-gas approach to GHG reduction can considerably reduce the costs of mitigating climate change.[14]

Endogenous Technological Change

Policy optimization and policy guidance models balance the risks at different levels of climate change with the costs of mitigating climate change. These costs are very sensitive to assumptions about technological development, related to energy efficiency and low-carbon energy. The Innovation Modeling Comparison project has shown that the consideration of induced technological change significantly reduces estimates of emissions abatement costs.[15]

Consideration of Future Learning

An increasing number of analyses have considered adaptive decision strategies, in which future policies depend on incoming information (dubbed "learn then act") that reduces or eliminates uncertainties in the future.[16] These studies have revealed that the relationship between key uncertainties of the climate problem, their resolution in the future, and effective climate policies is rather complex. The consideration of learning may increase or decrease near-term mitigation efforts, depending on the specific assumptions in a particular analysis.

Multi-Actor Analysis

Most optimizing IA models identify a globally optimal climate policy under the assumption of full cooperation of all relevant actors. Due to the heterogeneous distribution of climate impacts and mitigation costs, however, the global optimum may not be optimal from the perspective of individual actors. Game-theoretic analyses have explored the implications of this heterogeneity for global climate policy and have suggested measures to overcome barriers for cooperation.[17]

Second-Best Analyses

Most early optimizing IA models assumed a "first-best" world with perfect markets and without distortionary factor taxes and cost externalities. The context in which climate policies will actually be implemented, however, is full of such distortions. Recent analyses have found that optimal climate policies in a second-best world can differ significantly from those in a first-best world.[18] For instance, net costs of emission reductions are lower (and optimal abatement rates higher) when carbon taxes are used to replace existing distortionary taxes or when the reduction of health impairments associated with current energy production is considered.

Modular IA Models

IA models have traditionally been developed by individuals or by closed teams. Recently there have been various approaches for the development of modular IA models (e.g., CIAS) in which individual components can be easily replaced by alternative sub-models, either from the core modeling group or from the wider scientific community.[19] In this way, uncertainty assessment can be extended from individual model parameters to whole modeling approaches.

Illustrative Applications

This section presents two illustrative applications of IA models of climate change: a probabilistic assessment with a globally aggregated climate-economy model and a guard-rail analysis with a multiregional climate-economy-impact model.

Probabilistic Integrated Assessment

Probabilistic IA represents one or more key uncertain parameters of the coupled social and natural system in probabilistic terms. This information is used to generate probabilistic output, such as probability distributions for future temperature increase or the probability of specific climate impacts under different climate scenarios.

Figure 14.1 shows results from a probabilistic analysis of climate stabilization targets. The solid and dashed lines depict the probability of exceeding an equilibrium global warming of 2 degrees Celsius above preindustrial levels (the official climate stabilization target of the European Union) at GHG stabilization levels between 350 and 750 parts per million (ppm) CO_2-equivalent based on various probability distributions for climate sensitivity.

According to figure 14.1, ensuring that the probability of staying below the 2-degree-Celsius target is "likely" (defined as more than 66 percent probability) or "very likely" (greater than 90 percent probability) requires that the CO_2-equivalent stabilization level must be below 410 ppm for the majority of climate sensitivity distributions, with a range from 355 to 470 ppm across all models considered.

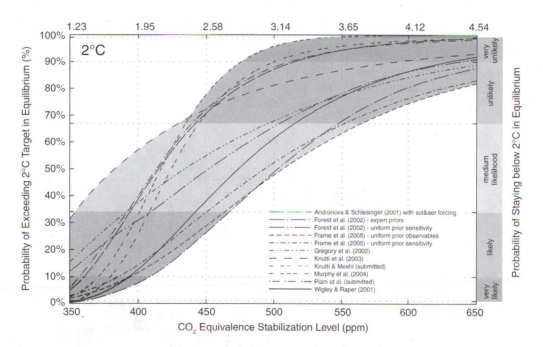

FIGURE 14.1. Probability of exceeding an equilibrium global warming of 2 degrees Celsius (above preindustrial levels) for different CO_2-equivalent stabilization levels. Source: M. Meinshausen. What does a 2°C target mean for greenhouse gas concentrations? A brief analysis based on multi-gas emission pathways and several climate sensitivity uncertainty estimates, in *Avoiding Dangerous Climate Change*, H. J. Schellnhuber, W. Cramer, N. Nakicenovic, T. Wigley, and G. Yohe, eds. Cambridge University Press, Cambridge, UK, 2006, chapter 28, pp. 265–79.

Conversely, exceeding the 2-degree-Celsius target is "likely" or "very likely" at 550 ppm CO_2 equivalence for almost all models considered. Another way to state this is to say that the probability of staying below the 2-degree-Celsius threshold is unlikely (less than 34 percent) or very unlikely (less than 10 percent) for CO_2 equivalent concentrations above 550 ppm. To put this in context, GHGs must be stabilized below *current* levels to make the probability of exceeding the 2-degree-Celsius target "very unlikely" for the majority of probability distributions.

Climate impacts depend not only on the concentration stabilization level but also on the path to reach this level. Several analyses have found significant differences in transient temperature change and warming rates between concentration paths that allow temporary overshoot of the stabilization concentration and paths that do not.[20] For example, the probability of at least temporarily exceeding the 2-degree-Celsius target until 2200 was determined to be 70 percent higher in an overshoot scenario rising to 600 ppm CO_2-equivalent and then stabilizing at 500 ppm CO_2-equivalent, compared with a non-overshoot scenario stabilizing at the same level.[21]

Guardrail Analysis

Guardrail analyses assess the implications for global or regional climate protection strategies of various normative guardrails specified in terms of climate, climate impact, and/or climate policy variables. This approach was first implemented in the ICLIPS model.[22]

Figure 14.2 shows results from a guardrail analysis with the ICLIPS model.[23] This example involves an illustrative "impact guardrail," which specifies that fundamentally transforming more than 30 percent of the Earth's ecosystems would be unacceptable. The pairs of dot-

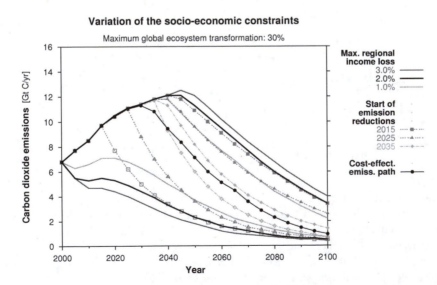

FIGURE 14.2. Sensitivity of global emission corridors to variations of two socioeconomic guardrails.

ted, solid thick, and solid thin lines without markers denote the upper and lower boundaries of the admissible global CO_2 emissions corridor, which observes this impact guardrail as well as a socioeconomic guardrail that limits mitigation costs for any generation in any world region to 1 percent, 2 percent, or 3 percent of per capita income, respectively. A lower willingness to pay for climate protection narrows the emissions corridor and limits the temporal flexibility for reducing emissions. The dashed lines marked with squares, triangles, and diamonds in figure 14.2 show the effects of delaying emission reductions until 2015, 2025, and 2035 for the 2 percent mitigation cost guardrail. Thus, implementing carbon emissions sooner preserves a greater margin of action later to stay within the ecological and economic guardrails. The cost-effective emissions path optimizes discounted intertemporal welfare for the given environmental and social targets.

Conclusion

Some of the robust findings of IA studies are that the following issues are highly important for climate policy analysis:

- non-CO_2 greenhouse gases and sulfate aerosols;
- assumptions on socioeconomic development;
- methods for aggregation across time and regions in policy optimization models;
- adjustment costs, learning by doing, and induced technological change;
- considering uncertainty explicitly rather than through best-guess values; and
- large-scale climate instabilities.

The following specific findings of IA are also robust:

- Where, when, and how flexibility can substantially reduce the total costs of emissions abatement.
- Expanding the consideration of climate impacts (e.g., nonmarket impacts and impacts from extreme weather events and large-scale climate instabilities) generally increases the level of mitigation recommended in policy optimization models.
- The calculation of mitigation activities for individual regions or nations requires explicit burden-sharing assumptions, such as rules for allocation of emission permits in emission-trading schemes.
- The complexity of the climate system and its response to increasing greenhouse gas concentrations, as well as uncertainties regarding the future trajectory of human development and adaptive capacity, generate significant uncertainties in the analysis of emissions pathways consistent with impact-based targets.
- Early mitigation activities preserve a broader range of stabilization options. In contrast, delaying mitigation activities increasingly makes achievement of low stabilization targets infeasible, except via overshoot scenarios.

Notes

1. M. G. Morgan, M. Kandlikar, J. Risbey, and H. Dowlatabadi. Why conventional tools for policy analysis are often inadequate for problems of global change. *Climatic Change*, 41:271–81, 1999; F. L. Toth, and M. Mwandosya. Decision-making

frameworks. In B. Metz, O. Davidson, R. Swart, and J. Pan, eds., *Climate Change 2001: Mitigation*, chapter 10, pages 601–88. Cambridge University Press, Cambridge, UK, 2001.

2. W. D. Nordhaus, ed. *Economics and Policy Issues in Climate Change*. Resources for the Future, Washington, DC, 1998; OECD. *The Benefits of Climate Change Policies: Analytical and Framework Issues*. Organization for Economic Cooperation and Development, Paris, 2004; D. Helm, ed. *Climate-Change Policy*. Oxford University Press, Oxford, 2005; A. Haurie and L. Viguier, eds. *The Coupling of Climate and Economic Dynamics*. Springer, Dordrecht, 2005; H. J. Schellnhuber, W. Cramer, N. Nakicenovic, T. Wigley, and G. Yohe, eds. *Avoiding Dangerous Climate Change*. Cambridge University Press, Cambridge, UK, 2006.

3. W. D. Nordhaus and J. Boyer. *Warming the World: Economic Models of Global Warming*. MIT Press, Cambridge, MA, 2000; J. Guo, C. J. Hepburn, R. S. J. Tol, and D. Anthoff. Discounting and the social cost of carbon: A closer look at uncertainty. *Environmental Science & Policy*, 9:205–16, 2006; A. S. Manne and R. G. Richels. An alternative approach to establishing trade-offs among greenhouse gases. *Nature*, 410:675–77, 2001.

4. S. J. DeCanio. *Economic Models of Climate Change*. Palgrave Macmillan, New York, NY, 2003.

5. M. Kainuma, Y. Matsuoka, and T. Morita, eds. *Climate Policy Assessment*. Springer, Berlin, 2002; J. Alcamo, R. Leemans, and E. Kreileman, eds. *Global Change Scenarios of the 21st Century: Results from the IMAGE 2.1 Model*. Pergamon Press, Oxford, 1998.

6. C. D. Kolstad. Learning and stock effects in environmental regulation: The case of greenhouse gas emissions. *Journal of Environmental Economics and Management*, 31:1–18, 1996.

7. F. L. Tóth. ICLIPS: Integrated assessment of climate protection strategies. *Climatic Change*, Special Issue 56(1–2), 2003.

8. M. G. Morgan and M. Henrion. *Uncertainty: A Guide to Dealing with Uncertainty in Risk and Policy Analysis*. Cambridge University Press, Cambridge, 1990.

9. E. Kriegler. *Imprecise Probability Analysis for Integrated Assessment of Climate Change*. PhD thesis, Institute of Physics, University of Potsdam, Potsdam, Germany, 2005. URN: urn:nbn:de:kobv: 517-opus-5611; J. W. Hall, G. Fu, and J. Lawry. Imprecise probabilities of climate change: Aggregation of fuzzy scenarios and model uncertainties. *Climatic Change*, 81:265–81, 2007.

10. S. Dessai and M. Hulme. Does climate policy need probabilities? Working Paper 34, Tyndall Centre for Climate Change Research, Norwich, UK, 2003.

11. G. Heal and B. Kriström. Uncertainty and climate change. *Environmental and Resource Economics*, 22:3–39, 2002; M. D. Mastrandrea and S. H. Schneider. Probabilistic integrated assessment of "dangerous" climate change. *Science*, 304:571–75, 2004; A. Ingham and A. Ulph. Uncertainty and climate-change policy. In D. Helm, ed., *Climate-Change Policy*, chapter 3, pages 43–76. Oxford University Press, Oxford, 2005.

12. D. C. Hall and R. J. Behl. Integrating economic analysis and the science of climate instability. *Ecological Economics*, 57:442–65, 2006.

13. T. Roughgarden and S. H. Schneider. Climate change policy: Quantifying uncertainties for damages and optimal carbon taxes. *Energy Policy*, 27:415–29, 1999; J. Gjerde, S. Grepperud, and S. Kverndokk. Optimal climate policy under the possibility of a catastrophe. *Resource and Energy Economics*, 21:289–317, 1999; K. Keller, K. Tan, F. M. M. Morel, and D. F. Bradford. Preserving the ocean circulation: Implications for climate policy. *Climatic Change*, 47:17–43, 2000; M. D. Mastrandrea and S. H. Schneider. Integrated assessment of abrupt climatic changes. *Climate Policy*, 1:433–49, 2001; C. Azar and K. Lindgren. Catastrophic events and stochastic cost-benefit analysis of climate change. *Climatic Change*, 56:245–55, 2003; A. Baranzine, M. Chesney, and J. Morisset. The impact of possible climate catastrophes on global warming policy. *Energy Policy*, 31:691–701, 2003; A. C. Fisher and U. Narain. Global warming, endogenous risk, and irreversibility. *Environmental and Resource Economics*, 25:395–416, 2003; M. Meinshausen. What does a 2°C target mean for greenhouse gas concentrations? A brief analysis based on multi-gas emission pathways and several climate sensitivity uncertainty estimates. In H. J. Schellnhuber, W. Cramer, N. Nakicenovic, T. Wigley, and G. Yohe, eds., *Avoiding Dangerous Climate Change*, chapter 28, pages 265–79. Cambridge University Press, Cambridge, UK, 2006; K. Keller, B. M. Bolker, and D. F. Bradford. Uncertain climate thresholds and optimal economic growth. *Journal of Environmental Economics and*

Management, 48:723–41, 2004; K. Kuntz-Duriseti. Evaluating the economic value of the precautionary principle: Using cost benefit analysis to place a value on precaution. *Environmental Science & Policy*, 7:291–301, 2004.

14. J. Reilly, R. Prinn, J. Harnisch, J. Fitzmaurice, H. Jacoby, D. Kicklighter, J. Melillo, P. Stone, A. Sokolov, and C. Wang. Multi-gas assessment of the Kyoto Protocol. *Nature*, 401:549–55, 1999; M. Meinshausen, B. Hare, T. M. M. Wigley, D. Van Vuuren, M. G. J. Den Elzen, and R. Swart. Multi-gas emissions pathways to meet climate targets. *Climatic Change*, 75:151–94, 2006.

15. O. Edenhofer, C. Carraro, J. Koehler, and M. Grubb. Endogenous technological change and the economics of atmospheric stabilisation. *The Energy Journal*, 57(Special issue), 2006.

16. C. D. Kolstad. Learning and stock effects in environmental regulation: The case of greenhouse gas emissions. *Journal of Environmental Economics and Management*, 31:1–18, 1996; R. J. Lempert, M. E. Schlesinger, and S. C. Bankes. When we don't know the costs or the benefits: Adaptive strategies for abating climate change. *Climatic Change*, 33:235–74, 1996; L. J. Valverde A. Jr., H. D. Jacoby, and G. M. Kaufman. Sequential climate decisions under uncertainty: An integrated framework. *Environmental Modeling and Assessment*, 4:87–101, 1999; R. J. Lempert and M. E. Schlesinger. Adaptive strategies for climate change. In R. G. Watts, ed., *Innovative Energy Strategies for CO$_2$ Stabilization*, chapter 3, pages 45–86. Cambridge University Press, Cambridge, UK, 2002; G. Yohe, N. Andronova, and M. Schlesinger. To hedge or not against an uncertain climate future. *Science*, 306:416–17, 2004.

17. C. Kemfert. Climate policy cooperation games between developed and developing nations: A quantitative, applied analysis. In A. Haurie and L. Viguier, eds., *The Coupling of Climate and Economic Dynamics. Essays on Integrated Assessment*, chapter 6, pages 145–71. Springer, Dordrecht, 2005.

18. R. B. Howarth. Second-best pollution taxes in the economics of climate change. In J. D. Erickson and J. M. Cowdy, eds., *Frontiers in Environmental Valuation and Policy*. Edward Elgar, London, 2006.

19. J. Schellnhuber, R. Warren, A. Haxeltine, and L. Naylor. Integrated assessment of benefits of climate policy. In *The Benefits of Climate Change Policies: Analytical and Framework Issues*, chapter 3, pages 83–110. Organization for Economic Cooperation and Development, Paris, 2004; M. Leimbach and C. Jäger. A modular approach to integrated assessment modeling. *Environmental Modeling and Assessment*, 9:207–20, 2005; C. Beltran, L. Drouet, N. Edwards, A. Haurie, J.-P. Vial, and D. Zachary. An oracle method to couple climate and economic dynamics. In A. Haurie and L. Viguier, eds., *The Coupling of Climate and Economic Dynamics: Essays on Integrated Assessment*, chapter 3, pages 69–95. Springer, Dordrecht, 2005.

20. B. O'Neill and M. Oppenheimer. Climate change impacts are sensitive to the concentration stabilization path. *Proceedings of the National Academy of Sciences*, 101:16411–16, 2004; S. H. Schneider and M. D. Mastrandrea. Probabilistic assessment of "dangerous" climate change and emissions pathways. *Proceedings of the National Academy of Sciences*, 102:15728–35, 2005; B. Hare and M. Meinshausen. How much warming are we committed to and how much can be avoided? *Climatic Change*, 75:111–49, 2006; M. Meinshausen, B. Hare, T. M. M. Wigley, D. Van Vuuren, M. G. J. Den Elzen, and R. Swart. Multi-gas emissions pathways to meet climate targets. *Climatic Change*, 75:151–94, 2006.

21. S. H. Schneider and M. D. Mastrandrea. Probabilistic assessment of "dangerous" climate change and emissions pathways. *Proceedings of the National Academy of Sciences*, 102:15728–35, 2005.

22. F. L. Tóth. ICLIPS: Integrated assessment of climate protection strategies. *Climatic Change*, Special Issue 56(1–2), 2003.

23. F. L. Tóth, T. Bruckner, H.-M. Füssel, M. Leimbach, G. Petschel-Held, and H.-J. Schellnhuber. Exploring options for global climate policy: A new analytical framework. *Environment*, 44(5):22–34, 2002.

Chapter 15

Risk, Uncertainty, and Assessing Dangerous Climate Change

STEPHEN H. SCHNEIDER AND MICHAEL D. MASTRANDREA

Introduction

Human activities are already changing the climate in many ways, with the potential to threaten natural and social systems. But how large and how fast will these changes be? And how can our policy choices reduce the threat they pose?

The global scale of climate change and its driving forces, and the subtly intensifying and persistent nature of many of its impacts, contrast uneasily with the short-term, local-to-national scales of most management systems, posing a major dilemma for managing planetary sustainability in the decades ahead. Furthermore, significant uncertainties plague projections of climate change and its consequences. Policy makers struggle with imprecise information in order to make decisions that have far-reaching and often irreversible effects on both the environment and society. Not surprisingly, efforts to incorporate uncertainty into decision making enter the negotiating parlance through catchphrases such as "the precautionary principle," "adaptive environmental management," "the preventive paradigm," and "principles of stewardship."

On the other hand, uncertainty can also be used as a justification for delaying action until a better understanding of the risks involved and the cost of reducing those risks becomes available. Phrases such as "wasted investments" and "burdensome regulations" permeate the rhetoric of those who invoke uncertainty as a reason to avoid or delay policies to adapt to or mitigate climate change risks.

Like many other complex socio-technical problems with large stakes and discordant stakeholder communities, uncertainty is an inherent component of the global climate science and policy debate.[1]

Uncertainties remain in scientific understanding of the scale and distribution of changes to the climate system in response to increasing greenhouse gas concentrations. And even if we could perfectly predict the response of the climate system, there is considerable uncertainty in the future trajectory of human development and the drivers behind greenhouse gas emissions, which is heavily dependent on policy choices worldwide. Simply put, if we wait to act until undesirable impacts occur, the inertia in the climate system and in the socioeconomic systems that produce greenhouse gas emissions will have committed us to even more severe impacts.

Understanding not only how much climate change is likely, but also how to characterize and analyze the effects of our policy choices, is essential. Decision making intended to address climate change must account for uncertainty, and response strategies—including deliberate inaction—must be determined before uncertainty is resolved.

Uncertainty and Risk Management

Scientists can help policy makers evaluate climate policy options by laying out the elements of risk, classically defined as *consequence times probability*. In other words: what can happen and what are the odds of it happening? Both of these factors are important in determining whether and how we address specific risks.

The plethora of uncertainties inherent in climate change projections clearly makes risk assessment difficult. Decision makers must weigh the importance of climate risks against other pressing social issues competing for limited resources. Some fear that actions to control potential risks may unnecessarily consume resources that could be used for other worthy purposes, especially if any hypothesized climatic impacts turned out to be on the benign side of the projected range. This can be restated in terms of "Type I" and "Type II" errors (essentially false positives or false negatives, respectively). If governments were to apply the precautionary principle and act now to mitigate risks of climate change, they would be said to be committing a Type I error if their worries about climate change proved exaggerated and their investments in hedging strate-

gies were considerable. A Type II error would occur if serious climate change materialized after insufficient hedging action. Deciding whether to be Type I versus Type II error averse can be informed by scientific risk assessment (what are the magnitudes of the risks on either side?), but this decision is a value judgment based on the risk aversion philosophy of the decision maker, one that is often made implicitly. The inertia in the climate and socioeconomic systems mentioned above, and the fact that emissions of greenhouse gases continue to rise given the absence of strong mitigation policies, indicate that globally most policy makers are currently acting as if they were more wary of making a Type I error.

To make informed policy decisions, policy makers must have information about potential impacts associated with different levels of climate change (consequences) and the range of future climate changes that could be induced by different levels of future emissions (the probability of those consequences occurring). Extensive literature has arisen in recent years linking social, geophysical, and biological impacts with changes in specific climate parameters such as global average temperature, and the Intergovernmental Panel on Climate Change Fourth Assessment Report (IPCC AR4) has specifically addressed impacts and thresholds that are of particular concern for policy makers by defining and identifying "key vulnerabilities." This assessment is based on seven criteria: magnitude of impacts, timing of impacts, persistence and reversibility of impacts, potential for adaptation, distributional aspects of impacts and vulnerabilities, likelihood (estimates of uncertainty) of impacts and vulnerabilities and confidence in those estimates, and the importance of the vulnerable system(s).[2] Many potential impacts of climate change are covered in chapters 3 through 12 in this volume and in the IPCC AR4.[3] To complete the risk equation, we also need to estimate the probability of these consequences occurring. This chapter focuses on how climate projections can inform and guide climate risk management.

Modeling Future Climate

It is frequently asserted that meteorologists' inability to predict weather accurately beyond about ten days bodes ill for any attempt at long-range climate projection. This misconception misses a key difference—the instantaneous state of the atmosphere (weather) versus its time and space averages (climate). It is indeed impossible, even in principle, to predict credibly the details of weather beyond about two weeks, and no amount of weather data collection, computing power, or model sophistication will improve predictions of weather beyond a few weeks. This is because the evolution of atmospheric conditions is an inherently chaotic process in which the slightest perturbation today can make a huge and unpredictable difference in the weather 1,000 miles away and weeks hence—a time known as the "weather predictability period." But large-scale climate shows little tendency to exhibit chaotic behavior (at least on timescales longer than a decade), and appropriate models therefore can make reasonable climate projections decades or even centuries forward in time when forced by both natural and anthropogenic drivers.

An illustrative scenario was chosen for each of the six scenario groups: A1B, A1FI, A1T, A2, B1, and B2. All should be considered equally sound.

The SRES scenarios do not include additional climate initiatives, which means that no scenarios are included that explicitly assume implementation of the United Nations Framework Convention on Climate Change or the emissions targets of the Kyoto Protocol.

Emissions Scenarios

In 2000, the IPCC released the Special Report on Emission Scenarios (SRES), which presented a suite of scenarios for greenhouse gas emissions based on different assumptions regarding economic growth, technological developments, and population growth, critical driving forces behind greenhouse gas emissions.[4] These scenarios, displayed in figure 15.1 and described in box 15.1, were designed to represent different possible "baselines" (also called business-as-usual scenarios) without explicit emissions-reduction policies. One of the most interesting contrasts across the SRES scenarios is that between the emission pathways of two variants of the A1 scenario family. The A1 family assumes relatively high economic growth, lower population growth, and considerable contraction in income differences across regions—a world that builds on current economic globalization patterns and rapid income growth in the developing world. One variant, A1FI, is a fossil fuel–intensive scenario, in which the bulk of the energy needed to fuel economic growth continues to be derived from burning fossil fuels, especially coal. Emissions grow from the current level of about 8 billion tons carbon of per year to nearly 30 billion, resulting in a tripling of CO_2 (from current levels) in the atmosphere by 2100 and implying at least a quadrupling of CO_2 as the twenty-second century progresses. The contrasting variant is A1T—the technological innovation scenario—in which fossil fuel emissions increase

BOX 15.1. THE EMISSIONS SCENARIOS OF THE SPECIAL REPORT ON EMISSIONS SCENARIOS (SRES)

A1. The A1 storyline and scenario family describe a future world of very rapid economic growth, global population that peaks in mid-century and declines thereafter, and the rapid introduction of new and more efficient technologies. Major underlying themes are convergence among regions, capacity building and increased cultural and social interactions, with a substantial reduction in regional differences in per capita income. The A1 scenario family develops into three groups that describe alternative directions of technological change in the energy system. The three A1 groups are distinguished by their technological emphasis: fossil intensive (A1FI), non-fossil energy sources (A1T), or a balance across all sources (A1B) (where balanced is defined as not relying too heavily on one particular energy source, on the assumption that similar improvement rates apply to all energy supply and end use technologies).

A2. The A2 storyline and scenario family describe a very heterogeneous world. The underlying theme is self-reliance and preservation of local identities. Fertility patterns across regions converge very slowly, which results in continuously increasing population. Economic development is primarily regionally oriented and *per capita* economic growth and technological change are more fragmented and slower than in other storylines.

B1. The B1 storyline and scenario family describe a convergent world with the same global population, that peaks in mid-century and declines thereafter, as in the A1 storyline, but with rapid change in economic structures toward a service and information economy, with reductions in material intensity and the introduction of clean and resource-efficient technologies. The emphasis is on global solutions to economic, social and environmental sustainability, including improved equity, but without additional climate initiatives.

B2. The B2 storyline and scenario family describe a world in which the emphasis is on local solutions to economic, social and environmental sustainability. It is a world with continuously increasing global population, at a rate lower than A2, intermediate levels of economic development, and less rapid and more diverse technological change than in the B1 and A1 storylines. While the scenario is also oriented towards environmental protection and social equity, it focuses on local and regional levels.

Intergovernmental Panel on Climate Change (IPCC), 2007(b). Climate Change 2007: The Scientific Basis, Contribution of Working Group I to the Fourth Assessment Report of the IPCC, Solomon, S., et al. (Eds.), Cambridge University Press: Cambridge, United Kingdom, and New York, NY, USA. IPCC, 2000, Emissions Scenarios—A Special Report of Working Group III of the Intergovernmental Panel on Climate Change Nakicenovic, N., and R. Swart (Eds.), Cambridge University Press, Cambridge, United Kingdom, 599 pp.

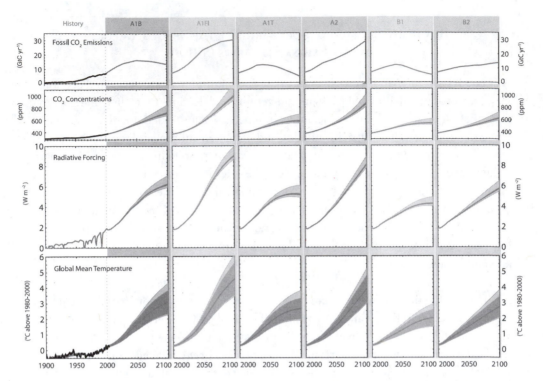

FIGURE 15.1. IPCC SRES emissions scenarios. Scenario family and subgroup (figure columns) are described in box 15.1, and for each, emissions and concentrations of CO_2, radiative forcing (including other greenhouse gases), and projected temperature increase up to 2100 are displayed. Dark shaded area in temperature panel represents mean ± one standard deviation of nineteen model projections. The lighter-shaded outer ranges of temperature increase represent higher or lower assumptions about the strength of carbon cycle feedbacks over the twenty-first century. Source: Intergovernmental Panel on Climate Change (IPCC), 2007(a). *Climate Change 2007: Impacts, Adaptation and Vulnerability, Contribution of Working Group II to the Fourth Assessment Report of the IPCC*, Parry, M., et al., eds. Cambridge University Press: Cambridge, United Kingdom.

from the present by about a factor of two until the mid-twenty-first century, but then, because of technological innovation and deployment of low carbon-emitting technologies, global emissions drop to well below current levels by 2100. Even so, as seen in figure 15.1, CO_2 levels roughly double by 2100. However, they would not increase much more in the twenty-second century because emissions rapidly approach zero by 2100.

Climate Projection Uncertainty

Note that for all scenarios in figure 15.1, despite their major differences in emissions, their paths for global temperature increase do not diverge dramatically until after the mid-twenty-first century. This has led some to declare that there is very little difference in climate change across scenarios and therefore that emissions reductions can be delayed many decades. That inference would be a serious misrepresentation of this result. What must be recognized is the considerable inertia in the economic systems that produce emissions and in the climate system response to increasing concentrations. It takes many decades to replace energy supply and end-use systems, thus delaying significant abatement for decades even if we started today (though we could realize some abatement benefits much faster with the use of performance standards like fuel economy rules or building efficiency codes). For example, the technological advances built into the A1T scenario lead to a sharp divergence in emissions from the fossil-intensive A1FI scenario after 2030. By around 2050, the temperature response diverges as well. This delay between the divergence in emissions and divergence in temperature is due to the inertia in the response of the climate system to forcing, caused by the large heat capacity of the oceans. After the mid-twenty-first century, there is a very large difference in cumulative emissions, leading to large differences in the projected temperature increases—and the risks of dangerous climate changes—for the late twenty-first century and beyond. For the A1FI scenario, for example, the models project a warming of 2.4 to 6.4 degrees Celsius by the year 2100 (including a component for carbon cycle feedbacks). For the A1T scenario, the models project a warming of 1.4 to 3.8 degrees Celsius.

These different ranges of warming imply very different risks of experiencing key impacts and vulnerabilities. While warming at the low end would be relatively less stressful, it would likely still be significant for some communities, sectors, and natural ecosystems, as some of these systems have already shown concerning responses to the approximate 0.75

degree Celsius warming during the last century. Warming at the high end of the range could have widespread catastrophic consequences. A temperature change of 5 to 7 degrees Celsius on a globally averaged basis is about the difference between an ice age and an interglacial period occurring in merely a century, not millennia like ice age interglacial transitions.

There is a further fundamental difference between the A1FI and A1T scenarios. At the end of the twenty-first century, greenhouse gas emissions in A1FI are still extremely high, implying further increases in atmospheric concentrations and significant temperature increase in the twenty-second century. In contrast, emissions in A1T have dropped considerably and are trending downward. Natural processes remove greenhouse gases from the atmosphere (see chapter 1 in this volume). If anthropogenic emissions drop below the level of natural removal (as would be possible in the twenty-second century after following the A1T path), atmospheric concentrations would begin to fall, which in turn would allow temperatures to fall as well. Such a scenario is often referred to as an overshoot or peaking scenario, in which concentrations peak and decline, and temperatures follow, and may very well be the most likely general trajectory for future concentrations and temperature change.

Probabilistic Climate Projection

Much of the uncertainty contributing to the ranges of temperature increase is represented by the so-called climate sensitivity, often stated as the equilibrium global mean surface temperature increase from a doubling of atmospheric CO_2. Based on the properties of complex climate models, the IPCC estimates that there is a 66 to 90 percent chance that the climate sensitivity is between 2 and 4.5 degrees Celsius and roughly a 5 to 17 percent chance that it is above 4.5 degrees Celsius (with the remainder being the chance it is less than 2 degrees Celsius). They also offered a "best guess" of 3 degrees Celsius climate sensitivity. The likelihood of a higher climate sensitivity is of particular concern in terms of risk management, because a high climate sensitivity will induce large temperature change and more severe climate impacts.

Many recent studies have produced probability distributions for climate sensitivity with a long "right-hand tail," meaning that high climate sensitivity values, while relatively unlikely, are still plausible at nontrivial probabilities of above a few percent. One such example is displayed in figure 15.2, for which there is a very uncomfortable 10 percent chance that the climate sensitivity is higher than 6.8 degrees Celsius. The median re-

sult—that is, the value that climate sensitivity is as likely to be above as below—is 2.0 degrees Celsius, while there is a 10 percent chance the climate sensitivity will be 1.1 degrees Celsius or less. Although it is not possible currently to confidently prefer this distribution to the estimates in IPCC, nor to determine the exact climate sensitivity, these three percentile values (6.8 degrees Celsius, 2.0 degrees Celsius, and 1.1 degrees Celsius) can be used as representative of high, medium, and low climate sensitivity estimates.

Each possible value can be used to produce a projection of temperature over time, using a climate model (in this case a simple mixed-layer climate model), once an emissions scenario is specified. In the example below, these three climate sensitivities are combined with two of the SRES storylines: the fossil fuel–intensive scenario (A1FI) and the high-technology scenario (A1T), where development and deployment of advanced lower carbon-emitting technologies dramatically reduces long-term emissions (see figure 15.1).

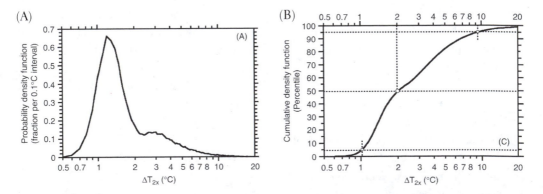

Figure 15.2. Probability density function (panel A) and cumulative density function (panel B) for climate sensitivity from Andronova and Schlesinger (2001) generated by scaling observed temperature trends against estimates of radiative forcing during the twentieth century. The distribution in panel A has a long "right-hand tail," leaving open the possibility of high values for climate sensitivity. Source: Andronova, N. G., and M. E. Schlesinger (2001). Objective estimation of the probability density function for climate sensitivity. *Journal of Geophysical Research* 106, 22605–12.

These make a good comparison pair since they almost bracket the high and low ends of the six SRES representative scenarios' range of cumulative emissions to 2100. And since both are for the "A1 world," the only major difference between the two is the technology component—an aspect that decision makers can influence with policies and other measures. Even though not originally constructed as a climate policy scenario, A1T is not a bad surrogate for a climate policy storyline and thus makes an instructive pairing with A1FI— a more business-as-usual projection. Figure 15.3 displays the results. We compare these temperature projections to a threshold of 3.5 degrees Celsius warming above 2000 levels, which is a very conservative interpretation of the information presented in the IPCC Working Group II report that projects many serious and potentially "dangerous" climate change impacts for temperature increase above 2 to 3 degrees Celsius.[5]

What is "dangerous" climate change? The term was introduced in the 1992 United Nations Framework Convention on Climate Change (UNFCCC), which calls for stabilization of greenhouse gases to "prevent dangerous anthropogenic interference with the climate system."[6] The UNFCCC further states that: "Such a level should be achieved within a time frame sufficient to allow ecosystems to adapt naturally to climate change, to ensure that food production is not threatened, and to enable economic development to proceed in a sustainable manner." While these criteria suggest that "dangerous anthropogenic interference" (DAI) is associated with the impacts of climate change on social and natural systems, it does not define a threshold with any precision. Although there may be some thresholds beyond which there would be abrupt climate impacts (e.g., ice sheet

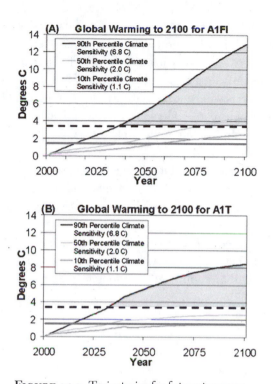

FIGURE 15.3. Trajectories for future temperature increases for two emissions scenarios, A1FI (panel A) and A1T (panel B). In each panel, three possible temperature trajectories for each emissions scenario are displayed, assuming different climate sensitivities. Shading indicates exceeding a very conservative threshold of 3.5 degrees Celsius for potentially "dangerous" climate impacts, whereas the dark horizontal lines are close to the EU threshold for "dangerous" warming (of 2 degrees Celsius above preindustrial temperatures, or around 1.4° above 2000 levels). The EU threshold is likely to be exceeded in all these calculations except for the tenth percentile climate sensitivity case with the A1T scenario. Schneider, S. H., and J. Lane (2006). An overview of "dangerous" climate change, in *Avoiding Dangerous Climate Change*, J. Schellnhuber, W. Cramer, N. Nakicenovic, G. Yohe, and T. M. L. Wigley, eds. Cambridge University Press, New York.

collapse and subsequent rapid or large sea-level rise) that most would deem "dangerous," there are other categories for which different points along a rising continuum of impact intensity will be seen as "dangerous" by different people (e.g., gradual sea-level rise threatening eventual abandonment of some islands and coastal regions).

What constitutes being "dangerous" is informed by scientific evidence and analysis, but it is ultimately a value-laden decision, that can differ significantly depending on geographical location, socioeconomic standing, and ethical value system. The EU, for example, has committed to a long-term target of limiting warming to 2 degrees Celsius above preindustrial levels.

The most striking feature of the two panels of figure 15.3 is the top ninetieth percentile line that rises very steeply above the other two lines below it. For the ninetieth percentile results, both the A1FI and the A1T temperature projections exceed the threshold of 3.5 degrees Celsius at roughly the same time (around 2040), but the A1FI warming not only goes on to outstrip the A1T warming, it is still steeply sloped upward at 2100, implying additional warming in the twenty-second century. For the fiftieth percentile results, the A1FI scenario exceeds the 3.5 degrees Celsius threshold around 2075, while the A1T scenario stays under the threshold. Nearly all calculations shown substantially exceed the EU estimate of DAI at 2 degrees Celsius above preindustrial or about 1.25 degrees Celsius above present, shown by the dark horizontal lines on figure 15.3.

Another important feature that is easy to visualize on figure 15.3 is that, as mentioned earlier, the inertia of both economic systems generating emissions and the climate system prevent a major separation of the temperature rise curves for both scenarios until about mid-century. Again, this should not be misinterpreted as suggesting that emissions reductions activities can be delayed until midcentury, for then the separation point of the two warming curves would be even further delayed, and the absolute level of warming would eventually be higher. This simple analysis shows, through a small number of curves, the significant variation in temperature change over time for three values of climate sensitivity within a climate sensitivity probability distribution (tenth, fiftieth, and ninetieth percentiles).

Treatment of Uncertainty and Risk Management

The difficulties of explaining uncertainty have become increasingly salient as society seeks policy advice to deal with global climate change. How can science be most useful to society when evidence is incomplete or ambiguous, the subjective judgments of experts about the likelihood of outcomes vary, and policy makers seek guidance and justification for courses of action that could cause significant environmental and societal changes?

As shown in figure 15.4, uncertainty can be decomposed into three ranges. The smallest range represents "well-calibrated" uncertainty, which is based on conventional modeling literature, with the Ms symbolically representing four different model results. The middle range represents "judged" uncertainty, which is based on expert judgments—including factors not well represented in models, processes that could, as the figure suggests, either enhance or reduce estimates made with models that exclude adequate treatments of such factors. However, even experts can ex-

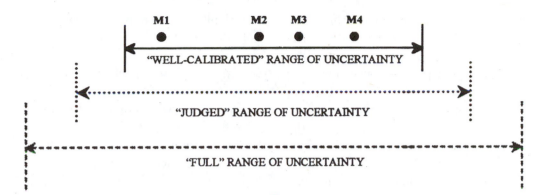

FIGURE 15.4. Three ranges of uncertainty for future projections, representing the "well-calibrated" range found in conventional modeling literature, the "judged" range based on expert judgment that includes factors not well-represented in models, and the "full" range that includes outlier possibilities that may not be considered within the "judged" range. Source: Schneider, S. H., and K. Kuntz-Duriseti (2002). Uncertainty and climate change policy, in *Climate Change Policy: A Survey*, S. H. Schneider, A. Rosencranz, and J. O. Niles, eds. Island Press, Washington, D.C.

hibit cognitive biases by anchoring on well-known results or available ideas, thus the judged range may not encompass the "full" and largest range of uncertainty, which takes into account the possibility of such cognitive biases that can lead to overconfidence. Underestimation of risky outlier possibilities is particularly problematic when potentially large, irreversible, or inequitable results are implied by outcomes connected with outlier risks.[7]

For example, the SRES scenarios attempt to span the range of possible business-as-usual future worlds. But, although it is disturbing to consider, pandemics or global economic stagnation could result in lower emissions than B1. Few would hope to mitigate climate change via economic or social collapse scenarios. Likewise, the A1FI scenario, a business-as-usual scenario with CO_2 tripling by 2100, might not even be the highest imaginable scenario. High gasoline prices and continued demand for heavy, inefficient personal vehicles could drive a large market for high CO_2-emitting fuel sources such as synthetic

fuels from tar sands or oil shale. SRES deemphasized a storyline on either of these outlier, highly disquieting possibilities.

Conclusion

In dealing with uncertainty in science, normal scientific practice is to strive to reduce uncertainty through standard science: data collection, research, modeling, simulation, and so forth. The objective is to overcome the uncertainty—to make known the unknown. New information, particularly reliable and comprehensive empirical data, may eventually narrow the range of uncertainty. In this way, further scientific research into the interacting processes that make up the climate system can reduce uncertainty in the response of the system to increasing concentrations of greenhouse gases. This uncertainty, however, is very unlikely to be reduced quickly, given the complexity of the global climate and the many years of high-quality data necessary to resolve such uncertainty. Meanwhile, even the most optimistic business-as-usual emissions pathway is projected to result in dramatic, and perhaps dangerous, climate impacts. Therefore, there is a clear need to make policy decisions before this uncertainty is resolved, rather than using it as a justification for delaying action.

That leaves an alternative: to manage the uncertainty rather than master it, to integrate uncertainty into climate research tools and policy making. This risk-management framework for decision making is often practiced in defense, health, business, and environmental risk situations. Policy makers must make the final decisions about investments and strategies that determine what climate change risks to face and what to avoid, but scientific research can provide essential information about how different mitigation or adaptation strategies could influence climate impacts and their probability of occurrence, and thus help to put risk management decisions on a firmer scientific foundation.

Is the Science "Settled" Enough for Policy? An Editorial Capstone

Sometimes critics claim that there should be no strong climate policy until the science is "settled" and major uncertainties resolved, whereas supporters of strong policies suggest the science is already "settled enough" and it is time to proceed with action to reduce risks. The science that demonstrates a significant warming trend during the last century is settled; moreover, it is virtually settled that the past few decades of warming have been largely of anthropogenic origin and that much more and potentially dangerous climate change is

being built into the emissions pathways of the twenty-first century. Sounds like the "settled already" side has won the debate: warming is occurring, and human activities are the primary driver of recent changes.

How severe will warming and its impacts be in the future, especially when projections for "likely" warming by 2100 vary by a factor of six? The magnitude of future climate change is dependent on two key uncertainties: what we do and how the natural climate system responds. Uncertainty about what we do incorporates a broad range of social factors: possible trajectories for economic development, population growth, utilization of low-carbon-emitting energy sources, and other societal factors that affect greenhouse gas emissions. The second component of uncertainty is the response of the climate system to increasing greenhouse gas concentrations—how much the climate will warm; how that warming will affect other processes like rainfall patterns, ice-sheet dynamics, and sea level and the natural uptake of carbon by the ocean and growing vegetation; how changes will be distributed across different regions; and so on. Even if we could predict exactly how emissions will play out in the future, we would still be faced with a range of possible climate changes and a range for the severity and distribution of impacts.

Policy decisions can strongly influence the first source of uncertainty (future emissions), but will have little influence on the second source (climate response to emissions). We cannot know precisely what the severity of impacts will be for a specific trajectory for future emissions, but we can confidently say that the severity will be reduced if emissions are reduced. In this case, the question of "settled or not" is really not very relevant. What is important is the degree of risk aversion of various groups and sectors given the prospects for key vulnerabilities—which multiply rapidly in number and intensify considerably with warming, which, in turn, increases with emissions. The lowest IPCC scenario projects an additional around 150 percent warming (1.1 degrees Celsius, on top of current amount of 0.75 degree Celsius). That is disquieting, though clearly more adaptive capacity exists for this lower end of the IPCC uncertainty warming range than for the 6.4 degrees Celsius warming at the upper end of the IPCC "likely" range for the A1FI scenario. That, recall, is about the magnitude of an ice age to interglacial transition, but occurring 10 to 100 times faster. The literature suggests this would very likely be considered a "dangerous" outcome by nearly all stakeholders and groups represented in the literature.[8] Remember, too, that a nontrivial chance remains that the actual climate sensitivity is above upper limit of the IPCC range of 4.5 degrees Celsius and therefore that temperature increase could be even greater for each scenario. Thus the answer to the question "is the science settled enough for strong hedging policies" is not primarily scientific, but involves value judgments about risk aversion—such as whether one is more Type I (committing resources unnecessarily) versus Type II (failing to prevent dangerous change) error averse. The role of science is to estimate the risks, not choose the level that should trigger strong policy.

In our personal value frames, we believe that it is already a few decades too late for having implemented some policy measures. Had we begun mitigation and adaptation investments decades ago the job of remaining safely below dangerous thresholds would be easier to accomplish and cheaper to implement. Similarly, beyond a few degrees Cel-

sius of warming—now considered at least an even bet if we remain anywhere near current course—it is likely in our view that many "dangerous" thresholds will be exceeded and the United Nations Framework Convention on Climate Change injunction to "prevent dangerous anthropogenic interference with the climate system," which has now been signed into law by 191 nations, will have been violated. Clearly, to lower the probability of that negative outcome, strong mitigation and adaptation actions are long overdue, even if there is a small chance that by luck climate sensitivity will be at the lower end of the uncertainty range and, at the same time, some fortunate, soon-to-be discovered, low-cost, low-carbon-emitting energy systems will materialize and keep the planet comfortably below most dangerous thresholds. For us, that is a high-stakes gamble not remotely worth taking with our planetary life-support system. Despite the large uncertainties in many parts of the climate science and policy assessments to date, uncertainty is no longer a responsible justification for delay. Finally, IPCC Working Group 3 has reinforced earlier analyses that even seemingly expensive mitigation actions costing trillions of dollars over the rest of the century are really only a small fractional cost out of the growth rate of the world's projected economy.[9] As Azar and Schneider argued in 2003, is it really such an expensive "insurance premium" for planetary climatic stability to be 500 percent per capita richer in 2102 with about 450 parts per million of CO_2 equivalent stabilized concentrations versus about 500 percent per capita richer in 2100 with more than 600 parts per million stabilization levels and many more key vulnerabilities triggered?[10] That is the value choice we face; and for us the answer is, indeed, "settled": hedge against dangerous climate change and lower the risks of many key vulnerabilities.

Notes

1. Funtowicz, S. O., and J. R. Ravetz, 1993. Three types of risk assessment and the emergence of post-normal science, in Krimsky, S., and D. Golden, eds., *Social Theories of Risk*, Westport, CT: Greenwood, 251–73.

2. Schneider, S. H., et al., 2007. Assessing key vulnerabilities and the risk from climate change, in Parry, M., et al., eds., *Climate Change 2007: Impacts, Adaptation and Vulnerability, Contribution of Working Group II to the Fourth Assessment Report of the IPCC*, Cambridge University Press: Cambridge, United Kingdom, and New York, NY, USA.

3. Intergovernmental Panel on Climate Change (IPCC), 2007(a). *Climate Change 2007: Impacts, Adaptation and Vulnerability, Contribution of Working Group II to the Fourth Assessment Report of the IPCC*, Parry, M., et al., eds., Cambridge University Press: Cambridge, United Kingdom, and New York, NY, USA.

4. Ibid.

5. IPCC, 2007(a) op. cit.

6. UNFCCC, 1992. United Nations Framework Convention on Climate Change. www.unfccc.de.

7. See Stern and colleagues (2006) for discussion of the possible magnitude of damages associated with such outliers. N. Stern, *The Economics of Climate Change: The Stern Review* (Cambridge Univ. Press, Cambridge, 2007).

8. IPCC, 2007(a) op. cit.

9. Azar, C., and S. H. Schneider, 2002. Are the economic costs of stabilizing the atmosphere prohibitive? *Ecological Economics* 42, 73–80.

10. Azar, C., and S. H. Schneider, 2003. Are the economic costs of stabilizing the atmosphere prohibitive? A response to Gerlagh and Papyrakis. *Ecological Economics* 46, 329–32.

Chapter 16

Risk Perceptions and Behavior

ANTHONY LEISEROWITZ

Risk Perception

Social scientists have found that public risk perceptions strongly influence the way people respond to hazards. What the public perceives as a risk, why they perceive it that way, and how they will subsequently behave are thus vital questions for policy makers attempting to address global climate change.

This chapter introduces the concepts of risk and risk perception, summarizes public opinion on climate change internationally and in the United States, and reports results from an in-depth study of public climate change risk perceptions, policy preferences, and individual behaviors in the United States.

The technical meaning of risk is "the probability that an outcome will occur times the consequence, or level of impact."[1] This technical definition is relatively neutral and allows for both gains and losses; thus, there are potentially both "good risks" and "bad risks." Among broader society, however, the term "risk" has come to predominantly mean "danger," especially among the lay public.[2] In everyday language, "risk tends to be used to re-fer almost exclusively to a threat, hazard, danger, or harm."[3]

As the field of risk analysis and assessment developed, it became evident that there were often great disparities between what experts and the lay public viewed as a risk—regarding natural hazards, new technologies, environmental protection, medical procedures, and so on—often with large individual, social, economic, or political consequences. Early studies compared expert versus lay assessments of the probabilities and severity of consequences (typically fatalities) for particular hazards and often found great discrepancies. Experts, for example, warned of the high-probability, high-consequence risks of living in floodplains or earthquake zones. The public, however, continued to build homes and live in these areas while remaining relatively unconcerned.[4] Conversely, researchers found that the lay public often were very concerned about low-probability risks, such as radiation from nuclear power plants, while ignoring higher-probability risks like radon in the home. These and many other expert-public conflicts over risk led some to decry the apparent "irrationality" of the public. As one scholar later described these critiques:

> Experts are seen as purveying risk assessments, characterized as objective, analytic, wise, and rational—based upon the real risks. In contrast, the public is seen to rely upon perceptions of risk that are subjective, often hypothetical, emotional, foolish, and irrational.[5]

In the early 1980s, however, researchers identified additional risk dimensions that are often more important for the lay public than just probabilities and the severity of consequences (the technical, yet limited, definition of risk). When combined, two dominant factors in public perceptions of risk were identified across eighty-one different hazards: *dread risk* (including the dimensions of "perceived lack of control, dread, catastrophic potential, fatal consequences, and the inequitable distribution of risks and benefits") and *unknown risk* (including the dimensions of "hazards judged to be unobservable, unknown, new, and delayed in their manifestation of harm").[6] For example, the public rated nuclear power very high on both the dread and unknown risk dimensions, while hazards like alcohol were rated very low. Laypeople subsequently rated nuclear power as a much greater risk than alcohol, despite the fact that nuclear accidents are low-probability events that have killed relatively few people. By contrast, alcohol directly and indirectly (e.g., drunk driving) kills thousands of people each year. Other studies found that nuclear power technology was highly stigmatized, with the public's risk perceptions driven by dimensions such as the catastrophic potential of an accident, the potential impact on future generations, and the perceived lack of control over and involuntariness of such an event. Further, in the wake of events like Three Mile Island and Chernobyl, the public came to deeply distrust the industry and government experts who promoted nuclear power. As a result, no new nuclear power plant has been constructed in the United States since 1979, despite the relatively low number of fatalities from nuclear power generation.[7] Thus, this research suggested that public risk perceptions are often only irrational when considered within the narrow confines of a technical analysis of probability and fatalities. They are rational, however, when considered within the more complex and comprehensive conception of risk used by the public, which includes a range of evaluative dimensions (e.g., catastrophic potential, dread, unequal distributions of costs and benefits, and so on) not included in technical assessments of probabilities and fatalities.

More recent research has focused on the broader social, cultural, and political context of risk perception, including sociodemographic factors like sex, race, income, and education and cultural factors like trust, social values, and worldviews.[8] For example, several studies have found that women generally perceive greater risk across a variety of hazards than men, leading some researchers to conclude that women are either biologically or culturally predisposed to worry more about risk. Other work, however, has demonstrated that most of this gender difference, at least in the United States, is due to a subpopulation of "low-risk white males" who perceive much less risk than all other demographic groups.[9] Likewise, other researchers have found that an egalitarian worldview is associated with greater concern about nuclear power, while people who hold strong individualism values are less concerned.[10] Thus, researchers are increasingly asking not just

"What does the public perceive as a risk and why?" but also "Who perceives risk and why?"

Risk Perceptions of Climate Change

Public risk perceptions are critical components of the sociopolitical context within which policy makers operate. Public risk perceptions can fundamentally compel or constrain political, economic, and social action to address particular risks. For example, public support or opposition to climate policies (e.g., treaties, regulations, taxes, subsidies, and so on) will be greatly influenced by public perceptions of the risks and dangers of climate change. To date, however, there have been only a few in-depth studies of public climate change risk perceptions.[11] Nonetheless, we can consider the limited findings of public opinion polling.

International Public Opinion

We still know very little about international public opinion on climate change, in large part because only a few multinational surveys have included even a single question on the issue.[12] A 2000 survey of thirty-four countries found that majorities in each country said that climate change was a somewhat to very serious problem (see figure 16.1).[13] This same survey, however, found that climate change ranked seventh out of eight environmental issues among the global public. These results suggest that the global public, in both developed and developing countries, does consider climate change to be a significant problem, albeit a lower priority than most other environmental problems. Meanwhile, concern about

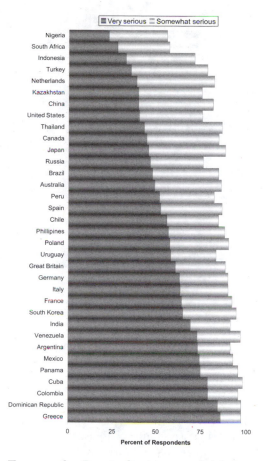

FIGURE 16.1. Perceived seriousness of global warming. Source: GlobeScan, Inc. (2000).

climate change may be increasing. A resurvey of these same countries in 2006 found that the percent of respondents saying climate change was a "very serious threat" had increased in most countries (see figure 16.2).[14]

American Public Opinion

In the United States, numerous public opinion polls demonstrate that large majorities of Americans are aware of global warming (92 percent), believe that global warming is real and already underway (74 percent), believe

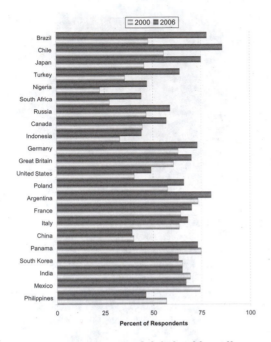

FIGURE 16.2. Percent of global public calling climate change a "very serious problem" in 2000 and 2006. Source: GlobeScan, Inc. (2000, 2006).

that there is a scientific consensus on the reality of climate change (61 percent), and already view climate change as a somewhat to very serious problem (76 percent).[15] The Gallup Organization has been tracking American levels of worry about global warming for nineteen years. They found that the percentage of Americans saying they were worried "a great deal" has fluctuated over the years, reaching a high of 40 percent in the year 2000, followed by a decline to 26 percent in 2004 (see figure 16.3). Since then, the percentage has risen, reaching a new high of 41 percent in March of 2007 ($n = 1,009$; margin of error ±3 percent).[16] It remains to be seen whether the recent increase in public worry about global warming represents the first signs of a social tipping point, leading to greater levels of public engagement and political pres-

sure, a new plateau, or the peak before another downward slide as other national and global issues take the spotlight.

At the same time, however, Americans continue to regard climate change as a relatively low national priority. For example, respondents to a national survey in 2004 ranked global warming tenth out of ten national issues (behind the economy, terrorism, health care, education, Social Security, the federal budget deficit, tax cuts, Medicare, and crime). Even among other environmental issues, global warming ranked below air and water pollution, loss of the ozone layer, and toxic waste as a priority. More recently, a survey conducted by Gallup and Yale University in 2007 found that the environment (including climate change) remained at the bottom of public priorities. Climate change may be moving slowly up the priority ladder in the United States, but clearly has a long way to go before it overtakes issues like the economy, health care, education, and so forth.

Thus, Americans seem generally concerned about global warming, yet view it as less important than nearly all other national or environmental issues. Why? Additionally, why do some Americans see climate change as an urgent, immediate danger, while others view it as a gradual, incremental problem, or not a problem at all? To better understand American climate change risk perceptions, policy preferences, and behaviors, an in-depth national survey study was conducted in 2003.[17]

American Risk Perceptions

This study found that Americans as a whole perceived climate change as a moderate risk. On average, Americans were somewhat con-

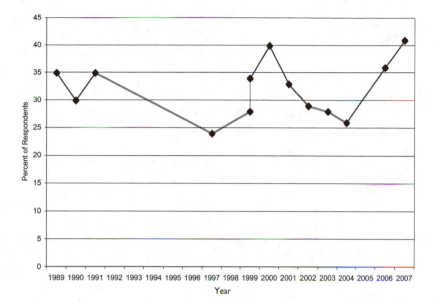

FIGURE 16.3. Percent of Americans worried "a great deal": Trend 1989–2007. Source: Gallup. (2007). *Did Hollywood's Glare Heat Up Public Concern about Global Warming?* Retrieved June 15, 2007, from www.galluppoll.com/content/?ci=26932.

cerned about global warming, believed that impacts of global warming on worldwide standards of living, water shortages, and rates of serious disease are somewhat likely, and believed that the impacts will be more pronounced on nonhuman nature. Importantly, however, they were less concerned about local impacts, which were rated as somewhat unlikely. The moderate level of public concern thus appeared to be greatly influenced by the perception that climate change is a danger to geographically and temporally distant people, places, and nonhuman nature.

This conclusion was supported by the results of a separate question. Respondents were asked, "Which of the following are you *most* concerned about? The impacts of global warming on . . . you and your family; your local community; the United States as a whole; people all over the world; nonhuman nature; or not at all concerned."

A large majority of respondents (68 percent) were most concerned about the impacts on people around the world and nonhuman nature. Only 13 percent were most concerned about the impacts on themselves, their family, or their local community. This helps explain why global climate change remains a relatively low national priority. Higher-ranking national issues (e.g., the economy, education, health care, and so on) and environmental issues (clean air, clean water, urban sprawl) are all issues that are more easily understood as having direct local relevance. Global climate change, however, is not yet perceived as a significant local concern among the American public. Former Speaker of the U.S. House of Representative Tip O'Neill once famously stated that "all politics is local." To the extent that this is true, climate change is unlikely to become a high priority until Americans consider themselves personally at risk.

American Images of Global Warming

This study also assessed the connotative meanings of global warming in the American mind using a simple form of word association. Respondents were asked to provide the first thought or image that came to mind when they heard the words "global warming." The results were then grouped thematically into eight categories. Associations to melting glaciers and polar ice were the single largest category of responses, indicating that this current and projected impact of climate change was the most salient image of global warming among the American public. This was followed by heat and rising temperatures, impacts on nonhuman nature, ozone depletion, images of devastation (Alarmists), sea-level rise and the flooding of rivers and coastal areas, references to climate change, and finally associations indicating skepticism or cynicism about the reality of climate change (Naysayers).

Thus, two of the four most prevalent associations (melting ice and nonhuman nature), provided by a third of all respondents, referred to impacts on places or natural ecosystems distant from the everyday experience of most Americans. Most of the references to "heat" were relatively generic in nature and likely indicated associations with the word "warming" in "global warming." Finally, 11 percent of Americans provided associations to the separate environmental issue of stratospheric ozone depletion, indicating that a substantial proportion of Americans continue to confuse and conflate these two issues. Thus, nearly two thirds of respondents provided associations to impacts geographically and psychologically distant, to generic increases in temperature, or to a different environmental problem. Critically, this study found that most Americans lack vivid, concrete, and personally relevant images of climate change, reinforcing the earlier finding that few Americans were concerned about the impacts on themselves, their families, or their local communities.

Policy Support

This study also measured American public support for a variety of policy proposals to mitigate global warming at the national and international levels. It found that of those Americans who had heard of global warming (92 percent):

- 90 percent thought the United States should reduce its greenhouse gas emissions.
- 88 percent supported the Kyoto Protocol, and 76 percent wanted the United States to reduce greenhouse gas emissions regardless of what other developed or less-developed countries do.
- 79 percent supported an increase in vehicle fuel economy standards (CAFE).
- 77 percent supported government regulation of carbon dioxide as a pollutant and a shift in subsidies from the fossil fuel industry to the renewable energy industry (71 percent).
- While a majority favored a tax on "gas guzzlers" (54 percent), large majorities opposed a 60-cent gasoline tax (78 percent) or a 3 percent business energy tax (60 percent) to reduce greenhouse gas emissions.
- Americans divided evenly (40 percent) regarding a market-based emissions trading system, while 18 percent were uncertain.
- Democrats and liberals expressed

stronger support for climate change policies than Republicans and conservatives.

- Independents and moderates showed levels of support more similar to Democrats and liberals than to Republicans and conservatives, yet:
- Majorities of Republicans and conservatives supported most climate change policies.

This research thus identified a contradiction in American climate change risk perceptions and policy preferences. On one hand, Americans expressed moderate levels of concern about global warming, strongly agreed that the United States should reduce its greenhouse gas emissions, strongly supported national regulation of carbon dioxide as a pollutant, and supported international treaties like the Kyoto Protocol. On the other hand, the public strongly opposed increasing energy and gasoline taxes—both direct pocketbook issues. A majority of Americans did support a tax on "gas guzzler" vehicles, but they were evenly split regarding an international market in emissions trading. Thus, the public largely supported policy action at the national and international levels, but opposed two tax policies that would directly affect them.

Individual Behaviors

This survey also asked respondents whether they had already taken actions to mitigate or reduce greenhouse gas emissions. Overall, approximately half said they had used energy-efficiency in past consumer choices (51 percent) or installed insulation or weatherized their home (45 percent). In addition, 46 percent said they had chosen not to buy an aerosol spray can, which provides further evi-

dence that many Americans continue to confuse or conflate global warming with stratospheric ozone depletion. The second highest response was planting a tree (49 percent), an action that has become perhaps the quintessential, symbolic "environmental act."

Only a quarter (26 percent) of the American public had used alternative forms of transportation, such as rail, carpools, walking, bicycling, and so on instead of driving. Only 4 percent of Americans reported purchasing alternative energy, which in part reflects limited access to renewable energy sources. Likewise, few Americans had engaged in political action on global warming. Only 15 percent said they had joined, donated money to, or volunteered with an organization working on global warming issues, and only 9 percent said they had contacted politicians to communicate their concerns about global warming. Finally, only a quarter (27 percent) of Americans reported talking to family members, friends, or colleagues about how to reduce or prevent global warming. Until global warming becomes a topic of common discourse, it will probably remain a low-priority issue for most Americans.

Interpretive Communities

The above aggregate results, however, gloss over substantial variation in risk perceptions within the American public (see figure 16.4).

In particular, this study identified several distinct "interpretive communities" within the American public—groups that share relatively homogeneous risk perceptions, imagery, values, and sociodemographic characteristics. Different interpretive communities are predisposed to attend to, fear, and socially amplify some risks, while ignoring, discounting, or attenuating others. For example, this study

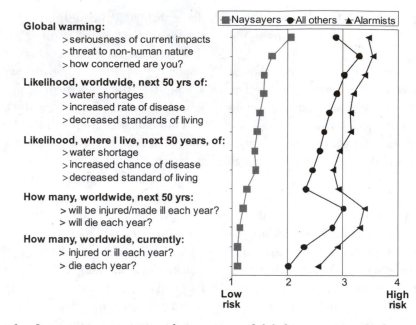

FIGURE 16.4. Interpretive community risk perceptions of global warming. For the first nine items, marked with ▶, mean *n*'s are Naysayers 35, All others 450, and Alarmists 55. For the last four items, marked with ◆, mean *n*'s are Naysayers 27, All others 274, and Alarmists 27; "don't know" responses are excluded. Source: Leiserowitz, A. (2005). American Risk Perceptions: Is Climate Change Dangerous? *Risk Analysis*, 25(6), 1433–42.

found that naysayers perceived climate change as a very low or nonexistent risk. This group was subsequently found to be predominantly white; male; politically conservative; holding pro-individualism, pro-hierarchism, and anti-egalitarian values and anti-environmental attitudes; distrustful of most institutions; and highly religious; and to rely on radio as their main source of news.[18] Further, the naysayer interpretive community articulated five distinct reasons why they doubted the reality of global climate change:

1. Belief that global warming is natural ("Normal Earth cycles"; "It is just the natural course of events"; "A natural phenomenon that has been going on for years").

2. Hype ("It is not as bad as the media portrays"; "The 'problem' is overblown"; "Environmentalist hysteria").

3. Doubting the science ("There is no proof it exists"; "Around ten years or so ago it was global cooling"; "Junk science").

4. Flat denials of the problem ("A false theory"; "There is no global warming").

5. Conspiracy theories ("Hoax"; "Environmentalist propaganda"; "Scientists making up some statistics for their job security").

While only 7 percent of the U.S. adult population were naysayers (or approximately 12 million people) according to these survey results, they are politically active, are significantly more likely to vote, and have strong representation in national government and powerful allies in the private sector.

By contrast, alarmists perceived climate change as an extreme risk and provided apocalyptic images of climate change far worse than scientific projections, such as: "Bad . . . bad . . . bad . . . like after nuclear war . . . no vegetation"; "Heat waves, it's gonna kill the world"; and "Death of the planet." Alarmists held pro-egalitarian, anti-individualist, and anti-hierarchist values; were politically liberal; strongly supported government policies to mitigate climate change (including raising taxes); and were significantly more likely to have taken personal action to reduce greenhouse gas emissions. It is also important to note, however, that the rest of the American public had climate change risk perception levels much closer to alarmists than naysayers. This demonstrates that most Americans are predisposed to view climate change as a significant danger, albeit not as extreme as alarmists, while climate change naysayers have substantially lower risk perceptions than the rest of American society.

Conclusion

Global climate change is perhaps the preeminent environmental risk confronting the world in the twenty-first century. As one survey respondent commented:

> We can't know where global warming is going, how far it will go, or whether we will be able to arrest it if it progresses past a certain point. All we know with reasonable certainty is that it's happening, it's big, and we're contributing to it. My impression, which I think is shared by most scientists (I'm not a scientist) concerned with the subject, is that there's a nontrivial risk of widespread death and destruc-

tion if we continue doing what we're doing.

The consequences will be global and especially severe for the poor, children, future generations, and diverse species. Thus, climate change raises deep questions of morality and social justice, as those who will likely suffer the greatest impacts are not the ones who have obtained the greatest benefits from the exploitation of fossil fuels. The reduction of emissions in the effort to limit global climate change will require significant social, political, and economic changes at all levels of society, including the individual. There will be costs and benefits, winners and losers, and consumers will have to make different choices in service of the local, national, and global effort to shift modern civilization from dependence on fossil fuels to noncarbon energy sources. The global public, in their collective actions as voters and consumers, will therefore play a decisive role in the future trajectory of global warming.

Overall, these and other research findings suggest that the global public is already predisposed to view climate change as a significant risk and to support local, national, and international action to reduce greenhouse gas emissions. What is lacking is a sense of public urgency, strong leadership, and political will.

Notes

1. Kammen, D. M., and Hassenzahl, D. M. (1999). *Should We Risk It? Exploring Environmental, Health, and Technological Problem Solving.* Princeton: Princeton University Press, p. 3.

2. Douglas, M. (1992). *Risk and Blame: Essays in Cultural Theory.* London: Routledge, p. 24.

3. Lupton, D. (1999). *Risk.* New York: Routledge, p. 8.

4. Cutter, S. L. (1993). *Living with Risk: The Geography of Technological Hazards.* New York:

Edward Arnold; Palm, R. I., and Carroll, J. (1998). *Illusions of Safety: Culture and Earthquake Response in California and Japan*. Boulder: Westview Press.

5. Slovic, P. (1997). Trust, emotion, sex, politics, and science: Surveying the risk-assessment battlefield. In M. Bazerman, D. Messick, A. Tenbrunsel, and K. Wade-Benzoni, eds., *Environment, Ethics and Behavior*, pp. 277–313. San Francisco: New Lexington Press, p. 278.

6. Slovic, P. (1987). Perception of risk. *Science*, 236: 280–85, p. 283.

7. Slovic, P., Layman, M., and Flynn, J. H. (1991). Perceived risk, trust, and the politics of nuclear waste. *Science*, 254: 1603–8; Flynn, J., Burns, W., Mertz, C. K., and Slovic, P. (1992). Trust as a determinant of opposition to a high-level radioactive waste repository: Analysis of a structural model. *Risk Analysis*, 12(3): 417–30; Flynn, J., Peters, E., Mertz, C. K., and Slovic, P. (1998). Risk, media, and stigma at Rocky Flats. *Risk Analysis*, 18(6): 715–27.

8. See, e.g. Leiserowitz, A. A. (2006). Climate change risk perception and policy preferences: The role of affect, imagery, and values. *Climatic Change*, 77: 45–72; Slovic, 1997, op cit.; Flynn et al., 1992, op. cit.; Flynn et al., 1998, op. cit.; Peters, E., and Slovic, P. (1996). The role of affect and worldviews as orienting dispositions in the perception and acceptance of nuclear power. *Journal of Applied Social Psychology*, 26: 1427–53; Finucane, M., Slovic, P., Mertz, C. K., Flynn, J., and Satterfield, T. (2000). Gender, race, and perceived risk: The "white male" effect. *Health, Risk and Society*, 2(2): 159.

9. Flynn et al., 1992, op. cit.; Flynn et al., 1998, op. cit.

10. Peters and Slovic, 1996, op. cit.

11. See, for example, Bostrom, A., Morgan, M. G., Fischhoff, B., and Read, D. (1994). What do people know about global climate change? *Risk Analysis*, 14(6): 959–70; Kempton, W. (1997). How the public views climate change. *Environment*, 39(9): 11; Bord, R. J., Fisher, A., and O'Conner, R. E., (1998). Public perceptions of global warming: United States and international perspectives. *Climate Research*, 11(1): 75–84; Poortinga, W., Pidgeon, N., and Lorenzoni, I. (2006). *Public Perceptions of Nuclear Power, Climate Change and Energy Options in Britain: Summary Findings of a Survey Conducted during October and November 2005*. Technical Report (No. Understanding Risk Working Paper 06-02). Norwich: Centre for Environmental Risk; Leiserowitz, A. (2004). Before and after *The Day after Tomorrow*: A U.S. study of climate change risk perception. *Environment*, 46(9), 22–37; Leiserowitz, A. (2005). American risk perceptions: Is climate change dangerous? *Risk Analysis*, 25(6), 1433–42; Leiserowitz, 2006, op. cit.

12. Leiserowitz, A., Kates, R. W., and Parris, T. M. (2005). Do global attitudes and behaviors support sustainable development? *Environment*, 47(9), 22–38; Brechin, S. R. (2003). Comparative public opinion and knowledge on global climatic change and the Kyoto Protocol: The U.S. versus the world? *International Journal of Sociology and Social Policy*, 23(10): 106–34.

13. GlobeScan, Inc. (2000). *Environics International Environmental Monitor Survey Dataset*. Kingston, Canada: Environics International. http://jeff-lab.queensu.ca/poadata/info/iem/iemlist .shtml (accessed October 5, 2004).

14. GlobeScan, Inc., 2000; GlobeScan, Inc. (2006). *Thirty-Country Poll Finds Worldwide Consensus that Climate Change Is a Serious Problem*. Toronto: GlobeScan, Inc. www.globescan.com/ news_archives/csr06_climatechange.pdf.

15. Leiserowitz, A. (2003). *Global Warming in the American Mind: The Roles of Affect, Imagery, and Worldviews in Risk Perception, Policy Preferences, and Behavior*. Unpublished Dissertation, University of Oregon, Eugene; Program on International Policy Attitudes. (2005). *Global Warming*. Retrieved October 10, 2003, from www.americansworld.org/digest/global_issues/global_warming/gw _summary.cfm.

16. Gallup. (2007). *Did Hollywood's Glare Heat up Public Concern about Global Warming?* Retrieved June 15, 2007, from www.galluppoll .com/content/?ci=26932.

17. For study details, see Leiserowitz, 2005, op. cit.; Leiserowitz, 2006, op. cit.; Leiserowitz, A. (2007). Communicating the risks of global warming: American risk perceptions, affective images and interpretive communities. In S. C. Moser and L. Dilling, eds., *Communication and Social Change: Strategies for Dealing with the Climate Crisis*. Cambridge: Cambridge University Press.

18. Ibid.

Chapter 17

What Is the Economic Cost of Climate Change?

Michael Hanemann

Introduction

Much of the economic analysis of climate change revolves around two big questions: What is the economic cost associated with the impacts of climate change under alternative GHG emissions scenarios? What is the economic cost of reducing GHG emissions?

The economic aspect of the policy debate intensified with the publication in the UK of the *Stern Review of the Economics of Climate Change*.[1] Stern stated that, if no mitigative action is taken, "the overall costs and risks of climate change will be equivalent to losing at least 5 percent of global Gross Domestic Product (GDP) each year, now and forever." This conclusion has been criticized by many economists, particularly in the United States, where Professor William Nordhaus of Yale, the leading expert on climate economics, estimates that the economically optimal policy involves only a modest rate of emission reduction in the near term, followed by larger reductions later.[2]

The disagreement between Stern and Nordhaus has aroused considerable interest. In this chapter I focus specifically on the limited issue of the assessment of the economic costs of climate change to the United States if there were a 2.5-degree-Celsius increase in average global temperature.[3] This serves as something of a benchmark for Nordhaus's well-documented damage function. Particularly problematic, most existing economic analyses rely on rather general information and are highly aggregated over space and time, which has important consequences for estimating the economic costs of climate change. Since the impacts of climate change are likely to vary both spatially and temporally, aggregation distorts economic assessments, sometimes quite dramatically.

The Framework for Damage Assessment

In an economic assessment the physical, biological, and social impacts of climate change—whether beneficial or adverse—are expressed in monetary terms. Given a particular future scenario, an impact assessment translates the resultant emissions first into changes in temperature, precipitation, and sea level; then into a suite of physical, environmental, and social impacts (e.g., changes in crop yield, water supply, disease incidence, species abundance, and so on); and, finally, into an aggregate monetary cost.[4] Each step is marked by uncertainty and some degree of scientific disagreement.

I refer to the physical, environmental and social impacts in the next-to-last step as "damages" and their economic valuation (the last step) as the "economic cost" of climate change. The impact assessment involves the use of damage functions, which express physical or environmental outcomes as a function of changes in climate variables; for example, there may be agricultural damage functions or health damage functions. The economic cost associated with damages is then represented through a valuation function, which translates the given set of damages into an economic cost. Alternatively, the impact may be expressed directly in monetary terms using a cost function, which represents the economic cost as a function of changes in climate variables, thereby combining the two steps into one.

Four points should be emphasized. First, the disagreement among damage and cost functions is significantly larger than that among climate change projections.[5] This should not be a surprise. Climate modeling has been going on for longer and at a higher level of activity than the damage and cost modeling, and is therefore in a more mature state. In addition, damage estimation is inherently more complex; it involves a high level of spatial disaggregation and a wide range of biological, chemical, hydrological, and physical phenomena, most of which are not yet well modeled.

Second, the paucity of the available data for many of the factors that determine the impacts of a given climate change scenario can

scarcely be overstated. Because of the spotty nature of the data available, analysts inevitably have to extrapolate the results from an analysis of one location to other locations that could be quite different. The resulting damage functions, therefore, depend very heavily on subjective judgments by the researcher. This accounts in part for the differences in conclusions regarding the economic impact of climate change.

Third, while there is some controversy both about the economic concept of value—representing the consequences of climate change through a monetary measure—and also about specific approaches to the measurement of economic value, this is *not* a major factor in the controversy about the economically optimal reduction in GHG emissions.[6] As explained below, the issues have more to do with whether the impacts are positive or negative, and how large, than with whether and how they should be monetized. In other words, the damage function is more problematic than the valuation function.

Finally, the precise representation of climate and its relationship to damages matters greatly to the outcome of the econometric analysis. Regrettably, many analyses use a quadratic functional form to represent the relationship between climate variables, such as temperature, and damages. This form assumes a smooth, symmetric, hill-shaped relationship with a unique optimum, around which losses and gains offset each other exactly for equal temperature changes. However, a quadratic function more likely than not turns out to provide a poor representation of this relationship. For example, Schlenker and Roberts demonstrate convincingly that the relationship between temperature and crop yields is highly asymmetric.[7] The improvement from temperature changes below the turning point is far smaller than the loss from increases in temperature beyond the turning point; that is, the losses at high temperatures greatly outweigh the gains at low temperatures for comparable temperature increases. It seems likely that a similar pattern would connect climate variables and damages from climate change, namely that the relationship is asymmetric and the slope becomes particularly steep for very high temperatures. While imparting tractability, functional simplifications such as the quadratic functional form distort climate damages.

The Issue of Aggregation

Aggregation over space and time when characterizing impacts matters greatly and accounts for some of the differences in assessed impacts. Using annually averaged temperatures masks the harmful effects of high temperature extremes on a daily timescale and obscures crucial differences in seasonal temperature. While warmer winter temperatures may be beneficial in some cases, including energy demand and health, warmer summer temperatures can have an even larger negative impact, especially if they involve more frequent heat waves.

Regional Differences

Consider the Hadley CM3 projection of a 2-degree-Celsius increase in *global* average temperature by 2100 under the B1 emissions scenario compared to the average from 1970 to 1999.[8] However, the temperature increase is distributed unevenly around the globe; the increase is smaller over the ocean and in lower latitudes and larger on land and at higher latitudes.

For example, by 2100 in California and much of the western United States under this scenario, there is a 3.3-degree-Celsius increase in average annual temperature. In addition, the increase is different at different times of the year. Statewide average winter temperature (December through February) in California rises by 2.3 degrees Celsius, while statewide average summer temperature (June through August) rises by 4.6 degrees Celsius. Moreover, there is spatial variation between the temperature increases along the coast versus inland. In the Central Valley, the main farming area in California, the increase in summer temperature reaches 5 degrees Celsius.[9]

Given the nonlinearity of damage function, it makes a substantial difference to the estimated impact on California agriculture whether one represents the climate change as an increase of 2 degrees Celsius (global average annual temperature), 3.3 degrees Celsius (statewide average annual temperature), or 5 degrees Celsius (Central Valley average summer temperature). While the effect on yield of a 2-degrees-Celsius temperature increase combined with carbon fertilization may or may not be positive, the effect of a 5-degree-Celsius increase during the growing season is likely to be negative. To the extent that there are significant seasonal and spatial differences in the rate of temperature increase, relying on globally or annually averaged temperatures can produce an excessively optimistic assessment of the consequences.

Agriculture

A similar distortion occurs when analyses include spatial aggregation to derive estimates of climate damages. Agriculture is both the most heavily studied sector in the climate economics literature and also the most controversial in terms of the range of divergent impact estimates. In most of the existing economic analysis the agricultural effects of climate change have been limited to the effects of temperature and CO_2 fertilization on crop yield, especially for the major grain crops. The effect on product quality and the impacts of climate on water need, weeds, pests, and ozone have largely been overlooked.

For example, Mendelsohn, Nordhaus, and Shaw conclude that adapting to climate change yields positive benefits to U.S. agriculture.[10] However, they fail to control adequately for the effects of irrigation by pooling both dry-land farming and irrigated land. The analysis assumes that local precipitation is the source of water supply for the county's agriculture. While this assumption is accurate in dryland farming areas, it does not apply to irrigated areas where the water supply is either imported or comes from groundwater. The richest farmland in the United States is found in irrigated areas of California and parts of Arizona; these are also the driest and hottest regions of the United States. If one does not control for irrigation, it looks as though being dry and very hot in the summer makes for richer farming. When Schlenker, Hanemann, and Fisher control for the effects of irrigation and repeat the analysis just for dry-land farming areas of United States, they reveal a loss in these areas of $11 billion, instead of the gain of $2.3 billion reported by Mendelsohn, Nordhaus, and Shaw.[11]

For irrigated areas, there is clear evidence that warmer temperatures will increase evapotranspiration, and thus crop demand for water, while also reducing the effective supply of surface water. Assuming an increase in temperature while maintaining effective water supply (which is an unduly optimistic assumption), Schlenker, Hanemann, and Fisher determine that farmland values in California toward the end of the century (2070 to 2099) could be reduced 24 to 52 percent, depending on the scenario assumptions regarding energy usage and technological advances.[12]

Taken together, the revised analysis that takes into account differences in water sources between rain-fed and irrigated land indicate that climate change will entail significant losses to U.S. agriculture in the range of $10 to $15 billion (1990$), if not more.

Similar problems with aggregation emerge in analyses of the effects of climate change on health, energy, and water supplies.

Health

Mortality rates, as one measure of the human health impacts of climate changes, are likely to decrease in the winter and increase in the summer as globally averaged annual temperatures rise. However, it cannot be assumed that increased mortality in the summer will be offset by reduced mortality from winter warming.

The precise shape of the relation between temperature and mortality varies by location, which complicates calculation of the net effect of projected temperature increases. The estimated impact depends on some key measurement issues, including whether one uses monthly or daily average temperature, whether one distinguishes between different levels of extreme temperature (e.g., treating all temperatures above 90 degrees Fahrenheit as being in the same category), whether one uses absolute or relative temperature maxima (e.g., days over 90 degrees Fahrenheit versus days over the ninety-ninth percentile for that location), and whether one allows for interactions between the temperature one day and that on previous days (e.g., three or more consecutive days over 92 degrees Fahrenheit) or between temperature and other factors such as humidity and air quality (e.g., ozone). In general, the more refined the measure of temperature extremes, the more likely one is to conclude that the increase in summer mortality outweighs the reduction in winter mortality.[13]

The approach currently being used in California seems to account well for variation in both air quality and daily mortality.[14] Applied to Los Angeles, for example, this approach projects heat mortality to increase by a factor of 3 toward the end of the century compared to the 1971 to 2000 average (about 300 deaths annually instead of about 100, assuming a constant population).[15] The existing evidence suggests that, in the United States, the increased mortality in the summer is likely to outweigh the reduced cold mortality in the winter, leading to an overall increase in mortality loss.

Energy

Similarly, climate change will produce regional and seasonal variation in energy demands for space heating and cooling, which represents 54 percent of all residential and commercial energy use in the United States. Although warmer temperatures lower wintertime demand for energy for space heating, they raise summertime demand for energy for space cooling. Again, there is no reason to believe that one effect offsets the other when temperature changes of this order of magnitude occur. In fact, residential winter space heating occurs mostly at night and involves baseload energy, while residential space cooling typically occurs in the late afternoon and involves peak power. From an economic perspective, the latter is considerably more costly than the former. In addition, climate change is expected to affect winter and summer temperature changes differently. For example, a 2.5-degree-Celsius increase in global average annual temperature translates into an increase of 2.3 degrees Celsius in California statewide average winter temperature, but an increase of 4.6 degrees Celsius in statewide average summer temperature.[16] This seasonal difference in temperature changes will amplify changes in energy consumption that will disproportionately affect energy demand during the summer months.

In addition to the effect on demand, climate change can also affect the *supply* of energy when extreme weather events occur. In the United States, power plants discharging cooling water often face restrictions of the temperature of the discharge water and sometimes have to limit operations when the ambient air and water temperature become too high, notably along the Gulf of Mexico. Extreme heat also lowers the carrying capacity of electricity transmission lines. Hurricanes, storms, and extreme weather conditions can sometimes disrupt the production and distribution of energy; in 2005, Hurricanes Katrina and Rita damaged 558 pipelines and destroyed more than 100 offshore oil platforms with one major platform remaining out of production for eight months.

Water

Finally, changes in patterns of precipitation due to climate change are likely to have a significant economic impact. Three quarters of the freshwater used in the United States comes from surface water rather than groundwater and, therefore, depends directly on precipitation. Not only is climate change likely to alter precipitation patterns, but increases in temperature will also affect water supply more profoundly than precipitation in some regions.

Temperature affects both the timing and volume of runoff. In areas that rely on the snow pack for natural water storage, like California, the Pacific Northwest, and the Colorado River Basin, winter warming reduces the snowpack available at the beginning of spring for use in the summer. In California, the Sierra snowpack accounts for about one third of the state's total surface water supply storage; by 2070 to 2099 it is projected to decline by 70 to 90 percent under the A1Fi emissions scenario.[17]

The main economic impact from this water supply reduction in California, for both agricultural and urban water users, occurs not in average years but in drought years, which occur with greater frequency under the climate change scenarios.[18] For example, for urban water users in Southern California in 2070 to 2099 in a scenario without the benefit of technological improvements, shortages that require rationing occur twice as frequently (in one third instead of one sixth of years) and are more intense (one third greater loss of consumer's surplus). The net impact of climate change to urban water users in Southern California is a loss of $4 billion per drought year.

Warmer winter weather also will cause precipitation to fall more often as rain instead of snow, leading to earlier and more intense snowmelt. Both changes increase the likelihood of flooding.[19] These changes are already underway: the 100-year, three-day peak flows on the American, Tuolumne, and Eel rivers have more than doubled between the period before 1955 and the period since then; in California during the period 1950 to 1997 the annual property damage from these floods averaged about $280 million per year (in 1998 dollars), with an average of almost six deaths per year.

Temperature also affects the ground cover in a watershed, which in turn influences runoff. In forested areas, for example, the combination of higher temperature and dryer soil could cause more frequent or intense wildfires, reducing forest cover, lowering moisture retention, and accelerating runoff. Because of the increase in water consumed by ground cover, Nash and Gleick find that, if the temperature in the Colorado River Basin increased by 4 degrees Celsius with no change in precipitation, this would reduce the mean annual runoff there by nearly 20 percent.[20]

The intensity of precipitation is also important since a greater intensity means less moisture is retained in the soil. Trenberth and colleagues hypothesize that, with warming, precipitation will tend on average to be less frequent, but more intense when it does occur.[21] At the same time, with warming, the soil dries earlier in the year so that there can be both more winter flooding *and* more summer drought. Increases in summer temperature also raise crop evapotranspiration, which would in turn increase the demand for agricultural and outdoor urban water use and exacerbate the effect of reduced summer water supply.

Sea level rise can exacerbate water supply constraints in coastal areas due to increased saltwater intrusion into freshwater aquifers, changes in the pattern of sedimentation that affects water utility operations, an increase in the tidal range in rivers and bays that brings saltwater closer to the intake points of water supply systems, and inundation of wetlands that provide a natural filter for water supply.

Conclusion

Damage estimates by Nordhaus and Boyer are typical of assessments that project modest climate change impacts.[22] Although Nordhaus and Boyer consider a more complete set of factors than those considered here, their overall conclusion is that climate damages to the United States from a 2.5-degree-Celsius increase in global average annual temperature would lower GNP (gross national product) by approximately 0.5 percent. I believe that these relatively optimistic climate damage projections underestimate climate damages once assessments take into account the localized effects of climate change. Based upon a more extensive analysis than that presented here, I argue that these climate damages would be closer to 1.5 to 2 percent of U.S. GNP, due in part to the distortion caused by aggregation. These revisions are necessarily speculative, but hopefully they provide a more nuanced accounting of what climate change might mean for the United States.

Three main factors account for the difference between my estimates and those of Nordhaus and Boyer: (1) a change in global average annual temperature, on which they tend to focus, translates into significantly larger changes in *seasonal* temperatures in the United States; (2) much of the impact is driven by extreme weather conditions occurring on the time scale of a few hours or a few days, not by the conditions that occur most of the time; and (3) damage functions are asymmetric, and the most severe damages are triggered when thresholds are crossed. The bias

introduced by aggregation of climate change diminishes assessments of the economic costs of climate damages. Aggregation — whether spatial or temporal — is *not* innocuous.

Notes

1. Nicholas Stern, *The Economics of Climate Change: The Stern Review*, Cambridge University Press, Cambridge, U.K., 2006.

2. W. D. Nordhaus, *A Question of Balance*, Yale University Press, New Haven, 2008.

3. In this volume, chapter 13 by Hallegatte and Ambrosi covers similar issues, but not specifically for the United States.

4. On a national basis, the aggregate monetary impact — the change in income plus the WTP equivalent of the other changes in well-being — is typically expressed as a percentage reduction in gross domestic product (GDP), the value of the market economy.

5. The disagreement among damage functions is an important factor in the controversy over the *Stern Review*: his critics have complained that Stern's damage estimates are "ten to twenty times higher" than estimates in the existing literature (Yohe, G. W., and R. S. J. Tol, 2008, The Stern Review and the economics of climate change: An editorial essay. *Climate Change* 89 (3–4), 231–40.

6. For example, Schneider, S. H., K. Kuntz-Duriseti, and C. Azar, 2000, Costing nonlinearities, surprises, and irreversible events. *Pacific and Asian Journal of Energy* 10 (1), 81–91. The authors recommend that five separate metrics be used to measure the impacts of climate change: monetary loss (by which is meant market impacts); loss of life; quality of life (including conflict over resources, cultural diversity, loss of cultural heritage, and forced migration); species and biodiversity loss; and distribution/equity.

7. Schlenker, W., and M. J. Roberts, 2006a, Nonlinear effects of weather on corn yields. *Review of Agricultural Economics* 28 (3), 391–98.

8. The emissions scenarios are from Nakicenovic, N., et al., 2000, *Special Report on Emissions Scenarios: A Special Report of Working Group III of the Intergovernmental Panel on Climate Change*, Cambridge University Press, Cambridge, U.K.;

also refer to chapter 15 of this volume for a summary of scenarios.

9. For further details, see Hayhoe, K., D. Cayan, C. Field, P. Frumhoff, E. Maurer, N. Miller, S. Moser, S. Schneider, K. Cahill, E. Cleland, L. Dale, R. Drapek, W. M. Hanemann, L. Kalkstein, J. Lenihan, C. Lunch, R. Neilson, S. Sheridan, and J. Verville, 2004, Emissions pathways, climate change, and impacts on California. *Proceedings of the National Academy of Sciences* 101 (34), 12422–27. For comparison, under the Hadley CM3 projection of the A1Fi emission scenario, the corresponding figures are: a 4.1-degree-Celsius increase in global average annual temperature; a 5.8-degree-Celsius increase in statewide average annual temperature, broken down into a 4-degree-Celsius increase in statewide average winter temperature and an 8.3-degree-Celsius increase in statewide average summer temperature; and a 10-degree-Celsius increase in average summer temperature in the Central Valley.

10. Mendelsohn, R., W. D. Nordhaus, and D. Shaw, 1994, The impact of global warming on agriculture: A Ricardian analysis. *American Economic Review* 84 (4), 753–71.

11. Schlenker, W., W. M. Hanemann, and A. C. Fisher, 2005, Will U.S. agriculture really benefit from global warming? Accounting for irrigation in the Hedonic approach. *American Economic Review* March, 395–406. Mendelsohn, Nordhaus, and Shaw, 1994 op. cit.

12. Schlenker, W., W. M. Hanemann, and A. C. Fisher, 2007, Water availability, degree days, and the potential impact of climate change on irrigated agriculture. *Climatic Change* 81, 19–38.

13. For example, Martens, W. J., 1998, Climate change, thermal stress and mortality changes. *Soc. Sci. Med.* 46 (3), 331–44 proposes a relation between monthly average temperature and mortality which is used by Tol et al. (2002) in the health component of their damage function (Tol, R. S. J., et al., 2002, Estimates of the damage costs of climate change: Part I: Benchmark estimates. *Environmental and Resource Economics* 21, 47–73). Two recent studies that have been influential among economists (Deschenes, O., and E. Moretti, 2007, *Extreme Weather Events: Mortality and Migration*, Working Paper 13227, and Deschenes, O., and M. Greenstone, 2007, *Climate Change, Mortality and Adaptation: Evidence from Annual Fluctuations in Weather in the U.S.* MIT Econom-

ics Department Working Paper 07-19) assume that the thresholds above or below which temperatures trigger excess mortality are the same across all counties in the United States, and they treat all days with a temperature above 90 degrees Fahrenheit as the same. This significantly understates the impact of extreme heat on mortality.

14. See, for example, Sheridan, S. C., and L. S. Kalkstein, 2004, Progress in heat watch-warning system technology. *Bulletin of the American Meteorological Society* 85 (12), 1931–41.

15. Drechsler, D., N. Motallebi, M. Kleeman, D. Cayan, K. Hayhoe, L. Kalkstein, N. Miller, S. Sheridan, J. Jin, and R. A. VanCuren, 2006, *Public-Health Related Impacts of Climate Change in California*, California Energy Commission, White Paper, CEC-500-2005-197-SF.

16. In California, the metric for identifying occurrences of peak power demand is the so-called T90 temperature—the historical 1961 to 1990 maximum temperature threshold for the 10 percent warmest days in June through September. Daily maximum temperatures above the T90 threshold occurred an average of twelve times a year during 1961 to 1990. By 2100, they are projected to occur three or four times more frequently under the B1 emissions scenario, and six or seven times more frequently under the A1Fi scenario. Miller, N. L., K. Hayhoe, J. Jin, and M. Auffhammer, 2008, Climate, extreme heat, and electricity demand in California. *Journal of Applied Meteorology and Climatology* 47, 1834–44.

17. Hayhoe, K., D. Cayan, C. Field, P. Frumhoff, E. Maurer, N. Miller, S. Moser, S. Schneider, K. Cahill, E. Cleland, L. Dale, R. Drapek, W. M. Hanemann, L. Kalkstein, J. Lenihan, C. Lunch, R. Neilson, S. Sheridan, and J. Verville, 2004, Emissions pathways, climate change, and impacts on California. *Proceedings of the National Academy of Sciences* 101 (34), 12422–27.

18. Hanemann, M., L. Dale, S. Vicuna, D. Bickett, and C. Dyckman, 2006, *The Economic Cost of Climate Change Impact on California Water: A Scenario Analysis*, California Energy Commission PIER Project Report, CEC-500-2006-003.

19. Besides flooding, another common hazard of wet winter storms in California is mudslides when hillside soils become saturated in urban areas. The risk of landslides is especially severe in areas that have recently been burned in wildfires.

For example, heavy rains from a winter storm in December 2003 caused mudslides resulting in fifteen deaths and widespread property damage in areas of San Bernardino County that had been burned in the Southern California wildfires of October 2003. As noted below, warmer summer temperatures will increase the occurrence of such fires.

20. Nash, L. L., and P. H Gleick, 1993, *Colorado River Basin and Climatic Change: The Sensitivity of Stream Flow and Water Supply to Variations in Temperature and Precipitation.* Pacific Institute Technical Report. The effect is likely to be nonlinear. A recent estimate for river basins in Australia by Young (personal communication) suggests that a 1 percent reduction in precipitation there will lead to a 3 percent reduction in reservoir inflows, but a 10 percent reduction in precipitation could lead to a 50 percent reduction in inflow.

21. Trenberth, K. E., A. Dai, R. M. Rasmussen, and D. B. Parsons, 2003, The changing character of precipitation. *Bulletin of the American Meteorological Society* 84, 1205–17. The Fourth IPCC notes a recent increase in the frequency of heavy precipitation events over most land areas. IPCC, 2007, *Climate Change 2007: The Physical Science Basis*. Contribution of Working Group I to the Fourth Assessment Report of the Intergovernmental Panel on Climate Change, Solomon, S., D. Qin, M. Manning, Z. Chen, M. Marquis, K. B. Averyt, M. Tignor, and H. L. Miller, eds. Cambridge University Press, Cambridge, U.K., 996 pp.

22. Nordhaus, W. D., and J. Boyer, *Warming the World: Economic Models of Global Warming*, MIT Press, Cambridge, 2000.

Chapter 18

Cost-Efficiency and Political Feasibility

Christian Azar

Introduction

Global CO_2 emissions from burning fossil fuels have surpassed 8 billion tons of carbon per year (GtC/yr), and they are on track to exceed 20 GtC/yr by the end of the century.[1] In order to stabilize atmospheric concentrations below 400 parts per million (ppm) CO_2, global carbon emissions may have to be reduced by some 75 percent by 2100, compared with today's levels.[2] Several energy scenarios have demonstrated the technical feasibility of meeting stringent climate targets.[3] Figure 18.1 shows a global energy scenario that meets a stringent atmospheric CO_2 concentration target. In the short run, it builds on intense use of existing technologies for biomass, wind, and energy efficiency. In the long term, this scenario relies more heavily on new and advanced carbon-free (or near-free) technologies: solar power, hydrogen, and both fossil fuels and biomass with carbon capture and storage. These technologies need to be developed further before they are employed large-scale.

In order to achieve low carbon emissions, climate policies are needed. In this chapter, I first present the climate policies that, in theory, help us protect the climate at the lowest possible cost. Then I discuss how political feasibility often seems to be inversely proportional to cost-efficiency.

Cost-Efficient Climate Policy

Three categories of policies are central to meeting climate targets cost-efficiently:

- *Economy-wide price incentives*: A carbon tax or a cap-and-trade system to correct for the externalities associated with emissions.[4]
- *Specific technology policies that enhance the development of advanced energy technologies through support to research and development (R&D) and to market diffusion programs*: To correct for myopic behavior in industries, uncertainty about future climate policies, and difficulties in keeping industry inventions in-house when long time scales are involved.
- *Standards for energy efficiency*: To cor-

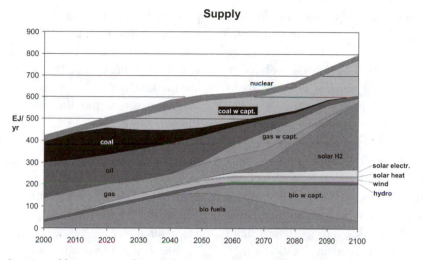

FIGURE 18.1. A world energy-supply transformation scenario under a CO_2 stabilization target of 350 ppm. Bioenergy with carbon capture and storage (BECCS) is one way to achieve negative carbon emissions, and this is the key reason why CO_2 concentrations as low as 350 ppm are feasible in this scenario (1 EJ = 1 exajoule = 10^{18} J). Azar, C., Lindgren, K., Larson, E., and Möllersten, K. 2006. Carbon capture and storage from fossil fuels and biomass: Costs and potential role in stabilizing the atmosphere. *Climatic Change* 74, 47–79.

rect for market failures associated with asymmetric information, skewed incentive structures, and other barriers to energy efficiency.

Economy-Wide Price Incentives

Policies that rely on price signals are in general more cost-efficient than policies that regulate individual companies and their technology choices.[5] Price signals, such as a carbon tax (see chapter 19), have four main advantages:

- They offer incentives to shift fuels toward less carbon-intensive alternatives (renewables; natural gas instead of coal; nuclear; and fossil fuels with carbon capture and storage).

- They lead to higher energy prices; this increases the benefits of improving energy efficiency, choosing less energy-intensive ways of life (public transportation instead of private cars, rail instead of trucks), saving energy through lifestyle changes, and choosing less energy-intensive materials (the benefits of recycling materials increase).

- They offer, at least if they are part of credible long-term plans to reduce CO_2 emissions, incentives for the private industry to develop new and more advanced CO_2-neutral energy technologies (on both the supply and demand side).

- They generate government revenue, which can be used to reduce other distortionary taxes, thereby reducing the cost the abatement policy.

Sweden introduced carbon taxes in 1991, and the experience deserves more widespread international attention. In the district heating system, oil and coal have been replaced by biomass, primarily residues from forests industries. The resulting reduction in carbon emissions has been significant. In fact, Sweden is one of only a few countries in the world that has reduced its carbon emissions since 1990 as a result of deliberate climate policies. Emissions have dropped by around 8 percent.

Other policies—including subsidies to renewables, renewable energy portfolios, or energy taxes (instead of carbon taxes)—are more expensive for a given emission reduction than a price on carbon since they do not provide incentives to reduce emissions on both the demand and the supply side of the equation. For instance, a subsidy to renewables does not lead to increases in the energy price, and for that reason options to reduce emissions through energy efficiency improvements are not exploited.[6] Further, it costs money for taxpayers to subsidize the use of an alternative technology, meaning that other taxes may be needed to restore the fiscal budget, and these taxes might distort the economy.

Policies to Bring Advanced Energy Technologies to the Market: R&D and Market-Diffusion Programs

Technological change comes in two varieties. In one, we simply choose a different technology already available. In this case, price incentives (and sometimes performance standards) are preferable.

The second kind of technological change amounts to bringing new technologies to the menu of available options. It requires not only research and development, but equally important—or possibly more important—are efforts by governments to develop bridges over what is sometimes called the "valley of death." In essence, there is a need for policies that create markets where "new" technologies may mature, may pass from the stage of invention, through innovation and full-scale diffusion.[7] The key is to set in motion a process of self-sustained growth, driven by dynamic learning and scale-effects, where cost reductions generate market growth that, in turn, generates investments and learning that lead to further cost reductions.

Investment in creating new technologies requires sustained support from the public sector because (1) it takes a very long time (thirty to fifty years) to take advanced technologies from invention through innovation to mass diffusion; (2) diffusion of knowledge at this time scale implies that private companies are not likely to reap the full benefits of in-house research on advanced technologies; and (3) uncertainty regarding whether stringent climate policies will be put in place and high internal discount rates in private companies entail that too little research—from society's point of view—is carried out on developing these technologies.

In addition, it is likely that policy makers will never dare set and implement stringent emissions targets unless they can be sure that these can be met without too much sacrifice. Therefore, policies that specifically aim to develop new advanced technologies will not only help us meet stringent climate targets at lower costs, but they are also of critical importance in the process of setting these stringent targets.

Possible policy instruments for this include investment subsidies, subsidies based on the amount of "green" electricity produced, feed in tariffs, green certificates, portfolio standards, and public procurement programs (where, for instance, governments decide to buy only hybrid cars, or municipal bus companies decide to run only buses using fuel cells powered by hydrogen). All these policies aim at creating niche markets that enable learning by doing, and there are pros and cons with all these approaches.[8] The efforts to support market development for wind energy in, for example, Denmark, Germany, Spain, and the United States, provide an example where policies to create growth and cost reductions for emerging carbon-neutral energy technologies have been successful. Global installed wind capacity has now surpassed 100 GW.

GOVERNMENTS SHOULD NOT PICK WINNERS: AN ELUSIVE OBJECTIVE?

It is often claimed that governments should not pick winners because they do not choose as capably as the market. This view has merits when it comes to already available technologies. However, when it comes to developing new technologies, governments have reason to be selective. For instance, supposedly technology-neutral biofuel policies (e.g., a policy that requires a certain share of biofuels in the transport system) may end up offering more support to existing poorly performing systems (such as corn-ethanol) than potentially better-performing second generation fuels (such as cellulose ethanol or methanol from woody biomass). If, on the other hand, the same policy were introduced a decade later, then more advanced, then better-performing second generation biofuels systems would likely win market share (given that these have been developed to a sufficiently mature level). The timing of the policy is critical. For this reason, a technology-neutral policy is an evasive objective when the goal is to develop more advanced carbon-free technologies. In these cases, explicit technology priorities should be reflected in government support. Key criteria include: (1) technologies that are expected to experience significant reductions in costs as a result of the policy and (2) technologies that have a reasonable likelihood of winning market share in the future when there is only a cap-and-trade or a carbon tax in place.

Efficiency Standards

Improving energy efficiency is central to meeting low-emission targets; it may in fact contribute as much or more to reduced CO_2 emissions as all renewables, carbon capture and storage, and nuclear combined.[9] Such improvements can be expected from higher carbon prices (which drive higher energy prices), technology policies that aim at developing more energy efficient products (e.g., fuel cells, hybrid cars, plug-in hybrids, zero-heating houses, and so on), as well as standards for energy efficiency, especially when split incentives exist between installers of the technology (e.g., insulation purchased by builder) and consumers of the resource (e.g., heating costs paid by residents).

Performance standards for energy efficiency are likely to enhance cost-efficiency—but they should be implemented with care. For instance, performance standards are in general preferable to technology specific standards. There is a risk that the standard itself freezes at the same level for too long. Standards might become impediments to a more dynamic development. An alternative might be to introduce dynamic efficiency standards that mirror the best available technology with a time lag that depends on the sector.

Price incentives, policies to speed up technology development, and performance standards for energy efficiency are complementary in the sense that they need one another to work properly. For example, if a consumer purchases a smaller and less expensive car, which is also more fuel efficient, s/he will save both money and emissions. The consumer could use the savings to drive additional miles or buy an airline ticket to a vacation destination, thereby negating some or all of the carbon emission reductions from the higher fuel efficiency—this is the so-called rebound effect. For this reason, a complementary carbon cap or tax is needed to maintain a nationwide pressure on the overall emissions.

Some Additional Reflections

So far I have only discussed three key categories of policy instruments for reducing CO_2 emissions: price incentives, technology policies, and efficiency standards. These three main categories must form the backbone of any climate policy aiming at cost-efficiency because they directly address the key reasons why we "over-emit" CO_2: the cost of climate damages are not included in the price of the product or activity that causes the emissions; incentives to develop new technologies when the incumbents are expected to remain cheaper are weak; and markets do sometimes fail to exploit cost-efficient options for energy efficiency.

However, other policies are also needed. One could mention the need for policies to initiate markets when there are chicken-and-egg-like problems (e.g., the establishment of infrastructures for alternative transportation fuels is particularly important for gaseous fuels), the establishment of standards for new fuels (e.g., what pressure should be the standard in hydrogen tanks and other distribution systems), and demand-side management (e.g., city planning and the provision of better-performing public transportation systems and railway networks), and an end to subsidies to fossil fuels would be both environmentally friendly and economically beneficial.[10]

We may also have to carefully consider how other policies affect energy use. For instance, implementing policies that reduce other pollutants could lead to some CO_2 reductions as a "co-benefit." So far, European fuel taxes may have been one of the more significant carbon abatement policies implemented in the world, even though they were not introduced in order to reduce carbon emissions.[11] Other policies might instead exacerbate carbon emissions—for example, promoting the use of coal to enhance energy security.

Finally, in the process of solving one problem we should be careful not to create others. For instance, with a higher carbon price, the profitability of well-performing biomass systems (e.g., forest residues used for heat generation) will increase, but this may drive up food prices and cause biomass plantations to supplant natural forests and land held by poor farmers in developing countries with poor property rights.[12]

Similarly, a higher carbon price will lead to higher electricity prices, and higher profitability for nuclear power. With nuclear power in place, and in particular with control of the fuel cycle, the step to nuclear weapons is much shorter (see chapter 46).[13]

The key issue illustrated by the biomass and nuclear examples is that energy policy is not only about selecting the lowest-cost option, but also affects more broad-ranging issues such as hunger, human rights, ecosystems, and international security. Choices go beyond merely opting for the cost-effective solution to the climate problem.

Political Feasibility versus Cost-Efficiency

The key reasons why so little has been done to reduce carbon emissions are a lack of political will in government circles, lack of sufficient support from the general electorate (although most voters, even in the United States, think that climate change is a serious issue), and, perhaps most important, resistance from industry groups and other special interests who think—rightly or not—that they have much to lose from climate policies.

Policy makers should spend as much time determining how to create public acceptance for climate policies as they spend on the cost-efficiency of policies. They need to think more about how to manage special interests and how to make feasible what is currently perceived as politically impossible. This involves agenda-building (i.e., information dissemination and explicit discussions with the general public) so as to generate support for climate policies.

Economists, too, need to carefully consider their role in the policy-making process. It is easy to repeat the mantra of "same carbon price in all sectors and all countries" but the real-world policy options are far from the economist's utopia. There is a risk that the perfect becomes the enemy of the good.

Consider, for instance, the discussions in early 1997, prior to negotiations in Kyoto. The real-world choice was not one between targets only for the industrialized countries and cost-efficient global reductions, but one between reductions in industrialized countries and a scenario with no abatement at all. Many economists argued that Kyoto was flawed because of its (somewhat) cost-inefficient features; this strengthened those who did not want to do anything about climate change. The economists' objections were used to justify the positions of energy-intensive companies and others who rejected Kyoto for less lofty reasons. In the end, the U.S. withdrawal from Kyoto led to an even less cost-efficient approach to the problem.

Thus, one of the more intellectually challenging problems for policy makers, policy advisers, and economists is to find so-called second-best solutions to the climate problem, where both cost-efficiency and political feasibility are addressed. Next, I consider a few examples highlighting the tension between these two objectives.

Protection of Energy-Intensive Industries

Should countries or regions with unilateral climate policies try to protect their energy-intensive industries? This can be done through tax differentiation (Sweden, for example, has a much lower carbon tax for industry than for other sectors), border-adjusted taxes (e.g., by implementing an implicit carbon tax on imported energy-intensive goods), through the free allocation of permits to energy-intensive industries (as the EU has done), or by subsidies to the negatively affected industries (e.g., based on production levels, so as to maintain the incentive to reduce emissions).[14]

The reasons for protection include:

- Reducing carbon leakage (without protection, steel manufacturing, to take a concrete example, may move out of the country into a region with, possibly, higher CO_2 emissions per ton of steel);
- Avoiding job losses in sensitive regions or sectors (although, in the long run, trade policies will not determine the number of jobs or the level of unemployment in the country as a whole);
- Avoiding loss of capital;
- Increasing the political feasibility of climate policy (large energy-intensive industries often make a very powerful lobby, and such groups can often block carbon policies);
- Sending signals to the rest of the world that the country is serious about its climate ambitions. This benefit only arises if the protection is carried out through border adjusted taxes. If they are designed so that they only apply to countries outside the carbon abating region, they also provide an incentive to introduce carbon policy in countries that currently do not have such policies.[15]

Arguments against protection include the fact that protection tends to *increase* the cost of reducing domestic emissions. Further, the protection may grow entrenched and difficult to get rid of even when no longer justified. We need only think of agricultural policies in Europe, cotton subsidies in the United States, and subsidies to coal in Germany. Finally, border tax adjustments are difficult to implement, may be introduced for sectors where they are not warranted, and may increase the risk of escalating trade conflicts.

Weighing these arguments is difficult, but it is clear that cost-efficiency cannot be the sole criterion.

The EU Emissions Trading Scheme (ETS): Which Sectors to Encompass?

Currently, the EU Emission Trading Scheme encompasses only large point sources, primarily large industries, plants, refineries, steel mills, and the like, and power generation. From a textbook economics perspective, this is economically inefficient. But would including the transportation and fuel use in households be desirable? Putting all sectors under the same cap implies that all sectors will face the same stringent policy, or, more specifically, they would face the same marginal price on carbon. Cost-efficiency would increase.

However, there are political risks. Transportation does not compete on international markets, whereas many energy-intensive industries (steel producers, for example) do, and they have already voiced their concerns. Governments may not dare introduce a stringent cap on the overall emissions for fear of losing the energy intensive industries to non-abating countries.

When assessing this policy option, and many others, one cannot only consider cost-efficiency arguments but also what the proposal means for the political feasibility of introducing a stringent cap on the emissions.

Emissions Trading: Grandfathering versus Auctioning

Another key issue in all emissions trading schemes is free allocation (currently carried out in the EU system through grandfathering) versus auctioning. Auctioning is preferable from an economic perspective (the revenues from auctioning can be used to reduce other taxes and thereby reduce the societal cost associated with the climate policy) and for reasons related to simplicity and fairness.

Clearly, by choosing to allocate allowances for free based on emissions history, it became possible to soften objections to the EU trading scheme. The main foes were thereby *bought* off. Paying those who do wrong in order that they do the right thing is objectionable, but perhaps this kind of thing is required in order to make certain difficult changes acceptable. But there are limits. Grandfathering permits to coal-fired power plants goes too far, turning the polluter pays principle on its head, especially since there is little trade in electricity across the EU-region border. The EU commission has now proposed that auctioning should be the main mechanism through which permits will be allocated.

Conclusion

In this chapter, I have stressed the need for (1) economy-wide price instruments for achieving low-cost emissions reductions; (2) technology-specific policies to drive technological development so that future targets can be met; and (3) performance standards for energy efficiency. A policy that *only* relies on, say, a technology push strategy like the Manhattan or Apollo project is not likely to succeed since such programs do not change the economic environment in which investments in new technologies are made. For instance, even a successful technology program resulting in solar cells at US$0.10/kWh would not in itself lead to major reductions in emissions, since coal-fired power plants cost even less.

In addition, I have indicated the need for policies that make sure that new problems are not created in the process of avoiding old ones (examples from biomass and nuclear energy were given). This multifaceted approach must form the backbone of all successful and cost-efficient climate policies.

In the real world the implementation of such policies faces obstacles. It is not easy to tax CO_2 emissions from households and industry at the same rate, since energy-intensive industries face international competition and households do not. As long as the rates with which countries implement climate policies are different, this will likely remain a problem. Lowering the tax on households would increase cost-efficiency in meeting a domestic CO_2 target—but the target would be lower! These are the trade-offs that have to be considered.

When assessing not fully cost-efficient approaches, the following key criterion should be used: the approach should pave the way for either more advanced technologies or more cost-effective policies. For instance, we may accept grandfathering of permits in the interest of getting a policy in place, but it is important that policy makers then clearly state that the long-term objective is auctioning (e.g., by auctioning a minor share of the permits right from the start). Further, subsidies to individual technologies may be acceptable as a way of bringing down costs but only if these technologies are expected to be able to compete without the subsidies one day. The challenge is to select policies that form a platform upon which it becomes easier to implement new, more stringent, and more cost-efficient policies.

Notes

I would like to express my gratitude to Paulina Essunger, Kristin Kuntz-Duriseti, Kerstin Åstrand, Paul Baer, Dean Abrahamson, Jonas Nässen, Frances Sprei, Julia Hansson, and Martin Persson for comments on the paper and to Björn Sandén, Tomas Kåberger, Thomas Sterner, and Per Kågeson for discussions about climate policy over the years that form the basis for this paper. Thanks to Formas and Swedish Energy Agency for financial support.

1. See www.globalcarbonproject.org/carbon trends/index.htm. See also Marland, G., Boden, T. A., and Andres, R. J., 2006. Global, Regional, and National Fossil Fuel CO_2 Emissions, database available at http://cdiac.ornl.gov/trends/emis/em_cont.htm.

2. Azar, C., and Rodhe, H., 1997. Targets for stabilisation of atmospheric CO_2. *Science* 276, 1818–19. Wigley, T. M. L., Richels, R., and Edmonds, J. A., 1996. Economic and environmental choices in the stabilization of atmospheric CO_2 concentrations. *Nature* 379: 240–43.

3. See, for example, Nakicenovic, N., and Riahi, K., 2003. Model runs with MESSAGE in the context of the further development of the Kyoto Protocol, German Advisory Council on Global Change, available at www.wbgu.de/wbgu_sn2003_ex03.pdf; Azar, C., Lindgren, K., and Andersson, B., 2003. Global energy scenarios meeting stringent CO_2 constraints: Cost effective fuel choices in the transportation sector. *Energy Policy* 31 (10), 961–76; Azar, C., Lindgren, K., Larson, E., and Möllersten, K., 2006. Carbon capture and storage from fossil fuels and biomass: Costs and potential role in stabilizing the atmosphere. *Climatic Change* 74, 47–79.

4. In textbook economics, the tax should equal the marginal cost of CO_2 emissions. In reality, it is both technically and ethically difficult to estimate this cost. For instance, what is the value of life, in particular in poor countries? What is an appropriate valuation of future damages? What is the risk of large-scale catastrophic events? For this reason, the tax may be set at a level that yields reductions that are in line with those believed to be required to minimize the risk or avoid dangerous climatic changes. Further, the implementation of a tax means that the consumer pays for the damages, but it does not necessarily follow that the damage is paid for (the Swedish carbon tax is recycled to consumers and not distributed according to the potential impact of climate change). Internationalizing external costs is not the same as equitably compensating the potential victims.

5. Cost-efficient incentives are ensured if the price on carbon is the same in all sectors, and for all countries. In real-world politics the price may be higher in one sector. Under those conditions, other policy incentives are not necessarily less cost-effective.

6. A subsidy to renewables has further disadvantages from a cost perspective: when the subsidized renewables replace other energy sources, the carbon intensity of the displaced energy sources are not considered. For instance, if natural gas is the electricity supply option on the margin, natural gas might very well be phased out instead of coal-based electricity (which under most circumstances would be more cost-efficient).

7. Sandéén, B. A., and Azar, C., 2005. Near-term technology policies for long-term climate targets: Economy-wide versus technology-specific approaches. *Energy Policy*, 33, 1557–76.

8. Haas, R., 2000. Promotion strategies for electricity from renewable energy sources in EU countries, Joint report by the cluster "green electricity." Available at www.tuuleenergia.ee/failid/reviewreport.pdf. Åstrand, K., 2006. Energy policy instrument: Perspectives on their choice, combination and evaluation. PhD dissertation, Lund University, Lund, Sweden.

9. The following back-of-the-envelope calculation may be illustrative: primary energy supply per capita in the OECD region is around 200 GJ/cap/year. Assume that 10 billion inhabitants of the world would enjoy that level on average by the year 2100. We would then end up at an energy supply of 2,000 EJ/year, which is five times higher than today. This level is in line with many global energy scenarios See, for instance, Nakicenovic, N., et al., 2003. *Special Report on Emissions Scenarios: A Special Report of Working Group III of the Intergovernmental Panel on Climate Change*. Cambridge University Press, New York.

Assuming that energy efficiency improves by an additional one percentage point per year over this period of time, by the end of the century we would be roughly 60 percent more efficient and the energy supply 800 EJ/year (which is in line with the levels shown in figure 18.1, 1 TW = 31 EJ/year). This is not a far-fetched objective; there are already cars being sold that are more than twice, or three times, as efficient as the current average; there are houses built in cold climates that do not need conventional heating systems; lighting can be made five times as efficient, and so forth.

10. Kammen, D. M., and Pacca, S., 2004. Assessing the cost of electricity. *Annual Review of Environment and Resources* 29, 301–44. Although these subsidies are large, and removing them is clearly worthwhile, it should be kept in mind that their removal would not likely change the picture fundamentally. Oil in the Middle East costs a few dollars per barrel, coal for electricity generation is still likely to be less costly than many renewables, and so on. Subsidies to liquefied petroleum gasoline might even be warranted from a CO_2 perspective if these facilitate the transition away from charcoal that is derived from non-renewable-biomass.

11. Sterner, T., 2007. Fuel taxes: An important instrument for climate policy. *Energy Policy* 35, 3194–3202.

12. Azar, C., 2005a. Emerging scarcities: Bioenergy-food competition in a carbon constrained world, in Simpson, D., Toman, M., and Ayres, R., eds., *Scarcity and Growth Revisited: Natural Resources and the Environment in the New Millennium*. Resources for the Future, Inc. John Hopkins University Press, pp. 98–120.

13. Abrahamson, D., and Swahn, J., 2000. The political atom. *Bulletin of the Atomic Scientists* 56 (4), 39–44.

14. For an overview of the pros and cons of different approaches to protecting the energy intensive industries, see Azar, C., 2005b. Post-Kyoto climate policy targets: Costs and competitiveness implications. *Climate Policy* 5, 309–28.

15. France proposed (in November 2006) that the EU introduce such a border adjusted carbon tax. The United States immediately criticized the proposal. See AFP, 2006. http://news.yahoo.com/s/afp/20061115/sc_afp/unclimateusfrancetax.

Chapter 19

Carbon Taxes, Trading, and Offsets

Danny Cullenward

Why Put a Price on Carbon?

This chapter will examine the structures of two policy responses to the problem of global warming: carbon taxes and carbon cap-and-trade. Both of these policies put a price on carbon, which is a necessary first step to controlling emissions.[1] Pricing carbon is particularly important in the energy sector, the dominant source of greenhouse gas emissions. By ignoring the damage done by greenhouse gases, society is in effect subsidizing fossil fuels. Low- or zero-carbon energy technologies are at a cost disadvantage with traditional fossil fuels as long as greenhouse gas emissions are free.

Setting a price on carbon helps level the playing field for alternative and low-carbon energy. But pricing carbon is a highly political exercise, since almost every person and company will be affected. The problem is vast: almost every facet of modern life, for both the poor and rich in this world, relies on relatively cheap supplies of carbon-intensive fossil fuels. So who will pay the costs of climate policy? On this question, the design of a climate policy can make or break its effectiveness and de-termine whether progress will be made or public trust squandered. As one might imagine, the devil is in the details. Knowing what has worked and what has not will be key in taking the first steps to reduce greenhouse gas emissions.

Cap-and-Trade versus Taxes

The most popular policy approach to date is a cap-and-trade system, modeled after a successful program used to control sulfur dioxide emissions in the United States. Under carbon cap-and-trade, also called carbon trading, a government sets a limit on the total allowed emissions. It then creates carbon credits—each representing the legal right to emit 1 ton of carbon dioxide per year—with the total number of credits equal to the emissions cap. Credits are either sold to companies at auction or given away freely.

The advantage of trading is based on the potential of the market to minimize costs of meeting a given emissions target, as each company decides how to comply with the policy at the lowest possible cost. Economic theory predicts that companies will optimize this process, buying and selling permits until the marginal cost of reducing emissions equals the market-clearing price for permits. Because credit supply is fixed, prices float according to how hard it is for the economy to meet the targets. Those companies that can reduce their emissions easily will profit by selling extra credits to those companies for which emissions reductions are not economically sound. And because carbon dioxide is not a local air pollutant, it does not matter whether emissions concentrate in particular geographic areas, so long as total emissions remain capped.[2] Hence, unlike with local pollutants such as mercury, a cap-and-trade system does not result in environmental justice issues from the pollution itself.

The primary alternative to a cap-and-trade system is a tax regime. In this system, the government sets a price on carbon and charges emitters throughout the economy. Then, most critically, it must decide what to do with the revenue. This is no small question. For example, at a price of $10 per ton of carbon dioxide, U.S. greenhouse gas emissions in 2007 were worth about $73 billion.[3] One could imagine that the taxes could be collected for energy research, development, and deployment; or spent to upgrade existing energy systems to reduce emissions. Others have proposed "recycling" the revenue to reduce payroll taxes.[4] With this approach, the overall impacts of the carbon tax are reduced, as the revenue is transferred to relieve economically inefficient income taxes.

Taxes and trading are the basic choices when designing a policy to put a price on carbon, and each comes with certain advantages and challenges. Academics have long framed the question of taxes versus cap-and-trade as a problem of managing uncertainty. In the economic literature, this is called the "prices versus quantities" trade-off.[5] Under a carbon-trading system, the level of emissions is fixed but prices are unknown and are determined by the market. In contrast, under a carbon-tax system, the price of carbon is set but the actual emissions reductions can't be predicted ahead of time.

The effect of this trade-off has immense importance in practice, because most of the technologies and sectors contributing greenhouse gases have long economic lives. Power plants, for example, can last thirty years or more. Oil and gas drilling operations occur over similar timescales. It is difficult to imagine a private company taking a risk on an expensive new technology—one whose economic value is contingent on a higher price of carbon—unless there is some kind of financial guarantee that it will be able to recoup its capital investment.

It is precisely for this reason that carbon taxes have an advantage over trading systems from an investment perspective. Price volatility in the largest cap-and-trade system in the world, the European Union's Emission Trading Scheme (EU ETS), illustrates the point.[6] At the beginning of 2006, EU ETS carbon prices approached $40 per ton of CO_2. But when market analysts realized that EU governments had over-allocated credits (swamping the market with extra permits to pollute), spot market prices crashed to under $1 per ton of CO_2.[7] The financial impacts are considerable: for a standard 500 MW coal power plant, the annual carbon liability would have fallen from $119 million to $3 million.[8] Perhaps with time and experience such events will become uncommon, but as carbon caps tighten over time, each adjustment will bring new challenges to the market. This possibility underlines an important point: the stability of carbon-trading systems is reliant on their institutional capacity to manage risk and information. Trading systems put their trust in the ability of markets to handle a complicated and uncertain future.

Tax systems face an enormous challenge, too. While they help solve the problem of investment certainty, one cannot guarantee any level of emissions reductions for a given tax. From a biophysical point of view, there is only so much carbon the atmosphere can absorb for a given climate change risk tolerance. But if we sustain a national carbon tax in the United States—arbitrarily, let's say $40 per ton of CO_2—there is no way to know with much certainty what national emissions will be in 2020. We can build models to predict behavior, and eventually there might be some empirical evidence about how carbon prices affect emissions levels, but the exact relationship is unknown at present. Hence, over time, a tax system would require adjustment—likely raising prices over time, just like a cap-and-trade system would have increasingly strict emissions limits over time.

It is also possible to hybridize the two policy frameworks in what is called a "safety valve" approach. To do so, government guarantees that the price of carbon will not exceed a certain threshold in its trading system. That price, or safety valve, is an otherwise arbitrary value reflecting the maximum price acceptable to the political community. So long as carbon prices stay below that level, the policy behaves like a trading system. However, when carbon prices approach the safety valve limit, the government offers to sell unlimited permits at that price, and the policy becomes a tax system. Safety valve policies have some appeal with private industry, as they offer the potential gains from trading alongside a government guarantee that the cost of compliance is capped. However, all the same caveats with

tax systems apply—it is impossible to know with precision what emissions will be for a given safety-valve price. And although the price ceiling allows investors to measure their maximum potential carbon risk, it does not offer a corresponding price floor to help low-carbon projects compete against established technologies.

In lieu of direct climate policy like taxes or trading, some politicians have suggested using greenhouse gas intensity targets. These goals are expressed in pollution per unit GDP, and are essentially a measure of how efficient the economy is in greenhouse gas terms. Unfortunately, such policies do not account for either total emissions or per capita emissions. Most proposals of this type are not serious options for reducing total emissions, because economic or population growth can outweigh the gains in efficiency that the policy mandates.[9]

At the international level, concerns about climate equity have led to discussions of per capita emissions convergence. A quick look at the disparity already in place helps make the case: Americans emit 20 tons of CO_2 per person per year, compared to 8.5 for the EU-27, 3.9 for the average Chinese, and 1.1 for the average Indian.[10] Some argue that the only fair solution is one that results in relative equality in per capita emissions, leading to ethical and political questions well beyond the scope of this chapter.[11] If such a solution were accepted at the international negotiating table, it would still require the participants to adopt mandatory emissions controls, such as carbon taxes or trading.

Permit Allocation and Carbon Offsets

For either a tax or trading system, the political challenge of changing the status quo will be difficult, as the price of carbon has a huge impact on many sectors of the economy. Vested interests will likely complicate matters, just as they have historically and continue to do today. These forces are often quite strong and can alter both the choice of policy framework and the details of its implementation.

Consider, for example, the problem of distributing carbon credits under a cap-and-trade regime. The government has two options: allocate permits for free or sell them at auction. By charging a price for pollution (or otherwise limiting emissions), government is creating a valuable new property right, and those to whom permits are awarded stand to gain significant financial worth. Economic studies suggest that if enough permits are given away for free to polluting industries, their value might outweigh the cost of complying with emissions limits.[12] These results indicate that policymakers could literally buy off opposition forces in order to create a cap-and-trade system.

The ability to award valuable permits allows politicians to grant favors to stakeholders and constituent groups, which in turn creates problems for policy efficacy and equity. The historical record of the EU trading system sheds some light on this problem. In the first compliance period of the EU ETS, which ran from 2005 through 2007, participating governments were required to freely allocate at least 95 percent of credits.[13] Some chose not to auction any at all. The choice of credit awardees was no accident, but instead reflected the protectionist interest of many countries' heavy industries. For example, the German utility RWE was granted ample credits for some of its coal power plants. The German government also helped finance the conversion of these plants to natural gas, which emits approximately half the carbon per unit of electricity. As a result, RWE had a surplus

of permits resulting in windfall profits, ultimately at consumers' expense.[14]

In addition to the problem of distributing valuable permits or tax revenues, the other major political challenge in implementing climate policy is dealing with offsets. Offsets are loosely defined as projects completed outside a climate policy system that produce emissions reductions. Put simply, offsets are the policy equivalent of outsourcing: in many cases, cheap options for greenhouse gas emissions reductions occur outside the tax or trading system. For example, companies and governments in the EU face high costs of compliance at home. As an alternative to reducing emissions in domestic sectors, some choose to buy offsets from the Kyoto Protocol's Clean Development Mechanism (CDM), which tenders credits from emissions reductions projects undertaken in the developing world. The challenge of determining a baseline—the estimation of business-as-usual, against which emission reductions are measured—is covered in detail in chapter 27 in this volume. These problems have led analysts (myself included) to criticize the use of offsets.[15]

Biological offsets present even more challenges for monitoring and enforcement. Projects to monetize carbon locked up in forests, ecologically friendly agricultural practices, and conservation programs are all potential sources of biological carbon credits. The same baseline concerns apply, as does the problem of reliably estimating carbon sequestration in highly heterogeneous natural and managed land systems.

Each sector has its own specific concerns, too. For example, halting deforestation is a noble goal, but difficult to guarantee with piecemeal protection of individual lands—deforestation can always move to a nearby region when one patch of land is protected. While many are optimistic about the future prospects for biological offsets, at present there is no way to reliably regulate these systems.[16] From a credibility point of view, biological offsets projects are still uncharted territory in climate policy.

The same can be said of domestic voluntary offsets markets. In the United States, for example, there is no mandatory climate policy, and yet there are companies willing to sell carbon credits to offset the footprints of concerned individuals and corporations. While many of the credit vendors have good intentions, the market is wholly unregulated and quality concerns dominate. Some analysts have accused participating firms of shuffling paper with few actual emissions reductions taking place.[17] Under either taxes or trading, offsets are a concern needing further attention from academics, government, and industry — at present there is no way to guarantee the reliability of these credits, and they will be the subject of debate for years to come.

Is a Carbon Price Enough?

Setting up a tax or trading system is only the first step in addressing global warming. Taxes and trading are designed to put a price on carbon, but prices alone are insufficient to encourage change. Even at the highest prices sustained in the European carbon market, most electric power companies would not have the financial incentive to switch away from coal.[18] In the transportation sector, things are even worse: consider that even at a hypothetical carbon price of $50 per ton CO_2, the additional price per gallon of gasoline is only 42 cents.[19] While not negligible, that cost is dwarfed by the volatility in gasoline prices during the last few years, which has not

led to a fundamental rethinking of transportation technologies, gasoline consumption, or urban planning.

Dealing with the big picture on climate necessitates a focus on technology, and especially those options that address emissions from coal and oil. In the long term, energy efficiency measures, renewable energy, and carbon sequestration technologies will need to out-compete today's cheap and dirty fossil options. However, because research benefits are hard for private companies to appropriate, the market tends to underinvest in technology relative to what would be best for society. Hence, government R&D support is a natural partner for climate change policy. Sadly, however, energy-related funding levels in the United States today are only half what they were in the early 1980s, and just a tiny fraction of the overall research budget.[20]

Conclusion

The human impact of greenhouse gases on the atmosphere is unsustainable, and because greenhouse gas emissions are unpriced, society has no financial incentive to change its behavior. Both commercial and developing technologies to help control carbon are at a disadvantage under the status quo. Two dominant policy options are available for governments to control their domestic emissions and set a price on carbon: carbon cap-and-trade and carbon taxes.

With carbon trading, the government mandates a total level of emissions reduction and allows the market to trade pollution permits. The resulting price is uncertain, and this volatility makes long-term investment planning difficult. Governments must also deal with the politically complicated question of how to allocate permits. Giving permits away for free subsidizes existing emitters of greenhouse gases and can lead to market manipulation. Auctioning permits helps alleviate these concerns, but is less appealing politically, as it imposes the full cost of greenhouse gas pollution on current emitters. Auctioning also results in a large sum of money being transferred to the government.

With carbon taxes, the price of emissions is set, but the resulting reduction in emissions is uncertain. Long-term investment planning is simplified with known costs. Like auctioning carbon permits under trading, taxes result in increased government revenues, which are subject to a political appropriations process. This money can be used for any number of useful programs: research and development, subsidies for renewable and low-carbon energy, reducing payroll taxes, preparation for climate adaptation, or general government funds. A tax system reduces the opportunities for industry manipulation of a given climate policy, but for the same reason meets with resistance from some established greenhouse gas emitters.

In either system, carbon offsets could play a role. However, this option should be exercised only after careful study. Offsets markets today, both international and domestic, are riddled with questionable methodology. Investments resulting from offsets have not targeted technologies and economic sectors that help manage emissions or climate change impacts in the long run. Reforming the way offsets programs are run will be key if they are to be part of climate policies.

Ultimately, establishing a carbon price is only the first step in managing global warming. Properly designed climate policy will incorporate a price signal (achieved by taxes or trading), coordinate research support, and as-

sist technology demonstration and deployment. But even a modest climate policy will involve many billions of dollars, so getting the details right is not just a problem for policy makers and economists. The fundamental question of who pays will depend on the allocation of credits or tax revenues, and has enormous political and equity dimensions. Finally, the efficacy of any climate program rests on the interaction of carbon price signals with energy markets and the broader world of technology policy—and so climate policy can never be considered in a vacuum.

Notes

1. One typically reads about taxing or capping "carbon," but this usually means setting a price on all greenhouse gases. Non-CO_2 gases are converted into CO_2-equivalents via the Global Warming Potentials (GWPs) published by the Intergovernmental Panel on Climate Change.

2. Interestingly, warming from greenhouse gases has important impacts on local air quality and is expected to increase illness and mortality from existing air pollution. See Jacobson, M. Z. (2008), On the causal link between carbon dioxide and air pollution mortality. *Geophysical Research Letters* 35: L03809, doi:10.1029 2007GL031101.

3. United States Energy Information Administration (US EIA) (2008), Emissions of Greenhouse Gases Report. December 3, 2008. Online at www.eia.doe.gov/oiaf/1605/ggrpt.

4. Goulder, L. H. (1995), Environmental taxation and the double dividend: A reader's guide. *International Tax and Public Finance* 2(2): 157–83.

5. Weitzman, M. L. (1974), Prices versus quantities. *Review of Economic Studies* 4: 477–91.

6. Kettner, C., A. Köppl, S. P. Schleicher, and G. Thenius (2007), Stringency and Distribution in the EU Emissions Trading Scheme: The 2005 Evidence. Fondazione Eni Enrico Mattei Working Paper No. 22. Online at: http://papers.ssrn.com/sol3/papers.cfm?abstract_id=968418.

7. While the spot market price crashed, the futures market remained relatively stable. Historical

data are available from the European Climate Exchange: www.europeanclimateexchange.com.

8. Assuming an 85 percent capacity factor and an emissions rate of 0.8 ton of CO_2 per MWh, a 500 MW coal plant produces 2,978,400 tons of CO_2 per year. At $1 per ton of CO_2, the carbon liability is $2,978,400. At $40 per ton of CO_2, the carbon liability is $119,136,000.

9. See, for example: Northrop, M., and D. Sassoon (2007), Cap and trade and more. *Environmental Finance*, June: 2–4.

10. World Resources Institute, Climate Analysis Indicators Tool, Version 6.0 (using 2004 data). Online at http://cait.wri.org. Please note that the reported emissions are for CO_2 only; data for non-CO_2 emissions are difficult to estimate with comparable accuracy.

11. See generally, Tóth, F. L., ed. (1999), *Fair Weather? Equity Concerns in Climate Change.* London, U.K.: Earthscan Publications Ltd.

12. Bovenberg, A. L., L. H. Goulder, and D. J. Gurne (2005), Efficiency costs of meeting industry-distributional constraints under environmetnal permits and taxes. *RAND Journal of Economics* 36(4): 950–70.

13. In the current phase of the ETS, Phase II, EU governments can auction up to 10 percent of total credits. See Hepburn, C., M. Grubb, K. Neuhoff, F. Matthes, and M. Tse (2006), Auction of EU ETS phase II allowances: How and why? *Climate Policy* 6: 137–60.

14. Kanter, J. (2008), EU carbon trading system brings windfall for some, with little benefit to climate. *The International Herald Tribune*, 9 December 2008. Online at www.iht.com/articles/2008/12/09/business/windfall.php.

15. Wara, M. (2007), Is the global carbon market working? *Nature* 445 (February 8, 2007): 595–96. Victor, D. G., and D. Cullenward (2007), Making carbon markets work. *Scientific American* 297(6): 44–51.

16. Daily, G., and K. Ellison (2002), *The New Economy of Nature.* Washington, D.C.: Island Press. Sedjo, R. (2006), Forest and Biological Carbon Sinks after Kyoto. Resources for the Future Background Paper. Online at http://www.rff.org.

17. Trexler Climate + Energy Solutions, Inc. (2007), A Consumer's Guide to Retail Carbon Offsets Providers. Clean Air-Cool Planet report. Online at www.cleanair-coolplanet.org/Consumers GuidetoCarbonOffsets.pdf.

18. Douglas, J. (2006), Generation technologies for a carbon-constrained world. *EPRI Journal*, Summer 2006.

19. This number accounts only for the CO_2 emissions from the fuel itself and does not include emissions from production, refining, or distribution. United States Energy Information Administration website. Fuel and Energy Source Codes and Emission Coefficients. Accessed January 8, 2009. Online at www.eia.doe.gov/oiaf/1605/coefficients.html.

20. Nemet, G. F., and D. M. Kammen (2007), U.S. energy research and development: Declining investment, increasing need, and the feasibility of expansion. *Energy Policy* 35: 746–55. American Association for the Advancement of Science (AAAS) (2007), From the Hill: Federal R&D funding stuck on hold. *Issues in Science and Technology* 23(2), Winter 2007.

Chapter 20

The Cost of Reducing CO$_2$ Emissions

Christian Azar

The cost estimates for reducing CO$_2$ emissions are central in the debate over what to do about climate change. Proponents of mitigation tend to state that there are plenty of low-cost options available, based on employing existing technology. Opponents claim that the cost of achieving deep reductions in carbon emissions during this century measure in the trillions, or even tens of trillions, of dollars, based on the opportunity cost of investment.

IPCC estimates that the cost of stabilizing the atmospheric concentration of CO$_2$ at 450 parts per million (ppm) is 2.5 to 18 trillion USD, depending on the model and the emission pathway.[1] For less ambitious targets, the costs fall; a stabilization target of 650 ppm is expected to cost around 1 trillion USD. The costs refer to the net present value of the abatement costs between 1990 and 2100, in 1990 USD.

Of course, estimating the cost of an almost complete transformation of the energy system requires several heroic assumptions. At the start of the twentieth century, you would almost certainly have been completely wrong about what the next 100 years would look like. Not only have technological advances been enormous, but the number of countries in the world has almost quadrupled as political systems have emerged and collapsed. I am not terribly convinced that we are much better at guessing the future now than we were then.

Still, cost estimates may give us some sense of the order of magnitude involved, and, perhaps more important, cost estimates are used in the public debate over what to do about climate change and in comparing policy alternatives. Therefore, it is important to understand what the numbers really mean.

Costs measured in trillions of dollars are often perceived to be prohibitively high, as if they threaten our welfare and our way of life. Most people do not even know how many zeroes there are in a trillion, and that in itself is a reasonable cause for concern. We worry that our income levels will become lower than at present, we will have to stop using cars, significantly reduce indoor temperature in cold climates, and so on.

In order to better understand the significance of these costs, it may be useful to view them in light of the expected overall global economic development. The upper line in figure 20.1 depicts global income under the

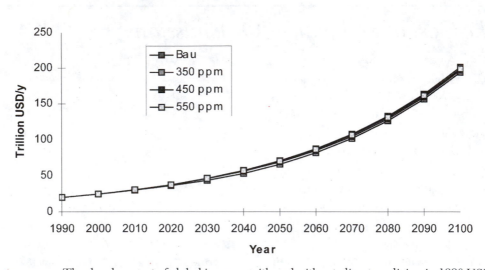

FIGURE 20.1. The development of global income, with and without climate policies, in 1990 USD. Climate change–related damages are not included in the graph. Source: Azar, C., and Schneider, S. H., (2002). Are the economic costs of stabilizing the atmosphere prohibitive? *Ecological Economics* 42: 73–80.

business-as-usual scenario, in which no policies to reduce CO_2 emissions are introduced. The growth rate in the business-as-usual scenario is assumed to be 2.1 percent per year.

In order to stabilize atmospheric concentrations, emissions have to be reduced from a baseline scenario in which emissions roughly increase by a factor of three during this century. To meet a 350-ppm scenario, for instance, emissions have to drop to essentially zero during the course of this century. A 450-ppm scenario would require a drop in emissions by roughly 80 percent compared to the baseline by the end of the century.[2] There are costs associated with those reductions, and those costs are deducted from the business-as-usual income level in the graph. The three lines below the income trajectory in the business-as-usual scenario show expected global income levels under 350-, 450-, and 550-ppm scenarios.[3]

Thus, the graph shows—loosely speaking—how much richer the world as a whole becomes with carbon abatement policies in place compared to how much richer we get in the baseline scenario.

The income levels in the graph are estimated without taking into account possible negative impacts from climate change. The reason is that monetarization of climate damages is a difficult, complex, and contentious process full of value-laden assumptions.[4] For instance, when estimating the cost of climate change, one needs to consider the value of losses of lives in different parts of the world. One way to do that is to look at how much governments are willing to spend to save a statistical life (e.g., by avoiding traffic accidents). One controversial issue arises from the fact that governments in poor countries tend to spend less to save a statistical live than governments in, say, Sweden. In many economic studies of the cost of climate change, this and similar observations have been evoked to justify the use of significantly lower values for people who die from climate change in poor countries than in rich countries. For this and similar reasons related to the evaluation of ecosystems and risks, it is very difficult to find an objectively correct estimate of the cost of climate change.

The difference in GDP levels between the business-as-usual scenario and the 350-ppm scenario represents a net present value cost of 18 trillion USD (the difference for the 450-ppm scenario is $5 trillion, and for the 550-ppm scenario, $2 trillion). Hence, the graph shows roughly the same costs as those reported in the IPCC estimates, but when we compare the cost figures to the expected future income, a completely different picture emerges.

Although trillion-dollar costs are large in absolute terms, they are minor compared to the expected perhaps tenfold increase in global income over the next 100 years. Instead of getting ten times richer by the year 2100, we become ten times richer in 2102 — a delay of two years.

This result can be explained as follows: top-down models typically suggest that the cost of a 50 percent reduction of global CO$_2$ emissions compared to the baseline would be some 1 to 4 percent of global GDP by 2050. A 75 to 90 percent reduction by 2100 would cost some 3 to 6 percent.[5] But since these studies also assume that global income grows by 2 to 3 percent per year, this abatement cost would be compensated for after a few years of income growth, with the abatement scenario catching up to what the GDP was in the business-as-usual scenario two or three years earlier. Thus, the cost of "climate insurance" amounts to a delay of a couple of years in achieving a very impressive growth in per capita income levels. If the cost by the year 2100 is as high as 6 percent of global GDP and income growth is 2 percent per year, then the delay time is three years, whereas the delay time is only one year if income grows by 3 percent per year and the abatement cost is 3 percent of GDP.

Interestingly, similar observations can be made regarding the cost of meeting near- and mid-term climate targets for different countries, regions, or the world as a whole. If the cost of meeting Kyoto would be 100 billion USD per year for the United States, or roughly 1 percent of GDP, then this amounts to a four-month delay in achieving a given income level (assuming a growth rate of 3 percent per year). Essentially all energy economy models suggest that we will become substantially richer even if stringent climate policies are adopted; the cost amounts to not becoming *as* rich *as* fast as was hoped for.

Further, a difference in global income of 4 percent by the end of the century amounts (assuming that the difference in income grows smoothly) to an average difference in growth rates during the century of roughly 0.04 percent per year. This means that stringent climate policies would lead to growth rates that are around a few hundredths of a percentage point lower than in the business-as-usual scenario. It may even in retrospect be difficult to disentangle the cost to the economy!

Finally, it would be wrong to conclude that the minor difference in growth rates between a stringent climate policy and business-as-usual implies that the low-carbon future will materialize by itself. On the contrary, major efforts are required to achieve the required almost complete transformation of the energy system. In chapter 18 in this volume, I discuss policies for effecting this transformation.

A Back-of-the-Envelope Calculation to Get a Sense of the Numbers

Very often results from complex models are difficult to understand, and that makes them somewhat less useful. However, understanding why the annual costs end up at a few percent of the GDP can be rather straightforward. Here, I will offer a back-of-the-envelope calculation, applicable to most industrial countries. Assume a GDP per capita of 25,000USD per year and carbon emissions of 2.5 tons per year. This is more like Europe than the United States.

Assume further that the cost of reducing CO_2 emissions is equal to 200 USD/ton of carbon. It would raise the cost of producing electricity from a coal-fired power plant by around 5 USc/kWh (more than double the current total production cost). This would make profitable bioelectricity, wind, nuclear, if one so wishes, and a huge range of energy-efficiency improvements. Further, carbon capture and storage would—most likely—be profitable under these circumstances.[6]

Under these assumptions, the maximum expenditure for completely phasing out most CO_2 emissions is equal to 500 USD per capita per year, which corresponds to 2 percent of the income. This simple calculation enables us to reproduce in rough terms the cost estimates given previously (a few percent of GDP).

One More Look at the Numbers

Although the costs reported above are typical for top-down models, there are nevertheless reasons why they may turn out to be both under- and overestimates of the real future costs.

Reasons why they may turn out to be overestimates include:

- Bottom-up studies, that is, studies that take their point of departure in detailed assessments of the existing energy system including or even with a focus on end-use technologies, often find large inefficiencies in the energy system. They point to the fact that there are refrigerators, cars, bulbs, houses, and industrial processes that are much more efficient than those used today, with comparable costs.[7] They also offer plausible reasons for why such technologies are not being used to the extent that seems economical.[8]

- The reaction of many economists to reports about upcoming scarcities is to claim that the economy is resilient, that human ingenuity will find new ways of managing, and so on. I would be surprised if those arguments do not hold true also when it comes to dealing with restricting the amount of carbon dioxide that we emit to the atmosphere—but for such ingenuity to be triggered, carbon policies are required. With these incentives, new advanced technologies and system solutions (logistics, city planning, and so on) might become better and less costly than what we can imagine today.

- Our values may change too, in a way not reflected by current models. In the energy modelling literature, we discuss how to endogenize technological change—that is, develop models so that the performance of a given technology depends not only on what is being prescribed for it but also on how prevalent that technology is assumed to be.[9] There is also a need, I think, to endogenize *values*. In economic models of climate change, a carbon tax is needed to raise the price of fuels so that people invest in a different car that consumes less fuel. Currently, I think that that is a correct description of reality. However, if people's values change in response to concerns about climate change, they might start to buy smaller and more efficient cars just because they want to contribute to "saving the climate." If so, they would save money and CO$_2$ emissions by buying a smaller car. In current models, without endogenized value changes, such a decision would be associated with a cost because this is not how consumers behave today.[10] But for our assumed future consumer, it would not amount to any sacrifice. Parallels can be drawn to vegetarians who choose not to eat meat for ethical reasons, and this decision is for most vegetarians not a cost but a benefit. It is, in my view, more likely than not that the more important the climate problem is perceived to be, the more willing people will be to save energy in their daily activities.

On the other hand, a number of factors suggest that costs may be underestimated:

- Top-down models often assume that policy instruments are implemented in a cost-effective way. In the real world, this is unlikely. For instance, politically it is very difficult to employ a single carbon price throughout the world. This is at least what we see today. This means that the cost of achieving a certain global reduction target becomes higher. Further, it also implies that govern-

ments that do act (e.g., in the EU), will likely implement tougher policies on households and transportation than on industries; this too is associated with higher costs.

- Often top-down models are criticized for being pessimistic about future technological development or for being very scarce in technological detail. Some models may have only a few technologies available for reducing emissions. On the other hand, in some cases rather optimistic assumptions are made, for instance, that carbon capture and storage, hydrogen storage, nuclear, and so on will become available at reasonable costs and accepted by the general public. In reality, further development of many technologies is needed, and we cannot know with certainty that some of these technologies will really be commercialized.

- Top-down models are often poor at capturing nonequilibrium processes which may be costly (e.g., rapid structural changes in the economy). In the real world, such changes may be associated with long-term unemployment, premature phase-out of capital in the energy sector and elsewhere, and so on. In many models, a farmer may shift into a computer modeler, or a steel worker into a flight attendant, overnight. This may not be a critical factor for climate targets close to 550 ppm, but for targets approaching 350 ppm, CO$_2$ emissions would have to be reduced rather rapidly, and that may lead to higher transition costs than those discussed here.

Finally, a remark on the way top-down models estimate future GDP levels: these are not only determined by the amount of labor and capital available in the model, but also by increases in productivity—that is, the amount of output produced per unit of input (capital, labor, energy). In the models, this increase in productivity is estimated based on historic data and does not depend on other variables in the models such as energy prices or population (for that reason it is referred to as an exogenous productivity parameter). Clearly, there is a lot of uncertainty about what value to use for the future, and the uncertainty about future productivity levels plays a key role in explaining the uncertainty in long-term estimates of future GDP levels. More important, the link between this exogenous productivity growth—or residual, in Solow's analysis—and energy prices and availability is poorly understood. The models assume that this exogenous growth in productivity will remain the same, regardless of energy prices; in reality, this is a matter of debate.

It is also of relevance to consider the real reasons for this exogenous productivity increase, one of which is of course technological development spurred by R&D. In a general equilibrium model, in which all resources are allocated optimally (i.e., production inputs are put to their best usage), higher R&D expenditures on renewables, nuclear, carbon capture and storage, and energy efficiency, driven by climate policies, would probably lead to less R&D in other sectors and cause a decline in productivity growth in nonenergy sectors. But in the real world, where R&D is most likely underfunded, climate policies and associated attempts to spur technology change in the energy sector might lead to an increase in overall R&D spending, and this might boost productivity growth. The way the exogenous productivity increase will be affected by climate policies may have a large impact on future GDP and

the estimated cost of climate abatement, but unfortunately we don't know to what extent it will be affected.

What Really Matters

Policy makers and industry leaders who, for various reasons, do not want to take action on the problem of climate change may use the "trillions of dollars" estimates to legitimize their position. However, arguments based on the total cost of reducing CO_2 emissions are, in my view, a red herring.

The majority of those who resist climate policies are not doing so for the sake of the total social costs but for reasons related to the costs that may affect their own industries. In a world where carbon policies may increase the costs for a carbon-intensive industry by tens of percent of revenues, opposition to unilateral climate policies is to be expected. For that reason, studies on the cost of CO_2 mitigation should perhaps focus more on understanding the fate of certain industries, how they may be affected by climate policies, and what may be done to protect their interests during periods when different countries act with different enthusiasm.[11]

Similarly, national costs may be substantial depending on how emission quotas are allocated across countries. A per capita distribution, which I think is the most ethical approach, may lead to large transfers of sums across national borders—sums that are larger than the total official aid transfers.

Conclusion

The main aim of this chapter is to show that the way costs are presented largely determines how they are perceived. In the graph, costs are presented in a way that makes them seem small, almost negligible. I am not trying to argue that they are. Absolute costs—as estimated by top-down models—are substantial.

Instead, the point is to reject the rather widespread perception that climate policies are not compatible with continued economic development. If policy makers and the general public understand that the cost amounts to a delay of a few years in becoming ten times richer by the year 2100—or a difference in growth rate of, on average, less than 0.05 percent per year—their willingness to accept climate policies would probably be greater.

Notes

The first part of this chapter is built on joint work with Steve Schneider published in *Ecological Economics* (2002). I wish to thank Paulina Essunger and Fredrik Hedenus for helpful comments on the manuscript and the Swedish Energy Agency for financial support.

1. Intergovernmental Panel on Climate Change (IPCC), 2001. *Climate Change 2001: Mitigation Contribution of Working Group III to the Third Assessment Report of the Intergovernmental Panel on Climate Change.* Metz, B., Davidson, O., Swart, R., and Pan, J. Cambridge: Cambridge University Press. See especially chapter 8.

2. Wigley, T. M. L., Richels, R., and Edmonds, J. A., (1996). Economic and environmental choices in the stabilization of atmospheric CO$_2$ concentrations. *Nature* 379 (6562): 240–43.

3. For details, see Azar, C., and Schneider, S. H., (2002). Are the economic costs of stabilizing the atmosphere prohibitive? *Ecological Economics* 42: 73–80.

4. Spash, C. L., (1994). Double CO$_2$ and beyond: Benefits, costs and compensation. *Ecological Economics* 10: 27–36. Schneider, S. H., (1997). Integrated assessment modeling of global climate change: Transparent rational tool for policy mak-

ing or opaque screen hiding value-laden assumptions. *Environmental Modeling and Assessment* 2: 229–48. Azar, C., (1998). Are optimal emissions really optimal? Four critical issues for economists in the greenhouse. *Environmental and Resource Economics* 11: 301–15.

5. IPCC, 2001 op. cit., p. 548.

6. Intergovernmental Panel on Climate Change (IPCC), 2005. Special report on carbon capture and storage, Summary for policy makers. Cambridge: Cambridge University Press, p. 15. Available at www.ipcc.ch. In reality there are lots of ways of reducing emissions that cost substantially less than 200 USD/ton of carbon, and for that reason it may be an overestimate. On the other hand, for very deep reductions we may have to move into the area of hydrogen production and fuel cells, required for cars if carbon capture should be applicable, and we might confront higher costs.

7. For interesting empirical evidence and theoretical analysis, see Ayres, R. U., (1994). On economic disequilibrium and free lunch. *Environmental and Resource Economics* 4: 435–54; Levine, M. D., Koomey, J. G., McMahon, J. E., Sanstad, A. H., and Hirst, E., (1995). Energy efficiency policy and market failures. *Annual Review of Energy and the Environment* 20: 535–55.

8. For an early but interesting analysis of the difference between top-down and bottom-up models, see Wilson, D., and Swisher, J., (1993). Exploring the gap: Top-down versus bottom-up analyses of the cost of mitigating global warming. *Energy Policy* 21: 249–63.

9. Grubb, M., et al., eds., (2006). Special issue on endogenous technological change and the economics of atmospheric stabilisation. *Energy Journal* 27.

10. A carbon tax would have been needed to induce them to behave in this way, and the new behavior would be associated with a loss in the consumer's utility.

11. Reinaud, J., (2004). Industrial competitiveness under the European Union Emissions Trading Scheme. IEA Information Paper. International Energy Agency, Paris. Azar. C., (2005). Post-Kyoto climate policy targets: Costs and competitiveness implications. *Climate Policy* 5: 309–28.

International Considerations

Chapter 21

International Treaties

M. J. Mace

Introduction

A number of legal frameworks are now in place at the international level to address the challenge of monitoring and reducing greenhouse gas (GHG) emissions, including the UN Framework Convention on Climate Change (UNFCCC), the Kyoto Protocol to the UNFCCC (Protocol), and the European Union's Emissions Trading Scheme (EU-ETS). This chapter highlights key provisions of these frameworks, explores key challenges for their evolution, and then describes the processes now in place at the international level for the further elaboration of these frameworks to address emissions in the post-2012 period.

Key International Frameworks

The UN Framework Convention on Climate Change

The 1992 United Nations Framework Convention on Climate Change, with more than 190 country parties, is the overarching international law framework for intergovernmental efforts to address climate change. This agreement between countries aims to stabilize GHG concentrations in the atmosphere at a level that will prevent dangerous man-made interference with the climate system, and aims to achieve this goal "within a time frame sufficient to allow ecosystems to adapt naturally to climate change, to ensure that food production is not threatened and to enable economic development to proceed in a sustainable manner." What constitutes "dangerous" is left undefined, and many developing countries particularly vulnerable to the impacts of climate change, including small island states, rightly assert that this level has already been exceeded for them.

The Convention contains a series of key principles to guide its implementation. Once central principle is found in Article 3.1, which provides that:

> Parties should protect the climate system for the benefit of present and future generations of humankind, on the basis of equity and in accordance with their common but differentiated responsibilities and respective capabilities. Accordingly, the developed country Parties should

take the lead in combating climate change and the adverse effects thereof.

This implicitly recognizes that developed countries have made the greatest historical contribution to GHG concentrations in the atmosphere and are also most capable of responding to climate change and its adverse effects. Importantly, the precautionary principle found in Article 3.3 provides that parties "should take measures to anticipate, prevent, or minimize the causes of climate change and mitigate its adverse effects. Where there are threats of serious or irreversible damage, lack of full scientific certainty should not be used as an excuse for postponing action."

Reflecting equitable principles, the Convention divides countries into two broad groups: (1) developed countries (Annex I Parties) and (2) developing countries (Non-Annex I Parties). The group of Annex I Parties is further broken down into a more advanced subset: Annex II Parties and countries with Economies in Transition (EITs).

Under the Convention, all parties (whether developed or developing) agree to share information on their GHG emissions, develop national and regional programs containing mitigation and adaptation measures, cooperate in preparing for adaptation to the impacts of climate change, cooperate in scientific research and systematic observation of the climate system, cooperate in the development and transfer of technology, and cooperate in promoting education, training, and public awareness on climate change.

Annex I Parties have a series of further mitigation commitments. They agree to adopt national policies and measures to limit GHG emissions that will clearly demonstrate that developed countries are taking the lead in modifying longer-term trends in emissions.[1]

The narrower group of Annex II Parties bear certain additional financial commitments. They agree to provide financial support to developing countries to assist them in reporting on their national GHG emissions and on Convention implementation, and financial resources for GHG mitigation measures, adaptation measures, and the transfer of environmentally sound technologies.[2] They further agree to provide assistance to developing countries particularly vulnerable to the adverse effects of climate change in meeting costs of adaptation.[3]

The Convention creates a set of institutions to facilitate implementation and monitor the Convention's implementation. These include a Subsidiary Body on Scientific and Technological Advice (SBSTA), a Subsidiary Body on Implementation (SBI), and a financial mechanism that has been entrusted to the Global Environment Facility (GEF). The financial mechanism is intended to provide new and additional financial resources for Convention implementation on a grant or concessional basis, including for the transfer of technology needed to assist developing countries achieve emission reductions.

The Kyoto Protocol

The 1997 Kyoto Protocol, which came into force in 2005, significantly strengthens the Convention by committing Annex I Parties to the UNFCCC to individual, legally binding targets to limit or reduce their greenhouse gas emissions relative to 1990 emission levels in a first commitment period, which runs from 2008 to 2012. These targets are set out in Annex B to the Protocol and function much like countrywide emissions budgets for the five-year first commitment period. Each country has an assigned amount of emissions allowable for the period based on its target.[4]

In the aggregate, first commitment period targets were expected to reduce GHG emissions by at least 5 percent below 1990 levels for Annex I Parties as a whole.[5] Targets cover emissions of the six main GHGs: carbon dioxide (CO_2); methane (CH_4), nitrous oxide (N_2O); hydrofluorocarbons (HFCs); perfluorocarbons (PFCs); and sulphur hexafluoride (SF_6). For the purpose of comparing emission reductions, each is measured in terms of carbon dioxide equivalents, using the estimated global warming potential of each gas.

The Protocol does not prescribe how Annex B Parties should meet their binding targets. Instead, each country has complete discretion on how to meet its target. Countries are permitted to "bubble" their efforts, fulfilling their commitments jointly under Articles 3 and 4, as the European Community has elected to do. In addition, countries may make use of the most unique feature of the Kyoto Protocol — its creation of three "flexible mechanisms." The flexible mechanisms allow Annex B Parties to lower the overall cost of meeting their targets in three ways: (1) by engaging in international emissions trading with other countries with Kyoto targets; (2) by undertaking emission reduction projects in other Annex B Parties (joint implementation); or (3) by undertaking emission reduction projects in developing country parties through the "Clean Development Mechanism."[6]

Through international emissions trading, Annex B Parties may transfer or acquire portions of their assigned amounts. They may also trade credits generated through joint implementation (JI) and Clean Development Mechanism (CDM) projects, denominated in metric tons of CO_2-equivalent, which may be used by Annex B Parties toward their targets. JI and CDM projects must create reductions in emissions that are additional to what would otherwise have occurred in the absence of these projects. To ensure that developed countries take the lead in reducing emissions, the Protocol requires that Annex B countries' use of the flexible mechanisms only supplement domestic actions.[7] The flexible mechanisms give countries flexibility in how they achieve compliance. They also give countries an incentive to achieve reductions below their Kyoto targets, because any excess reductions have a market value.

After the Kyoto Protocol was agreed, an extensive set of rules for the flexibility mechanisms were negotiated and adopted as the Marrakesh Accords at the Seventh Conference of the Parties to the UNFCCC in 2001 and formally adopted by the Kyoto Protocol parties in 2005, after the Protocol entered into force. As part of the same package of decisions, the parties also adopted procedures and mechanisms on compliance, containing penalties for parties that fail to meet their Kyoto targets.[8] If at the end of the first commitment period a country's emissions have exceeded its assigned amount, that country will be found in noncompliance. As a penalty, it will have deducted from its assigned amount for the second commitment period a number of metric tons equal to 1.3 times its excess emissions in the first period. This multiplier is intended to deter noncompliance and delayed compliance. The party will also be required to develop a compliance action plan, and its eligibility to make transfers of tradable units will be suspended until certain requirements are met. These penalty provisions may become significant for negotiations on second commitment targets, as some Annex B Parties are on track to miss their Kyoto targets by a substantial margin.

The EU Regional Framework

In 2003, the EU established an internal cap-and-trade system for CO_2—the EU Emissions Trading Scheme (EU-ETS)—as a tool to assist EU member states and the European Community as a whole in achieving Kyoto commitments.[9] When the EU-ETS was launched in 2005, it was the first international trading system for CO_2 emissions in the world. It covered more than 11,500 energy-intensive installations across the fifteen EU member states, representing 45 percent of Europe's CO_2 emissions from combustion plants, oil refineries, coke ovens, and iron and steel plants, as well as factories making cement, glass, lime, brick, ceramics, pulp, and paper.[10] The EU-ETS had a first three-year phase (2005 to 2007). It is now in a second five-year phase (2008 to 2012), which corresponds to the Kyoto Protocol's first commitment period. An eight-year third phase will run from 2013 to 2020.

From January 2005, all regulated installations have been required to hold a greenhouse gas permit. A fixed number of greenhouse gas allowances are allocated to these installations, denominated in metric tons of carbon dioxide equivalent, entitling them to emit a corresponding quantity of GHGs. Regulated installations must monitor and report CO_2 emissions; at the end of each year, they must surrender sufficient allowances to cover their emissions for that year.[11]

Allowances are freely transferable. Operators that succeed in reducing emissions below their allocations may sell their excess allowances into the market. Operators with excessive emissions must either acquire additional allowances from other regulated installations or from the market. They may also use a limited number of emission reduction credits from Kyoto JI and CDM projects to cover their emissions. Operators holding an insufficient number of allowances in the first phase of the trading programme were required to pay a €40 penalty for each tonne of excess emissions (2005 to 2007); this penalty increased to €100 per tonne in the second phase (2008 to 2012).[12]

Cap-and-trade systems aim to reduce emissions by creating a scarcity in allowances that drives emission reductions. Unfortunately, the first phase of the EU-ETS saw an excessive allocation of allowances in some member states and some sectors, due to reliance on projections and a lack of verified emissions data. The number of allowances issued actually exceeded 2005 emissions by a substantial margin. For the second trading period, the European Commission has been far more strict, and allocation decisions should secure a 6 to 7 percent reduction compared to 2005 verified emission levels.[13] In the third trading period, the Commission will centrally determine the overall number of allowances to be allocated across the now 27 EU member states.[14]

For EU member states to achieve their Kyoto commitments, substantial emission reductions will be needed from sectors not yet covered by the EU-ETS. In January 2008, the European Commission proposed amendments to the EU-ETS that would widen the trading program's coverage to include CO_2 emissions from petrochemicals, ammonia, and aluminium, as well as certain N_2O and PFC emissions. A proposal was also made to bring the aviation sector within the EU-ETS.[15] Both of these proposals have now been adopted, together with a legislative framework for sharing effort across the EU in reducing emissions in sectors not covered by the EU-ETS.[16]

How successful have the EU's measures been? At the end of 2006, aggregate GHG emissions of the EU-15 were 2.2 percent below their 1990 level, with a drop of 0.8 percent between 2005 and 2006 alone; EU-27 emissions were 7.7 percent below 1990 levels, with a drop of 0.3 percent between 2005 and 2006.[17] However, in 2007, projections of the impacts of existing policies and measures for the EU-15 showed that GHG emissions might only be 0.6 percent below 1990 levels by 2010, which would miss the collective reduction target of 8 percent agreed for 2008 to 2012 for these countries by a significant margin.[18] In 2007 the EU made a firm unilateral commitment to achieve a 20 percent reduction of GHG emissions by 2020 compared to 1990 levels.[19] In the international negotiating process, the EU has repeated an offer to increase this effort to 30 percent provided that other developed countries commit themselves to comparable emission reductions and economically more advanced developing countries contribute adequately according to their responsibilities and respective capabilities.[20] The accession of new EU member states with troubled economies is likely to assist the European Community greatly in achieving its Kyoto target.

Post-2012 Framework Challenges

Neither the UNFCCC nor the Kyoto Protocol sets out a specific long-term GHG reduction target or a timeframe for meeting that target through a sequence of shorter-term milestones. This raises some fairly daunting challenges for the negotiation of future commitments under these instruments.

At what concentration level should GHGs be stabilized in the atmosphere, and over what timeframe? Different stabilization concentrations (e.g., 350 parts per million by volume (ppmv) or 450 ppmv) and different timeframes for achieving these concentrations will have vastly different impacts on the climate system and consequently on vulnerable populations and ecosystems. Moreover, the opportunity to stabilize concentrations at certain levels will be lost if sufficient emission reductions cannot be secured rapidly enough. Atmospheric concentrations of CO_2 had already risen from 280 ppmv before the industrial revolution to approximately 385 ppmv by the end of 2008.[21] When all GHGs are included, concentrations are already close to 430 ppmv and increase yearly by about 2 ppmv.[22]

How should the principle of "common but differentiated responsibilities and respective capabilities" be applied to developed and developing countries? Kyoto targets now apply only to developed countries. Should some or all developing countries be asked to take on commitments, in view of the rapidly increasing emissions from this group of countries? Which countries, when, and what kind of commitments? What kinds of economic incentives and opportunities might be needed to expand participation?

Should the Protocol's second commitment period be longer than the first five-year period? A longer period may provide regulatory certainty to industry and guide long-term investment decisions; however, it might also limit flexibility if it becomes clear that still stricter measures are needed to stabilize GHG concentrations.

What types of commitments could be taken for a second commitment period? If commitments other than fixed Kyoto-like targets are to be permitted or encouraged for some countries (e.g., carbon intensity targets, sectoral targets, energy efficiency targets, renewable energy targets, policies, and measures), how can countries' different efforts from different kinds of commitments be compared? How can overall progress be measured?

Can sufficient technology transfer occur through the flexible mechanisms or other market-based mechanisms? Or, should a supplemental technology agreement be negotiated that builds upon the Convention and the Kyoto Protocol? What form of agreement might drive the development and deployment of clean technologies and energy-efficient goods?

How can equitable burden sharing for adaptation be achieved? The Convention requires developed countries to assist particularly vulnerable countries in meeting the costs of adaptation, but provides no detail on how this is to be done. An automatic revenue stream for adaptation has been created under the Kyoto Protocol, where a 2 percent share of the proceeds of CDM project activities is contributed to an Adaptation Fund. Perversely, in this way the Adaptation Fund is linked to emission *reductions* undertaken in developing countries rather than to GHG emissions in developed countries. This leaves developing countries in some ways paying for their own adaptation needs, though they have contributed little to the global emissions that are now impacting them. Wouldn't it be more appropriate to link adaptation funding directly to emissions, through a share of Annex B parties' assigned amount units or through a direct levy on GHG emissions? Moreover, the United States—a major industrialized country—remains outside the Protocol and does not invest in CDM projects. How can a secure and predictable revenue stream for adaptation be generated that draws upon the resources of Annex I parties equitably? How can the adaptation needs of vulnerable countries be satisfactorily addressed?

Should anything be done to address the impacts of mitigation efforts on developing countries heavily dependent on fossil fuel production or consumption? The Convention and the Kyoto Protocol require parties to consider the impacts of their GHG reduction measures on developing country economies that are highly dependent on fossil fuel production or consumption (e.g., lost revenue for producing countries or higher fuel prices for consuming countries). What action, if any, is needed to address adverse impacts of these response measures in a time of increasing demand? In the context of increasing oil prices? In the context of decreasing oil prices? What are the roles of OPEC, large petroleum producing countries, and major oil corporations?

What should be the role of the flexible mechanisms in a second commitment period? The Kyoto Protocol does not resolve the scope of activities that can be included in the CDM in the second commitment period. The possible inclusion of certain land-use change and forestry activities, including measures to help avoid deforestation, and the proposed inclusion of carbon capture and geological storage activities are both controversial. Can the flexible mechanisms be used to create additional opportunities for cost-effective emission reductions and support sustainable development, without jeopardizing the environmental integrity of the Kyoto Protocol? How can it be ensured that tons of emissions reduced or avoided through CDM projects actually represent additional reductions, or tons kept permanently out of the atmosphere?

Architectural Options under Consideration

Many approaches have been suggested to secure deeper emission reductions by more countries in the international climate regime after 2012.[23] Many of these would build upon or complement existing Kyoto commitments and offer opportunities to engage non-Kyoto Parties and developing countries in GHG reduction efforts. Examples include:

- Absolute targets—Kyoto-like numerical targets that reflect emission limitations or emission reductions compared to emissions in a country's base-year (e.g., a limitation of X percent over 1990 levels, or a reduction of X percent below 1990 levels). Targets for most developing countries could represent an increase during the base year chosen, to allow for ongoing development. Absolute targets have the advantage of leading to measurable overall emission reductions.
- Carbon intensity targets—limitations or reductions of emissions per unit of output, relative to GDP or another indicator, with targets applied to sectors or to economies as a whole. A decrease in carbon intensity does not, however, necessarily indicate a reduction in overall emissions.
- Sectoral targets—measures to be undertaken in specific sectors in an economy (e.g., energy generation, cement, steel, or transport), with the type of target differing with the characteristics of the sector.[24] Sectoral targets recognize that the bulk of GHG emissions come from a limited number of industrial sectors and may assist in addressing competitiveness concerns across different countries.
- Renewable energy targets—a targeted level of generation or use of renewable energy, or increase in the generation or use of renewable energy. Many countries have already adopted renewable energy targets. For example, the EU's

Renewables Directive, adopted in 2001, aimed to achieve a 22 percent share of electricity from renewable energies by 2010. In 2004, China adopted a target for a 10 percent share of its energy to be generated from renewables by 2020.[25]

- Energy efficiency targets—targets for energy-saving that require improved energy efficiency in industry, housing construction, or the design of energy-using products.[26]
- A global emissions trading scheme—linking the EU-ETS with compatible mandatory emissions trading schemes in other countries (e.g., those emerging in Australia and the United States).[27]
- Sustainable development policies and measures (SD-PAMs)—measures that make the development path of a country more sustainable by lowering GHG emissions as a side benefit, such as energy-efficient building materials.[28]
- Incentives for reducing emissions from deforestation—positive incentives to encourage the conservation of forests to protect carbon sinks. An estimated 20 percent of global CO_2 emissions come from land-use change emissions, mainly deforestation.[29]
- Technology agreements—explicit agreements to support energy-efficient technologies, renewable energy technologies, or carbon capture and storage, to complement or support commitments taken under the UNFCCC on technology development, diffusion, and transfer.

In reaching agreement on a comprehensive post-2012 framework, top-down, bottom-up, or a combination of approaches is possible.[30] The international community could agree on a top-down, overarching target (e.g., a percentage reduction for the global community to achieve, or a concentration of GHGs not to be exceeded) with responsibility then divided among countries through multilateral negotiations. This could ensure that all countries are moving together toward a single, measurable goal. Alternatively, countries might decide what types of commitments they are prepared to take through a bottom-up approach and then pledge to achieve these commitments. However, given the urgency of the action needed, it is increasingly clear that bottom-up approaches alone will not produce sufficient GHG reductions in the necessary timeframe. Hence a combination of top-down and bottom-up approaches will be needed to secure both measurable results and global participation.

To increase developing country participation, various multistage approaches to mitigation have been proposed. Most would differentiate among groups of developing countries based on objective criteria to enable different groups of developing countries to undertake different levels or types of participation in GHG reduction efforts based on their national circumstances.[31] Countries might be encouraged to take on greater commitments when they cross one or more thresholds, with incentives offered to assist countries in moving through stages and increasing their reduction efforts.[32] Criteria for differentiating groups of countries could include historic GHG emissions, capacity to reduce emissions, GDP per capita, emissions per capita, emissions per unit of GDP, human development index, emission growth rates, or some combination of these indicators. A number of proposals have been put forward to show ways in which a multistage regime could be envisaged.[33] (See chapter 24 in this volume on equity.)

Commitments on adaptation will also have to form a key pillar of the post-2012 regime. Some options that have been discussed to augment available funding for adaptation and address adaptation needs include: a levy directly on GHG emissions to implement the "polluter-pays" principle; an increase in the 2 percent levy now imposed on the share of proceeds from CDM projects for the Kyoto Protocol's Adaptation Fund; an extension of the Adaptation Fund levy across all three flexible mechanisms, including on assigned amount units (representing GHG emissions); a per-ton levy on all current global GHG emissions, with levels set differently for developed and developing countries; a levy on aviation emissions; a share of the revenue from auctioning allowances in an increasingly global carbon market; support for the preparation of comprehensive adaptation strategies by developing countries with clear channels for funding these strategies; use of climate risk assessments by donors and international finance institutions to ensure that all investment projects actually reduce vulnerability and increase resilience; and government supported, insurance-type approaches as a means of helping vulnerable populations contend with climate impacts.[34]

Negotiations on Commitments for the Post-2012 Period: Montreal, Nairobi, Bali, and Copenhagen

Negotiations over the scale and shape of intensified efforts to address the causes and impacts of climate change have proceeded along two parallel tracks—one under the Convention and one under the Kyoto Protocol.

The Convention makes provision for periodic reviews of the obligations of parties in light of the objective of the Convention, experience gained in implementation, and evolving scientific knowledge. See, for example, articles 4.2(a), 7.2, and 10.2(a). However, because no timeframe is given for these reviews, and because Annex I parties have failed to demonstrate that they are taking the lead in reducing emissions under the Convention and Protocol, it has been challenging to use these provisions to discuss the need for greater ambition and new efforts by all Convention parties.

In contrast, article 3.9 of the Kyoto Protocol firmly requires Kyoto parties to initiate consideration of second commitment period targets for Annex I Parties to the Protocol at least seven years before the first commitment period ends (in 2012). Hence in 2005, shortly after the Protocol entered into force, developed country parties with first commitment period targets for 2008 to 2012 faced the prospect of beginning negotiations on further binding emission reduction targets for themselves without the largest developed country emitter—the United States—at the table. At that time, both the United States and Australia had elected to remain outside the Protocol. Meanwhile, emissions from major-emitting developing country parties (including China, India, and Brazil) had risen to levels that could not be ignored if efforts by developed countries to address emissions were to have any likelihood of global impact. Faced with this predicament, Annex I Kyoto parties with targets were understandably wary of proceeding without a mechanism to engage all parties in broader discussions on future mitigation efforts.

After a very challenging negotiation in Montreal at COP 11 and COP/MOP 1 in December 2005, parties agreed on a two-track process going forward. Kyoto Protocol parties agreed by decision 1/CMP.1 to establish an "Ad Hoc Working Group on Further Commitments for Annex I Parties under the Kyoto Protocol" (AWG-KP) to consider targets for the period beyond 2012, which would aim to complete its work in time to ensure no gap between the first and second commitment periods.[35] At the same time, Convention parties agreed by decision 1/CP.11 to establish a parallel "Dialogue on long-term cooperative action to address climate change by enhancing implementation of the Convention" (Dialogue).[36] To manage concerns of non-Kyoto parties and developing country parties without binding targets, it was agreed that the Dialogue would proceed "without prejudice to any future negotiations, commitments, process, framework, or mandate under the Convention" as an "open and nonbinding exchange of views, information, and ideas" that would "not open any negotiations leading to new commitments."[37] Both processes were to report back to the parties on their progress.

The AWG-KP met for the first time in May 2006.[38] The AWG-KP's work program was agreed in Nairobi in November 2006 at COP/MOP 2.[39] A timeframe for the completion of this work program was subsequently agreed in Bali in December 2007 at COP 13/MOP 3 and further elaborated at COP 14 in Poland.[40] The AWG-KP's agreed tasks included:

1. Analysis of the mitigation potentials and ranges of emission reduction objectives of Annex I Parties and their environmental, economic, social, and sectoral consequences;

2. Analysis of possible means to achieve these mitigation potentials, including emissions trading and the project-based mechanisms under the protocol, land-use and

land-use change activities, sectors and gases to be covered, and possible sectoral approaches; and

3. Consideration of the scale of emission reductions to be achieved by Annex I Parties in aggregate, the allocation of the corresponding mitigation effort, and agreement on further commitments of Annex I Parties and the duration of the commitment period.[41]

Without agreeing on a specific range of mitigation effort for Annex I Parties, reports of the AWG-KP to the COP/MOP have noted the IPCC's Fourth Assessment Report finding that achieving the lowest GHG stabilization concentration range analyzed to date (445 to 490 CO_2-equivalent ppmv), and its corresponding potential damage limitation, would require Annex I Parties as a group to reduce emissions in a range of 25 to 40 percent below 1990 levels by 2020.[42] These reports have also noted the concerns of small island developing states and other developing countries with the lack of more ambitious stabilization scenarios and the need for the review of this range in light of further scientific information.[43]

The parallel Dialogue established in Montreal succeeded in creating a vehicle to engage non-Kyoto parties and developing countries in discussions on long-term cooperative action.[44] Four planned workshops were convened in 2006 to 2007 and enabled parties to express views on possible actions in the four agreed thematic areas: (1) advancing development goals in a sustainable way; (2) addressing action on adaptation; (3) realizing the full potential of technology; and (4) realizing the full potential of market-based opportunities. The Dialogue co-facilitators reported back on these events at COP 12 and COP 13. By COP 13, however, pressure had grown to convert this process into a formal negotiating track,

not just for the benefit of Kyoto parties facing targets but also for the benefit of vulnerable countries needing firm commitments from all Annex I Parties on adaptation support and technology transfer.

At COP 13 in Bali in December 2007, after extremely difficult negotiations, Convention parties agreed to move beyond the Dialogue format into an "Ad Hoc Working Group on Long-term Cooperative Action under the Convention" (AWG-LCA) through the Bali Action Plan (decision 1/CP.13).[45] The Bali Action Plan launched "a comprehensive process to enable the full, effective and sustained implementation of the Convention" in order to "reach an agreed outcome" for adoption at COP 15 in Copenhagen, focused on four building blocks: (1) mitigation; (2) adaptation; (3) technology; and (4) financing and means of implementation.[46] A series of elements were agreed for consideration under each of these headings, which have guided the work of the AWG-LCA since Bali. At the same session, the AWG-KP adopted a detailed work program and agreed to conclude its work by 2009 to enable the adoption of an amended Annex B to the Protocol by COP/MOP 4 in Copenhagen.[47]

The AWG-KP and AWG-LCA streams are parallel, but have become more and more closely linked. For example, both the AWG-KP and AWG-LCA have addressed the overall scale of emission reductions to be achieved in the post-2012 period: the AWG-KP in the context of the contribution of developed countries to the global reductions needed, as identified in scenarios presented in the IPCC's Fourth Assessment Report; and the AWG-LCA in the context of a "shared vision for long-term cooperative action, including a long-term goal for emission reductions, to achieve the ultimate objective of the Convention" under paragraph 1(a) of the Bali Action Plan. Both the AWG-KP and the AWG-LCA have addressed mitigation efforts for different groups of parties. The AWG-KP, for example, despite its narrow mandate under article 3.9, has considered mitigation achievable in developing countries through discussions on possible improvements to the Clean Development Mechanism and possible sectoral approaches for developing countries linked to crediting schemes. Meanwhile the AWG-LCA has considered "enhanced national/international action on mitigation of climate change" for all parties, including "[m]easurable, reportable and verifiable nationally appropriate mitigation commitments or actions, including quantified emission limitation and reduction objectives, by all developed country Parties, while ensuring the comparability of efforts among them" under paragraph 1(b)(i) of the Bali Action Plan, and "[n]ationally appropriate mitigation actions by developing country Parties in the context of sustainable development, supported and enabled by technology, financing and capacity-building in a measurable, reportable and verifiable manner" under paragraph 1(b)(ii).[48]

The expectation is that the two parallel work streams created under the Convention and Protocol will join up in Copenhagen, through a comprehensive agreement on future commitments for all parties, or through a package of complementary agreements on future commitments. Three and possibly four meetings of the AWG-KP and AWG-LCA will take place in 2009 in advance of the Copenhagen COP. These multiple meetings provide an opportunity for the new United States administration to engage, but they also provide time desperately needed to close a wide gap between parties on fundamental issues.

For example, the EU, a relatively progressive voice among developed country parties, has taken the position that global average temperatures should not exceed 2 degrees Celsius above preindustrial levels, that developed countries should aim to reduce their collective emissions by 30 percent below 1990 levels by 2020, and that global emissions (including those from developed countries) should peak and then decrease in the next two decades to have even a 50 percent chance of staying below a 2-degree-Celsius temperature increase.[49] Some developed countries have expressed the view that these goals are too ambitious. At the same time, developing countries that are particularly vulnerable to the impacts of climate change have been adamant that these goals are clearly insufficient to meet the Convention's objective. Small island states, for example, have called for a limitation of global average surface temperature to as far below 1.5 degrees above preindustrial levels as possible and a stabilization of GHG concentrations as far below 350 ppmv CO_2-equivalent as possible, with impacts on particularly vulnerable developing countries serving as a benchmark for the effectiveness of the post-2012 package.[50] Clearly, 2009 will be a busy year.

Conclusion

Recent scientific studies plainly indicate that global emissions must peak as soon as possible and then decline as rapidly as possible to avoid potentially catastrophic impacts on the climate system. Meanwhile, global emissions continue to increase and it is estimated that by 2020 developing countries will account for more than 50 percent of these emissions. This dynamic only underscores the enormous challenge facing the global community in trying to craft an equitable and effective global agreement.

Against this backdrop of increasing emissions and accelerating climate change impacts, to be credible, any agreement for the post-2012 period will have to include: binding commitments toward substantial emissions reductions from all developed country parties; meaningful incentives for major developing countries to participate in the development and uptake of clean technologies; incentives for developing countries to address the challenge of deforestation; tools to monitor, report, and verify mitigation efforts by developing countries and funding and technology flows by developed country parties to support these efforts; a mechanism for adaptation funding that is substantial and directly tied to GHG emissions; and a means to address loss and damage experienced by developing countries from the impacts of climate change.

Notes

1. See UNFCCC articles 4.2(a) and (b).

2. UNFCCC articles 4.3, 4.4, 4.5.

3. UNFCCC article 4.4.

4. See Annual compilation and accounting report for Annex B Parties under the Kyoto Protocol (FCCC/KP/CMP/2008/9/Rev.1), available at www.unfccc.int.

5. Kyoto Protocol article 3.1.

6. See Kyoto Protocol article 6 (joint implementation), article 12 (clean development mechanism), and article 17 (emissions trading).

7. Ibid.

8. See UNFCCC decision 24/CP.7. These procedures and mechanisms were subsequently formally adopted by the parties to the Kyoto Protocol at COP/MOP 1 by decision 27/CMP.1.

9. Directive 2003/87/EC of the European Parliament and of the Council of 13 October 2003 establishing a scheme for greenhouse gas emission allowance trading within the Community and amending Council Directive 96/61/EC.

10. Communication from the Commission to the Council, the European Parliament, the European Economic and Social Committee and the Committee of the Regions, Limiting Global Climate Change to 2 degrees Celsius: The way ahead for 2020 and beyond, Brussels, 10.1.2007, COM(2007) 2 final, p. 6. Questions and Answers on Emissions Trading and National Allocation Plans, Memo 04/44 at 5, 4 March 2004 available at http://europa.eu/environment/climat/pdf/m06_452_en.pdf

11. See articles 4 and 6(2)(e) of the directive; see also Decision 280/2004/EC of the European Parliament and of the Council of 11 February 2004 concerning a mechanism for monitoring Community greenhouse gas emissions and for implementing the Kyoto Protocol [Official Journal L 49, 19.2.2004].

12. Article 16 of the directive.

13. Proposal for a Directive of the European Parliament and of the Council amending Directive 2003/87/EC to improve and extend the greenhouse gas emission allowance trading system of the Community, Brussels, 23.1.2008, COM(2008) 16 final, p. 2.

13. Ibid.

14. European Parliament legislative resolution of 17 December 2008 on the proposal for a directive of the European Parliament and of the Council amending Directive 2003/87/EC so as to improve and extend the greenhouse gas emission allowance trading system of the Community (COM(2008) 0016 – C6-0043/2008 – 2008/0013(COD)).

15. Proposal for a Directive of the European Parliament and of the Council amending Directive 2003/87/EC so as to include aviation activities in the scheme for greenhouse gas emissions trading within the Community, 20.12.2006, COM(2006) 818 final {SEC(2006)1684}{SEC(2006)1685}.

16. European Parliament legislative resolution of 17 December 2008 on the proposal for a decision of the European Parliament and of the Council on the effort of Member States to reduce their greenhouse gas emissions to meet the Community's greenhouse gas emissions reduction commitments up to 2020 (COM(2008)0017 – C6-0041/2008 – 2008/0014(COD)).

17. Annual European Community greenhouse gas inventory 1990–2006 and inventory report 2008 ("EC National Inventory Report"), p. 11, available at http://unfccc.int/national_reports/

annex_i_ghg_inventories/national_inventories_submissions/items/4303.php. Also see p. 10.

18. Commission Staff Working Document, Limiting Global Climate Change to 2 degrees Celsius: The way ahead for 2020 and beyond, Impact Assessment Summary, Brussels, 10.1.2007, {SEC(2007) 7}{SEC(2007 8}, p. 2. See also EC National Inventory Report, pp. 11–12. ("Under the Kyoto Protocol, the EC agreed to reduce its emissions by 8% by 2008-12, from base year levels. Assuming a linear target path from 1990 to 2010, in 2006 total EU-15 GHG emissions were 3.7 index points above this target path . . .")

19. EC National Inventory Report, p. 11.

20. Proposal for a Directive of the European Parliament and of the Council amending Directive 2003/87/EC to improve and extend the greenhouse gas emission allowance trading system of the Community, Brussels, 23.1.2008, COM(2008) 16 final, pp. 2–3, 13; Communication from the Commission to the European Parliament, the Council, the European Economic and Social Committee, and the Committee of the Regions, Towards a comprehensive climate change agreement in Copenhagen, Provisional version, {SEC(2009) 101} {SEC(2009) 102} COM(2009) 39/3, p. 2. (Hereafter, Towards a Comprehensive Climate Change Agreement in Copenhagen.)

21. At December 2007, CO_2 concentrations were measured at the Mauna Loa Observatory in Hawaii at approximately 383.5 ppmv. As of December 2008, this figure had increased to approximately 385 ppmv. See http://cdiac.ornl.gov/ftp/trends/co2/maunaloa.co2 and http://www.esrl.noaa.gov/gmd/ccgg/trends/.

22. Ibid.; Commission Staff Working Document, Limiting Global Climate Change to 2 degrees Celsius: The way ahead for 2020 and beyond, Impact Assessment Study, Brussels, 10.1.2007, SEC(2007) 7, p. 4.

23. See *International and EU Climate Change Policies after COP 12/MOP 2: Challenges and Opportunities for the New Member States and Candidate Countries*, Workshop in Prague, April 12, 2007, Background Information (Ecologic, FIELD, ISD, IVM, DIW Berlin), p. 14.

24. See, e.g., CCAP 2006, p. 13.

25. Directive 2001/77/EC of the European Parliament and of the Council of 27 September 2001 on the promotion of electricity produced from renewable energy sources in the internal elec-

tricity market; China Passes Renewable Energy Law, March 9, 2005, at http://renewableenergyaccess.com/rea/news/story?id=23531. ("In 2003, China's renewable energy consumption accounted for only 3 percent of the country's total energy consumption. The government plans to lift up the figure to 10 percent in 2020.")

26. See EU Commission Staff Working Paper, Winning the Battle Against Climate Change, Background Paper (February 2, 2005), p. 41 (noting that globally it is estimated that 50 percent of future emissions could be eliminated through efficiency gains).

27. See Communication from the Commission to the Council, the European Parliament, the European Economic and Social Committee and the Committee of the Regions, Limiting global climate change to 2 degrees Celsius: The way ahead for 2020 and beyond, Brussels, 10.1.2007, COM(2007) 2 final, {SEC(2007) 7}{SEC(2007) 8} p. 6.

28. Ibid., p. 44.

29. Council Conclusions on Climate Change, 2785th Environment Council Meeting, Council of the European Union, Brussels, February 20, 2007, p. 4 (emissions from deforestation in developing countries amount to about 20 percent of global carbon dioxide emissions); Commission Staff Working Document, Limiting Global Climate Change to 2 degrees Celsius: The way ahead for 2020 and beyond, Impact Assessment Summary, Brussels, 10.1.2007, SEC(2007) 7 (land-use change emissions, mainly from deforestation, are responsible for around 20 percent of global emissions); Reducing emissions from deforestation in developing countries: Aapproaches to stimulate action (FCCC/CP/2005/MISC.1).

30. See *International and EU Climate Change Policies after COP 12/MOP 2: Challenges and Opportunities for the New Member States and Candidate Countries*, Workshop in Prague, April 12, 2007, Background Information (Ecologic, FIELD, ISD, IVM, DIW Berlin).

31. Ibid.

32. EU Commission Staff Working Paper, Winning the Battle Against Climate Change, Background Paper (February 2, 2005), p. 45.

33. See Blok, K., N. Hohne, A. Torvanger, and R. Janzic, Towards a Post-2012 Climate Change Regime: Final Report, June 2005; see also N. Höhne, et al., Options for the Second Commit-

ment Period of the Kyoto Protocol, ECOFYS, Federal Environmental Agency, Berlin, February 2005, p. 14.

34. Many years ago, AOSIS proposed the creation of an insurance fund to compensate low-lying developing countries and small island developing states for the impacts of sea-level rise from climate change. The fund was to be sourced by contributions from industrialized countries, through a formula based on their GHG emissions (responsibility) and their GDP (ability) (see A/AC.237/Misc.1/Add.3). See Linnerooth-Bayer, J., M. J. Mace, and R. Verheyen, Insurance Related Actions and Risk Assessment in the Context of the UNFCCC, background paper prepared for UNFCCC workshops (May 2003), pp. 3–5. A similar proposal was presented to the AWG-LCA in 2008 (Ideas and proposals on the elements contained in paragraph 1 of the Bali Action Plan (FCCC/AWGLCA/2008/Misc.5/Add.2 (Part I) submission by Alliance of Small Island States on a Multi-Window Mechanism to Address Loss and Damage from Climate Change Impacts)). See, for example, Blok, K., N. Hohne, A. Torvanger, and R. Janzic, Towards a Post-2012 Climate Change Regime: Final Report, June 2005, pp. 6, 10; Report on the workshop on risk management and risk reduction strategies, including risk transfer and sharing mechanisms such as insurance (FCCC/AWGLCA/2008/CRP.7). For a compilation of ideas and proposals by Parties on adaptation, see FCCC/AWGLCA/2008/16/Rev.1 (Assembly Document).

35. Decision 1/CMP.1 (FCCC/KP/CMP/2005/8/Add.1).

36. Decision 1/CP.13 (FCCC/CP/2007/6/Add.1).

37. Ibid.

38. Ibid. and FCCC/KP/AWG/2006/L.2/Rev.1.

39. FCCC/KP/AWG/2006/4.

40. FCCC/KP/AWG/2007/5 (Bali) and Annex I; FCCC/KP/AWG/2008/8 (Poznan).

41. FCCC/KP/AWG/2006/4.

42. FCCC/KP/AWG/2007/4, para. 19 (Vienna), FCCC/KP/AWG/2007/5, para. 16 (Bali), FCCC/KP/AWG/2008/8, para. 18 (Poznan). For stabilization scenarios, see *Contribution of Working Group III to the Fourth Assessment Report of the Intergovernmental Panel on Climate Change, Summary for Policymakers*, Table SPM.5, available at www.ipcc.ch/pdf/assessment-report/ar4/wg3/ar4-wg3-spm.pdf.

43. FCCC/KP/AWG/2007/4, para. 20 (Vienna), FCCC/KP/AWG/2007/5, para. 17, (Bali), FCCC/KP/AWG/2008/8, para. 19 (Poznan).

44. Decision 1/CP.11 (FCCC/CP/2005/5/Add.1).

45. Decision 1/CP.13 (FCCC/CP/2007/6/Add.1).

46. Ibid.

47. FCCC/KP/AWG/2007/5 (Bali), Annex I.

48. Decision 1/CP.13 (FCCC/CP/2007/6/Add.1).

49. Communication from the Commission to the Council, the European Parliament, and Social Committee and the Committee of the Regions, Limiting global climate change to 2 degrees Celsius: The way ahead for 2020 and beyond, Brussels, 10.1.2007 COM(2007) 2 final; see also Towards a Comprehensive Climate Change Agreement in Copenhagen.

50. FCCC/AWGLCA/2008/Misc.5/Add.2 (Part I) (Alliance of Small Island States); see also FCCC/AWGLCA/2008/16/Rev.1, pp. 21–22.

Chapter 22

EU Climate Policy

TOM R. BURNS IN COLLABORATION WITH MIKAEL ROMAN

Introduction

The processes leading up to and involving the EU in mediating and adopting the Kyoto Protocol reveal how rapidly the EU changed its political position regarding emission trading mechanisms. Initially, the EU was skeptical about the market-oriented core of the Kyoto Protocol introduced by the United States. Ironically, the EU now appears as if it were the founder of the Kyoto Protocol's emission trading scheme (ETS).

The Path Leading to the Kyoto Protocol

The EU in Climate Change Initiatives

At the Rio Earth Summit in 1992 (but also earlier in the 1980s), the European Union stressed the urgency of the climate change issue and sought to establish policies that would accomplish direct cuts in emissions. In general, the EU has favored central measures such as the coordination of member-state policies, including harmonization of energy taxes across the EU (to eliminate competitive advantages based on differences in such taxes among EU countries), adoption of energy efficiency measures, establishment of speed limits on motor vehicle traffic, and improvements in transport management.[1] In the 1980s, plans for an EU-wide carbon or energy tax were vehemently opposed by certain industries (transport, agriculture, petroleum, and chemical) and several member states, particularly the UK and Ireland. This initiative faced a serious barrier: according to EU constitutional rules, the introduction of an EU tax would require unanimous support of the member states.

After Rollback Came a Process of Groping Ahead

After the collapse of the carbon tax initiative prior to Kyoto, the EU focused on determining climate change goals within the EU, both collectively and by individual member states. The EU devoted much effort to establish binding "common and coordinated" policies and measures (PAMs), despite early indications that few EU members shared this enthusiasm or approach (some aspiring to EU membership were sympathetic, but never fully endorsed the EU proposals). Part of the EU's motivation apparently rested on the hope that the United States might eventually also agree to binding PAMs.

Although the EU was unsuccessful in developing concrete mechanisms for reducing emissions, it managed to establish target-sharing agreements for the EU zone, which was a major accomplishment, and it played a key mediating role in bringing about the Kyoto Protocol itself.[2] The EU had political will and a powerful rhetoric about dealing with climate change, but was relatively weak when it came to action and implementation. Some industrialized countries and NGOs pointed to the "credibility gap" between the EU's political goals and rhetoric, on the one hand, and the lack of a coherent implementation strategy on the other hand (especially after the collapse of the carbon tax initiative). Nevertheless, the EU has been relatively successful in limiting its share of global carbon emissions in 2000 to 14 percent (10.5 tons per capita) compared to 20.6 percent (24.5 tons per capita) for the United States.[3]

Kyoto Process

By the time of the Kyoto meetings (December 1997) the EU had little to show for its efforts over two decades at trying to achieve major emission reductions in the EU area, despite some headway from individual countries. The EU Commission worked hard to get the member states to agree to country-specific emission reductions and made considerable progress at this—although there was no EU-wide mecha-nism. Apparently, the EU hoped that other countries, particularly the developed countries (Annex I countries in UN terminology) and, above all, the United States, would agree to substantial reductions and support binding policies and mechanisms. The United States was reluctant to establish goals for emissions, but was determined that "the agreement should include an array of flexible, market-based approaches for reducing emissions."[4] Those approaches were expected to keep the net costs relatively modest. In order to "involve" the United States as a major player, the EU supported the American proposals for market mechanisms without really wanting (or knowing precisely how) to buy into them.

Prior to and throughout the Kyoto process, the EU played a mediating role with respect to the United States, increasing the likelihood (but without guaranteeing) that the United States would sign on. In order to keep the United States in the bargaining process, it went along with supporting the emission trading mechanism. The United States succeeded in gaining not only the inclusion of market mechanisms in the Kyoto protocol (article 3 [10], article 3 [11], and article 17), but also a more comprehensive basket approach of greenhouse gases (six gases instead of three). The EU played a key role in mediating the involvement of the United States, which led ultimately to the increased engagement and commitment of the EU to ETS.

The Aftermath of Kyoto

After Kyoto, and after some initial lethargy and passivity, the EU established a bold ETS—arguably, the most important ETS for CO_2 on a global level.[5]

The Immediate Years After Kyoto: 1997–2000

Prior to and following Kyoto, the ETS was generally viewed with skepticism in the EU. In the months immediately following Kyoto, the EU showed passivity toward the emission mechanism.[6] At the grassroots level in Europe, however, several initiatives were launched. A major private driver, British Petroleum (BP), established in 1998 an ET pilot scheme for a limited number of its business units (eventually 150 were involved) with an aim to cut their emissions by 10 percent by 2010 from their level in 1990; this initiative was partly the result of the U.S. pressure group Environmental Defense Fund (EDF), the most active supporter for many years of emissions trading in the United States.[7] On the public side, the Danish government initiated legislation in May 1998 and set up the first member state trading scheme for the Danish power sector. More attention-getting was the establishment in June 1999 of the UK Emission Trading Group (ETG) and the publication in November 2000 of *A Greenhouse Gas Emissions Trading Scheme for the United Kingdom* by the UK Department for the Environment, Transport, and the Regions.

This general trend was crowned at the end of 2000 when the German Federal Environment and Economics Ministries established a national stakeholder group to investigate the possibility of domestic emissions trading. These developments brought wider recognition of such trading as a realistic policy alternative. In addition, some small EU Member States such as Austria, Finland, and Ireland acknowledged that trading in GHG permits was something they would like to pursue, but

believed that their own trading volumes would be relatively small. National trading schemes would not make sense for them (although this did not stop Denmark).

During this period (1997–1999), the EU remained relatively passive. However, in the face of a number of grassroots movements, the EU became concerned about the fragmentation of European markets. The EU took several initiatives in 2000 and 2001. A green paper titled "Greenhouse Gas Emissions Trading within the European Union" was brought out in March 2000 by the European Commission, highlighting a "bottom-up pathway" to an emission trading scheme. The EU established a multi-stakeholder working group in the European Climate Change Programme, meeting ten times between July 2000 and May 2001; group discussion focused on the "bottom-up" approach (already taking place) and a mid-level approach stressing coordination and harmonization of design features.[8] Already, the EU was on the road that would lead it to establish the first major world trading system in GHGs. European Commission staff members Peter Zapfel and Matti Vainio reported:

> In May 2001 the stakeholder group concluded its work with the clear recommendation that European trading in GHG permits should be established "as soon as practicable". Astonishingly, the group—bringing together diverse interests with about thirty representatives from several Member States, industry, and environmental pressure groups—achieved a high degree of consensus and failed to adopt a consensual recommendation on very few issues. These were highlighted as areas of diverging opinions in the final report.[9]

In October 2001, the European Commission advanced the issue to a new level by adopting a proposal for a directive on EU-wide trading in GHG permits.[10] The proposal foresaw the mandatory introduction of trading in GHG permits in all EU Member States by 2005. It proposed to apply consistently across Europe a number of actions in particular sectors (power and heat generation, iron and steel, oil refining, pulp and paper, and cement and other building materials) but focused initially on CO_2 emissions only. The details of the initial allocation decisions would be largely left to member states, according to the proposal.

George W. Bush's repudiation of Kyoto in March 2001 arguably also played a role in the EU mobilization and commitment to establishing a world trading system in GHGs. As Zapfel and Vainio point out, "These events united Europe and triggered further interest in the establishment of an EU-wide market in GHG permits and other measures to rebuild momentum in international climate mitigation efforts."[11] Once the United States was out of the picture, the EU and some NGOs solidified their commitment to the ETS paradigm, and improved the EU contacts and discussions with Group 77.[12]

As sociologist Don MacKenzie stresses, "a significant current of opinion amongst European policy makers galvanized, [and] momentum built up behind the project of building a European-wide trading scheme."[13] By December 2002, the design of the European scheme was agreed and completed. Less than a year later (October 2003), the directive establishing the EU ETS was published, and it was followed in October 2004 by the "linking directive," which allows countries of the EU to invest in projects that reduce emissions in developing countries as an alternative to reducing emissions (typically more expensive) in their own countries.[14] In January 2005, the EU's carbon trading began operations.

In a certain sense, the EU stumbled into the role of global leadership of emission trading. The EU became the major promoter of the Kyoto "flexibility mechanisms" in setting up the first large-scale international ETS system (2005) for greenhouse gases. The EU's behavior today might give the impression that it is the true parent of this regime. However, most of the intellectual basis of the ETS came from U.S. experience, competence, and forceful proposals. Moreover, the EU was adopting systems that were very different from anything it had advocated or known about earlier. It was not until much later that the EU gained basic knowledge and expertise about the market-based mechanism embodied in the Kyoto Protocol.

Conclusion: Public Policy Shift?

In what sense can one claim that a major public policy shift took place with the EU adoption and implementation of the ETS?[15] We have suggested that the properties of the ETS were very different from earlier EU climate change public policy initiatives and proposals. Going back at least to the 1980s, the EU was oriented toward centralized policies and measures relating to mechanisms of coordination, regulation, and finance.[16] The most characteristic of these was the idea of introducing a carbon/energy tax for the entire EU, which ended in failure (there remains even today interest and potential in such taxation schemes).

The ETS is not the entire EU climate change policy profile.[17] Decisions beginning in 2003 led to the establishment and funding of a comprehensive initiative, Intelligent Energy Europe, including such programs as (1) energy saving (SAVE); (2) new and renewable energy sources (ALTENER); (3) energy aspects of transport (STEER); and (4) renewable energy sources and energy efficiency in developing countries (COOPENER).[18] The EU has adopted a number of other measures that directly or indirectly affect the emission of greenhouse gases, such as a 1999 directive on landfills that tries to reduce methane gas generation through better monitoring of landfills and the reduction of land filling of biodegradable waste; a 2002 directive on the promotion of electricity from renewable energy sources; a 2003 directive on the promotion of the use of biofuels or other renewable fuels for transport; a 2004 directive establishing a system to monitor the emission of greenhouse gases and the implementation of the Kyoto Protocol; and other measures, in particular in the areas of transport, energy, and agriculture policy.[19] There have also been numerous EU voluntary agreements with auto producers relating to CO_2 emissions from cars, and with producers of electronic equipment concerning energy efficiency. And there is still a possibility—with the changes in public opinion—that the EU will reconsider a carbon or energy tax. If such a system of EU taxation were established, then two contrasting economic approaches, taxation and emissions trading, would be operative at the same time, but with the potential for conflicting interactions and unintended consequences.

In spite of the setbacks since the 1990s, the EU climate change achievements are substantial. As environmental law professor Ludwig Krämer points out, "There does not seem to be any developed State which has, with the same political determination, invented a climate change policy and put it into operation."

Notes

We are grateful to Bo Kjellen, Stephen Schneider, and Matti Vainio for their comments and suggestions on an earlier draft of this article. A more elaborated and theoretically grounded chapter on this case is appearing in M. Carson, T. R. Burns, and D. Calvo (eds.) forthcoming in 2009, *Public Policy Paradigms: Theory and Practice of Paradigm Shifts in the EU*, Berlin/New York: Peter Lang Publishers.

1. Wettestad, J. 2000. The complicated development of EU climate policy: Lessons learnt, in *Climate Change and European Leadership: A Sustainable Role for Europe?* ed. J. Gupta and M. Grubb (Boston: Kluwer Academic).

2. Yamin, F. 2000. The role of the EU in climate negotiations, in *Climate Change and European Leadership: A Sustainable Role for Europe?* ed. J. Gupta and M. Grubb (Boston: Kluwer Academic).

3. Baumert, K. 1998. Carbon Taxes vs. Emissions Trading: What's the Difference, and Which is Better? (available online at http://env.chass.utoronto.ca/env200y/ESSAY2001/tax.htm). Baumert points out that, because the EU is already relatively energy efficient, carbon taxes would be less of a burden than in the United States. Improvements have been made steadily since the late 1980s through electricity deregulation, tax coordination, and agreements with industrial sectors.

4. MacKenzie, D. 2008. Constructing emissions markets, in *Material Markets: How Economic Agents Are Constructed*, ed. D. MacKenzie (New York: Oxford University Press).

5. We note that ETS schemes are established or are being established in Japan, in Australia, and on the East and West Coasts of the United States, among other places.

6. EU bureaucrats and politicians as well as EU NGOs with a concern about climate change felt more comfortable with established policy instruments such as a general tax or a directive setting constraints on polluting behavior.

7. This goal is comparable in magnitude to salient Kyoto targets (Zapfel, P., and M. Vainio. 2002. *Pathways to European Greenhouse Gas Emissions Trading History and Misconceptions*. Nota di Lavoro 85.2002 Fondazione Eni Enrico Mattei. www.feem.it/web/activ/_activ.html; MacKenzie, 2008: 13). The Environmental Defense Fund

(EDF) viewed emissions trading as a way to end the polarity between business and environmental groups over the climate change issue. EDF forged a partnership with BP that, in 1997, under its new chief executive John Browne, broke ranks with the other major oil companies, announcing that it accepted that global warming was a major threat and that action was needed. EDF's president, Fred Krupp, "lobbied John Browne to adopt a cap and trade system" (Mackenzie, 2008: 13; Victor, D. G., and J. C. House. 2005. BP's emissions trading scheme. *Energy Policy* 34:2102.

8. According to Zapfel and Vainio (2002 op cit.), an important factor in the EU development was the active involvement of experts from the United States in the European debate, as the country not only with the biggest interest, but also with the broadest experience in applying the instrument.

9. See European Commission 2001 Final Report: ECCP Working Group I "Flexible Mechanisms," available at http://europa.eu.int/comm./environment/climat/final_report.pdf.

10. European Commission 2001 Proposal for a framework Directive for greenhouse gas emissions trading within the European Community—COM(2001)581, October.

11. Zapfel and Vainio, 2002 op. cit.

12. According to sociologist Don MacKenzie (2008 op cit.), the initial belief in Europe had been that a system of international emissions trading, based around the United Nations but led by the United States, was on its way. As it became clear that this was not to be case, it was replaced by a sense that "we have to make it happen at home first."

13. At the heart of emission market construction—an *in vivo* economic experiment—was a small team of officials of the European Commission, led by economist Jos Delbeke of the Commission's DG Environment (MacKenzie, 2008 op. cit.: 16).

14. European Commission 2003 "Directive 2003/87/EC of the European Parliament and of the Council of 13 October 2003 establishing a scheme for greenhouse gas emission allowance trading within the Community and amending Council Directive 96/61/EC." Official Journal of the European Union, L 275/32, 25.10.2003; European Commission 2004 Directive 2004/101/EC of the European Parliament and of the Council of 27 October 2004 amending Directive 3003/87/EC establishing a scheme for greenhouse gas emissions allowance trading within the Community, in respect of the Kyoto Protocol's project mechanisms. Official Journal of the European Union, L338/18, 13.11.2004.

15. Concerning major policy shifts, see Carson, M., T. R. Burns, and D. Calvo (eds.). Forthcoming, 2009. *Public Policy Paradigms: Theory and Practice of Paradigm Shifts in the EU*, Berlin/New York; Peter Lang Publishers.

16. Baumert (1998 op. cit.) stresses, "The European Union has traditionally been in favor of strong coordinated policies and measures, such as energy/carbon taxes, among countries."

17. A range of EU initiatives includes promotion of electricity from renewable energy sources ("green electricity"), renewable energy growth in the EU from less than 7 percent today to 20 percent by 2020, the promotion of co-generation (combined heat and power), the promotion of energy performance of buildings (all new buildings as well as large existing buildings undergoing innovation), energy labeling of domestic appliances, the landfill directive, research funding for energy conservation and new and alternative energy sources, energy certification prior to sale, support of energy crops and voluntary agreements with auto companies to cut CO_2 emissions.

18. Intelligent Energy Europe, available online at http://ec.europa.eu/energy/intelligent/projects/index_en.htm.

19. L. Kramer, 2006. Some reflections on the EU mix of instruments on climate change, in *EU Climate Change Policy: The Challenge of New Regulatory Policy*, ed. M. Peeters and K. Deketelaere, 279–96.

Chapter 23

Population

Frederick A. B. Meyerson

Introduction

Human population growth is closely linked to both the causes and effects of climate change. Global population is projected to increase by an average of 76 million people per year through 2020, by which time the population is projected to be 7.6 billion.[1] A slow decline in the rate of growth is projected thereafter, assuming that improvements in reproductive healthcare and fertility decline trends continue. Even with those assumptions, in the 2040s growth is still expected to be more than 50 million people per year.

Average per capita carbon dioxide emissions from fossil fuels (the major cause of global warming) have barely changed at the global scale since 1970, remaining in a narrow range around 1.2 metric tons of carbon (mt) per capita (see figure 23.1).[2] The minor fluctuations have followed global economic cycles, with recessions causing temporary decreases in average per capita emissions, and economic booms causing short-term increases. The effect is that population growth (more than 80 percent globally since 1970 or 3 billion people) has been a major factor in greenhouse gas

(GHG) emission trends and will be in the future. While hope remains that per capita emission rates can be decreased in major emitting countries, we have had little success on that front in recent years even among countries committed to Kyoto Protocol reductions.

Many developed countries and regions have reached per capita emissions plateaus. The *level* of the plateau has varied substantially—for example, consider the United States (around 5.5 mt per capita) and Western Europe (around 2.1 mt per capita).[3] This plateau phenomenon and the level reached by particular countries and regions is a function of many variables, including the establishment and subsequent inertia of housing, industrial, transportation, and fuel-delivery infrastructure and systems. With longer life expectancies and increasing median age, the consumption and behavior patterns of developed country populations also become more static and less dynamic.

The effect of this "stickiness" of per capita emissions is that, for developed countries, a major determinant of national emissions increases and decreases has been population growth and decline. For example, U.S. total

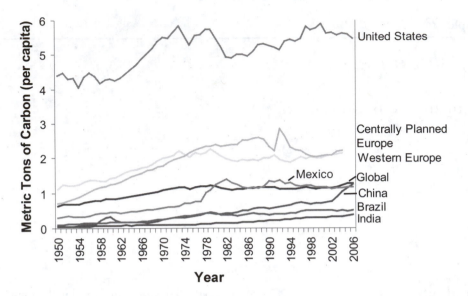

FIGURE 23.1. Per capita emission trends, 1950–2006.

emissions have grown dramatically in recent decades compared to those of Western Europe, in large part as a result of demographic differences rather than technological or behavioral trends. Between 1970 and 2004, the U.S. population increased by 43 percent (88 million people), and total annual emissions also increased by 43 percent (almost 500 million metric tons). In the same period in Western Europe, population and emissions each increased by less than 15 percent.[4]

To be sure, many factors other than demography and population growth affect emissions of countries, including climate, the availability of renewable energy sources, and other geographical factors. However, the patterns described in figure 23.1 suggest that it will not be easy to change the historical inertia of per capita emission trends. The rest of this chapter will dissect the historical and projected population and emissions trends and briefly explore other demographic aspects of climate change.

Population and Greenhouse Gas Emissions Trends and Plateaus

During the last two centuries of our 2-million-year history, humanity has rapidly shifted from a society based on agriculture and the use of domestic animals to an economic system propelled by and dependent on fossil fuels. The energy and industrial revolutions and the nineteenth- and twentieth-century population explosions are inextricably intertwined. It is unlikely that the human population could have increased from 1 billion in 1800 to its present (2008) level of 6.7 billion in the absence of the discovery and exploitation of fossil fuels.[5] The population growth rate peaked during the 1960s at nearly 2.2 percent per year and declined to 1.2 percent by the early twenty-first century. However, the 1980s and 1990s saw the greatest numbers of added people, about 80 million per year, with only a slight decrease occurring since that time.[6]

Global emissions of carbon dioxide from

fossil fuel combustion grew from 8 million to 7,910 million metric tons (of carbon) between 1800 and 2004, an almost 1,000-fold increase.[7] Per capita carbon dioxide fossil fuel emissions increased by 60 percent from 1950 to 1970. However, since 1970, global per capita emissions have fluctuated in a narrow range, averaging 1.15 mt from 1970 to 2004, the most recent year for which final global estimates are available.[8] Notably, there has been no change in per capita emissions since 1990, the benchmark year for the Kyoto Protocol, or since 1997, when it was signed.

In the United States, per capita annual emissions also plateaued around 1970, but at a much higher level (around 5.5 mt per capita). Recessions, economic booms, and changes in energy prices and fuel sources have caused small variations, but U.S. per capita emissions have been very "sticky" (5.6 mt in 2004). The same is true for other developed countries and regions (see figure 23.1). For instance, per capita emissions of Western Europe exceeded 2 mt per person for the first time in 1970 and have remained within 10 percent of that level, before and after the region's adoption of the Kyoto Protocol.[9] The same apparent plateau phenomenon occurred in "centrally planned Europe," which includes the former Soviet Union and now Russia, in the early 1970s when emissions first exceeded 2 mt per person, with some fluctuation in the 1980s and 1990s as a result of the disintegration of the Soviet Union.

The mechanisms that lead to stalling of per capita emissions are complex and deserve more research. In many developed countries, it appears that greater consumption, or affluence, has effectively canceled out technological gains in energy efficiency. For example, we have built larger new houses, but with more efficient energy systems and insulation. In the United States, while the emissions intensity per unit of GDP has improved, energy use (and emissions) per capita have remained essentially stable.[10] One additional complicating factor is the tendency for developed countries to "export emissions" by importing goods produced by energy-intensive activities in developing countries. One study estimated that U.S. emissions would be 3 to 6 percent higher if the country produced the goods that it imports from China.[11]

A few developed countries have substantially reduced per capita emissions. These fall into two major categories: political and economic upheaval, and fuel-switching. For example, per capita emissions in Russia and the Ukraine decreased after the Soviet Union collapsed, and economically and energy-inefficient industries were shut down or replaced. France and the United Kingdom significantly reduced per capita emissions by adopting new energy sources (nuclear power and North Sea gas and oil, respectively), which replaced coal and other high-emissions energy sources.[12] However, in each of the above cases, it appears that a new per capita emissions plateau has emerged, albeit slightly lower than the previous one.

There is somewhat more variation among developing countries. Mexico, for instance, experienced the same kind of plateau, but not until about 1980, when per capita emissions first exceeded 1.1 mt and have remained there since (1.1 mt in 2004). At about the same time, Brazil's per capita emissions first exceeded 0.4 mt, and they remain close to that level twenty-five years later, partly because of the country's development of a large biomass to ethanol fuel industry, whose emissions are not counted as fossil fuel emissions. The commonality of the per capita emissions plateau phenomenon among countries is striking, as is the wide range among those plateaus.

The world's two most populous developing countries, India and China, are important exceptions in that they have not yet reached per capita plateaus. India's emissions tripled from 0.1 to 0.3 mt per capita between 1970 and 2004, though growth has been much slower in the last decade (only about 10 percent). In the same period, China's per capita emissions quadrupled from 0.26 to 1.05 mt per person (nearly the global average), although recent wild swings in reported annual emissions create questions about data accuracy. Suffice it to say, that end of the story has not been written for either of these countries, which combined already contain more than a third of the world's population and are expected to grow by almost 1 billion people between 2000 and 2050.

Population growth or decline will continue to be a key determinant of future emissions increases.[13] Among the top ten carbon dioxide–emitting countries, which accounted for about 65 percent of global fossil fuel emissions in 2004, only India, China, and the United States are still experiencing significant population growth (see figure 23.2).[14] They are projected to add 804 million, 155 million, and 138 million people, respectively, between 2000 and 2050.[15] However, by around 2030, China's population is projected to peak and begin a slow decline, due in part to the effects of the "one-child" family planning policy, which severely penalizes many families who have more than one child. By 2050, among the top ten carbon dioxide–emitting countries, only the United States and India are projected to have fertility rates above the replacement level (around 2.1 children per woman) and significant continued population growth.[16]

Demographic Projections

Population projections more than a few decades into the future are quite speculative, since they necessarily depend on estimates of the fertility, mortality, and family size preference of people who have not yet been born.

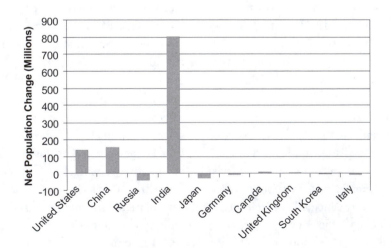

FIGURE 23.2. Top ten CO_2-emitting countries, 2004.

For example, demographic projections by the United Nations indicate that by 2050, the global population could be as low as 7.8 billion and declining, or as high as 10.8 billion and continuing to increase rapidly, though neither of these extremes is considered likely. The medium variant scenario, often incorrectly referred to as a prediction, projects a population of 9.2 billion by 2050, which would be an almost 40 percent increase over the current level.[17] As of 2006, the U.S. Census Bureau, using slightly different data and assumptions, projects a global population of 9.4 billion in 2050 (see figure 23.3). Both the UN and U.S. census projections for 2050 are revised every year or two, often increasing or decreasing by half a billion people.

Long-range demographic projections and trends are highly sensitive to small changes in fertility and mortality assumptions.[18] The relatively small difference in the fertility assumption by 2050 between the high and low UN variants—less than one child per woman in

2050—results in a net difference of 3 billion people, or the equivalent of the entire global human population in the mid-1960s. The current underlying assumption for the UN projections is that fertility rates in each of the world's countries will eventually converge to 1.85 children per woman (below the replacement level of 2.1), regardless of historical trends, and that most of that homogenization will happen by 2050.

However, there is limited evidence that this convergence will actually occur. Fertility rates in many European countries remain near or below 1.5 and have shown few signs of increasing. The U.S. fertility rate has hovered at or above 2.0 for years and is projected by the Census Bureau to increase slightly over the next few decades. Fertility rates in several developing countries, such as Egypt, Kenya, and Guatemala, may be stalling at 3 or above, despite predictions of decline. Some developing countries still have fertility rates of 5 children per woman or more. In short, the prospects

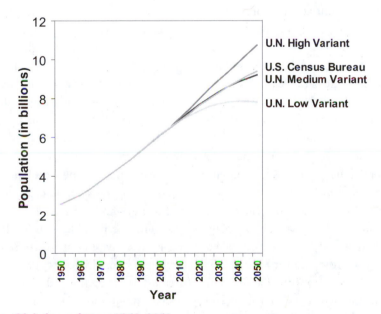

Figure 23.3. Global population, 1950–2050.

seem more likely for a future similar to the past, with divergent rather than convergent fertility rates.

Population Policy

The future size of the global human population will be critical in terms of both mitigation and adaptation to climate change. There may be increasingly difficult trade-offs among uses of land and solar radiation for food production, biofuels, forests, and other energy sources and carbon sinks. Population policy may therefore be an essential component of long-term climate policy.

One cost-effective long-term strategy for reducing greenhouse gas emissions and therefore climate change would be to provide reproductive health services to the hundreds of millions of people who still do not have adequate access.[19] In general, the fertility rate tends to reach or decline below replacement level (around 2.1 children per woman) in countries in which contraceptive prevalence rises to about 70 percent. Currently, only about 54 percent of the world's women use modern contraception, and the rate is still below 10 percent in some of the least-developed countries, including much of Africa.[20]

The climate- (emissions-) related costs associated with each additional birth are estimated to be in the range of several hundred to several thousand dollars.[21] These amounts are roughly comparable to the average costs for providing access to contraception (calculated on a per-birth-averted basis), which has many other social benefits. It has been estimated that the achievement of universal access to family planning by 2015 (the Cairo goal) could result in the avoidance of 100 to 200 million unwanted births by that date.[22] How-

ever, attainment of those family planning goals is uncertain due to stagnation in international reproductive health care funding in recent years.

The composite data for developing countries suggest that women on average have had about one more child than their stated family size preference.[23] Even in the United States, almost half of pregnancies are unintended, in part because contraception effectiveness is imperfect.[24] The demographic implications are significant. The relatively high failure rate of contraception in actual use means that women continue to have more pregnancies than they would otherwise choose, and unless they have access to both effective contraception and abortion, are having unintended children. There are an estimated 46 million abortions per year globally (both legal and illegal), and therefore, as a practical matter, abortion also continues to be a substantial global demographic factor, given that the current net annual world population increase is around 76 million per year.[25]

Reproductive health care remains one of the great success stories of international development aid and human rights. The percentage of women and families in the developing world with access to and using modern forms of contraception has increased slowly but steadily since U.S. A.I.D., the United Nations, and other organizations established programs in the 1960s. As of 2006, more than 50 percent of the women in the developing world use modern forms of contraception.[26] Improvements in reproductive health care and family planning are possible with a relatively small economic investment (often $20 to $30 per couple per year in developing countries), particularly when compared to the costs of other social expenditures. Yet hundreds of millions of women still have unmet family planning needs, and international aid, particularly U.S. funding, has been flat or decreasing in recent years.[27]

Because of below-replacement-rate fertility and stable or even slightly declining populations in many developed, donor countries, there is a false perception that global population growth issues have essentially been resolved. In some developed countries, there is serious concern that "birth dearth" (below replacement fertility and population decline) poses a threat to economic stability—a few governments have even instituted tax and other incentives to promote childbearing. At the global scale, however, the largest generation of young people in history is entering their reproductive years in the developing world.[28] Whether or not the international community makes a commitment to meet global family planning needs will have a major impact on whether the population in 2050 most resembles the high, low, or medium UN scenario.

Future Demographic Emissions and Climate Challenges

At the global scale, assuming continued static per capita, attaining something close to the UN low projection for 2050 (7.8 billion people) could result in almost 30 percent lower emissions compared to the high projection (10.8 billion). Meeting the reproductive health care goals of the 1994 UN Cairo agreement would go a long way toward achieving this end. However, even that result would only be a partial solution, because total emissions would still rise above the current levels and result in a greater than doubling of atmospheric CO_2 concentrations. Greater than 30 percent global per capita emission reductions

(in addition to limiting population to the low projection for 2050) would also be necessary just to stabilize emissions at 1990 levels, and this presents a much more complicated demographic, social, and political challenge.[29]

It has often been convenient to descriptively divide the world into developed and developing countries, but the reality for GHG emission purposes is a complex spectrum of demographic and development experiences.[30] Most developed countries have relatively stable or declining populations, and, even after net immigration, their populations are likely to decline in the aggregate between now and 2050. Stable or declining population is a double-edged sword in terms of GHG emissions, however. Developed countries with slow or zero population growth and longer life expectancies have a higher average age, and individual per capita emissions tend to rise through middle age, with only a small decline thereafter.[31] Lower growth also creates fewer opportunities for changes in infrastructure and therefore limits opportunities for efficiency gains, since housing stock and transportation systems turn over very slowly.

By the same token, a few developed countries, notably the United States, Canada, and Australia, still have growing populations and therefore significant opportunities for per capita emissions reductions through new infrastructure efficiency improvements. In particular, the United States, which leads the world in fossil fuel use and GHG emissions, is projected to experience a population increase of almost 50 percent between 2000 and 2050, as a result of a combination of immigration, fertility, and increased longevity. This growth will require major new construction of homes, businesses, roads, and other infrastructure.

Likewise, in developing countries with

high population growth, there is the opportunity to establish and maintain a lower energy use and emissions pathway. Particularly in areas where little infrastructure exists, it may be possible for a "technological leapfrog"—to effectively jump over the high emissions technology employed by the developed world to a state of high development with much lower fossil fuel use and emissions. China and India, the most populous countries in the world with a combined projected population of more than 3.2 billion by 2050, of which 1 billion have not yet been born, could accomplish this feat, though to date their per capita emissions continue to rise. China in particular is rapidly building a large infrastructure of coal-fired power plants. An increasing number of developing countries, including China, Mexico, and South Korea, have per capita emissions that already match or exceed the global average. The window of opportunity to change their per capita emissions trends may be short.

Migration trends will also play a significant role in future emissions, since people not surprisingly tend to migrate from countries with lower per capita emissions and GDP to those with higher emissions and GDP. For example, Mexico exports about 25 percent of its population growth (several hundred thousand people) to the United States each year, where per capita emissions are about five times those in Mexico. If fertility rates continue to decline as projected, migration is likely to become the dominant demographic factor affecting emissions. Note also that significant international migration complicates any reduction system that depends exclusively on national, rather than per capita, targets. Climate change itself may also begin to drive human migration, either by attracting people toward areas with improved conditions as a result of warming, or by driving them away from

areas negatively affected by sea level rise, storm frequency, changes in precipitation, and desertification.[32]

Conclusion

The world's ability to control emissions over the next few decades is likely to depend in large part on global population trends, particularly those of the United States, India, and China, which together account for more than 40 percent of the world's population and almost 45 percent of its CO_2 emissions.[33] The level at which per capita emissions eventually plateau in China and India is likely to be far more significant than the mitigation efforts of the group of developed countries that agreed to take action under the Kyoto Protocol, whose combined population is only a few hundred million, about 10 percent of the world total.

While global per capita emissions have been essentially flat for almost four decades, there is now more risk that they will rise rather than fall in the near future. It may be difficult to maintain the pace of energy-intensity improvements and decarbonization.[34] The supply of coal (with its higher CO_2 emissions per unit of energy) is much less constrained than that of oil or gas, and the promise of biofuels appears to be complicated. Coal production and combustion have been growing substantially in recent years, particularly in China and the United States. Total Chinese CO_2 emissions matched or exceeded those of the United States in 2007, and China's per capita emissions have also surpassed the global average.

With little near-term relief in sight in terms of per capita emissions, stabilizing global population remains a critically important objective to address both the causes and the effects of climate change. Lower population growth could also have a major impact on our ability to adapt to global warming. Curiously, while the 2007 Intergovernmental Panel on Climate Change report acknowledges population growth as a major driver of emissions, it does not include population policy in its lengthy list of policies, measures, and instruments available to mitigate climate change.[35] We clearly have the technological and health care capacity to reduce unintended births and therefore population growth, but our scientific and political will is in question.

Notes

I thank Gregg Marland, Holly Pearson, Laura Meyerson, several anonymous reviewers, and the editors of this volume for comments and suggestions that greatly improved this chapter. I also thank Tom Boden and Gregg Marland for their early provision of emissions data and preliminary estimates, which allowed me to develop per capita emissions estimates for 2005 and 2006.

1. U.S. Bureau of the Census. 2006. International Database. Available online at www.census .gov/ipc/www/idbnew.html (updated August 24, 2006).

2. Marland, G., T. A. Boden, and R. J. Andres. 2007. Global, regional, and national fossil fuel CO_2 emissions, in *Trends: A Compendium of Data on Global Change*. Carbon Dioxide Information Analysis Center, Oak Ridge National laboratory, U.S. Department of Energy, Oak Ridge, Tenn. Meyerson, F. A. B. 2002. Population and climate change policy, in S. H. Schneider, A. Rosencranz, and J. Niles, eds., *Climate Change Policy: A Survey*, 251–73. Island Press, Washington, D.C. References to emissions throughout this chapter will be in metric tons (mt) of carbon from fossil fuel carbon dioxide emissions, the largest and most accurately measured greenhouse gas source.

3. Marland et al., 2007 op. cit.

4. Ibid. U.S. Bureau of the Census, 2006 op. cit.

5. Meyerson, 2002 op. cit.

6. U.S. Bureau of the Census, 2006 op. cit.

7. Marland et al., 2007 op. cit. A similar though less dramatic trend has occurred in net carbon fluxes from land-use change, where average annual emissions have quadrupled during the last 150 years but per capita emissions have not increased (see Houghton, R. A., and J. L. Hackler. 2000. Carbon Dioxide Information Analysis Center. Available online at http://cdiac.esd.ornl.gov/trends/trends.htm). Land-use emissions can vary greatly from year to year, due in part to the effect of climate variations and cycles such as El Niño on natural and human-induced fires.

8. Marland et al., 2007 op. cit. Meyerson, 2002 op. cit.

9. This implies that to the extent that the Kyoto Protocol has been effective to date, this effect is primarily a product of the demographic stability or decline of countries that chose to adopt the protocol, rather than the result of policy or per capita behavioral changes.

10. Energy Information Administration. 2006. U.S. Department of Energy. www.eia.doe.gov/environment.html.

11. Shui, B., and R. C. Harris. 2006. The role of CO_2 embodiment in U.S.-China trade. *Energy Policy* 34: 4063–68. Likewise, a growing export industry may cause a country's per capita emissions to rise even though consumption patterns and population within the country are relatively stable. In a world where there are extreme differences in trade balances among countries, a significant climate policy question may be whether the emissions associated with production of goods and services should be considered as belonging to the exporting or the importing nation (see Marland, G. 2006. The human component of the global carbon cycle. Testimony before the Committee on Government Reform, Subcommittee on Energy and Resources. September 27, 2006.

12. With current technology, the combustion of coal produces almost twice the CO_2 emissions as natural gas and about 25 percent more emissions than oil to produce the same amount of energy.

13. Meyerson, 2002 op. cit.

14. Almost half of U.S. population growth during this period is projected to be the result of net annual immigration exceeding 1 million people

per year (see U.S. Bureau of the Census, 2006 op. cit).

15. Note that at 2006 per capita emissions levels, the addition of each U.S. citizen produces almost fifteen times the emissions as each Indian citizen (5.47 mt versus 0.37 mt of carbon per capita, respectively), so that population growth in the United States (3 million) corresponded to far more emissions than the population growth of India (18 million) in 2006 (see also Meyerson, F. A. B. 2003. Population, biodiversity, and changing climate, in L. Hannah and T. E. Lovejoy, eds., *Climate Change and Biodiversity: Synergistic Impacts, Advances in Applied Biodiversity Science* 4: 83–90).

16. U.S. Bureau of the Census, 2006 op. cit.

17. United Nations Department for Economic and Social Information and Policy Analysis Population Division. 2006. *World Population Prospects: The 2006 Revision*. United Nations, New York.

18. Population growth rates are also affected by the age at which women have children. Women in more developed countries tend to have children at older ages, increasing the time between generations and decreasing the growth rate.

19. United Nations Population Fund. 1999. *The State of World Population 1999: 6 Billion—A Time for Choices*. UNFPA, New York.

20. United Nations Population Fund. 2006. *The State of World Population 2006*. UNFPA, New York. http://www.unfpa.org.

21. O'Neill, B. C., F. L. MacKellar, and W. Lutz. 2000. *Population and Climate Change*. Cambridge University Press, Cambridge, England.

22. United Nations Population Fund, 1999 op. cit.

23. Pritchett, L. H. 1994. Desired fertility and the impact of population policies. *Population and Development Review* 20: 1–55.

24. Henshaw, S. K. 1998. Unintended pregnancy in the United States. *Family Planning Perspectives* 30: 24–29. In the United States, for instance, 9 percent of all women using contraception have unintended pregnancies each year. Trussell, J., and B. Vaughan. 1999. Contraceptive failure, method-related discontinuation, and resumption of use: Results from 1995 National Survey of Family Growth. *Family Planning Perspectives* 31: 64–72, 93.

25. Henshaw, S. K., S. Singh, and T. Haas. 1999. The incidence of abortion worldwide. *International Family Planning Perspectives* 25 (Supplement): S30–38. U.S. Bureau of the Census, 2006 op. cit.

26. Population Reference Bureau. 2006. World Population Data Sheet. Washington, DC. www.prb.org.

27. United Nations Population Fund. 2001. *State of World Population: Population and Environmental Change*. UNFPA, New York.

28. Ibid.

29. Meyerson, F. A. B. 1998a. Population, carbon emissions, and global warming: The forgotten relationship at Kyoto. *Population and Development Review* 24: 115–30.

30. Meyerson, F. A. B. 1998b. Toward a per capita–based climate treaty: Reply. *Population and Development Review* 24: 804–10.

31. See, for example, O'Neill, B. C., and B. Chen. 2002. Demographic determinants of household energy use in the United States, in *Methods of Population-Environment Analysis: A Supplement to Population and Development Review* 28, 53–88.

32. Meyerson, F. A. B., L. Merino, and J. Durand. 2007. Migration and environment in the context of globalization. *Frontiers in Ecology and the Environment* 5(4): 182–90. Ecological Society of America.

33. For a similar analysis, see Raupach, M. R., G. Marland, P. Ciais, C. Le Quéré, J. G. Canadell, G. Klepper, and C. B. Field. 2007. Global and regional drivers of accelerating CO_2 emissions. *Proceedings of the National Academy of Sciences of the United States of America* 104: 10288–93; published online before print as 10.1073/pnas.0700609104.

34. Pielke, R., T. Wigley, and C. Green. 2008. Dangerous assumptions. *Nature* 452: 531–32.

35. Intergovernmental Panel on Climate Change (IPCC). 2007. *Climate Change 2007: Impacts, Adaptation and Vulnerability*, Contribution of Working Group II to the Fourth Assessment Report of the IPCC, Parry, M., et al., eds. Cambridge University Press: Cambridge, United Kingdom.

Chapter 24

Inequities and Imbalances

Ambuj Sagar and Paul Baer

Introduction

The global nature of climate change is often highlighted as one of the key aspects of the issue that makes it such a thorny problem to solve. What makes the issue particularly problematic is the wide variation among countries of their contribution to climate change, the likely climate change impacts that they will experience, their vulnerability to these impacts, and the availability of resources to mitigate greenhouse gas (GHG) emissions and adapt to climate change. Adding to the complexity is the likelihood that many of the countries that have contributed and are contributing the most to climate change will not suffer the most extreme impacts. Climate models suggest that the greatest climate change impacts may be borne by countries that have contributed little to climate change and, to make matters worse, have limited resources to cope. These inequities and imbalances lie at the heart of the ongoing struggle to develop a practical approach to prevent major disruption to the global climate system.

While these international imbalances have contributed to the current stalemate in

the climate arena, the high costs of mitigating GHG emissions, of adaptation, and of residual damages further hinder progress. Fair ways of allocating these costs are critical to avoiding dangerous climate change and protecting the poor. It is essential to better represent, incorporate, and resolve these equity issues in the decision-making process. This, in turn, will require the exploration of novel and thoughtful policy approaches, as well as the strengthening of the appropriate analytical and participatory capacities in the north and south.

Who Pollutes and Who Suffers?

The responsibility of Annex I for the accumulation of fossil-fuel CO_2 (the largest greenhouse contributor) in the atmosphere is estimated to be about 70 percent (taking into account emissions from the years 1850 to 2004). This picture will change somewhat as developing countries industrialize—China, for example, increased its carbon emission by more than 5 percent per year between 1980 and 2004 and was the second-biggest CO_2

emitter, with 17.5 percent of the global fossil-fuel CO_2 emissions in the year 2004.[1]

As the IPCC has stated,

> The impacts of climate change will fall disproportionately upon developing countries and the poor persons within all countries. . . . Populations in developing countries are generally exposed to relatively high risks of adverse impacts from climate change. In addition, poverty and other factors create conditions of low adaptive capacity in most developing countries.[2]

Thus, climate change may exacerbate inequities in health status as well as in access to other basic needs such as food, energy, and water.

Agricultural production in many African countries and regions is projected to be severely compromised by climate change and increased variability; some countries may see a decline in yields from rain-fed agriculture of up to 50 percent, with serious implications for food security and food availability.[3] As another example, freshwater availability in Central, South, East, and Southeast Asia is projected to decrease due to climate change even as population growth and higher standards of living will contribute to greater demand which could adversely affect more than 1 billion people by the 2050s. Exposure to the effects of projected climate change is likely to affect the health status of millions of people, particularly those with low adaptive capacity.

In light of these steep imbalances in responsibility for, and potential impacts from, climate change, the issue of allocation of costs of responding to the problem takes on tremendous significance. While countries and other actors want to minimize the costs associated with climate change, a fair resolution of this issue will be critical to any progress.

Who Should Pay?
Who Should Pay for Mitigation?

Intimately linked to mitigation costs is the issue of targets and timetables of GHG-emissions reduction commitments. The tighter the targets and timetable, the greater and faster reductions are needed, which in turn increases the cost burden of mitigation. Thus the questions of what is the appropriate burden for mitigation and how should (and will) these burdens be shared go to the very heart of the debate and international negotiations on climate change. Estimates of mitigation costs vary, but are substantial. IPCC's third assessment report estimated that staying within an atmospheric CO_2 concentration ceiling of 550 parts per million by volume (double the preindustrial concentration, and the most widely discussed target) could require up to 800 billion USD (in 1990 dollars) during the next century, while meeting a tighter target of 450 ppmv would require between 350 and 1,750 billion USD.[4] More recent projections from a series of scenarios reviewed in IPCC's fourth assessment report indicate that net present value of cumulative abatement costs (for the tenth to the ninetieth percentile of the estimates) by 2100 ranges from nearly zero to 11 trillion USD (in 2000 dollars).[5]

Developing countries do not have the resources required to develop and deploy the advanced energy technologies that will be required to undertake large-scale GHG-emissions mitigation. Thus two major questions in the climate discussions have been the extent to which industrialized countries will be willing to share the burden of GHG emissions mitigation in developing countries and on what basis should this burden be allocated. So far, the industrialized countries have shown little interest in accepting any responsibility for the burden of emissions reductions in developing countries. Even if financial support were forthcoming, there are at least two issues of concern: whether these flows would be in addition to overseas development flows or merely replace them and whether the same historical pattern of "tied aid" that has bedeviled development aid will continue to operate here, thus hobbling the long-term effectiveness of mitigation programs.[6]

Who Should Pay for Adaptation?

While mitigation is clearly important for the long-term future to reduce the possibility of a highly disrupted climate system, it is clear that an era of a changed climate cannot be avoided. This means that there will be varied climate impacts in different parts of the world. While it is impossible to comprehensively assess the economic, social, and human costs of climate change, these have the potential to be enormous. Once again, poorer countries and the poor within them are particularly vulnerable—some, such as the Alliance of Small Island States (AOSIS) and other low-lying coastal countries, may suffer widespread impacts in absolute terms as well as relative to their GDP (and, in some cases, loss of their very country).

Given this potential—and, some might say, highly likely—future scenario, adaptation to climate change has been (belatedly) receiving increasing attention as a way to manage the impacts of climate change on economic and social systems as well as humans. Unfortunately the very countries and populations who are most vulnerable and most at risk do not have the capacity and resources to start putting in place the institutions and mechanisms for adaptation to climate change. Once again, this raises the issue of who should bear the costs of underwriting these kinds of activities. Importantly, this set of activities should not be seen as a substitute for mitigation (or vice versa)—the two actually complement rather than substitute for each other.[7]

Adaptation consists of a spectrum of options, from inaction to reaction to proaction, and from the individual to the global. Since responses at higher levels (e.g., global) affect the options and choices at lower levels (e.g., national and local), with inaction at high levels shifting full responsibility to the lower levels, "justice is always implicit in the choice of adaptive responses."[8] Thus the choice, prioritization, and design of adaptation activities must explicitly ensure that it redress existing imbalances in vulnerability and adaptive capacity.

Who Should Pay for Damages?

There is a third, and less-discussed (and certainly less considered), issue: liability. Adaptation will help prepare for coming climate impacts, with the intention of reducing the physical, economic, and human effects of climate-related phenomena on individuals, communities, and nations. While this is an important issue that needs further attention, at the same time it must be realized that despite all practical adaptation efforts, some damages from climate change are unavoidable. An important question then arises: Who should pay compensation for the damages? If one adheres to the "polluter pays" principle, GHG emitters should be held accountable, but this principle has not been tested in a court of law in the context of climate change.

Still, if it does become possible to seek damages for climate-related impacts, it will not only provide compensation for any harm but also serve as a deterrent to activities that lead to harm.

Lawsuits for climate damages will no doubt be complex propositions, given the intricacies of the issues, especially in attributing and apportioning damages to climate change, and the application of existing legal practice and theory to such cases. This will require scientific and technical expertise that will likely not be available in most developing countries. Still, it is likely that such challenges could appear through the development of coalitions of transnational groups that believe that such an approach has the potential to tackle the intransigence of major polluting states and corporations.

Inequity in Decision-Making Processes

There are two dimensions to inequity in the climate change arena: consequential inequity and procedural inequity. While the first category refers to inequity in outcomes, the second pertains to inequity in the process surrounding decision making.[9] Without ameliorating the latter, it will be difficult to achieve the former.

The UNFCCC is an agreement among countries; therefore states are the main actors in the climate policy processes. Thus, in a stylized representation, much of the decision making can be thought of as formation of national positions and policies, which, through interactions in the international arena, lead to multilateral agreements.[10] Given that this international negotiation process has mostly degenerated into positioning to protect national interests, the possibility of a fair outcome very much hinges on weaker states developing suitable decision-making capacity and strategy.

There are a number of imbalances between developing and industrialized countries in terms of their participation in the decision-making process. Some of the key issues are

- *Inequity in representation, analytical capabilities, and resources* to suitably inform and shape national policy and actions (i.e., defining and representing "national needs and interests")

In the case of an issue as complex as climate change, defining a "national interest" is an arduous task. It requires developing a good grasp of the science of climate change since that can have a bearing on determining the responsibility of the country in terms of emissions and on the mitigation and adaptation options that are relevant and available within the country. A better understanding of the economic, technical, institutional, and other resources necessary for undertaking actions becomes equally critical. This requires marshalling a whole host of capabilities—scientific, technical, and policy—and, equally importantly, linking up the research, analysis, and assessment process to the policy-making processes in the country.[11]

The reality, however, in most developing countries is very different. Often (and especially in the case of smaller countries), there is only limited relevant expertise in the country and limited resources to carry out the kind of work needed. At the same time, policy makers are often distracted by more pressing issues (such as economic and political crises). This stands in contrast to the enormous resources being directed toward climate science and policy, with the concomitant buildup of scientific and policy research enterprises, in industrialized countries. As an illustration, table 24.1 shows the distribution of publishing activity on topics related to climate change and climate policy for

TABLE 24.1

Percentage of published papers from Annex I and non–Annex I country authors on topics related to (A) climate change, (B) climate change policy, and (C) climate change and equity. Only the top publishing countries in each category are shown. Keywords search terms: (A) climate change OR global warming; (B) (climate change OR global warming) AND policy; (C) (global warming OR climate change) AND (equity OR justice OR fair OR fairness). Search carried out on March 29, 2008. The total percentages add up to more than 100 because of authors from different countries on the same paper.

(a)

Country/Territory	Record Count	% of Total
United States	9,781	41.5%
England	3,317	14.1%
Canada	1,981	8.4%
Germany	1,694	7.1%
Australia	1,123	4.8%
France	938	4.0%
Netherlands	872	3.7%
Japan	791	3.4%
China	704	3.0%
Sweden	680	2.9%
Switzerland	608	2.6%
Scotland	485	2.1%
Norway	408	1.7%
Italy	401	1.7%
Spain	370	1.6%
Russia	349	1.5%
Finland	344	1.5%
Denmark	341	1.5%
New Zealand	327	1.4%
India	271	1.2%

TABLE 24.1 (*Continued*)

(b)

Country/Territory	Record Count	% of Total
United States	767	46.4%
England	301	18.2%
Canada	120	7.3%
Germany	112	6.8%
Netherlands	108	6.5%
Australia	69	4.2%
France	45	2.7%
Austria	38	2.3%
Sweden	35	2.1%
Switzerland	34	2.1%
Norway	29	1.8%
Finland	26	1.6%
Italy	26	1.6%
Belgium	22	1.3%
Brazil	22	1.3%
Scotland	22	1.3%
Spain	20	1.2%
India	19	1.1%
Japan	17	1.0%

(c)

Country/Territory	Record Count	% of Total
United States	78	41.6%
England	31	16.8%
Netherlands	16	8.6%
Germany	15	7.6%
Canada	13	7.0%
India	11	5.9%
France	9	4.9%
Australia	7	3.8%
Sweden	6	3.2%
South Africa	5	2.7%
Switzerland	5	2.7%
Brazil	4	2.2%
Japan	4	2.2%
Norway	3	1.6%
Denmark	2	1.1%
Hungary	2	1.1%
China	2	1.1%
Russia	2	1.1%
Spain	2	1.1%
Taiwan	2	1.1%

major Annex I and non–Annex I countries. Not surprisingly, Annex I countries account for a much greater portion of papers published.

A related issue, that of participation and representation in policy deliberations, becomes enormously important in the climate domain. The poor, who are the most vulnerable and therefore suffer the greatest impacts, often remain unrepresented in policy discussions. But not incorporating their needs and concerns into climate policy may ultimately hobble its effectiveness. Issues such as the distribution of risk, the aggregation of impacts, and the dynamics of vulnerability are still not very well studied, but are critical, for example, to the development of appropriate adaptation strategies.

- *Inequity in negotiation capability* (i.e., protecting national needs and interests)

The lack of scientific, technical, and policy analytic capacity hobbles preparation for negotiations. There is even a wide gap between delegation sizes of industrialized and developing countries, and resulting institutional memories, as Muller has shown.[12] Developing countries do try to compensate for the limited size and clout of their negotiating teams through coordination and strategic interest groups (of which the G77/China is perhaps the most effective example). At the same time, NGOs also play an effective role in supporting developing countries either through analytical help to their negotiators or by putting pressure (individually or as part of transnational environmental networks) on industrialized-country negotiators.[13] It also must be recognized that climate change is but one issue that is part of the foreign policy agenda of countries and it may be given short shrift by developing countries themselves, given competing attention across issues such as strategic linkages, national security, trade, and so on.[14]

- *Discussion and representation of inequity in literature*

Equity issues, despite their importance, do not receive much attention in the climate literature—the number of equity-related papers is less than 1 percent of total (see table 24.1). Again, the major non–Annex I countries are represented on only a small fraction of these papers (although the disparity is less than in the climate change and climate policy categories). Thus academics and analysts in most developing countries are not heavily engaged in putting forward perspectives and proposals that would project equity concerns into the climate debate. There are a number of possible reasons for this: limited research and analysis capacity; other, more immediate and pressing, issues; and research agendas driven by foreign funding and priorities.[15]

Governments in developing countries have consistently tried to project the equity issue into the international negotiations, but they have been hobbled in their effort by the lack of cogent and creative proposals that advance this issue. Even in cases where the injustice is quite clear—for example, the possible submergence of small islands by sea-level rise—it has been almost impossible to get traction in the negotiation regime. While the UNFCCC does reflect a number of key fairness- and justice-related concerns (for example, see article 3, "Principles"), these have been translated into few concrete steps. Perhaps the most obvious example is the Kyoto Protocol, which does commit Annex I countries to take the first steps in reducing their GHG emissions, but even this has been foundering. The real manifestations of consequential inequity, though, might become clearer only as climate impacts start becoming more significant and obvious.

Conclusion

The question of how the costs of mitigation should be shared has received a relatively large share of attention in the climate debate and indeed can be characterized as "the equity question." Most analysts have concluded that fairness would seem to require acknowledgment of a fundamental equal right to make use of the global common sinks for greenhouse gas pollution.[16] Some have proposed a straight per capita allocation of emissions rights or (more commonly) convergence to an equal per capita allocation over time.[17] Other analysts have suggested that there are other principles of distributive justice—notably responsibility, capacity, and need—which are as relevant as the principle of equality and, at a minimum, require that some account be taken of the distinction between "luxury emissions" and "subsistence emissions" as well as historical emissions and accumulated economic benefits.[18] However, the practical consequence is that any defensibly fair system that protects the *right to development* of poor countries will require the industrialized countries to pay the vast majority of costs of mitigating climate change.[19]

It seems highly unlikely that industrialized countries will accede to such a proposal any time soon, given the large financial transfers

implied. Unfortunately this puts the developing countries in an unenviable position: holding out for such a scheme on the basis of principle, while completely morally justified, is already being portrayed as their slowing down progress on GHG emissions abatement. At the same time, since these countries generally are more vulnerable to climate change, it is also important for them that there be progress on a global agreement for serious GHG reduction commitments.

On the adaptation front, there is already some momentum building with the establishment of the Least Developed Countries (LDC) Fund and the Special Climate Change Fund (SCCF) under the UNFCCC.[20] The Kyoto Protocol also channels funds toward adaptation, although amazingly these are derived from clean development mechanism (CDM) activities, which effectively means that developing countries are being taxed to finance adaptation activities in their own or other poor countries (see chapters 27 and 28 in this volume).[21] Still, adaptation activities will need to be greatly scaled up, which will require larger allocation of resources, which in turn might require negotiations on Annex I country contributions toward this end. It might be worth considering some approach that apportions these costs among Annex I countries in a logical and transparent manner instead of the horse-trading that underpinned the Kyoto targets.[22] Given the magnitude of the problem, though, it will be important to leverage funds beyond those forthcoming under the UNFCCC. Mainstreaming climate change into other development and disaster management strategies may be particularly fruitful.[23]

In terms of appropriate design and implementation of adaptation activities, the experience of the disaster management community has highlighted that it is critical to focus on vulnerability as the key issue in determining, and hence responding to, climate risks (see chapter 26 in this volume). Hence reduction in vulnerability by enhancing disaster management capacity and system resilience must take primacy.[24] Such an approach will go a long way in reducing the significant risks posed for the poor by climate change.

Similarly, there have been efforts to enter the mostly uncharted territory of liability for climate damages. The prime minister of Tuvalu, a Pacific Island country that will likely be fully submerged by sea-level rise, has declared his intention to sue the United States and/or Australia for damages. Analysts have already started exploring the legal frameworks under which state or firm liability could be imputed for climate damages.[25] Others have proposed different approaches such as allocating climate refugees to countries on the basis of their cumulative emissions or developing a contingent permit scheme that would result in a compensation fund under specific climate scenarios.[26] We believe this is an extremely important issue to explore for two reasons: one, no practical level of adaptation can avoid all damages, and, two, torts can send a strong signal to emitters to reduce their emissions since that is the only way of reducing their exposure to possible claims.[27] A successful litigation in a climate tort case will do much to advance equity in the climate arena (see chapter 12 in this volume on the value of place).[28]

At the same time, there remains a pressing need for enhancing analytical and negotiating capacity within developing countries, which requires developing countries themselves to proactively take the lead and start devoting greater human, financial, and institutional resources toward this. In effect, this will require

a greater attention to the climate issue by policy makers in these countries (as is already beginning to happen in some countries) and, indeed, giving it the kind of primacy that it deserves (e.g., especially by focusing on the twin goals of sustainable development and climate change; see also chapter 25 in this volume).[29] At the same time, more creative approaches toward coordination of southern (or southern and northern) researchers (rather than activists, as is mostly the case so far) and of southern negotiators should be helpful. In fact, greater collaborations among southern and northern researchers may also help reshape the research agenda of both groups to focus more on issues of equity.

Channeling "untied" research funds to a broader array of southern researchers and institutions would also be helpful. In the spirit of equity, perhaps funders of climate justice work could pledge that for every dollar (or euro or yen) given to northern researchers and activists, an equal amount should be directed toward supporting independent research in the south.

Last, but not the least, the links between developmental and climate inequities need to be emphasized.[30] Developmental inequities exacerbate climate inequities and vice versa. Therefore, integration between sustainable development and climate policies (mitigation and adaptation) makes eminent sense, and this is increasingly being recognized by experts from various communities.[31] This is consistent with the UNFCCC, which states that policies and measures to address climate change "should be integrated with national development programmes" (article 3.4). Positive and sustainable changes in development trajectories can also have significant positive benefits for GHG-emissions mitigation.[32]

Thus, sustainable development should be seen as both a mitigation and adaptation strategy (see chapters 25 and 42 in this volume).

Notes

1. To be fair, though, the Chinese economy grew at almost 10 percent per year over that period (in purchasing-power-parity terms); therefore the energy intensity of its economy has shown a remarkable improvement. (Data from World Bank, *World Development Indicators* online, www.world bank.org/data.) The emissions data in this paragraph are from the Climate Analysis Indicators Tool (CAIT) version 4.0 (Washington, DC: World Resources Institute, 2007). Available at http://cait .wri.org.

2. Intergovernmental Panel on Climate Change, *Climate Change 2001: Synthesis Report. Contribution of Working Groups I, II, and III to the Third Assessment Report of the Intergovernmental Panel on Climate Change*, ed. R. T. Watson, et al. (Cambridge, UK: Cambridge University Press, 2001).

3. These projected impacts and those discussed in the remainder of this paragraph are from Intergovernmental Panel on Climate Change, "Summary for Policymakers," in *Climate Change 2007: Impacts, Adaptation and Vulnerability. Contribution of Working Group II to the Fourth Assessment Report of the Intergovernmental Panel on Climate Change*, ed. S. Solomon, et al. (Cambridge, UK: Cambridge University Press, 2007).

4. Intergovernmental Panel on Climate Change, *Climate Change 2001: Synthesis Report*. While one can argue about the accuracy of predicting costs during the next century (and we ourselves don't give much credence to such numbers), the fact remains that such assessments do shape policy.

5. B. Fisher, et al., Issues related to mitigation in the long-term context, in *Climate Change 2007: Mitigation of Climate Change. Contribution of Working Group III to the Fourth Assessment Report of the Intergovernmental Panel on Climate Change*, ed. O. Davidson and B. Metz (Cambridge, UK: Cambridge University Press, 2007).

6. "Tied aid" refers to aid with conditionalities, of which the most common is the longstanding practice of requiring the money be used to procure goods and services from the donor country. The practice of tied aid remains widespread, despite efforts to the contrary. Only a few donor countries have shown any significant progress in untying their overseas development aid. See United Nations Economic and Social Council, *Economic Report on Africa 2004: Unlocking Africa's Potential in the Global Economy* (New York: United Nations, 2004).

7. If there is no mitigation, it will not be possible to adapt to all the climate change that would occur. And even if mitigation activities are undertaken, there will still be some climate change; hence some level of adaptation will be unavoidable.

8. J. Paavola and W. N. Adger, Justice and adaptation to climate change, Working Paper 23, Tyndall Centre for Climate Change Research (2002).

9. T. Banuri, et al., Equity and social considerations, in *Climate Change 1995: Economic and Social Dimensions of Climate Change*, ed. J. P. Bruce, et al. (Cambridge, UK: Cambridge University Press, 1996).

10. Chasek and Rajamani suggest that there are four stages of multilateral negotiation: issue definition, fact-finding, bargaining, and strengthening of agreement. P. Chasek and L. Rajamani, Steps towards enhanced parity: Negotiating capacity and strategies of developing countries, in *Providing Global Public Goods: Managing Globalization*, ed. I. Kaul, et al. (New York: UNDP, 2003).

11. See, for example, M. Kandlikar and A. D. Sagar, Climate change research and analysis in India: An integrated assessment of a south-north divide, *Global Environmental Change* 9, no. 2 (1999): 119–38.

12. B. Muller, Framing future commitments: A pilot study on the evolution of the UNFCCC greenhouse gas mitigation regime, Oxford Institute for Energy Studies EV32 (2003).

13. See, for example, M. E. Keck and K. Sikkink, *Activists beyond Borders: Advocacy Networks in International Politics* (Ithaca, NY: Cornell University Press, 1998).

14. For example, there have been questions regarding the motives for the Indian government's willingness to join the U.S.-led Asia-Pacific Partnership for Climate Change, especially since such a program may undercut the multilateral negotiation

process (S. Narain, The new dirty deal, *Down to Earth*, September 15, 2005).

15. Kandlikar and Sagar, 1999 op. cit.

16. See, for example, M. Grubb, *The Greenhouse Effect: Negotiating Targets* (London: Royal Institute of International Affairs, 1989); A. Agarwal and S. Narain, *Global Warming in an Unequal World: A Case of Environmental Colonialism* (New Delhi: Centre for Science and the Environment, 1991); P. Singer, *One World: The Ethics of Globalization* (New Haven, CT: Yale University Press, 2002); P. Baer, Equity, greenhouse gas emissions, and global common resources, in *Climate Change Policy: A Survey*, ed. S. H. Schneider, A. Rosencranz, and J. Niles (Washington, D.C.: Island Press, 2002); for a rare counterargument see W. Beckerman and J. Pasek, The equitable international allocation of tradable carbon permits, *Global Environmental Change* 5 (1995): 405–13.

17. See, for example, A. Meyer, *Contraction and Convergence: The Global Solution to Climate Change* (Devon, UK: Green Books, 2000) or the website of the Global Commons Institute (www.gci.org.uk) for a discussion of the classic "Contraction and Convergence" proposal.

18. L. Ringius, A. Torvanger, and A. Underdal, Burden sharing and fairness principles in international climate policy, *International Environmental Agreements: Politics, Law and Economics* 2, no. 1 (2002): 1–22. See A. Agarwal and S. Narain, 1991 op. cit., or H. Shue, Subsistence emissions and luxury emissions, *Law and Policy* 15, no. 1 (1993): 39–59.

19. In addition to contraction and convergence, there are several such equity-based proposals for allocating emissions rights or emissions reductions currently under discussion. Among these are "Common but Differentiated Convergence" (N. Höhne, M. den Elzen, and M. Weiss, "Common but differentiated convergence [CDC]: A new conceptual approach to long-term climate policy, *Climate Policy* 6 [2006]: 181–90); the Climate Action Network's "Viable Framework" proposal (Climate Action Network, A viable framework for preventing dangerous climate change, [2002] available at www.climnet.org/pubs/CAN-DP_Framework.pdf); the "South-North Dialogue" proposal (H. Ott, et al., South-north dialogue on equity in the greenhouse: A proposal for an adequate and equitable global climate agreement, Deutsche Gesellschaft für Technische Zusamme-narbeit [GTZ] GmbH [2005]); and "Greenhouse Development Rights" (P. Baer, et al., The greenhouse development rights framework: Rationales, mechanisms, and initial calculations, EcoEquity and Christian Aid [2007], available at www.ecoequity.org/docs/TheGDRsFramework.pdf). For a review and comparison of these and some others, see P. Baer and T. Athanasiou, "A brief, adequacy- and equity-based evaluation of some prominent climate policy frameworks and proposals, Heinrich Böll Foundation Global Issue Paper #30 (2007), available at www.boell.de/index.html?http://www.boell.de/en/04_thema/5055.html&lang=en).

20. S. Huq, The Bonn-Marrakech agreements on funding, *Climate Policy* 2 (2002): 243–46.

21. Under article 12.8 of the Kyoto Protocol, a share of the proceeds from CDM activities is directed toward adaptation activities in particularly vulnerable countries.

22. See, for example, A. Najam and A. D. Sagar, Avoiding a COP-out: Moving towards dystematic decision-making under the climate convention, *Climatic Change* 39 (1998): iii–ix; and P. Baer, Adaptation to climate change: Who pays whom? in *Fairness in Adapation to Climate Change*, ed. W. N. Adger, et al. (Cambridge, MA: MIT Press, 2006), who suggests allocating adaptation funds on the basis of "net liability."

23. L. M. Bouwer and J. C. J. H. Aerts, Financing climate change adaptation, *Disasters* 30, no. 1 (2006): 49–63.

24. G. O'Brien, et al., Climate change and disaster management, *Disasters* 30, no. 1 (2006): 64–80.

25. D. J. Grossman, Warming up to a not-so-radical idea: Tort-based climate change litigation, *Columbia Journal of Environmental Law* 28, no. 1 (2003): 1–61; R. S. J. Tol and R. Verhayen, State responsibility and compensation for climate change damages: A legal and economic assessment, *Energy Policy* 32 (2004): 1109–30; M. R. Allen and R. Lord, The blame game: Who will pay for the consequences of climate change? *Nature* 432 (2004): 551–52.

26. S. Byravan and S. Chella Rajan, Immigration could ease climate-change impact, *Nature* 434 (2005): 435. S. A. Adamson and A. D. Sagar, Managing climate risks using a tradable contingent security approach, *Energy Policy* 30, no. 1 (2002): 43–51.

27. The two basic goals of tort law are (a) re-

ducing the overall costs of environmental harm and (b) providing corrective justice. D. J. Grossman, 2003 op. cit.

28. And climate-related damages could indeed be enormous. As an example of the costs of a major disaster, it is estimated that in addition to the attendant and widespread human and social misery, Hurricane Katrina may well cost the United States more than $150 billion (M. L. Burton and M. J. Hicks, Hurricane Katrina: Preliminary estimates of commercial and public sector damages, Marshall University, Center for Business and Economic Research, September 2005).

29. China, for example, released its first national plan on climate change in early June of 2007.

30. T. Athanasiou and P. Baer, *Dead Heat: Global Justice and Global Warming* (New York: Seven Stories Press, 2002).

31. See, for example, M. Munasinghe, Development, equity, and sustainability in the context of climate change, in *Proceedings of the IPCC Expert Meeting on Development, Equity and Sustainability, Colombo, 27–29 April, 1999*, ed. M. Munasinghe and R. Swart (Geneva: IPCC and World Meteorological Organization, 2000), and L. Schipper and M. Pelling, Disaster risk, climate change and international development: Scope for, and challenges to, integration, *Disasters* 30, no. 1 (2006): 19–38.

32. See Banuri, et al., Setting the stage: Climate change and sustainable development, in *Climate Change 2001: Mitigation. Contribution of Working Group III to the Third Assessment Report of the Intergovernmental Panel on Climate Change*, ed. O. Davidson and B. Metz (Cambridge, UK: Cambridge University Press, 2001). Note that this is very much related to issues of behavior, lifestyle, and consumption patterns. It is suggested often that developing countries should take a different development trajectory ("they should not make the same mistakes we made"). These expectations might be viewed as being inequitable by developing countries in that they would not have full control over their development trajectory. If the richer countries set an example by adopting sustainable (or climate-friendly) behavioral and consumption patterns, that might go a long way in persuading developing countries to follow, but unfortunately this has not happened so far.

Chapter 25

Ethics, Rights, and Responsibilities

PAUL BAER AND AMBUJ SAGAR

Introduction

The tension between the need to avoid significant climate impacts and the worry over the costs of doing so leads to two inescapably ethical policy questions: on what basis should limits on greenhouse pollution be set (we will call this the question of *precaution*)? And who should pay the costs of achieving the necessary reductions (the question of *allocation*)?[1]

We will argue in this chapter that, due to the priority of the general ethical principle that one should do no harm, it is necessary to rapidly reduce global greenhouse gas (GHG) emissions so as to minimize climate impacts and that it is an unjust imposition of harm and risk to advocate further delay in reducing emissions.[2] Furthermore, we argue that, as a consequence of the also widely held principle of equal concern, the costs of the necessary mitigation must be borne by those with the greatest responsibility and capacity. Finally, we argue that because individuals have little reason to believe that reducing their own pollution will solve the problem, there is an obli-gation to actively support the creation of the necessary institutions.

These arguments are not novel. Indeed, at one level we are simply asserting that the language of the United Nations Framework Convention on Climate Change should be taken seriously, that we are obligated to prevent "dangerous anthropogenic interference with the climate system," and that we should accept our obligations in proportion to "common but differentiated responsibilities and respective capabilities."[3] Our goal is to give a rather simple introduction to a very complex problem: what ethical reasons are there for choosing one climate policy goal or mechanism over another? In doing so we will not survey the literature on ethics and climate change, but we hope to tell a story that is comprehensible to those with no exposure to ethical theorizing as practiced by philosophers, yet also does at least partial justice to the scholarly tradition.[4] Thus before we return to our claim that the "do no harm" principle is a sufficient justification for a very stringent mitigation policy we will digress briefly and discuss our approach to ethical argument.

Our claim is that it is not necessary to study philosophy and decide what ethical theory is "best" in order to make reasoned ethical arguments.[5] A very wide range of approaches coexist in both academic and lay ethical reasoning, and individuals and communities can quite reasonably hold ethical premises that come into conflict in real situations. Equally important, different causal premises about how the world works can lead parties who share ethical views to reach different practical conclusions. While we acknowledge that there are true ethical dilemmas and that the relationship between ethics and interests is complicated, we maintain that there exist shared norms and experiences—however complicated and context-sensitive—that allow us to tell the difference between the legitimate and illegitimate pursuit of self-interest, and thus to distinguish the just from the unjust. We offer no new tricks—only an effort to engage the key questions with transparent arguments, sharing our assumptions and attempting to persuade our readers.

Climate Change, Pollution, and the Right to Protection from Harm

To make the matter concrete: as we suggested above, we believe that the ethical principle that it is wrong to cause harm to others for one's own benefit is widely held, and indeed effectively universal. By this we mean that some version of it appears in all cultures; this is not to say that anyone holds that it applies to all situations.[6] Quite plainly, cases of self-defense and self-preservation provide justification for overriding the basic principle; equally important, utilitarian arguments that advocate maximizing the common good at the expense of particular individuals are also accepted in many circumstances. The processes through which this tension between individual rights and the common good is resolved in general and in particular cases are complex, involving the assertion of rights, the use of persuasion, the establishment of legitimate decision procedures, and of course the exercise of power. Debates about pollution—which is a useful way to think about the climate threat—are paradigmatic examples of these processes.

Key to our argument in the case of climate policy is the claim that there is a general hierarchy of harm, in which death is at the apex, with physical suffering taking priority over the deprivation of property.[7] Equally important, imposing risk matters as well as causing actual harm; in other words, the right not to be harmed implies a duty not to impose risk of harm.

Because the risks from climate change involve the likelihood of widespread death, injury, and disease while the costs of mitigation are first and foremost the reduction of growth in income and consumption, we claim that a moral weighing of the alternatives strongly favors very steep emissions reductions, as we proposed above, even at the cost of what are currently considered impractically large economic costs. Furthermore, because of the asymmetry between rich and poor, between those who are most responsible and those who are most vulnerable (see chapter 24 in this volume), we suggest that it is appropriate to think of the problem as a question of justice, not merely a question of maximizing economic gains for an abstract global community. Those who benefit most from the highest levels of pollution and are frequently most insulated from its effects—namely, the wealthy and industrialized countries—are certainly not well positioned to judge the "value" of the harms that their actions impose on others without their consent.

The "do no harm" principle is of course not the only reason why one might support stringent climate mitigation. One might in fact believe that purely utilitarian reasons also justify stringent mitigation.[8] One might frame the question as a matter of sustainability, of preserving the opportunity of future generations to enjoy undiminished life chances, or as a requirement to provide stewardship of the climate system.[9] Or one might have an aesthetic preference for a future in which massive species extinctions or ecosystem transformations were prevented.[10] We choose here to emphasize the "do no harm" principle because we feel it is the strongest argument for stringent mitigation and appeals to the most universal ethical principles. Indeed, we suggest that many of these alternative framings draw their moral weight from the same kinds of prioritization of risks of bodily harm that we take as primary.[11] Furthermore, focusing on the "do no harm" principle makes it clear that what is at stake is not simply a matter of preferences (do we, or will people in the future, prefer a more stable climate or greater amounts of consumption?), but rather a right to protection and a corresponding moral duty that takes precedence over both interests and preferences.

Important arguments against stringent mitigation draw on the sanction against causing harm in a different way, by arguing from a utilitarian perspective that more harm would be caused by stringent mitigation than by the climate change caused by moderately increased GHG concentrations and that overall welfare would improve by maximizing economic growth. Typically these arguments are based on claims that there is effectively no risk from climate change (which we believe is no longer plausible), or on cost-benefit analyses (CBA), which allege to show that, even taking into account the prospective loss of life and other morally significant impacts, overall social well-being would be reduced by rapid and steep emissions cuts. These results are often an artifact of the choice of discount rate, which is itself a complex issue with important ethical considerations.

The applicability of CBA to problems like the threat of anthropogenic climate change has been challenged in many contexts on many grounds.[12] To begin with, because neither the likelihood of the most serious impacts of climate change nor their consequences in "lost utility" are well defined, no quantification of the benefits of a given mitigation policy can be robust to even an order of magnitude. Equally important, but less often discussed, the welfare impacts of a given mitigation policy are dependent on the distribution of the costs; thus the harm from the pure economic expense of mitigation depends on much more than just the stringency of the mitigation effort. For these reasons and more, we reject the claim that CBA can show the "optimal" level of emissions reductions.

Having said this, it is plain that there must be some point at which more rapid emissions reductions are unjustified; in the most extreme case, cutting CO_2 emissions to zero in one year would cause such unimaginable economic and social disruption (and indeed, likely loss of life on a large scale) as to be plainly unwarranted by any hypothetical climate benefits. Our argument, however, is the likely costs of reducing climate risk by steeply reducing GHG emissions are not unacceptably high, in a qualitative sense, at least if those costs are fairly distributed.

A related argument against stringent mitigation is that it is an *inefficient* way to benefit the potential victims.[13] This argument highlights the opportunity costs of mitigation and suggests that many more lives could be saved (or more broadly, a much greater increase in welfare provided) in poor countries through other mechanisms than through climate mitigation. This is an important argument and deserves much more space than we can give it here. At a high level, we would simply note that there is a tension between the utilitarian intuition that the sum of all suffering is what matters versus the intuition that it matters precisely who is exposed to harm or risk and why. On one hand, we as a society do accept that some amount of pollution and associated risk will be accepted in exchange for the benefits of the polluting activities; yet we do not then allow a company to increase its pollution beyond the legal level in exchange for an equivalent investment in, say, famine relief. With climate change, we believe two arguments are decisive: first, no one is actually suggesting that money not spent on mitigation be spent on other forms of development aid; and second, any such trade-off must surely be made with the consent of those who are being put at risk, not those who are benefiting from the risk-producing activities.

The Right to Sustainable Development and the Obligation to Pay for Mitigation

We previously suggested that a reasonable interpretation of the right to protection from harm warrants emissions reductions sufficient to ensure a high likelihood of avoiding dangerous climate change. The nature of the global GHG cycles and GHG lifetimes in the atmosphere in turn implies that GHG emissions need to peak soon and thereafter decline steeply. What are the consequences of staying within such a tight GHG budget? In addition, what would be a fair way of allocating the costs associated with reducing emissions to this level?

Figure 25.1 demonstrates the scope of the problem. The upper (solid) line shows a precautionary pathway in which CO_2 emissions peak in 2015 and fall to 50 percent of 1990 levels by 2050.[14] As an illustrative scenario, we show CO_2 emissions in Annex I countries dropping by 90 percent below 1990 levels in 2050, as proposed by Al Gore. Non–Annex I emissions are calculated as the difference between the global pathway and Annex I emissions, and reach a peak in 2018, just three years after the global peak.

With a few calculations, it is easy to show that, along such a precautionary trajectory, poor countries will have to reduce their absolute emissions long before their average per capita emissions or energy use reach even the levels of today's most efficient industrialized countries, and while they are still on average much poorer than the Annex I countries today.

Clearly, if we are to rapidly reduce global greenhouse pollution, developing countries will not be able to fuel their economies the same way that the rich countries have, and thus they will face costs for protecting the global commons that the industrialized countries did not. Given the importance of increased energy use for the development and improved well-being of the poorest people

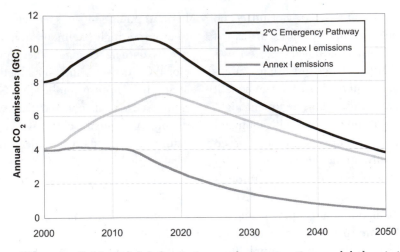

Figure 25.1. Illustrative division of global emissions under a precautionary global emissions trajectory. Global CO_2 emissions peak in 2015 and fall to 90 percent below 1990 levels in 2050. Annex I emissions decline to 90 percent below 1990 levels in 2050. Non–Annex I emissions peak in 2018.

and countries, we suggest that it would be unfair for them to bear all these costs themselves. The rich countries have both the greatest *responsibility* for the problem and the *capacity*—based on wealth in part derived from unrestricted greenhouse pollution—to pay for emissions reductions without a significant decrease in their welfare (see chapter 18).

Put differently, the ability to emit GHGs into the atmosphere represents a subsidy to persons and countries from the global commons, and it would be unfair for wealthy countries to continue to receive a larger per capita subsidy than poor countries. We suggest that minimally the principle of "equal concern"—which we believe underpins the idea of "equal opportunity" that is generally endorsed across the political spectrum—provides a strong justification for equal per capita access to global common resources. Moreover, we suggest that ultimately it is the cumulative benefit from the use of the global com-

mons that is the global resource to be fairly shared, not merely annual emissions.

Indeed, we suggest that the wealthy, industrialized countries have accumulated a kind of ecological debt.[15] While it is certainly true that CO_2 emissions were not historically recognized to be a damaging pollutant (at least not widely), and thus that their use would someday have to be greatly restricted, this absence of intent to cause harm (or even knowledge of harmful consequences) doesn't simply negate moral responsibility. The benefits of unrestricted GHG emissions contributed to the accumulation of capital that now supports the high income and consumption of the industrialized countries; if we accept these assets, presumably we should also accept the liabilities that come with them.[16] And since plausibly sustainable CO_2 emissions levels are on the order of half a ton of carbon per capita or less annually, while all today's wealthy societies have reached levels of 2 tC per capita annually (or much more), it's easy to see why even equal per capita emissions rights going forward will not be seen as fair by poor countries.[17] Given the very small remaining carbon budget, even an immediate transition to equal emissions rights would not deliver developmental equity.[18]

Finally, we argue that the underlying principle of equal concern or equal opportunity fundamentally addresses persons, not countries. Thus both ethically and politically it would be inappropriate for wealthy countries to subsidize unsustainable emissions levels for the wealthy people in poor countries. How institutions might manage this is beyond our scope, but we suggest that in giving form to this right to sustainable development, developing-country elites will have to also accept obligations appropriate to their responsibility and capacity (see chapter 26).

Conclusion

We have focused on the necessary global emissions trajectory, but individuals do not choose global emissions trajectories. They make various choices that increase or decrease their own production of pollution within a framework of laws and social norms that they only weakly influence. Thus climate is a classic commons problem. Such problems are difficult precisely because the individual incentive structure undermines the collective good. Reducing my CO_2 emissions to zero today would have a huge impact on my quality of life, but would have a negligible effect on carbon concentrations.[19] The cumulative effect of individual actions leads to the global crisis. This is the ethical dilemma at the heart of the so-called tragedy of the commons; even ethically motivated actors may continue to overexploit a commons because acting ethically brings no identifiable benefits to others.[20]

Commons problems require collective solutions, but our understanding of the relationship between ethical obligations—which must ultimately bind individuals—and practical institution building is limited.[21] For practical purposes we accept a de facto ethical division of labor, in which social and political leaders articulate moral choices, which individuals assent to through voting, purchasing decisions, and other means. It is essential for moral progress that some individuals choose these roles as moral leaders, but we have no obvious mechanism for judging whether enough people, or the right people, decide to take on such roles.

Moreover, politics represents a domain for pursuing both self-interest and the common good, and, arguably, the actual structures of representative government and capitalism favor self-interested over ethical behavior. Put simply, the people who profit most from the current institutional structures are in a position to provide incentives to political and moral leaders to support their interests, and to reward those who do. Furthermore, the politically and economically dominant actors in the most powerful countries are also able to shape global institutions to their interest.

In conclusion, therefore, we suggest that an ethics based on the right to protection from harm not only justifies the policies necessary to prevent the worst impacts from climate change, but also suggests that it is not enough for us as individuals to reduce our carbon footprint—we need to be part of the political change that will be required if the relevant policies are to be put in place.

Perhaps what is required is effective ethics of virtue, in which working for justice and the welfare of others is both socially rewarded and seen as intrinsically rewarding.[22] Such an ethics does in fact exist to some extent in our culture and worldwide; millions of people are to some degree involved in activities in pursuit of justice or charity. Yet injustice and vast unnecessary suffering still exist, and the climate problem, which may be one of the biggest challenges facing humanity this century, seems to require far more activism and leadership than we are already seeing. Where is it to come from? If some of us do not do much more to address the ethical and political issues, there seems to be little hope of solving the problem.

Notes

1. An equally important question is who should pay for adaptation to limit harm from unavoided climate change and liability for damages. We argue elsewhere that the obligation to pay for adaptation should also be allocated on the basis of responsibility and capacity; see chapter 24 in this volume, and also P. Baer, Adaptation to climate change: Who pays whom? in *Fairness in Adaptation to Climate Change*, ed. W. N. Adger, et al. (Cambridge, MA: MIT Press, 2006), 131–53.

2. Growing scientific evidence suggests that there are significant risks of dangerous and even catastrophic impacts at as little as a 2-degree-Celsius increase in global mean surface temperature above preindustrial levels; a high likelihood of keeping warming below this level requires global GHG emissions to peak within about ten years and fall to less than half of today's levels by 2050. (P. Baer and M. Mastrandrea, *High Stakes: Designing Emissions Pathways to Reduce the Risk of Dangerous Climate Change* (London: Institute for Public Policy Research, 2006, available at www.ippr.org).

3. See the UNFCCC website at http://unfccc.int/2860.php.

4. An excellent recent survey is S. M. Gardiner, Ethics and global climate change, *Ethics* 114, no. 3 (2004): 555–600. Pioneering work in ethics and climate change was done by Dale Jamieson and Henry Shue; their numerous relevant works are listed in a comprehensive ethics and climate change bibliography maintained by the Rock Ethics Institute, http://rockethics.psu.edu/climate. Both have interesting essays in the recent collection *Perspectives on Climate Change: Science, Economics, Politics, Ethics*, ed. W. Sinnott-Armstrong and R. B. Howarth (Amsterdam: Elsevier, 2005).

5. Philosophers have tried for many years to systematize the field of ethics, but it remains devoutly pluralistic, and there are a variety of taxonomies even of the most basic groupings. One common grouping is into deontological (rule-based), consequentialist (outcome-based, typically utilitarian), and virtue theories, but it is also possible to identify contractarian theories, common-sense theories, and many more classes and subclasses that overlap in various ways. The online *Stanford Encyclopedia of Philosophy* (http://plato.stanford.edu/) is a good source of definitions and accessible discussions.

6. We don't base this claim on our own anthropological expertise, but we've seen no counterexamples.

7. This hierarchy is independent of any formal ethical theory, but we suggest it is common enough to provide a reasonable consensus on most simple practical choices. Obviously, it doesn't by itself give guidance on complex comparisons.

8. See the *Stern Review* (at http://www.hm-treasury.gov.uk/stern_review_report.htm); but note that Stern and colleagues also suggest that cost-benefit analysis should set the *lower limit* on mitigation targets.

9. R. B. Howarth, Towards an operational sustainability criterion, *Ecological Economics* 63 (2007): 656–63. P. G. Brown, Stewardship of climate, *Climatic Change* 37 (1997): 329–34.

10. J. S. Risbey, Some dangers of "dangerous" climate change, *Climate Policy* 6 (2006): 527–36.

11. Note that the precautionary principle, which is also often invoked as a reason for stringent climate mitigation, has no content without some additional moral specification of what harms and risks warrant taking precaution against. See S. M. Gardiner, A core precautionary principle, *Journal of Political Philosophy* 14, no. 1 (2006): 33–60.

12. See, for example, F. Ackerman and L. Heinzerling, *Priceless: On Knowing the Price of Everything and the Value of Nothing* (New York: New Press, 2003); N. Hanley and C. Spash, *Cost-*

Benefit Analysis and the Environment (Aldershot, UK: Edward Elgar Publishing Ltd., 1993); M. Sagoff, *The Economy of the Earth: Philosophy, Law, and the Environment* (Cambridge: Cambridge University Press, 1988); or C. L. Spash, *Greenhouse Economics: Values and Ethics* (London, Routledge: 2002).

13. See, for example, T. C. Schelling, The cost of combating global warming: Facing the tradeoffs, *Foreign Affairs* 76, no. 6 (1997): 8–14; B. Lomborg, ed., *How to Spend $50 Billion to Make the World a Better Place* (Cambridge: Cambridge University Press, 2006).

14. These graphs are based on CO_2 emissions only, but including other GHGs would not change things significantly. Historical CO_2 data include an estimated 1,500 MtC of emissions from land-use change annually between 2000 and 2005.

15. K. Smith, Allocating responsibility for global warming: The natural debt index, *Ambio* 20 no. 2 (1991): 95–96; J. Martinez-Alier, *The Environmentalism of the Poor: A Study of Ecological Conflicts and Valuation* (Cheltanham, UK: Edwin Elgar, 2002), chapter 10; A. Simms, *Ecological Debt: The Health of the Planet and the Wealth of Nations* (London: Pluto Press, 2005).

16. Note that there are two separate issues here; first, responsibility for the "using up" of available carbon sinks and, second, liability for the harms that will be caused by climate change. On the first see E. Neumayer, In defence of historical accountability for greenhouse gas emissions, *Ecological Economics* 33 no. 2 (2000): 185–92; on the second, P. Baer, 2006 op. cit.

17. Uptake of CO_2 by oceanic and terrestrial sinks is generally estimated to be about 4 GtC annually; if global emissions were held to this level, CO_2 concentrations would (for a while at least) be roughly stable. With a population growing toward about 6.5 billion and expected to reach 8 billion in the next few decades, this is about one-half ton of carbon per person annually.

18. One of us has spent a great deal of his career arguing for equal per capita rights before coming to this perspective—see for example P. Baer, Equity, greenhouse gas emissions, and global common resources, in *Climate Change Policy: A Survey*, ed. S. H. Schneider, et al. (Washington, DC: Island Press, 2002) 393–408. For our newer perspective on developmental equity, see P. Baer, T. Athanasiou, and S. Kartha (2007), *The Right to Development in a Climate Constrained World: The Greenhouse Development Rights Framework*, available at www.ecoequity.org/docs/TheGDRs Framework.pdf.

19. This assumes that I reduce my actual emissions to zero, as opposed to paying somebody else to "offset" my emissions, which in theory can be done for perhaps a few hundred dollars.

20. W. Sinnott-Armstrong, It's not my fault: Global warming and individual moral obligations, in *Perspectives on Climate Change: Science, Economics, Politics, Ethics*, ed. W. Sinnott-Armstrong and R. B. Howarth (Amsterdam: Elsevier, 2005): 285–307.

21. Good discussions of the relationship between individual moral obligations and the institutions required to give substance to human rights can be found in H. Shue, *Basic Rights: Subsistence, Affluence and U.S. Foreign Policy* (Princeton, NJ: Princeton University Press, 1996), and T. Pogge, *World Poverty and Human Rights* (Cambridge, UK, Polity Press, 2002).

22. On virtue ethics and climate change, see D. Jamieson, When utilitarians should be virtue theorists, *Utilitas* 19, no. 2 (2007): 160–83.

Chapter 26

Developing Country Perspectives

Jayant Sathaye

Introduction

The climate change issue is part of the larger challenge of sustainable development.[1] Implementing climate policies can be more effective when consistently embedded within broader strategies designed to make national and regional development paths more sustainable. The impact of climate variability and change, climate policy responses, and associated socioeconomic development will affect the ability of countries to achieve sustainable development goals. The pursuit of these goals will in turn affect the opportunities for, and success of, climate policies. In particular, the socioeconomic and technological characteristics of different development paths will strongly affect emissions, the rate and magnitude of climate change, climate change impacts, the capability to adapt, and the capacity to mitigate.

Climate Change Problems and the Controversy

Developing countries, however, have widely varying views and perspectives on the strategies that should be followed for climate change mitigation and adaptation. Most developing countries oppose restrictions on their carbon emissions; all will need to adapt to climate changes. Small-island states and low-lying countries, like Bangladesh, are rightly concerned about projected sea-level rise; their focus is almost entirely on adaptation. Larger developing countries, such as Brazil, China, India, and Indonesia, already contribute significantly to carbon concentrations; their participation in reducing carbon emissions is necessary to stabilize greenhouse gas concentrations.

What is the best method to justly and equitably distribute the burden of stabilizing climate change among the countries? Most countries usually propose burden-sharing formulas that favor their economies; some have suggested schemes based on a combination of inherited and future emissions, a country's contribution to temperature change, GDP, and land area and other resource endowments. India, the fifth-largest emitter of greenhouse gases from fossil fuel in the 1990s, has suggested that the right to pollute the atmosphere be apportioned to all countries on the basis of their population. Using this gauge, China and India, the only countries with populations in excess of 1 billion each, could legitimately emit greenhouse gases for decades to a greater extent than other countries with smaller populations. Since their greenhouse gas emissions today are less than this proposed allocation, they could sell some of the rights to the industrialized countries.

Impacts of Climate Change: Implications for Developing Countries

Developing countries are already faced with immediate concerns that relate to forest and land degradation, fresh water shortage, food security, and air and water pollution, all of which will probably be exacerbated by climate change.[2] For example, by 2025, as much as two-thirds of the world population, much of it in the developing world, may be subjected to moderate to high water stress. Moreover, the populations of the developing world are more vulnerable as their infrastructure is not adequately developed to cope with additional stresses from climate change.

Although the ability to project regional differences in impact is still emerging, the consequences of climate change are projected to be more drastic in the tropical regions. Estimates of the effects of climate change on crop yields are predominantly negative for the tropics, even when adaptation and direct effects of CO_2 on plant processes are taken into consideration.

Role of Developing and Industrialized Countries in Addressing Climate Change: Mitigation and Adaptation

Regardless of the ranking of individual development strategies, the issue of greatest importance to developing countries is reducing the vulnerability of their natural and socioeconomic systems to projected climate change. Mitigation and adaptation strategies can both, if appropriately designed, reduce vulnerability to climate change and advance sustainable development and equity both within and across countries and between generations.

One approach to deciding among various adaptation and mitigation strategies is to compare the costs and benefits of each strategy. If adaptation of climate change could be carried out at negligible cost, then it may be the most feasible option, at least in the short term. Of course, there are complications in establishing the benefits of adaptation policies and consequent avoided damages.[3] Further, most mitigation and adaptation measures have significant co-benefits which need to be estimated.

The impacts or benefits of adaptation measures are more immediate and felt by the implementers of the measures. However, the climate benefit of mitigation will only be felt in the long run by the future generations. The regions implementing the mitigation measures could be different from the regions experiencing their impacts. The current generation of industrialized countries may invest in mitigation measures, and the main beneficiaries may be the next generation in the developing countries.

Certain mitigation measures, however, have been demonstrated to reduce greenhouse gas emissions while also yielding immediate economic and environmental benefits. The vast majority of energy efficiency measures fall in this category of "no-regrets" measures. Some renewable energy measures may not be cost-effective today but they too can reduce local pollution. While the climate benefits of mitigation may be long term, measures that yield economic and environmental benefits today are worth pursuing in developing countries. The Asia Pacific Partnership (APP) is one example of an initiative to promote such measures. Australia, Canada, China, India, Japan, South Korea, and the United States are partners in this venture.

An optimal mix of mitigation and adaptation strategies may elude the climate negotiations due to their different spatial and temporal dimensions, as well as the differing perceptions of industrialized and developing countries. Under the Kyoto Protocol and UNFCCC, developing countries have insisted that Annex I countries demonstrate commitment by promoting mitigation measures domestically and provide resources for adaptation measures in developing countries.[4] However, an overemphasis on adaptation might inhibit concerted mitigation actions by the Annex I governments, since adaptation measures are implemented and rewarded locally. Consequently, there is no incentive to participate in international negotiations if a country considers itself to be able to fully adapt to climate change.[5]

The future regime architecture can reduce the climate burden by giving greater emphasis to adaptation—for example, through an adaptation protocol—whereby mandatory funding by industrialized countries could support adaptation activities in developing countries. Additional policy options like support for adaptation planning and implementation, creation of a public-private insurance mechanism, and alignment of climate funds and development assistance can be deployed for gaining added benefits.

Climate Change and Sustainable Development

Sustainable development (SD) has become part of all climate change policy discussions at the global level. Three critical components in promoting sustainable development are economic growth, social equity, and environmental sustainability. Policy makers in developing countries often perceive a trade-off between economic growth and environmental sustain-

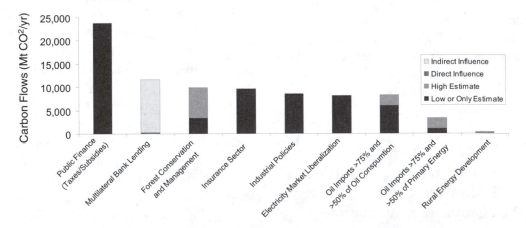

FIGURE 26.1. CO_2 emissions associated with sectors that could be targeted for sustainable development actions (2002). Source: Price L., S. de la Rue du Can, J. Sinton, E. Worrell, J. Sathaye, and N. Zhou, 2006. *Sectoral Trends in Global Energy Use and Greenhouse Gas Emissions*. Berkeley, CA: Lawrence Berkeley National Laboratory (LBNL-56144).

ability. However, there is growing evidence showing that environmental conservation for sustainability of natural resources is not a luxury but a necessity when considering long-term economic growth and development, particularly in the least developed countries. The decline and degradation of natural resources such as land, soil, forests, biodiversity, and ground water resulting from current unsustainable use patterns are likely to be aggravated due to climate change in the next twenty-five to fifty years. Africa, South Asia, and some regions of Latin America are already experiencing severe land degradation and fresh water scarcity problems.

Sustainable development strategies and actions that are directly targeted at reducing GHG emissions will be needed to achieve a low long-term GHG emissions trajectory. Both climate mitigation and nonclimate policies will need to contribute to stabilization of climate change. Nonclimate policies, such as electricity privatization, can increase emissions if, for example, it results in construction of natural gas power plants in place of hydro-electric power. However, it can also reduce emissions if the construction of additional coal power plants is avoided. Thus, judicious and informed choices will be needed when pursuing policies of sustainable development in order to ensure that GHG emissions are reduced and not increased. Sustainable development policies may be aimed at sectors that have larger mitigation potential. On the other hand, others that have low mitigation potential but a much larger societal benefit such as poverty alleviation may not need to take climate mitigation directly into consideration.

How should such SD actions be ranked? Ranking nonclimate SD actions requires estimating the current and future associated emissions of the targeted sector and the miti-gation potential of the SD action. Policy makers can then weigh the emissions reduction potential against other benefits of the SD action in choosing the appropriate policy to implement. In the discussion below, we present the associated emissions for selected sectors in which SD actions may be pursued, which provides some helpful guidance in ranking SD actions. A more complete analysis is needed.

Figure 26.1 shows selected examples of CO_2 emissions associated with sectors where sustainable development actions could be implemented. These examples are as follows.

- Through fiscal tax and subsidy policies, public finance can play an important role in reducing emissions. Rational energy pricing based on long-run-marginal-cost principle can level the playing field for renewables, increase the dissemination of energy-efficient and renewable-energy technologies, and improve the economic viability of utility companies, ultimately leading to greenhouse gas emissions reduction. For example, an OECD study showed global carbon dioxide emissions could be reduced by more than 6 percent and real income increased by 0.1 percent by 2010 if support mechanisms on fossil fuels used by industry and the power generation sector were removed.[6] Fuel prices/taxes/subsidies can impact the entire global fossil fuel emissions of CO_2, which amounted to 23.7 gross tons of CO_2 in 2002.

- Multilateral development banks (MDBs) can directly influence their own lending and indirectly influence the emissions of borrowing countries. The annual emissions from World Bank–funded energy activities alone, for instance were estimated to range from 268 to 323 Mt CO_2.[7] MDBs could directly influence more than the afore-mentioned amounts once emissions associated with all lending activities of all MDBs are counted. Indirectly, through policy dialogue and condition-ality, MDBs could influence the emis-sions from developing countries, which amounted to 11.2 Gt CO_2 in 2002.

- Adoption of forest conservation and sus-tainable forest management practices can contribute to conservation of bio-diversity, watershed protection, rural employment generation, increased in-comes to forest dwellers, and carbon sink enhancement. The forestry sector emissions show a high and low range to signal the uncertainty in estimates of de-forestation. An IPCC report noted that the emissions amounted to 6.6 Gt CO_2 with a +/–50 percent uncertainty for the 1990s.[8]

- Some insurers are beginning to recog-nize climate change risks to their busi-ness.[9] Examples of actions include pre-miums differentiated to reflect vehicle fuel economy; liability insurance exclu-sions for large emitters; improved terms to recognize the lower risks associated with green buildings; and new insur-ance products to help manage techni-cal, regulatory, and financial risks associ-ated with emissions trading.[10]

- Adoption of cost-effective energy effi-ciency technologies in electricity gener-ation, transmission, distribution, and end-use reduces costs and local pollu-tion in addition to reduction of green-house gas emissions. Building and in-dustrial energy efficiency improvements are being widely utilized, and policies

and programs are in place in many countries to promote their use. Indus-trial sector emissions amounted to 8.5 Gt CO_2, and all of it can be directly in-fluenced by actions taken to reduce intensity and switch to less-GHG-intensive fuels. Electricity deregulation or privatization can be practiced in any country and can impact global electric-ity-related emissions, which amounted to about 8.2 Gt CO_2 in 2003.

- Diversification of energy sources can enhance energy security and ensure a reliable supply of fuels and electricity. If low-carbon energy technologies replace oil imports, then reducing oil imports as a strategy to improve energy security of-fers a significant global opportunity to reduce emissions.

- Rural development policies, such as irri-gation and water management, or en-ergy and transportation infrastructure installation, can both promote sustain-able development and ensure long-term greenhouse gas emissions reduction.[11] For example, replacing traditional bio-mass fuels with fossil fuels would in-crease associated emissions; on the other hand, substituting inexpensive solar ovens would lower carbon emissions.

While we don't have data on mitigation potential, the associated emissions in figure 26.1 provide some guidance on prioritiz-ing sectors and actions that can yield larger benefits.

As indicators and measurement tools be-come available, the pursuit of sustainable development is moving out of academic dis-courses and being put into practice increas-ingly by institutions and private industry. The trend is likely to strengthen globally as nations come to recognize the limits on access to and development of natural resources.

Conclusion

The first commitment period of the Kyoto Protocol ends in 2012. Given the relatively short period to its termination, participating countries have been engaged in several dialogues, both within the UNFCCC auspices and elsewhere, about post-2012 commitments on emissions reductions and adaptation measures. The discussion at these dialogues ranges from mandatory economy-wide targets to sector-specific targets across all nations, to bilateral and/or multilateral agreements to voluntarily reduce GHG emissions. Industrialized countries, with the notable exception of the United States, already have agreed to adhere to economy-wide targets and seem keen to continue such an approach post-2012. Others have proposed sector-based approaches that require adoption of voluntary carbon intensity targets for the energy and other major industrial sectors in all countries. Key questions include how sectors are defined, how the voluntary target setting process unfolds, are there separate benchmark targets for new and existing facilities within a sector, when and how are reductions generated that can be sold, and how will sectoral benchmarks be part of an Annex I country target. A key to making a sector-based approach attractive to developing countries is the need for financial incentives to adopt such a target. A combination of technology finance and CDM/trading revenues could serve as one basis for making such targets attractive to developing countries.

Addressing adaptation in a post-2012 international climate regime could be done through the use of mainstreaming approaches such as insurance and innovative financing mechanisms. There is a growing interest in evaluating the role that innovative insurance mechanisms and other risk-spreading activities may offer in addressing adaptation needs.[12] These options can be structured so that they both help address impacts after the fact, expediting recovery efforts, or they can be set up so they encourage participants to take anticipatory actions that help reduce their vulnerability. Mainstreaming adaptation—where adaptation responses are considered and integrated into sustainable development and poverty reduction processes—would increase effectiveness of responses.

For developing countries, enhancing the economic well-being of their citizens remains an urgent and pressing goal. To the extent the new climate architecture was perceived as a barrier to this, it would be resisted and would fail to garner the wide support necessary for economic efficiency and coordination to derive multiple benefits. In the coming decades, the GHG emissions per capita in most developing countries will remain significantly below those in industrialized countries. Yet, for most many developing countries, this is the century when the majority of their citizens are likely to first experience economic prosperity. The next climate regime would succeed to the extent it would create instruments that align with sustainable development goals, activities, and processes in these nations.

Notes

1. The concept of sustainable development as adopted by the World Commission on Environment and Development launched sustainability into political, public, and academic discourses. It defined the concept as "development that meets

the needs of the present without compromising the ability of future generations to meet their own needs."

2. Ravindranath, N., and J. Sathaye, 2002. *Developing Countries and Climate Change*. Kluwer Academic Publishers, Dordrecht, Netherlands, p. 286.

3. Jepma, C. J., and M. Munasinghe, 1998. *Climate Change Policy: Facts, Issues and Analyses*. Cambridge University Press, Cambridge, UK.

4. Jacob James, 2005. The science, politics and economics of global climate change: Implications for the carbon sink projects. *Current Science* 89, no. 3, 10 August 2005.

5. Pielke, R. A. J., 1998. Rethinking the role of adaptation in climate policy. *Global Environmental Change* 8, no. 2: 159–70.

6. OECD (Organization of Economic Co-operation and Development), 2002. *Reforming Energy Subsidies: UN Environmental Programme and Organisation for Economic Cooperation and Development*. OECD/IEA, Oxford, UK.

7. World Bank, 1999. *The Effect of a Shadow Price on Carbon Emission in the Energy Portfolio of the World Bank: A Carbon Backcasting Exercise*. Report No. ESM 212/99.

8. IPCC, 2000. *Land Use, Land-Use Change, and Forestry*. Ed. R. Watson, I. Noble, B. Bolin, N. Ravindranath, D. Verardo, and D. Dokken. Cambridge University Press for the Intergovernmental Panel on Climate Change, Cambridge, UK.

9. Vellinga, P., E. Mills, G. Berz, L. M. Bouwer, S. Huq, L. A. Kozak, J. Palutikof, B. Schanzenbächer, C. Benson, J. Bruce, G. Frerks, P. Huyck, P. Kovacs, X. Olsthoorn, A. Peara, S. Shida, and A. Dlugolecki, 2001. Insurance and other financial services, in *Climate Change 2001: Impacts, Adaptation, and Vulnerability*. Ed. J. J. McCarthy, et al. Cambridge University Press, Cambridge, UK, 417–45; Mills, E. 2005. Insurance in a climate of change. *Science* 309, no. 5737: 1040–44.

10. Mills, E. 2003. The insurance and risk management industries: New players in the delivery of energy-efficeint and renewable energy products and services. *Energy Policy* 31, no. 12: 1257–72.

11. OECD, 2002 op. cit.

12. Vellinga, P. V., et al., 2001 op. cit.; Mills, E., 2003 op. cit.; Mills, E., 2005 op. cit.

Chapter 27

CDM *and Mitigation in Developing Countries*

David Wolfowitz

Introduction

If developing countries continue to industrialize based on energy from fossil fuels, then their future impact on the global climate will be enormous. This is especially true for the large and growing economies of countries like China, India, Mexico, Brazil, and South Africa, the so-called plus-five nations of the G8 climate dialogue. Mitigating emissions from rapidly developing economies is therefore a critical aspect of climate change mitigation. The Kyoto Protocol, however, does not include caps for developing countries in its current iteration, in effect through 2012. The countries assigned binding targets under Annex I of the agreement are industrialized countries that are responsible for the vast bulk of anthropogenic greenhouse gas emissions. The developing (non–Annex I) countries refused to take on binding targets before Annex I countries took meaningful action.

While it is politically sensitive to restrict emissions from developing countries, their limited existing fossil-fuel infrastructure presents two practical advantages. First and foremost, shifting to clean energy in the absence of existing infrastructure often presents a huge cost advantage, because it does not require retiring extensive capital assets (power plants, fueling stations, and so on) that are nowhere near the end of their operating life. Financing low-cost emission reductions in developing countries is therefore an important lever for Annex I nations to control the cost of meeting their Kyoto obligations. Secondary to cost minimization, there is a widely held view that industrialization based on fossil-fuel technologies creates path dependence and that early incentives for clean energy in developing countries have multiplicative benefits for the long-term cost and efficacy of mitigating climate change.[1]

The global climate regime primarily engages developing countries in mitigation activities through the Kyoto treaty's Clean Development Mechanism (CDM). The following analysis of CDM begins with an explanation of the origin and structure of the CDM, showing that certain design flaws reflect what was an essential political compromise at the time. Subsequent examination of CDM performance shows major success in rapidly manifesting enormous financial flows based on a climate-oriented commodity and major shortcomings with respect to clean development, climate change mitigation, and regulatory design. The last section enumerates a number of options to address these shortcomings, including reforms to CDM itself and solutions beyond CDM.

The Kyoto Surprise

Engaging developing countries in a mitigation regime was a central concern during Kyoto Protocol negotiations. But there was a considerable gulf between the goal of Annex I countries for a comprehensive mitigation regime and the firm view from non–Annex I countries that those nations that had principally contributed to the climate problem must take action first. As described by Matsuo, the problem of how to include non–Annex I countries in the Kyoto treaty had become "one of the deadlocks without any possible solution."[2]

CDM was the innovative resolution to the impasse. It allows projects carried out in non–Annex I countries to generate emissions reduction credits, but it does not subject non–Annex I countries to binding emissions targets. Annex I countries are then able to purchase those credits in lieu of reducing the same quantity of emissions domestically.

Because CDM allows participants to sell emissions credits without taking on carbon emission caps, it poses some immediate conceptual difficulties. First, the system must produce *additional* reductions that would not have occurred under business-as-usual conditions, so that Annex I countries aren't excused from domestic action without ensuring climatically equivalent mitigation measures elsewhere. Second, quantifying the credits attributable to an individual project requires calculating what emissions would have been without the project (i.e., under the *baseline* scenario). These two related issues—demonstrating additionality and quantifying baselines—have proved to be the core challenge of CDM implementation.

Determining additionality and quantifying baseline scenarios pose a huge challenge for CDM. These complex conceptual hurdles have few analogous predecessors. Furthermore, there is a staggering array of project options available to reduce GHG emissions—including substituting natural gas or renewables for coal, capturing methane emissions from landfills, flaring gaseous industrial by-products, and many, many others—each subject to different technical and economic considerations. These challenges have left little room for metrics that quantify the overarching goal of promoting clean development.

A look at the first few years of the CDM shows enormous growth in spite of these obstacles, indicating that the private sector has placed considerable confidence in this new market mechanism. However, there are also core doubts about whether CDM promotes clean energy development (due to a preponderance of credits from nonenergy projects) or mitigates climate change (due to the inherent subjectivity of project additionality). Whether CDM bolsters the global climate regime or undermines it in the long term depends on how these doubts are addressed.

CDM in Adolescence

At the end of 2007, with an unofficial transacted volume of almost one billion tonnes, the CDM market was valued at €12 billion, which was almost double the volume and more than triple the value reported the previous year.[3] Broadly speaking, this means that CDM constitutes roughly a third of the global carbon market. To realize this level of private-sector interest in transnational trading of an unfamiliar commodity that derives its value solely from global climate policy is no small achievement and must be recognized as such.

Aside from its magnitude, however, there are serious criticisms of the quality of this achievement. The first is that CDM growth has encouraged very little adoption of clean energy technology, because the strongest market activity has been focused on potent industrial gases that are not part of the energy system (see chapter 28 in this volume). Even credits from clean-energy projects face questions as to whether project-level additionality has been accurately evaluated. And doubts about CDM's efficacy are complicated by a regulatory process that offers limited transparency and has had difficulty keeping pace with surging growth in project activity.

Early experience created an expectation that CDM projects would focus on sustainable development in the energy sector. Under the UNFCCC's program of activities implemented jointly (AIJ)—a precommercial pilot phase for CDM implementation—project-based emissions reduction activities did in fact center on renewable energy and energy efficiency projects. However, as the CDM has emerged from infancy, low-cost emissions reductions of industrial gases like HFC-23 became the primary focus.

Wara (see chapter 28) discusses the HFC-23 issue in detail, illustrating how this type of project—based on turning extraordinarily potent industrial waste gases into far less potent carbon dioxide—in some ways contravenes the intention of CDM. For the purposes of

this chapter it is enough to note that these projects are certainly motivated by CDM incentives, which is a modest success, but in hindsight these reductions could have been achieved far more cheaply by other means.

Projects destroying industrial gases HFC and nitrous oxide dominated the early rounds of CDM registration, but they have not kept pace with other project types. After the CDM executive board disallowed credits from facilities commissioned after 2001, only a handful of preexisting facilities were eligible to register for CDM, and all of them have now done so.[4] Credit volume from industrial gases has plateaued while other project types continued to grow, such that industrial gases made up less than 25 percent of all expected credits by the end of 2007.[5] Figure 27.1 illustrates the declining share of industrial gas projects over time.

Existing industrial gas projects will continue producing credits, but current predictions indicate that finite industrial gas potential will be overtaken by strong growth in other project types.[6] Renewable energy and energy efficiency made up half of the expected volume by the end of 2007.[7] Nevertheless, the market forces that created an early advantage for straightforward, low-cost credits have not disappeared. Projects that meet higher environmental and sustainable development expectations—the NGO community's Gold Standard, for example—face obstacles due to higher cost or greater complexity (or both).[8] At the time of this writing only 88 out of almost 4,500 projects in the CDM pipeline qualified as Gold Standard, and only 39 of those projects were capable of generating more than a few thousand credits per year.[9]

The silver lining of industrial gas projects is that one can make a strong argument that these emissions reductions are financially

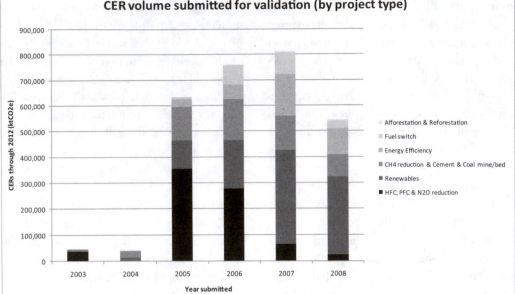

FIGURE 27.1. CDM project types. Source: UNEP-Risoe database, available at www.uneprisoe.org.

additional. With no source of revenue other than emissions credits they must have been motivated by CDM. This is a trickier question for most other projects, where carbon credits are just part—often a small part—of the overall return on investment. The regulatory core of CDM is largely geared toward making subjective assessments of financial additionality for projects where carbon credits are just a sliver of their total revenue. Due to the conceptual difficulty of the additionality issue, the constant struggle to ensure that CERs from each CDM project represent one-to-one reductions against a counterfactual scenario has yielded middling results.

Any discussion of additionality is likely to include a disclaimer that no approach can assess with full certainty whether a given project would have been initiated without carbon credits. In practice, establishing the additionality of an individual project under CDM is based on methods of extrapolating counter-

factual outcomes that are all highly subjective.[10] Among other issues, critics point to additionality determinations that are based more on presentation than on substance, the effect of host nation politics in CDM project approvals, and the nearly universal submission of projects in mature sectors—notably wind, hydro, and natural gas power in China—as evidence that additionality determinations under CDM are not reliable.[11]

Simply pushing for a more stringent examination of each project, however, is not a viable option. Even extreme levels of oversight would not eliminate the inherent subjectivity of additionality determinations, and project developers already complain that the CDM approval process is burdensome and fraught with delays and uncertainty.[12] Uncertainties due to the potential for false-negative additionality assessments create risks for investors. These risks discourage investments that rely on carbon credits to be viable (projects that are truly additional), while projects that don't need carbon revenue are largely unaffected.[13] Pushing for stricter project-level examination of additionality using the existing highly subjective approach would increase these risks and would have the largest negative impact on precisely those marginal projects that CDM is supposed to encourage.

Making the current system more stringent is also likely to worsen already serious problems with transaction costs and regulatory delays. Experts warned early on that high transaction costs would diminish interest in CDM projects.[14] While surging market growth would seem to disprove that prediction, transaction costs have strongly affected the distribution across project types. The most dramatic example is demand-side energy-efficiency projects, which require a large number of small improvements at many different locations, complicating their baseline and monitoring methodologies.[15] Although demand-side energy-efficiency projects include many win-win investments with strong benefits for sustainable development, high transaction costs have thus far limited their CDM presence to less than 0.1 percent of the registered project pipeline.[16] Further development of programmatic CDM methodologies may improve their representation, but this remains to be seen.[17]

Aside from discouraging certain project types, implementing the existing standards has also taxed the capacity of the regulatory authorities. Surging CDM submissions in 2006 and 2007 caused a backlog in registration requests such that at one point the CDM executive board required more than five months to review individual submissions.[18] Expanding capacity to accelerate registration has been difficult, and this is not the final hurdle. The final step in the CDM process, issuance of actual credits, appears to present an even more overwhelming logistical challenge.[19] Increasing the oversight of individual projects would compound the capacity problem even further.

Whether or not the CDM has lived up to its promise is largely a matter of how that promise is viewed. CDM has not become an engine for sustainable development in non–Annex I countries, but it has kept them attached to the global carbon market. CDM regulators have had difficulty guaranteeing environmental integrity and maintaining market efficiency, but they have built significant institutional understanding of emissions reduction projects that did not exist before. Nevertheless, the shortcomings of CDM under the Kyoto Protocol are almost certainly untenable in any post-Kyoto agreement. The next section will elaborate some options for fixing the holes in CDM without sacrificing its accomplishments, along with options beyond CDM for engaging developing countries in climate change mitigation.

What's on the Table

Measures to address the issues described previously can't just be focused on CDM itself. If the core question is how to engage developing countries in climate change mitigation in a way that is practicable, cost-effective, and environmentally sound, then CDM is only part of the answer. The concepts under discussion for meeting the broader challenge fall broadly into three categories: (1) ways to fix CDM; (2) complementary tools to reduce emissions from countries without GHG limits; and (3) ways to design binding GHG targets that might be acceptable to non–Annex I countries.

Reforming the CDM

The previous section described CDM short-comings with respect to clean energy development and additionality determination, issues that are made more serious by regulatory institutions that are not adequate to the task at hand. Measures to address these issues are all oriented toward building focus, certainty, and clarity into the regulatory process.

The first step is to draw the acceptable range of project types more tightly around clean energy. To prevent Annex I countries from substituting industrial gas abatement for clean energy investments, some suggest breaking up the Kyoto "basket of gases" into completely separate markets.[20] Alternatively, the strict atmospheric equivalence behind global warming potentials (GWPs) could be revised to a relative valuation that better reflects the perceived importance of these gases in the UNFCCC climate regime. Similar market restrictions and revaluations could be targeted at particular sectors or even project types to steer CDM toward clean energy development.[21]

If the CDM market were constrained to focus on a narrower range of projects, it would also be appropriate to deemphasize project-specific *financial* additionality in favor of broader guidelines on *environmental* additionality.[22] Despite meaningful effort, the CDM system has produced very few fixed guidelines on financial additionality. Environmental additionality, by contrast, has been folded into the various methodologies for calculating baseline and project emissions, and these methodologies constitute a significant body of knowledge related to greenhouse gas accounting. This asymmetrical success has made a persuasive case for considering tools like benchmarks and positive project lists that would not be evaluated at the individual project level.[23]

Measures that simplify and focus CDM will also set precedents that affect a wider range of projects, with greater potential to cause economic harm if they are applied without sound governance. In the current system, the CDM executive board has become the primary market authority—a de facto legislator and adjudicator combined into one body.[24] If reforms to the CDM system are to endow market regulators with greater top-down authority, these regulators—including the executive board, subsidiary panels, and designated operational entities (DOEs)—must also have clearly delineated roles, with due process to protect the rights of market participants.[25] Instituting this kind of administrative law might also optimize regulatory resources, improving overall confidence in the environmental effectiveness of the market.

None of these points can be adequately addressed by individual countries hosting CDM projects, or by the various domestic trading systems that constitute the demand market for CDM credits. Those entities are all driven by powerful incentives in the competitive global economy, and implementing any of the measures above is either against their individual interests or beyond their authority in the market. Furthermore, Annex I nations that do tighten CDM standards unilaterally (in spite of the competitive disadvantage) will find that more lenient rules elsewhere in the market continue to work against them.[26] The CDM is part of a global system to mitigate climate change, and as such it needs global reform by the UNFCCC parties as a whole.

Supplementing CDM

There are some mitigation measures that will never make sense as CDM projects even if they create cost-effective climate benefits. Agreements between Annex I and non–Annex I countries to address certain inputs to the climate problem (transport infrastructure, energy production, building codes, and so on) could have the potential to motivate mitigation efforts entirely outside the realm of emissions trading. Input-based measures are fundamentally different from the output-based emissions targets of the Kyoto Protocol, and there are very few existing mechanisms in the international community that provide instructive models. An input-based measure could address development priorities directly while providing a low-carbon pathway to economic growth, using financial incentives offered by individual Annex I parties or the multilateral climate regime as a whole.

Input-based measures might resemble the climate fund put forward by the finance ministers of the United States, United Kingdom, and Japan, or they might be direct commitments on specific actions by major developing countries.[27] Winkler and colleagues have proposed a sustainable development policies and measures (SD-PAMs) approach that would be based on development priorities put forward by non–Annex I countries, as opposed to the climate change commitments of industrialized nations.[28] Jackson and colleagues take this a step further by analyzing in depth the climate benefits from energy deals with China and India that prioritize economic development goals.[29] While such novel approaches are more difficult to implement, venues like the nascent Major Economies Process may prove to be more conducive than

the UNFCCC for moving input-based measures forward.

Alternative Target Design

As described above, CDM is a compromise. Ultimately, the most robust solutions to engage developing countries in the climate regime require non–Annex I countries (at least the major emitters) to adopt emissions limits of some kind. The Kyoto approach—directly negotiating absolute, economy-wide targets—raises issues of equity, capacity, and negotiating power for developing countries. Alternative concepts like equity-based targets and flexible compliance terms may offer potential for long-term convergence of Annex I and non–Annex I commitments.

The most straightforward approach to equity-based targets is based on the principle of "equal rights to the atmospheric commons for every individual."[30] If all people have an equal right to global atmospheric services (including both climate regulation and greenhouse gas disposal), then it follows that the most equitable emissions targets would be allocated on a per capita basis. Per capita allocations run counter to the interests of several key actors in climate treaty negotiations—notably the United States and Russia, among others.[31] A global climate regime cannot be effective without their participation, so per capita targets are out of the question for the foreseeable future. However, the principle is persuasive, and it favors major emitters like China and India where mitigation efforts will be critical in the long-term.

Another option to make targets more palatable to non–Annex I countries—one that faces less Annex I opposition—is to build flexibility into the terms of compliance using "no-lose" targets. A no-lose target would allow the relevant non–Annex I country to sell credits for emissions below the cap, but it would not carry any penalties if the cap were exceeded. By offering the opportunity to sell credits, no-lose targets create a positive incentive to reduce emissions, and they also have the potential to engage non–Annex I countries more fully in the global climate regime through emissions trading. Certain issues would require additional provisions—for example, forward credit sales under a no-lose target create a potential avenue for noncompliance if expected reductions are not achieved—however, no-lose targets could be feasible if they were designed with sufficient precision.[32]

Targets could also be intensity-based, stated in terms of emissions per unit of economic output. Historically, economic growth has been strongly correlated to proportional growth in greenhouse gas emissions. Intensity-based targets would encourage measures to reduce this correlation, while eliminating some of the economic uncertainties associated with mitigation commitments. Kim and Baumert even suggest dual-intensity targets, under which a modest binding target is hybridized with a more ambitious no-lose target, similar to the kind described above.[33] No-lose targets and intensity targets are mutually compatible in a number of ways that may make them more readily acceptable to all parties.

Conclusion

The potential impact of developing country emissions is huge. However, non–Annex I countries are still unwilling to take on binding emissions targets of any kind and are wary of measures that might lead to binding targets. UNFCCC discussions of sectoral approaches officially exclude mitigation commitments for non–Annex I countries because sector-specific targets are a commonly discussed strategy for phasing in caps on major developing country emitters.[34] As long as binding targets face this level of opposition, the global climate regime will have to make the best use of less robust measures like CDM and input-based mitigation to keep developing countries engaged.

CDM is the only one of these mechanisms that has an established presence. The market's rapid growth indicates enormous private-sector interest; to maintain this interest, policy makers must be careful not to alter the incentive in ways that seem opaque or arbitrary. Efforts to correct deep flaws in CDM's focus, implementation, and governance must be mindful of trade-offs between environmental certainty and commercial practicability. Focusing too strongly on one or the other would imperil the strongest existing link between developing countries and the global climate regime, and there may not be a viable alternative for a long time yet.

Notes

1. Unruh, G. (2000). Understanding carbon lock-in. *Energy Policy* 28: 817–30.

2. Matsuo, N. (2003). CDM in the Kyoto Negotiations: How CDM has worked as a bridge between developed and developing worlds? *Mitigation and Adaptation Strategies for Global Change* 8: 191–200.

3. Unless otherwise specified, "tonnes" refers to "equivalent tonnes of carbon dioxide" (tCO2e). The Kyoto Protocol applies the tCO2e metric to six gases based on their global warming potential (GWP). GWP is a conversion factor determined by the IPCC and is meant to measure how powerfully

a given GHG affects global climate. The six Kyoto gases are carbon dioxide (CO_2, GWP = 1), methane (CH_4, GWP = 21), nitrous oxide (N_2O, GWP = 310), the group of perfluorocarbons (PFCs, GWPs from 6,200 to 9,200), the group of hydrofluorocarbons (HFCs, GWPs from 140 to 11,700), and sulfur hexafluoride (SF6, GWP = 23,400). GWP values are currently used to equate emissions (and CDM credits) across different gases. For example, abating 1 ton of N_2O would generate 310 tCO_2e in CDM credits. The IPCC revises GWP values periodically. PointCarbon (2008). *Carbon 2008: Post-2012 Is Now*. Copenhagen. PointCarbon (2007). *Carbon 2007: A New Climate for Carbon Trading*. Copenhagen.

4. CDM Methodology AM0001 v3, available at: http://cdm.unfccc.int/UserManagement/File Storage/AM0001_version3 percent20.pdf. New Carbon Finance, Easy carbon credits coming to an end. Press release, July 29, 2008. Available at http://www.newcarbonfinance.com/pdf/2008-07-29_PR_Carbon_Credits.pdf.

5. Unless otherwise noted, percentages of total credit volume refer to forecasts of total credit production through 2012 for the entire CDM project pipeline. These forecasts vary from source to source, and variation can be significant depending on when they are made and what stages of project maturity they include. Various sources produce CDM pipeline forecasts; the most easily accessed is the UNEP-Risoe database, available at www.uneprisoe.org. PointCarbon, (2008) op. cit.

6. Forward-looking predictions of credit output indicate a growing share of credits from projects other than industrial gases. It should be noted that HFC-23 and N_2O projects still constitute more than 70 percent of approximately 185M CDM credits already issued at time of writing. For up-to-date CER issuance history, see http://cdm.unfccc.int/Issuance/cers_iss.html.

7. PointCarbon, (2008) op. cit.

8. The Gold Standard for CDM credits was created by a consortium of NGOs to certify premium credits from renewable energy and energy efficiency projects with sustainable development benefits, allowing these credits to fetch a higher price depending on market demand. Gold Standard is not an official UN certification. More information is available at www.cdmgoldstandard.org.

9. The Gold Standard project database, accessed January 9, 2009. Available at www.cdmgold standard.org/projects.php.

10. At present, all CDM projects must establish that they would not have taken place without carbon credits. This is a test of investment additionality or financial additionality, so this section focuses on this type of additionality determination. The complementary concept known as environmental additionality—establishing and quantifying climate benefits—is discussed in chapter 25.

11. Michaelowa, A., and P. Purohit (2007). Additionality determination for Indian CDM projects. Climate Strategies. Flues, F., A. Michaelowa, and K. Michaelowa (2008). UN approval of greenhouse gas emission reduction projects in developing countries: The political economy of the CDM Executive Board. Zurich, Center for Comparative International Studies (CIS). Wara, M., and D. Victor (2008). A realistic policy on international carbon offsets. Program on Energy and Sustainable Development (PESD).

12. (May 7, 2008). Time to rethink CDM additionality—Newcombe. *Carbon Finance*. Available at www.carbon-financeonline.com/index.cfm ?section=lead&action=view&id=11223.

13. Trexler, M., D. Broekoff, and L. Kosloff (2006). A statistically driven approach to offset-based GHG additionality determinations: What can we learn? *Sustainable Development Law and Policy* 6 (2): 30–40.

14. Michaelowa, A., M. Stronzik, F. Eckermann, and A. Hunt (2003). Transaction costs of the Kyoto mechanisms. *Climate Policy* 3 (3): 261–78.

15. IEA (2001). An view on methodologies for emission baselines: Energy efficiency case study. Paris.

16. Ellis, J., H. Winkler, J. Corfee-Morlot, and F. Gagnon-Lebrun (2007). CDM: Taking stock and looking forward. *Energy Policy* 35: 15–28. UNEP-Risoe Centre on Energy, Climate, and Sustainable Development (2008). *CDM Pipeline Overview*. Accessed January 9, 2009. Available at http://cdmpipeline.org/publications/CDMpipeline.xls.

17. Programmatic CDM methodologies allow a single set of baseline and monitoring protocols to be applied to large number of small, widely dispersed activities that can be submitted as a single project.

18. UNEP-Risoe, 2008 op. cit.

19. Wara and Victor, 2008 op. cit.

20. The primary demand market for CDM credits is the EU-ETS, Europe's domestic trading system. The EU-ETS regulates carbon dioxide emissions from fixed sources like power plants and heavy industry. This kind of domestic action to reduce fossil-fuel emissions addresses the core of the climate problem. By contrast, global emissions of industrial gases are quite small and could be considered peripheral to long-term climate change mitigation. Wara, M. (2008). Measuring the clean development mechanism's performance and potential: Program on Energy and Sustainable Development (PESD). *UCLA Law Review* 55 (6): 1759–1803.

21. Some critics of CDM advocate measures like quotas or discounts on all CDM credits across the board. It's important to note that because these measures don't differentiate credits (by greenhouse gas, sector, project type, country of origin, and so on) they don't do anything to improve the "quality" of CDM credits. These measures favor the cheapest, easiest credits, which are the most heavily criticized. Wara and Victor (2008 op. cit.) point out that Gresham's Law applies to carbon credits as well; when they are not distinguished from each other, "bad" credits drive out "good" ones.

22. Early discussions of CDM design distinguished between financial additionality (whether or not a project would happen without carbon credits) and environmental additionality (quantifying the project's environmental benefit relative to an appropriate baseline). For a full discussion of the distinction, see Baumert, K. (1999). Understanding additionality, in *Promoting Development While Limiting Greenhouse Gas Emissions: Trends and Baselines*. Ed. J. Goldemberg and W. Reid. UNDP/WRI.

23. Cosbey, A., D. Murphy, J. Drexhage, and J. Blaint (2006). *Making Development Work in the CDM: Phase II of the Development Dividend Project*. International Institute for Sustainable Development (IISD).

24. Streck, C. (2007). The governance of the clean development mechanism: The case for strength and stability. *Environmental Liability* 2007(2): 91–100.

25. For more on how stronger administrative law would improve CDM governance, see Streck, C., and J. Lin (2008). Making markets work: A review of CDM performance and the need for re-form. *European Journal of International Law* 19 (2): 409–42.

26. Wara and Victor (2008 op. cit.) point out the applicability of Gresham's Law to credit quality in the carbon market.

27. Paulson, H., A. Darling, and F. Kunaga (February 7, 2008). Financial bridge from dirty to clean energy. *Financial Times*.

28. Winkler, H., R. Spalding-Fecher, S. Mwakasonda, and O. Davidson (2002). Sustainable development policies and measures: Starting from development to tackle climate change, in *Building on the Kyoto Protocol: Options for Protecting the Climate*. Ed. K. A. Baumert, O. Blanchard, S. Llosa, and J. F. Perkaus. Washington, DC: World Resources Institute.

29. Jackson, M., S. Joy, T. Heller, and D. Victor (2006). *Greenhouse Gas Implications in Large Scale Infrastructure Investments in Developing Countries: Examples from China and India*. (PESD).

30. Baer, P., J. Harte, et al. (2000). Equity and greenhouse gas responsibility. *Science* 289 (5488): 2287.

31. Aslam, M. A. (2002). Equal per capita entitlements: A key to global participation on climate change, in *Building on the Kyoto Protocol: Options for Protecting the Climate*. Ed. K. A. Baumert, O. Blanchard, S. Llosa, and J. F. Perkaus. Washington, DC: World Resources Institute.

32. For more on no-lose targets, see Philibert, C. (2000). How could emissions trading benefit developing countries? *Energy Policy* 28: 947–56.

33. Kim, Y. G., and K. Baumert (2002). Reducing uncertainty through dual-intensity targets, in *Building on the Kyoto Protocol: Options for Protecting the Climate*. Ed. K. A. Baumert, O. Blanchard, S. Llosa, and J. F. Perkaus. Washington, DC: World Resources Institute.

34. UNFCCC (August 25, 2008). Report on the workshop on cooperative sectoral approaches and sector-specific actions, in order to enhance implementation of Article 4, paragraph 1 (c), of the Convention. FCCC/AWGLCA/2008/CRP.4. Associated Press (2008). Climate change conference makes progress on key dispute. *International Herald Tribune*, August 22, 2008. Available online at www.iht.com/articles/ap/2008/08/22/africa/AF-Ghana-Climate-Change.php.

Chapter 28

Measuring the Clean Development Mechanism's Performance and Potential

Michael Wara

Introduction

The original intent of the Clean Development Mechanism (CDM) was to spur development of low-carbon energy infrastructure in the developing world both because it would achieve sustainable development goals and because it would reduce the cost of developed-country compliance with the Kyoto Protocol by substituting for early retirement of expensive high-carbon energy infrastructure in the developed world (see chapter 27). It comes as a great disappointment to find then that the CDM pipeline bears little if any relationship to this vision. Instead, the subsidy provided by purchase of CERs will largely ensure that high global warming potential (GWP) industrial gases such as HFC-23 and N_2O as well as CH_4 emitted by landfills and confined animal feeding operations (CAFOs) in non–Annex I nations are captured and destroyed.

The CDM was designed around the insight that the marginal cost of emissions reductions in developing, and especially rapidly developing, countries would be less than for developed ones.[1] In fact, something far different has occurred: the CDM has primarily proffered CO_2 reductions in the developed world in exchange for reductions of various non-CO_2 gases in the developing world.

Goals of the CDM

Three major goals of the CDM are to prevent dangerous interference with the climate system, to promote sustainable development, and to lower costs of compliance for Annex I parties.[2] The first goal is accomplished by assisting developing countries in reducing their emissions of GHGs. By providing non–Annex I nations with financial incentives for low-carbon intensity development, these nations are encouraged to adopt more climate-friendly development paths and remain engaged for the long haul, thus accomplishing the second goal.

The third goal, lowering the cost of compliance, is achieved by subsidizing the new clean power capacity in the developing world, where rates of energy-sector growth are the highest and energy infrastructure is least

developed, as a substitute for premature retirement of old dirty power capacity in the
developed world.[3] This will lower the cost of
treaty compliance for Annex I parties while
still achieving the same environmental outcome since the overall decline of atmospheric
GHG concentrations is unrelated to the location where reductions occur.[4]

Unfortunately, the vast majority of emissions reductions generated by the CDM are
not of this type. To the extent that they do
achieve overall improvements in environmental outcomes, they do so at extremely
high cost, and so they create a positive ratio of
environmental benefits to subsidy costs. Furthermore, there are significant questions
about what fraction of the environmental benefits claimed on paper represent real changes
in GHG emitting activities in developing
countries.

Patterns of Development in CDM Projects

As of this writing, there are 702 projects in the
CDM pipeline, representing approximately 5
percent of Annex I 1990 GHG emissions.[5]
However, only a small number of large projects dominate the potential supply that has
reached the validation or registration phase of
the CDM project process (see figure 28.1). Indeed, more than 40 percent of the CERs
promised for delivery by all potential projects
can be accounted for by just ten large projects.[6] This pattern is repeated for registered or
soon-to-be registered projects, which account
for 221 out the approximately 700 total projects (32 percent). As can be seen from figure
28.1, a small number of very large projects
dominate the projected supply of CERs from
registered projects. In fact, the ten largest

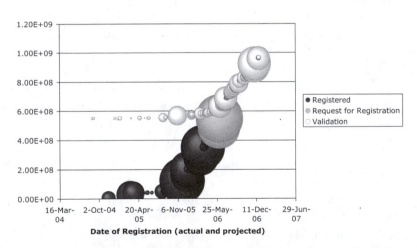

FIGURE 28.1. The total CER supply to December 2012 of all projects in the official CDM pipeline.
Total supply is based on the assumption that all projects in the CDM pipeline are registered and deliver CERs as promised in their PDDs is 966 MT CO_2e. An estimate of when validation stage projects will be registered is derived from the average time taken by currently registered projects to complete the process. Shown are projects in the CDM pipeline as of April 10, 2006.

projects (of the 221 registered projects shown in figure 28.1) represent 71 percent of the supply.

This indicates a bias toward registration of large projects, consistent with the impact that the relatively high transaction costs that the CDM's project-based system imposes on project proponents. It also suggests that an unknown number of smaller projects will in fact never be registered. However, since total supply of CERs is dominated by the larger projects, this possibility is unlikely to significantly impact the supply of CERs to Annex I parties.[7]

Supply of CERs in the CDM Pipeline by Project Type

The very large projects dominating the supply of CERs are confined to two relatively obscure industries—adipic acid and HCFC-22 manufacture—that represent nearly 55 percent of the supply of CERs in the CDM to date. Adipic acid is the feedstock for the production of nylon-66 and produces abundant N_2O as a production by-product.[8] HCFC-22 has two applications, first as one of two major refrigerants that were phased in to replace the CFCs under the auspices of the Montreal Protocol to Protect on Substances that Deplete the Ozone Layer.[9] Second, HCFC-22 is the primary feedstock in the production of PTFE, more commonly known by its Dupont brand name, Teflon.[10]

Contrary to ex-ante predictions, CO_2-based projects, including renewable low-carbon energy, energy efficiency, and cement process modification projects, account for just 29 percent of the CER supply to 2012. Renewable energy projects alone account for just 18 percent. Eleven HFC-23 capture proj-

ects at HCFC-22 production facilities make up 37 percent of projected supply while three projects that capture the N_2O made as a by-product of adipic acid or nitric acid production account for another 11 percent. Finally, 140 CH_4-capture and flaring projects, mostly located at large landfills and CAFOs, account for another 24 percent (see figure 28.2).

The bottom line is that the non-CO_2 gases dominate the supply of CERs to the carbon market, accounting for more than 70 percent of the predicted supply. Moreover, because the HFC-23 and N_2O (and to a lesser extent, CH_4) projects are typically of larger size than the renewable energy projects, they are more likely to overcome the transaction costs associated with registration and production of CERs than the smaller hydro-, wind-, and biomass-based energy projects that compose the CDM's renewable portfolio.[11]

Contrary to theory and expectation, the CDM market is not a subsidy implemented through a market mechanism by which CO_2 reductions that would have taken place in the developed world take place in the developing world. Rather, CDM subsidies are paying for the substitution of CO_2 reductions in the developed world for reductions in developing world emissions of industrial gases and methane.

HFC-23 Abatement Projects in the CDM
By-product of HCFC-22 Manufacture

There are eleven HFC-23 abatement projects currently participating in the CDM.[12] These projects consist of the capture and destruction of HFC-23 produced as a by-product of HCFC-22 manufacture.[13] The primary use of

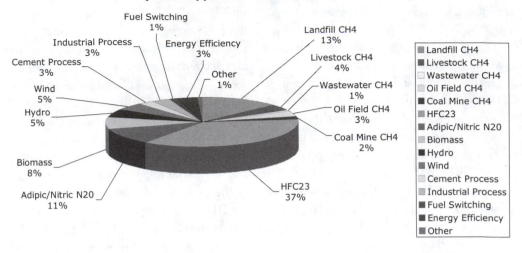

FIGURE 28.2. The fraction of CERs supplied by different project types.

HCFC-22 is as a refrigerant, although its use as a feedstock for fluoroplastics such as PTFE is also significant.[14] For every 100 tons of HCFC-22 produced, between 1.5 and 4 tons of HFC-23 is produced.[15]

HFC-23 is an extremely potent and long-lived greenhouse gas. Its 100-year GWP is 11,700.[16] As a consequence of this high GWP and the rules of the CDM, which allow conversion of the other six Kyoto Protocol gases to CO_2 equivalents and hence CERs, using their GWPs, 1 ton of HFC-23 abated is considered equivalent to 11,700 tons of CO_2.

Although approximately half of HCFC-22 production occurs in the developed world, there are essentially no by-product emissions of HFC-23 there because major producers have voluntarily adopted measures to capture and destroy it in order to burnish their environmental credentials.[17] Participation in voluntary abatement programs was substantial but not universal by 2004.[18]

The situation in the developing world was, prior to the CDM, quite different. There, HCFC-22 producers vented all HFC-23 produced to atmosphere.[19] One market analyst predicts that HCFC-22 production will grow by 6 to 7 percent per year to 2020 and by 16 percent per year in the developing world.[20] Thus, reducing non–Annex I emissions of HFC-23 should be a goal of any treaty aimed at curbing GHG emissions.

The Economics of HFC-23 Abatement as a CDM Project

The economics of HFC-23 projects create strong and perverse incentives for strategic behavior that, if left unchecked, undermine the environmental efficacy of the CDM. At current market prices for CERs, the production of HCFC-22 actually produces a subsidy far in excess of the price for HCFC-22.[21] Indeed, a developing-world producer of HCFC-22 can earn nearly twice as much from its CDM subsidy than it can gross from sale of its primary product.

The economics of HFC-23 CDM projects were a point of controversy from an early stage.[22] The CDM executive board decided to approve only those projects involving previously existing HCFC-22 production capacity.[23] No new plants or added capacity are currently allowed into the CDM.[24] In order to qualify as preexisting under the rule the executive board adopted, a plant must have been in operation and able to supply both HCFC-22 and HFC-23 production data for at least three years in the 2000 to 2004 period.[25]

Even with these restrictive rules on eligibility, there is strong circumstantial evidence that HCFC-22 producers participating in the CDM have behaved strategically to direct a greater share of the subsidy to themselves by artificially inflating their base year production in two ways: (1) by allowing the production of HFC-23 by-product to approach the 3 percent limit approved under the CDM mechanism, despite existing ability to produce less than half that amount; and (2) ramping up production at existing plants during the baseline period (2000 to 2004) far beyond the expected growth in the sector (15 percent) (see figure 28.3).

The CDM provides perverse economic incentives to HCFC-22 producers that have led to a large fraction of the CER supply being produced by HFC-23 abatement. Even if some fraction of these reductions are real and additional, they still may not be the best use of Annex I party resources for addressing non–Annex I GHG emissions. To abate all developing-world HFC-23 emissions would cost approximately $31 million per year.[26] Instead, by means of a CDM subsidy, the Annex I nations will likely pay between €250 and €750 million to abate 67 percent of non–Annex I HFC-23 emissions.[27]

This is a remarkably inefficient path to an environmental goal. It would seem that much more could have been done to combat global warming with these funds. The difference between the cost of abatement and the subsidy provided by the CDM also strongly suggests that a market-based mechanism that allows inter-convertibility of HFC-23 and CO_2 generated reductions may be a poor choice of treaty architecture for any post-2012 protocol to the UNFCCC whatever other form it may take.

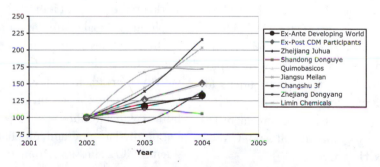

FIGURE 28.3. Percentage increases at HCFC-22 plants reporting multiple years of baseline data relative to ex-ante analyst predictions for the interval. 2002 = 100. Ex-ante developing world growth rate = 16.5 percent. Ex-post CDM participant growth rate = 25 percent. Thick lines show ex ante (filled circles) and average CDM participant (filled diamonds) rates of production growth.

Conclusion

Analysis of the financial incentives created for HFC-23 and N_2O emitters suggests that it is worth considering whether any future carbon trading program should be limited to CO_2 rather than including the other greenhouse gases covered by the protocol. Experience has shown that the other non-CO_2 gases can be abated at very low cost and at a relatively small number of facilities. Given these realities, it makes little sense for the developed world to subsidize their abatement at a price far in excess of cost. More sensible would be to pay developing-world emitters something closer to the actual cost of abatement.

The international community already has significant experience in compensating developing countries for the reduction of dangerous atmospheric emissions by paying for the cost of abatement rather than for a market determined price. The multilateral fund of the Montreal Protocol has been very successful at accomplishing the phaseout of the most harmful ozone depleting substances (ODSs) using this principle.[28] The fund facilitates developed nations' payment of any additional costs associated with the transition away from ODSs to new chemicals.[29] Under a future climate change protocol, this model could be adopted for the purposes of HFC-23 and N_2O abatement in the developing world with resulting emissions reductions applied to Annex I countries based on their contributions to the fund or some other agreed upon metric. A multilateral fund–style program might be worth considering for landfill and CAFO emissions of CH_4 as well.

Adopting a multilateral fund model for industrial emissions of GHGs has other advantages beyond just efficiency. Despite extremely strong financial incentives to do so,

the CDM's market incentives have only been able to produce about two-thirds of the abatement that is possible from the developing world HCFC-22 industry. The Montreal Protocol experience suggests that a multilateral fund aimed at abating emissions of N_2O and HFC-23 would allow better coverage of both industries so that maximum use of this low-cost emission reduction strategy could be realized, both in the developing and the developed world.

In addition, a strategy for these gases modeled on the multilateral fund would have the possibility of inducing a broader if perhaps shallower participation in a climate treaty. Nonparticipation is a fundamental problem with the Kyoto Protocol. Currently, the two largest emitters of GHGs, the United States, and China, are not participants in the components of the Kyoto Protocol that involve binding commitments. The United States refuses to take such an action without a similar commitment from China and India. These developing nations refuse to undertake binding emissions limits for fear that such a commitment will limit their economic development.

Lack of meaningful participation by the key global players in the most important climate change regime is unlikely to set the world on the path that it must take in order to resolve this global commons problem. Inducing these nations to enter into a series of protocols to the UNFCCC that dealt individually with the more tractable, less expensive aspects of the climate change problem might well build a global institution more capable of taking on the far more difficult challenge of reducing CO_2 emissions from power generation. Such a gradualist strategy has been successful in two of the great diplomatic successes of the twentieth century, the WTO and the European Union. Neither attempted to solve all of the difficult dilemmas at the beginning, but rather attempted to draw parties into a gradual process of accommodation that over time has produced substantial results.[30] New protocols to the UNFCCC that allow for rapid reductions in these trace gases offer the promise of building a solution to the problem of global warming in a piecemeal way that may encourage broader participation than has the grand bargain attempted by the Kyoto Protocol.

If, on the other hand, negotiators wish to stick with the Kyoto Protocol framework in the years after the first commitment period, even without American, Australian, and perhaps Canadian participation, there is yet another way to improve on the CDM. Nations are not required to purchase CERs or to allow private entities to purchase them; Europe has chosen to do so as a part of its compliance strategy. It is possible that Europe or the United States (if it participates in a future climate agreement) could modify its purchasing so as to encourage the kind of CDM that all had hoped for and discourage the accounting gimmicks and oversubsidization that have come to dominate the current market. It would be an easy matter for the European Commission to specify that post-2012, HFC-23 and N_2O project CERs would be either unexchangeable for European Union Allowances, the GHG currency of the Europe's Emissions Trading Scheme, or taxed at a very high level. Either action would go a significant distance toward leveling the playing field for energy efficiency and renewable energy projects.

The CDM's final disappointment is that among the CO_2 projects actually in the current program, there are virtually no large-scale power projects.[31] This is a major failure of the program in that the majority of GHG emissions from the developing world do and will in the future come from emissions from highly inefficient and carbon-intensive large energy projects. The small renewable energy projects currently typical in the CDM portfolio are unlikely to be more than marginal players in the energy market of the major developing countries. The major driver behind the inclusion of global carbon trading in the Kyoto Protocol was the insight that it was far cheaper to build new low-carbon emission energy infrastructure in the developing world than to replace it prematurely in the developed world. The CDM has not even begun to accomplish this goal. Resolving this failure of the current CDM market should be a central goal of any future trading program. Under current rules for the calculation of additionality it is unlikely to be possible. Even with a modification of these rules to encourage cleaner energy, it is unlikely given the marginal costs of building low- versus high-carbon–intensity large-scale power plants. Other workers have suggested that because of the politically determined nature of the energy-sector business-as-usual baseline, the best course of action in addressing this critical component of development may be to directly address the politics of energy-sector decision making in critical developing countries.[32] Whether or not the right incentives can be built into a market mechanism to drive low-carbon power development organically or such change must be precipitated by national policies and measures is beyond the scope of this chapter. In any case, the CDM as currently constituted is not doing the job.

Moving forward, the challenge for the international community will be to maintain the active participation in the CDM seen so far while honestly facing up to the flaws in the current system. By doing so, a far more efficient and therefore environmentally effective trading system could be fashioned.

Notes

1. See M. A. Toman, R. D. Morganstern, and J. Anderson, 1999. *The Economics of "When" Flexibility in the Design of Greenhouse Gas Abatement Policies.* Resources for the Future Discussion Paper 99-38-REV, pp. 2–3.

2. Conference of the Parties to the Framework Convention on Climate Change: Kyoto Protocol, 37 I.L.M. 22, at article 12, available at http://unfccc.int/resource/docs/convkp/kpeng.pdf (last visited April 3, 2006).

3. Energy Information Administration, U.S. Department of Energy, *International Energy Outlook 2005*, pp. 65, 67, available at www.eia.doe.gov/oiaf/ieo/index.html (last visited May 4, 2006).

4. W. Nordhaus, 2005. *Life after Kyoto: Alternative Approaches to Global Warming Policies*, NBER Working Paper No. 11889, p. 6, available at www.nber.org/papers/W11889 (last visited May 4, 2006). Because CO_2 is a well-mixed atmospheric gas with a long residence time, the extent to which it causes environmental harm is a function of its concentration in the atmosphere rather than the rate at which it is being added at any one time.

5. UNFCCC, *Greenhouse Gas Emissions Data for 1990-2003 Submitted to the United Nations Framework Convention on Climate Change*, 15 (2005), available at http://unfccc.int/resource/docs/publications/key_ghg.pdf (last visited April 3, 2006).

6. As of April 2006, the CDM supply to December 31, 2012, of the ten largest projects that have entered the CDM pipeline is estimated to be 409 Mt CO_2e while the pipeline as a whole has 966 Mt CO_2e. Thus more than 42 percent of CER supply is produced by less than 1.5 percent of the projects.

7. For example, were the smallest third (233 of 700) of projects in the pipeline to fail to achieve registrations, total CER volume produced to December 31, 2012, would be reduced by only 2.7 percent.

8. R. A. Reimer, C. S. Slaten, M. Seapan, T. A. Koch, and V. G. Triner, 1999. Adipic acid industry—N_2O abatement: Implementation of technologies for abatement of N_2O emissions associated with adipic acid manufacture, in *NON-CO_2 Greenhouse Gases: Scientific Understanding, Control and Implementation, Proceedings of the Second International Symposium, Noordwijkerhout, The Netherlands*, September 8 to 10, p. 347, ed. J. Van Ham, A. P. M. Baede, L. A. Meyer, and R. Ybema, 2000.

9. A. McCulloch, 2005. *Incineration of HFC-23 Waste Streams for Abatement of Emissions from HCFC-22 Production: A Review of Scientific, Technical and Economic Aspects*, p. 2, at http://cdm.unfccc.int/methodologies/Background_240305.pdf (last visited May 4, 2006).

10. Ibid.

11. E. Haites, 2004. *Estimating the Market Potential for the Clean Development Mechanism: Review of Models and Lessons Learned*, p. 45, available at http://carbonfinance.org/docs/EstimatingMarketPotential.pdf (last visited May 27 2005).

12. As of April 10, 2006.

13. INEOS Fluor Japan Limited, Foosung Tech Corporation Co., Ltd., and UPC Corporation Ltd., 2005. Revision to approved baseline methodology AM0001. *Incineration of HFC-23 Waste Streams*, p. 1, at http://cdm.unfccc.int/methodologies/DB/128DEJUIF08LJESQJHJKBTK86E1HKB/view.html (last visited May 4, 2006).

14. McCulloch, 2005 op. cit.

15. Haites, 2004 op.cit.

16. Ibid., p. 21.

17. Nordhaus, 2005 op. cit.

18. IPCC, 2001. *Climate Change 2001: The Scientific Basis*, p. 397, available at www.grida.no/climate/ipcc_tar/wg1/248.htm (last visited April 6, 2006).

19. McCulloch, 2005 op. cit.

20. Ibid.

21. McFarland, M., 2005. Environmental Fellow, DuPont Fluoroproducts, personal communication to Professor Tom Heller, Stanford Law School.

22. Letter from Thomas R. Jacob, senior adviser, June 3, 2004, Global Affairs, Dupont, to Mr. Jean-Jacques Becker, Chair, CDM Methodology Board, at http://cdm.unfccc.int/methodologies/inputam0001 (last visited May 4, 2006).

23. INEOS Fluor, 2005 op. cit.

24. Toman, 1999 op. cit.

25. Ibid.

26. Ibid.

27. 50 Mt $CO_2e \times €5 = €250,000,000$; 50 Mt $CO_2e \times €15 = €750,000,000$.

28. R. E. Benedick, 1998. *Ozone Diplomacy*, pp. 265–68.

29. Ibid., pp. 254–65.

30. But note that both of these treaty structures began with, and at least in the case of the EU continue to have, relatively small memberships when compared with the UNFCCC.

31. Out of 700 projects, only three will produce more than 250 MW of power. Only twenty-three will generate more than 100 MW. Meanwhile, China's electricity sector is growing at 4.8 percent or 100 billion kilowatt-hours per year. See Energy Information Administration, op. cit. 4, p. 98.

32. T. C. Heller and P. R. Shukla, 2003. *Development and Climate: Engaging Developing Countries, in Beyond Kyoto: Advancing the international Effort Against Climate Change*, pp. 111, 117.

Chapter 29

Understanding the Climate Challenge in China

Joanna I. Lewis, Jeffrey Logan, and Michael B. Cummings

Introduction

China's climate change mitigation strategy is of particular interest due to its role in global greenhouse gas emissions. Surging energy demand in China is having significant impacts within China and across the globe. Inside China, energy shortages, a resource-intensive industrial structure, and environmental pollution have incited debate on a wholesale reevaluation of Chinese energy and industrial policies, and their role in meeting the country's economic development objectives. Internationally, concerns over climate change, energy security, commodity trade, geopolitical conflict, and the larger global economic architecture have accompanied the rise in Chinese energy demand.

Though energy growth rates in China have slowed in the global economic downturn that began in 2008, Chinese coal demand reached an estimated 2.7 billion tons that year, more than twice the level of the United States. Perhaps surprisingly, Chinese petroleum consumption was still only one-third that in the United States despite very strong growth in recent years. While per capita

greenhouse gas emissions remain distinctly below those in the United States and other industrialized countries, China is now the world's largest national emitter of carbon dioxide. Over the past few years, China has constructed hundreds of pulverized coal-fired power plants that will lock in emissions for fifty years or more, and account for a significant share of the remaining global carbon budget needed to stabilize atmospheric concentrations of greenhouse gases. In 2006 and 2007 alone, China added more than 160 gigawatts (GW) of new pulverized coal capacity, equal to about half of the total U.S. coal-fired capacity. Clearly, China's energy and climate decisions during the coming decades will play a dominant role in influencing global technology, geopolitical, trade, and environmental outcomes.

The relationship between economic growth and energy use in China matters greatly. Between 1980 and 2000, China quadrupled its gross domestic product while only doubling its energy demand. This income elasticity of energy demand of roughly 0.5 during this period set an unprecedented mark for developing countries and saved

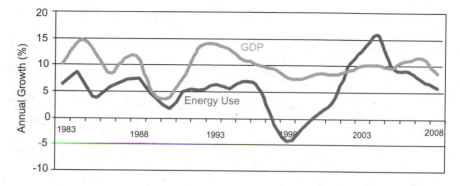

FIGURE 29.1. China's Energy Consumption and GDP, Annual Growth. Source: China Statistical Yearbook and authors' estimates. Data for 2007 and 2008 are preliminary.

China hundreds of millions of tons of coal combustion. Between 2002 and 2005, however, energy elasticity rose to well over 1.0 and erased some of the earlier benefits (figure 29.1). In response to the surge in energy growth, China set the extremely ambitious target of reducing energy intensity by 20 percent between 2006 and 2010, and mobilized much of the nation to accomplish it. Initial results show that these efforts may slowly be working; following increases in energy intensity each year from 2003 to 2005, intensity declined 1.23 percent in 2006, marking a reversal in the trend of increasing intensity, but falling short of the goal for that year of a 4 percent decline.[1] Preliminary figures for 2007 and 2008 appear to reinforce declining energy intensity. Still, the target will be challenging to meet.

Why did the Chinese economy suddenly become more energy intensive starting in 2002? Clearly, investment in energy-intensive cement, steel, and petrochemical production grew dramatically. But this does not explain everything. While Chinese energy and economic statistics are often unreliable, researchers have developed methods to evaluate and improve upon official statistics.[2] Many of these analyses have focused on the reported

decline in energy consumption during the late 1990s, a decline accompanied by an increase in economic growth. Some of this decline was real and resulted from tough economic reforms pushed by then-Premier Zhu Rongji. Some of the decline was illusory, however, and resulted from unreported coal production and consumption.[3] The recent surge in intensity, due in part to the recent surge in coal consumption for electricity and building material production, also conveniently corrects, at least partially, some of the earlier unreported coal use.[4]

According to the International Energy Agency (IEA) and other researchers, China surpassed the United States to become the largest carbon dioxide emitter in the world in 2007, much sooner than earlier estimates had predicted (see figure 29.2).[5] From a historical perspective, however, it will take many decades for China to surpass the United States in cumulative emissions. Data from 1930 to 2030, for example, indicate that the United States will have emitted 30 percent more energy-related carbon dioxide than China over the period: 416 gigatons CO_2 for the United States compared to 320 for China. These values correspond to the respective areas under the curves shown in figure 29.2. Viewing

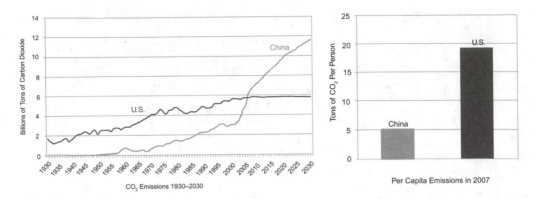

FIGURE 29.2. Chinese and U.S. Carbon Dioxide Emissions. Source: *World Energy Outlook 2008*. IEA; Carbon Dioxide Information Analysis Center, Oak Ridge National Lab.

cumulative emissions over a century or more timescale is a useful perspective because that is roughly how long carbon dioxide remains in the atmosphere.

Coal is likely to be the fuel of choice for many decades in China, despite the severe economic, social, and environmental dislocations it creates. Even with a reported 11 percent of the world's proven coal reserves, at current consumption rates China's coal supply may not last as long as once predicted, especially if coal liquefaction (using coal to make liquid fuels, e.g., for transportation) is practiced on a significant scale.[6] To what degree China can shift away from coal, or achieve decarbonized fuel supply, is the central issue.

For the time being, China's climate strategy remains centered on its energy development strategy, which in turn seems dominated by macroeconomic development policy. Like other large economies, China will eventually need to deploy a broad array of energy strategies if "greenhouse gas concentrations in the atmosphere are to be stabilized at a level that would prevent dangerous anthropogenic interference with the climate system," as the United Nations Framework Convention on Climate Change (UNFCCC) requires.[7]

China will face increasing international pressure to de-carbonize its energy sector in the years ahead. The level and speed at which the pressure builds will depend on many factors, but U.S. leadership on climate may be the most important.

China's "Wait and See" Climate Policy

A Domestic Scientific and Political Consensus

Scientists in China have made strong statements about the existence and the urgency of the climate change problem. It is also apparent that leading Chinese research organizations that often provide analytical input to shape government policy decisions are scaling up their work in this area.[8] Many Chinese scientists are engaged in research investigating the potential impacts of climate change, and the government recently released an interagency climate change assessment report, similar in structure to the reports of the Intergovernmental Panel on Climate Change (IPCC).[9] The increasing body of scientific knowledge concerning China's vulnerability to climate impacts has no doubt contributed to elevating the significance with which this issue is treated at the government level.[10]

In the early 1980s, the Chinese government treated climate change as a primarily scientific issue and gave the China Meteorological Administration (CMA) the responsibility of advising the government on policy options. However, the start of the international climate negotiations meant the engagement of the Ministry of Foreign Affairs (MFA) and the gradual politicizing of the climate change issue. As both political awareness and sensitivity surrounding the climate change issue increased, the primary role of representing the Chinese government was shifted to the more powerful National Development and Reform Commission (NDRC), the main government agency responsible for studying and formulating policies for economic and social development.[11] The move indicated a shift in the relative importance given to the issue, as well as perhaps a shift in perspective; initially viewed as primarily a scientific issue, climate change had now become recognized as predominantly a development issue.[12] The NDRC also serves as the primary energy policy decision-making authority in China, along with the new National Energy Administration (NEA), and this move may have reflected the clear need for climate priorities to be coordinated better with energy decisions. It is also home to the National Coordination Committee on Climate Change, which oversees climate activities within the NDRC, the MFA, the Ministry of Science and Technology (MOST), and the Ministry of Environmental Protection Administration (MEP).

Further institutional change came recently with the release of China's national climate change plan in June 2007.[13] The plan, later supplemented with the October 2008 climate change white paper, lays out an array of actions that China is taking to mitigate its greenhouse gas emissions, as well as the reported impacts it is likely to face in a changing climate.[14] The plan also announced the establishment of a high-level leading group on climate change chaired by Premier Wen Jiabao and reporting to the State Council. Subsequently, the Foreign Ministry announced that it had also established a leading group in charge of international work on climate change, headed by Foreign Minister Yang Jiechi. In September 2007, Ambassador Yu Qingtai was appointed to be China's new special representative of the Foreign Ministry for climate change negotiations. The role of this new special representative is to help implement China's domestic action plan to respond to climate change and to demonstrate "the government's active participation in international cooperation on responding to climate change."[15] At the end of 2007, China's lead climate negotiator moved from the Ministry of Foreign Affairs to a new leadership position within NDRC's National Coordination Committee on Climate Change. NDRC vice chairman Xie Zhenhua, former head of the MEP, has provided China's ministerial-level representation at the climate change talks in recent years. This institutional restructuring has both increased government capacity to work on climate change issues and has elevated the level of government at which they are being addressed.

Apprehension in the International Negotiations

Officials at the highest levels of China's government believe that climate change is a real and serious issue, and admit both their responsibility in and vulnerability to the problem.[16] However, this relatively advanced understanding of the problem has not contributed to a proactive policy stance on climate change mitigation—particularly at the international level.

Developing-country solidarity has been used as a tool since the early days of climate change negotiations, despite the growing economic differentiation within the developing world and often disparate interests with respect to climate policy issues. Aware of their limited weight of acting in isolation, developing countries attempt to build common positions in the framework of the Group of 77 (G-77).[17] The G-77 is the largest intergovernmental organization of developing states in the United Nations, providing a means for these countries to articulate and promote their collective economic interests and enhance their joint negotiating capacity on all major issues within the United Nations system.[18] China has historically associated itself with the G-77, despite the fact that it does not have the problem of limited weight in acting alone. In fact, China's motivation for siding with the G-77 is somewhat different—rather than having a fear of not being heard if acting alone, it can use the G-77 block as protection against being singled out. As the largest developing-country emitter, this is certainly a rational fear. However, its size does allow it to take a leadership role in formulating the positions of the G-77; China consequently has a hand in crafting its position while ensuring that a large contingent of countries will stand at its side when it is presented before the world.

In recent years, China's alliance with the G-77 has not waned, and its willingness to step out of the pack has declined even further as its fear of being singled out increases due to growing economic growth and energy use. In June 2005, Xie Zhenhua essentially announced China's plans to take a "wait and see" approach on climate change, stating that he hoped

> that some countries would, according to the obligations which are provided for in the Kyoto Protocol, implement in a substantive way their obligations and take up their commitments. . . . On the Chinese side, the Chinese government would make its own decision after making some assessments of the implementation by other countries.[19]

Some of China's hesitancy stems from concerns about energy data quality and transparency. China submitted its long-awaited initial National Communication to the UNFCCC in November 2004, in which it reported national emissions data through 1994—emissions estimates that at that point were ten years old and therefore revealed little about China's current emissions situation and likely future trajectory. Although developing countries are given complete discretion in submitting these communications, the unwillingness of many to submit official emissions inventories is often not due to a lack of financial or other resources, but rather a fear of subjecting emissions inventories to international scrutiny.[20]

One area in which China has clearly shaped the G-77 position is on the topic of technology transfer. With China's leadership, the issue of technology transfer has begun to transform from a rhetorical debate to a more concrete (though highly controversial) discussion involving specific funding mechanisms and dollar amounts. China has most recently proposed the formation of a new Multilateral Technology Acquisition Fund under the UNFCCC to support technology transfer and deployment in developing countries, and has called for developed countries to spend 0.7 percent of their GDP on helping developing nations to address climate change.[21]

In the past few years of international climate negotiations, countries within the G-77 have shown signs of diverging somewhat in their positions, which could leave China in a more isolated negotiating position. For example, some tropical forest countries—including Brazil and a coalition of thirty-two rainforest countries including Costa Rica and Papua New Guinea—have stated a willingness to take on voluntary avoided-deforestation targets in return for compensation.[22] Historically, voluntary international targets of any form have not been part of the G-77 position.

This evolution in the position of the G-77 was evident in the Bali Action Plan decision reached at COP13 in December 2007. The decision frames the upcoming negotiations on the form that the international climate framework will take after 2012—the year in which countries' current targets under the Kyoto Protocol expire. In addition to calling for "measurable, reportable, and verifiable nationally appropriate mitigation commitments or actions . . . by all developed country Parties," the decision also calls for "nationally appropriate mitigation actions by developing country Parties in the context of sustainable development, supported and enabled by technology, financing and capacity-building, in a measurable, reportable and verifiable manner," as well as "[p]olicy approaches and positive incentives on issues relating to reducing emissions from deforestation and forest degradation in developing countries."[23] The reference to measurable, reportable, and verifiable mitigation actions being taken by developing-country parties in particular marked a historic turning point in the international climate negotiations.

China in the Carbon Market

China has already ratified the primary international accords on climate change—the UNFCCC and the Kyoto Protocol—but as a developing country, China has no binding emissions limits under either accord. It is, however, an active participant in the Clean Development Mechanism (CDM) established under the protocol, which grants emissions credits for verified reductions in developing countries that can be used by developing countries toward meeting their Kyoto targets (see chapter 27 in this volume). China was initially skeptical about the introduction of the Kyoto mechanisms under the UNFCCC, viewing joint implementation (JI) and the CDM as instruments for developed countries to avoid their own responsibility to reduce emissions. China was also concerned about the potential for foreign exploitation surrounding rights to ownership of emission credits.[24] This position has changed drastically in recent years, however, as China began to realize the economic and political benefits the CDM could provide. The CDM has become a vehicle that not only stimulates investment in projects that mitigate greenhouse gas emissions, but also allows China to be viewed internationally as being proactive on the climate issue. Now the world leader in terms of CDM-induced GHG reduction credits on the books through 2012, China has learned how to use the CDM to its advantage.

The Chinese government has approached the CDM somewhat more cautiously and has taken a more involved role in the project approval process than other developing countries, getting China off to a relatively late start in the carbon market. Although CDM projects became eligible for crediting in 2000 (five years before the Kyoto Protocol entered into force), China did not ratify the treaty until August 2002, its Designated National Authority (DNA) overseeing CDM projects was not established until June 2004, and the State Council did not adopt rules for the management of CDM projects until October 2005.[25] These rules governing the CDM in China are viewed as "carefully crafted . . . to heavily favor Chinese interests and control, and to ensure Chinese 'resources' are protected," and have become a cause for complaint by many potential foreign investors—particularly the stipulation that only majority-owned Chinese enterprises may serve as project owners.[26] Investors have also complained that an unofficial price floor for Certified Emissions Reductions (CERs—credits issued under the CDM) has been set, illustrated by the fact that the DNA has been unwilling to approve CDM projects valuing CERs below €7 per ton.[27]

Despite these restrictions and complaints, China has emerged as the leading CDM host country, with more than 1.5 billion tons of carbon dioxide equivalent credits scheduled to be issued by the end of the Kyoto Protocol's first commitment period in 2012.[28] This means that more than half (about 53 percent) of total emissions reductions under the CDM are taking place in China. Although many CDM projects are being developed (see table

TABLE 29.1

CDM Projects and Associated Emissions Reductions in China (as of January 1, 2009). Source: UNFCCC CDM Statistics: http://cdm.unfccc.int/Statistics/index.html; UNEP Risoe CDM Pipeline, updated January 1, 2009. See also www.cd4cdm.org/Publications/CDMpipeline.xls.

	Number of Projects[a]	Number of Credits (ktons CO_2 by 2012)[b]	Percent of China-Based Reductions
HFC destruction	11	375,611	24.1%
Hydropower	764	329,259	21.2%
Energy efficiency[c]	262	216,236	13.9%
Wind energy	303	164,729	10.6%
Fossil fuel switching	31	127,455	8.2%
Coal bed methane	60	125,963	8.1%
N_2O capture	28	106,339	6.8%
Biomass energy	61	49,678	3.2%
Landfill gas	51	38,283	2.5%
Other[d]	3	10,043	0.6%
Biogas	25	6,172	0.4%
Cement	6	5,565	0.4%
Solar	4	695	0.0%
Forestry[e]	6	476	0.0%
Total	1,615	1,556,504	100.0%

[a]Includes projects that have been submitted but not yet approved (registered)
[b]Credits are estimates based on projected project performance
[c]Energy efficiency category includes industrial, own, and supply-side energy efficiency projects
[d]Other projects includes 1 transportation project, 3 energy distribution projects, and 1 fugitive emissions project
[e]Forestry includes afforestation and reforestation projects

29.1), a notable—and somewhat controversial—characteristic of these reductions is that just eleven HFC-23 projects are responsible for about one-quarter of these reductions. HFC-23 is released during the production of refrigerants (HCFC-22) and is a potent greenhouse gas, trapping 11,700 times more heat per unit than CO_2 (see chapter 28 in this volume).[29] Although the Kyoto Protocol states that CDM projects must contribute to sustainable development, and the Chinese CDM regulations explicitly prioritize energy efficiency, renewable energy, and methane capture and utilization projects, China (working with the World Bank) is leading the way in developing HFC-23 projects, and in doing so is generating very large amounts of credit sales with arguably no sustainable development benefits.[30] China has, however, signed a memorandum of understanding with the World Bank stating that it will reinvest a significant portion of this revenue toward meeting national sustainable development objectives, and subsequently began taxing CERs from HFC-23 projects at a rate of 65 percent, with the revenue being placed into a fund to be invested in climate change capacity building and other related activities.[31] At a price of $10 per ton, sales of the estimated $1.5 billion tons of reductions currently in the pipeline would represent a total investment in China of about $15 billion.

How Energy Policy Overshadows Climate Policy

In the absence of an explicit, national-level climate change mitigation strategy, China's energy policy becomes its de facto substitute. Today's energy policies not only impact short-term emissions growth, but its future emissions trajectory for decades to come. China is currently experiencing a dramatic expansion of its energy infrastructure that will remain operational for decades.

Energy Policy in Theory: Targets and Plans

Chinese energy policies come in many forms: policy statements, documents, plans, and pronouncements from the central government or high-ranking officials; laws passed by the national legislature; and governmental and quasigovernmental diplomatic and international business transactions motivated by energy concerns. The most recent comprehensive energy policy for China was released in 2003 in the form of the National Energy Strategy and Policy for 2000 to 2020 (the NESP).[32] China also released its first energy white paper, largely retrospective in nature, in December 2007 outlining the country's energy conditions and policies.[33] The overarching goal of the NESP calls for maintaining growth in energy use at a level half that of GDP, for a doubling of energy use between 2000 and 2020 while GDP quadruples (or a repeat of the growth in GDP and energy in China between 1980 and 2000). Yet even to maintain this relatively impressive GDP/energy growth intensity through 2020, the Chinese energy sector is poised to continue its breathtaking expansion—with serious implications for the country's greenhouse gas emissions.

In addition to establishing an overall energy-intensity target, the NESP also sets a goal to reduce coal to less than 60 percent of total energy use. To enable a relative decrease in the percentage of coal in the Chinese energy sector, the NESP calls for a dramatic expansion of nuclear, large-scale hydro, and small-scale- and non-hydro renewables. Nuclear capacity is to expand more than four times by 2020, large-scale hydro is to more than double (requiring the building of a dam the size of the Three Gorges Project every two years), and non-hydro renewables are to grow more than 100-fold.[34] While past growth rates indicate that the renewables targets are within reach, growing citizen opposition and government hesitation toward continued expansion of large-scale hydro plants may result in targets not being met.[35] In addition, China's ability to grow its nuclear fleet to levels contemplated by the NESP has been questioned by energy experts.[36] If large hydro and nuclear power facilities are not developed as outlined in the NESP, it is quite likely that any missing capacity will be met through additional coal-fired plants, leading to an even greater increase in China's greenhouse gas emissions than is currently projected in future scenarios.

Notwithstanding the aggressive targets for low-carbon electricity listed above, coal currently makes up 69 percent of total energy consumption in China, prompting another call by the NDRC to reduce the percentage of total coal consumption to just over 66 percent by 2010.[37] In addition, the NESP also calls for Chinese oil demand to be between 9 and 12 million barrels per day (mb/d) in 2020—but in 2008, China's oil consumption was estimated to already be at 7.9 mb/d, and annual oil demand growth was more than 5 percent. Demand for oil is expected to reach 8.2 mb/d in 2009.[38] Thus, the continued overwhelming dominance of coal in the Chinese energy sector, in combination with rapidly growing oil consumption, makes the targets set out in the NESP, and targets more recently announced by the government, look increasingly unrealistic.

Energy Policy in Practice: Decentralized Decision Making and Regionally Unbalanced Demand Growth

Chinese government reforms that have increased the autonomy and the wealth of local governments are contributing to the decreasing effectiveness of the central government in controlling the implementation of desired policies. Even if China decided to create an economy-wide climate policy in the future, there is increasing doubt that the central government would be able to guide or control it. China's partially reformed economy lies in an uncomfortable middle ground where neither the invisible hand of the market nor the iron fist of the centrally planned economy can exert effective control. To illustrate this point, the central government had little success in slowing overinvestment and growth in certain sectors of the economy—namely energy-intensive sectors—between 2003 and 2008.

Despite attempts to increase enforcement of environmental laws and policies at the national level, decisions pertaining to the development of China's energy sector are increasingly made at the local and regional level.[39] Consequently, the ability to have or enforce a national energy or climate policy has been doubted by some.[40] Key regulatory institutions—NDRC's Energy Bureau, for example—are woefully understaffed. In some cases, provincial and local-level leaders are lax in their enforcement of central government laws and regulations; and in other cases, local officials have effectively blocked implementation of central government initiatives. China's difficulty in implementing national energy policy goals is complicated by an unwillingness to allow energy prices to rise, as well as by a lack of incentives for efficient energy use at the local level. In an attempt to more effectively implement the 2010 national energy intensity goal, the NDRC is allocating the target among provinces and industrial sectors, and energy efficiency improvement is now among the criteria used to evaluate the job performance of local officials.

Even more worrisome than the rate at which coal power plants are being built in China is the central government's apparent lack of control over their being built. Regional power shortages have spurred a wave of new plant construction, often completed without central government approval or recognition of how such plants fit into these overarching national goals (see fig. 29.3). Government officials have recently signaled the need for an overarching energy law, along with a strong independent administrative agency to enforce such a law, but the establishment of such an agency is yet uncertain.[41] China installed a reported 93 GW of coal power capacity in 2006 and more than 80 GW in 2007.[42] The power plants added in 2006 alone will add 500 millions tons of carbon dioxide to China's annual emissions, or about 5 percent of total global emissions, and will lock in significant greenhouse gas emissions for decades to come. Also problematic is the fact that building more coal-fired power plants will not necessarily keep the lights on, due to pricing distortions and a mismatch between where China's coal resources are located and where the country's economic growth is concentrated. Government officials and experts alike have reported that many power shortages are not caused by a lack of power capacity, but by pricing distortions that create bottlenecks in the delivery networks.[43]

All of these factors combined call into question the Chinese central government's ability to begin down a different development

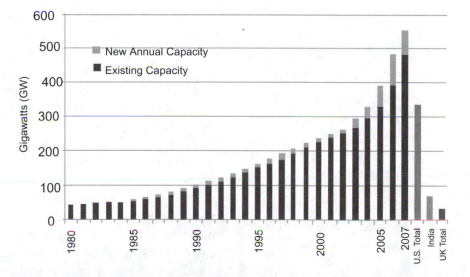

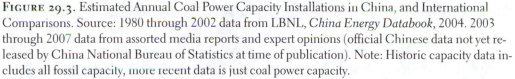

FIGURE 29.3. Estimated Annual Coal Power Capacity Installations in China, and International Comparisons. Source: 1980 through 2002 data from LBNL, *China Energy Databook*, 2004. 2003 through 2007 data from assorted media reports and expert opinions (official Chinese data not yet released by China National Bureau of Statistics at time of publication). Note: Historic capacity data includes all fossil capacity, more recent data is just coal power capacity.

trajectory, and to reduce its greenhouse gas emissions, without meaningful international engagement within the next decade—a critical time period in the development of the Chinese energy sector.[44]

Opportunities for Engaging China on Climate Change

Although China has made significant advances in some areas, the vast majority of energy development in the next few decades is likely to be fossil-fuel based and consequently will have profound implications for global greenhouse gas emissions. China is showing increasing recognition of a responsibility to engage with the rest of the world on issues related to climate change, but this transition is likely to be a gradual one. Additionally, the ability of the central government to effectively achieve its goals in certain economic and energy-sector reforms remains questionable. It is therefore critically important for the international community to ratchet up bi- and multilateral collaboration with China in order to address shared energy and environmental concerns before extensive carbon-intensive infrastructure is locked in place for a half century or more. Five areas are particularly well suited for further engagement and offer strong opportunities to expand global benefits: energy efficiency; energy security with climate benefits; clean coal and carbon sequestration; safe and secure nuclear power; and research, development, and demonstration (RD&D) efforts for renewable energy technologies.

Energy Efficiency

Efforts to reduce the wasteful use of energy are the most powerful and affordable measures China can take to meet economic development goals and reduce greenhouse gas emissions. Benefits of improved energy efficiency accumulate over time and can have a significant impact on energy security, local and regional environmental quality, and levels of investment for new energy supply infrastructure.

Just as China's increasing "car culture" is attracting fears about increasing its oil demand and imported oil dependence, China recently approved regulations for new fuel-economy standards for its rapidly growing passenger vehicle fleet that are more stringent than those currently in place in Australia, Canada, California, and the United States, but less stringent than those in the European Union and Japan.[45] Continuing its tradition of implementing relatively impressive energy efficiency policy measures, the target to cut energy intensity by one-fifth by 2010 is a dramatic example of how serious key decision makers have become about this issue.[46] In addition, several additional energy-efficiency programs were announced by the NDRC at the end of 2006, including a program to improve the energy efficiency of China's 1,000 largest enterprises (which together consume one-third of China's primary energy) and programs to shut down inefficient power and industrial plants.[47]

Yet, there is still room for additional collaboration with international partners, and there are several areas where bi- and multilateral collaboration to improve end-use and supply-side efficiency in China can be strengthened. There is a need for further capacity-building focusing on the business, financial, and regulatory skills needed to promote energy efficiency projects and standards, and to reform policies and regulations that impede market-driven efficiency projects. Developing incentives for accelerated technology and knowledge transfer, particularly be-tween Chinese and foreign energy suppliers and end users, are also crucial. Special efforts should be focused on improving energy, eco-nomic, and greenhouse gas data gathering, with the goal of improving the transparency of the energy-economic relationship in China.

Many of these efforts are already under-way, and Chinese government officials have proven open to international collaboration proposals that can help them meet their tar-gets. Foreign partners need to show an open and flexible approach so that efforts are not duplicative and can leverage maximum impact.

Energy Security Drivers and Climate Implications

China's booming economy has required a huge expansion in imported raw material in-puts, especially oil, since 2001. Chinese na-tional oil companies have embarked on a strat-egy to purchase oil and gas assets around the globe as a way to increase energy security. Some nations view these actions with alarm as there are potentially destabilizing military, political, and economic implications that stem from them. Whether this is a long-term policy course or just experimental first steps remains to be seen, but there is clearly the need for China and other major energy con-sumers to discuss more seriously their shared concerns. The fall in oil prices beginning in late 2008 allowed China to accelerate stock-piling of oil, but it is almost certainly not the end of energy insecurity.

To address the sense of liquid fuels insecu-rity, one option China has considered is coal liquefaction. While some Chinese policy makers find this a particularly attractive way to boost old-style self-sufficiency, it is not clear that China can solve its energy problems sim-ply by insulating itself from global reliance. One need look no further than the electricity crisis of 2004 and 2005, an entirely home-grown phenomenon in which ineffective planning led to severe electricity shortages, to realize this.[48] More important to the global system, however, is the climate impact that such a fuel substitution program would cre-ate. Coal liquefaction results in approximately twice as much CO_2 entering the atmosphere as standard petroleum use. While carbon cap-ture and storage could displace some of these emissions, it would add substantially to the cost. Already, it appears that China's water scarcity in the northern coal belt region may prevent coal liquefaction from taking off.

There is a clear need to better integrate China into the global energy system. Joining the IEA, G-8, and other global bodies in-volved in energy dialogues would give China greater ownership in the global energy system, though there still remain major challenges to bringing China in as a member.[49] The United States and other key nations should accelerate high-level dialogues with China to ensure that each others' concerns are understood and dis-cussions of opportunities and options occur regularly. The December 2007 announce-ment of U.S.-Chinese efforts to cooperate on biofuels development and use of strategic pe-troleum reserves and the passage of higher fuel-economy standards in the United States (to take effect by 2020) are both positive and long-overdue developments. Many additional opportunities exist for the United States to sig-nal to global oil markets a serious commit-ment to reducing long-term oil use, and such actions could facilitate more meaningful co-operation from China in areas of shared con-cern over energy security. As part of a more regular dialogue on issues of shared concern, China and international partners could also hold deeper technical collaboration on vehicle technologies, alternative fuels, and associated policies. The partnerships first need to focus on a dramatically improved atmosphere of trust and sincerity.

Advanced Coal Technologies and Carbon Capture and Storage

China's fleet of coal-fired power plants has undergone rapid expansion—with the equivalent of one to two large plants being constructed each week for the last few years. The total installed coal power plant capacity at the end of 2007 was about 554 GW, and nearly an additional 900 GW is projected to be brought on line in China by 2030.[50] These are anticipated to be almost universally pulverized coal (PC) plants and therefore represent a carbon commitment of more than 130 billion tons.[51]

Provided that it can overcome technical, cost, regulatory, and social barriers, carbon dioxide capture and geological sequestration (CCS) may become a critical option for reducing greenhouse gas emissions from fossil-burning plants throughout the world, but especially in coal-intensive countries such as China and the United States (see chapter 47 in this volume). While China is unlikely to invest in CCS systems for coal plants at a large scale in the next decade or two due to the added costs, it is beginning to expand its use of high-efficiency coal gasification technology and may initiate a few CCS demonstration projects in the near future. Integrated gasification combined cycle (IGCC) plants have the potential to produce significantly less pollution than pulverized coal alternatives and can

offer an efficient base from which to produce a number of petrochemical feedstocks using domestic coal. Experts currently believe that carbon dioxide emissions from these plants can also be captured for relatively low incremental cost if designed to do so in advance. Although retrofitting any type of coal plant later to capture emissions is not a simple or inexpensive option, some current estimates indicate that retrofitting IGCC plants for CCS would be cheaper than retrofitting supercritical pulverized coal plants for the same use.[52] At this time, such estimates and associated savings have not yet been tested in a commercial situation.[53] China is also interested in enhanced oil and methane recovery technologies that could use carbon dioxide in the process. The United States and Canada have extensive experience in this field. Building collaborative interest and capacity on these two topics could lay important groundwork for more rapid future deployment of CCS in China.

There are several international partnerships aimed at bringing advanced coal power technologies to China, including a new UK-led initiative as part of the China-EU partnership on climate change.[54] This partnership could lead to a CCS demonstration project starting between 2010 and 2015.[55] Instruments to finance such a demonstration plant will need to be resolved, as will potential technology transfer issues. China's Huaneng Power Company, in cooperation with the Australian Commonwealth Scientific and Industrial Research Organization, has also launched a post-combustion carbon capture project (without sequestration).[56] In 2007, the International Energy Agency in Paris started a separate but related set of collaborative studies on clean coal in China.

Huaneng is one of ten international energy companies that had been participating in the U.S. FutureGen "clean coal" project, attempting to become the world's first integrated sequestration and hydrogen-production power plant.[57] The future of this alliance is uncertain after the U.S. DOE announced in late January 2008 that it was withdrawing its financial support for the project, but it is possible that the Obama administration may revive it.[58] Huaneng is also leading its own version of FutureGen in China called "GreenGen." The project has the ambitious goal of demonstrating a near-zero emissions, 400-megawatt plant with entirely homegrown technology by 2020.[59]

In addition to FutureGen, China is also collaborating with international partners on coal and CCS technologies through the Asia Pacific Partnership on Clean Development and Climate (APP).[60] Officially launched in January 2006, the APP brings together China, the United States, Australia, India, Japan, and the Republic of Korea in an agreement based on clean energy technology cooperation. Some have criticized the APP as an attempt to further weaken the Kyoto Protocol and point out that voluntary technology cooperation partnerships offer little more than a distraction to the Bush administration's failed climate leadership.[61] At the time of this writing, it is uncertain how the new administration and Congress will support the APP.

China is also a member of the Carbon Sequestration Leadership Forum, an international initiative of twenty-two countries to promote the development of improved, cost-effective technologies for the capture and geological sequestration of carbon dioxide. The forum, initiated in 2003 by the U.S. DOE, is slowly beginning to open its process to new participants and initiatives, but to date has been heavy on process and light on action.

Safe and Secure Nuclear Power

China has expressed its intention to quadruple installed nuclear capacity by 2020. Leaders and commentators in other countries have also called for, and predicted, a new growth era for the nuclear industry motivated in part by concerns about climate change. For many countries, challenges related to cost, long-term waste disposal, and public acceptability must be adequately addressed before nuclear capacity can expand significantly. Even in countries with more competitive electricity markets and more transparent governance than China, considerable challenges remain (see chapter 46 in this volume).

Although China may have difficulty reaching its ambitious goals for nuclear power in the coming decades, it is still likely to appreciably increase its nuclear fleet. Considering these intentions, the international community should engage China and other nuclear countries in developing and enforcing an enhanced international waste and proliferation safeguards regime. If successful, such an enhanced international regime could help ensure an acceptable role for nuclear power to contribute to long-term global efforts to address climate change.

Research, Development, and Demonstration for Renewables

Motivated by the economic and environmental benefits these technology industries provide, China is committed to developing indigenous renewable energy technology industries and has accordingly set aggressive targets for renewables. China's national renewable energy law went into effect in January 2006, offering financial incentives for renewable energy development.[62] Targets that have been announced in conjunction with the renewable energy law and subsequent government documents include 15 percent of primary energy from renewables by 2020 (includes large hydropower—which would place the current share at about 7 percent today) and 20 percent of electricity capacity by 2020, which includes 30 GW of wind power, 20 GW of biomass power, and 300 GW of hydropower capacity. The wind target will likely be far exceeded, and the government has recently increased the 2020 target to 100 GW. Policies to promote many renewable energy technologies in China also aim to encourage local technology industry development; China is already producing commercial large wind turbines that sell for approximately 30 percent less than similar European and North American technology, and 35 million homes in China get their hot water from solar collectors—more than the rest of the world combined.[63] China also has a burgeoning solar photovoltaic manufacturing industry.

Nevertheless, non-hydro renewables will continue to make up a relatively small fraction of the energy mix in China during the next few decades. The further commercialization of new renewable energy technologies offers an important opportunity for international collaboration with China. The entry of Chinese manufacturers into these rapidly expanding global markets may drive cost reductions and increase the viability of renewable energy technology utilization worldwide, in both developing and developed countries. Assistance with commercializing several Chinese tech-

nologies that show promise now but are still in the demonstration stage will help push these technologies into the marketplace. Especially in light of the challenges facing the widespread deployment of CCS and nuclear power over the next few decades, a focus on renewables by the major energy-consuming countries of the world is warranted (see chapter 44 in this volume). Cooperatively harnessing China's growing manufacturing prowess with the innovation and commercialization experience of other industrialized countries to enable widespread deployment of solar photovoltaics and utility-scale wind turbines should be a high priority of the international community. Many existing international forums, such as the UNFCCC and the WTO, are being underutilized as opportunities to discuss key issues surrounding renewable energy technology transfer, including the role governments can play in facilitating the sharing and the protection of intellectual property rights.

There are clearly other options to collaborate with China on energy issues of shared concern; these are provided as initial examples. To regain trust with China on market-based energy and climate collaboration, the Obama administration will need to restore confidence in market fundamentals. Table 29.2 summarizes some of the key issues, impacts, players, and barriers related to collaboration. However, significant barriers to engagement in each of these areas remain, including a lack of political leadership, financing mechanisms, and intellectual property rights' concerns, among others. The bilateral and international partnerships mentioned above, along with the first set of EcoPartnership projects announced under the U.S.–China Strategic Economic Dialogue, begin to address some of these topics.[64] However, the time required to make progress in such in-

TABLE 29.2

Summary of energy and climate collaboration options
Notes: RD&D-Research, development, and demonstration; IFI-International Financial Institutions;
ESCOs-Energy service companies; IEA-International Energy Agency; IAEA-International Atomic
Energy Agency; IPR-Intellectual property rights.

	Focus	Potential Climate Impact	Primary Partners
Efficiency	Deployment	Moderate	Governments, IFIs, industry, ESCOs, NGOs
Energy security	Dialogue	Low-moderate	IEA, governments
Cleaner coal	RD&D	Moderate-high	IEA, IFIs, bilaterals, industry, NGOs
Nuclear safety	Dialogue	Low-moderate	IAEA, governments
Renewables	RD&D	Moderate	Industry, governments, IFIs, NGOs

ternational efforts in combination with the path-dependency of the global energy sector make efforts announced to date by the international community seem woefully inadequate to the meet the challenge.

Conclusion

Providing modern energy services for 1.3 billion people in a climate-friendly manner is a profoundly daunting challenge. Fortunately, the Chinese central government is demonstrating increasing awareness of the problems posed by climate change and interest in altering China's current energy development trajectory. However, the government's ability to significantly alter this trajectory without meaningful international engagement during the critical time period of the next one to two decades is questionable. Clearly, there is new urgency and opportunity to address linked climate protection, energy security, and economic development issues, especially among the world's largest energy consumers. The UNFCCC and Kyoto follow-on processes are important drivers that need continued support from host governments. Recent efforts by

the G-8, APP, and others offer opportunities to supplement the UN process, although it remains uncertain if they can add constructively to the broader challenges that to date have been outside the scope of climate change discussions.

U.S. leadership to address energy and climate issues at home and in international forums is essential to expand cooperation with China and other large developing countries. The new U.S. administration has promised far greater action in addressing climate change, but it remains to be seen if these promises are tempered by concerns with the domestic economy. The Obama administration must launch a serious, high-level dialogue with China that builds on the limited progress achieved in the Bush administration's Strategic Economic Dialogue. Such a dialogue could lead to energy security–enhancing initiatives with climate benefits and could lead the way toward climate-focused dialogue between the major energy consumers of the world.

Although sector-specific technical collaboration with China on the commercialization and expanded deployment of low-carbon options are ultimately crucial, initial efforts must begin with a demonstration of stronger political commitment. The International Energy Agency is one international forum in which serious engagement with China on energy and climate issues can be further developed. The world's major economies should expand their bilateral and multilateral efforts that seek to engage China on its core energy and climate challenges. There are also many opportunities for private-sector cooperation that facilitates the use of more advanced technologies in China.

There are a wide range of opportunities for collaboration within the five technology areas noted earlier in this chapter. Technology, finance, and policy issues need to be properly balanced to ensure full deployment of each. Enhanced cooperation on cleaner coal and energy efficiency is ripe for international cooperation and may produce the biggest results in the immediate future—particularly given recent growth in coal-fired power plants in China; the relative lack of geopolitical constraints surrounding options to collaborate in these areas; and the shared Chinese and international interest in helping clean and decarbonize Chinese industry. Recently initiated EU, U.S., and Australian efforts related to coal may be an important step in the right direction, but are insufficient as currently conceived considering the scale of the challenge. Finally, giving China greater understanding of and ownership in the industrial world's energy dialogue and security system will help minimize global tension and ensure more optimal outcomes for all.

Clearly, there are more opportunities for collaboration available than political willpower currently supports. But change can come quickly, and those prepared to engage will benefit first.

Notes

1. Nation unlikely to meet energy efficiency goal. *China Daily*, December 18, 2006. Available online at www.chinadaily.com.cn/bizchina/2006-12/18/content_761668.htm; China fails to meet this year's energy saving targets. *People.com.cn*, March 5, 2007. Available online at http://finance.people.com.cn/GB/1037/5440859.html.

2. Sinton, J. E., and D. G. Fridley, 2002. A guide to China's energy statistics. LBNL-49024 and *Journal of Energy Literature*, June. Available online at http://china.lbl.gov/publications/jel-china_stats-300402.pdf; Streets, D. G., X. Hu, J. E. Sinton, X.-Q. Zhang, D. Xu, M. Z. Jacobson, and J. E. Hansen, 2001. Recent reductions in China's greenhouse gas emissions. *Science* 294. Available online at http://china.lbl.gov/publications/streets_et_al_science-301101.pdf.

3. The reported decline in energy use in China during the mid- to late 1990s is discussed in Sinton, J. E., and D. G. Fridley, 2000. What goes up: Recent trends in China's energy consumption. *Energy Policy* 28(10): 671–87; Logan, J., 2001. Diverging energy and economic growth in China: Where has all the coal gone? *Pacific and Asian Journal of Energy* 11(1): 1–13; Lewis, J. I., D. G. Fridley, J. E. Sinton, and J. Lin, 2003. Sectoral and geographic analysis of the decline in China's national energy consumption in the late 1990s. *Proceedings of the American Council for an Energy Efficient Economy Summer Study on Energy Efficiency in Industry 2003.* Available online at http://china.lbl.gov/china_pubs-ind.html.

4. The 2006 *China Statistical Yearbook* (China National Bureau of Statistics) shows revised energy consumption data for the years 1999 through 2002.

5. International Energy Agency (IEA), 2007. *World Energy Outlook 2007: China and India Insights.* IEA, p. 191.

6. China to produce liquid fuel from coal. *China Daily*, March 20, 2007. Available online at www.chinadaily.com.cn/china/2007-03/30/content_840641.htm.

7. Article 2, United Nations Framework Convention on Climate Change. Available online at http://unfccc.int/essential_background/convention/background/items/1349.php.

8. CAS outlines strategic plan for China's

energy development over next 40 years. *Xinhua*, September 24, 2007. Available online at www .bjreview.com.cn/science/txt/2007-09/25/content_ 77642.htm.

9. China released its first *National Assessment Report on Climate Change* in late 2006, conducted as a collaborative effort between nine government departments that included the Ministry of Foreign Affairs, the National Development and Reform Commission, the State Environmental Protection Administration, the Ministry of Science and Technology, the China Meteorological Administration, and the Chinese Academy of Sciences, and took four years to complete. The Chinese assessment consists of three parts: the past and future of climate change, the impact and adaptation of climate change, and the socioeconomic impact from the slowdown of climate change. See National assessment report on climate change released. Press release, China Ministry of Science and Technology. Available online at www.most.gov.cn/eng/ pressroom/200612/t20061231_39425.htm.

10. Intergovernmental Panel on Climate Change in Climate Change, 2007. *Climate Change 2007: Impacts, Adaptation and Vulnerability. Working Group II Contribution to the Intergovernmental Panel on Climate Change*. Fourth Assessment Report, Summary for Policymakers, April 2007.

11. The NDRC reports directly to the State Council, China's lawmaking body, while the China Meteorological Administration is an additional administrative level below.

12. Bang, G., G. Heggelund, and J. Vevatne, 2005. Shifting strategies in the global climate negotiations. CICERO Report 2003:08. Available online at www.cicero.uio.no/media/3079.pdf.

13. National Development and Reform Commission, People's Republic of China, June 2007. China's National Climate Change Programme. Available online at www.ccchina.gov.cn/WebSite/ CCChina/UpFile/File188.pdf. The press widely reported that China would release its National Climate Change Plan on April 24, 2007; however, the release of this plan was delayed likely due to political concerns surrounding wide media reports that China was about to become the world's largest greenhouse gas emitter. See Exclusive: China preparing National Plan for Climate Change. *Planet Ark*, February 6, 2007, www.planetark.com/

dailynewsstory.cfm/newsid/40197/story.htm; McGregor, R., 2007. China delays climate change plan indefinitely, *Financial Times*, April 23, www .ft.com/cms/s/be763e8c-f1d6-11db-b5b6-000b5df 10621.html.

14. State Council of China, White Paper on *China's Policies and Actions for Addressing Climate Change*. Available online at www.gov.cn/english/ 2008-10/29/content_1134544.htm.

15. Chinese Foreign Ministry sets up climate change int'l working group. *Xinhua*, September 5, 2007. Available online at http://news.xinhuanet .com/english/2007-09/05/content_6667432.htm.

16. Liu, J., 2005. The challenge of climate change and China's response strategy. Keynote speech at the Round Table Meeting of Energy and Environment Ministers from Twenty Nations, vice chairman (minister level), National Development and Reform Commission of China.

17. Oberthur, S., and H. E. Ott, 1999. *The Kyoto Protocol: International Climate Policy for the 21st Century*. Springer, p. 24.

18. Group of 77 website available online at www.g77.org/.

19. See *China to Watch Others on Climate Change Action*, Reuters/Environmental News Network, June 15, 2005. Available online at www.enn .com/today.html?Id=7959.

20. The UNFCCC stipulates that each developing country (non–Annex I) party shall submit its initial national communication within three years of the entry into force of the convention for that party, or of the availability of financial resources (except for the least developed countries, who may do so at their discretion). The Framework Convention entered into force in China on March 21, 1994, but as a developing country China may submit its communication at its discretion. The only developing countries to have submitted a second national communication as of August 2006 were Mexico, the Republic of Korea, and Uruguay. In contrast, most developed country (Annex I) parties submitted their first communications before 1995 and were to have submitted their third national communications by January 2006.

21. UNFCCC Official Submission from China, September 28, 2008. China's views on enabling the full, effective, and sustained implementation of the convention through long-term cooperative action now, up to, and beyond 2012.

Available online at http://unfccc.int/files/kyoto_protocol/application/pdf/china_bap_280908.pdf. China urges developed nations to spend more on tackling climate change. *Xinhua*, October 29, 2008. Available online at http://news.xinhuanet.com/english/2008-10/29/content_10272172.htm.

22. See Coalition of Rainforest Nations, www.rainforestcoalition.org/eng/; Reducing emissions from deforestation in developing countries: Approaches to stimulate action, January 30, 2007. Available online at http://unfccc.int/files/methods_and_science/lulucf/application/pdf/bolivia.pdf (submission of views of seventeen parties to the 11th Conference of the Parties to the United Nations Framework Convention on Climate Change [UNFCCC]); Department of Environment and Special Affairs, Brazilian Ministry of External Relations, Brazilian perspective on reducing emissions from deforestation, February 26, 2007. Available online at http://unfccc.int/files/methods_and_science/lulucf/application/pdf/brazil.pdf.

23. The Bali Action Plan, December 2007. Available online at http://unfccc.int/files/meetings/cop_13/application/pdf/cp_bali_action.pdf.

24. Bang et al., 2005 op. cit.

25. China Office of the National Coordination Committee on Climate Change (NCCCC), 2005. Measures for operation and management of clean development mechanism projects in China. Available online at http://cdm.ccchina.gov.cn/english/NewsInfo.asp?NewsId=905.

26. Szymanski, T., 2006. China's take on climate change. *Sustainable Development, Ecosystems and Climate Change Committee Newsletter of the American Bar Association* 9 (1), May.

27. Point Carbon, 2006. China rejects CDM projects due to lower prices, claim buyers. August 2.

28. This number includes projects in the validation stage as well as registered projects, as of January 1, 2009. Source: UNEP RISOE CDM Pipeline, January 2009. Available online at www.cdmpipeline.org/publications/CDMpipeline.xls.

29. EPA, March 8, 2006. *High Global Warming Potential Gases: Hydroflurocarbons*. Available online at www.epa.gov/highgwp/scientific.html#hfc.

30. China Office of the National Coordination Committee on Climate Change (NCCCC), 2005.

31. Enav, P., 2005. Chinese companies, World Bank sign US$930 million pollution credit deal. *Associated Press*, December 19.

32. Development Research Center of the State Council, 2003. *China's National Energy Strategy and Policy 2000–2020* [Hereinafter the NESP Report], November 15–17. Available (in English) online at www.efchina.org/documents/Draft_Natl_E_Plan0311.pdf.

33. See www.china.org.cn/english/environment/236955.htm.

34. Sinton et al., 2002 op. cit.

35. See, for example, Yardley, J., 2005. Rule by law: Seeking a public voice on China's "Angry River." *New York Times*, December 26. Available online at www.nytimes.com/2005/12/26/international/asia/26china.html; Yardley, J., 2006. China proposes fewer dams in power project to aid environment. *New York Times*, January 12; Marsh, P., 2006. Power companies predict return of coal. *Financial Times*, January 16; Batson, A., 2007. Dissent slows China's drive for massive dam projects: Local critics scuttle hydropower plants; the right to say "no." *The Wall Street Journal*, December 19.

36. For instance, the U.S. EIA projects that China will have 23 GW of installed nuclear capacity in 2020 in its "reference case scenario" (or just over half of the NESP projections). U.S. EIA, *International Energy Outlook 2006*, p. 167. Available online at www.eia.doe.gov/oiaf/ieo/pdf/0484(2006).pdf.

37. Hong Guan, 2007. Energy plan: Reliance on coal and oil to be eased. *China Daily*, April 11. Available online at www.chinadaily.com.cn/bizchina/2007-04/11/content_847714.htm.

38. International Energy Agency (IEA), 2008. Oil market report. December 11. Available online at http://omrpublic.iea.org/.

39. See, for example, Batson, A., 2007. China eco-watchdog gets teeth. *Wall Street Journal*, December 18; French, H. W., 2007. Beijing seeks energy cuts; Localities find loopholes. *New York Times*, November 24.

40. See, for example, Lewis, S., 2004. *The Northeast Asia Energy Cooperation Workshop Program Conference Report: The Future of Energy Security and Energy Policy in Northeast Asia: Cooperation Among China, Japan, and the United States*, The James A. Baker III Inst. for Pub. Policy, Rice University, pp. 16–18. Available online at www

.rice.edu/energy/research/asiaenergy/docs/UFJ_ conferencereport_web.pdf; *MIT Study on the Future of Coal*, 2007. Chapter 5: "Coal Consumption in China and India." Cambridge: MIT Press.

41. See, for example, Yu, W., 2007. Need for an energy law overseer, says drafter. *China Daily*, April 28. Available online at www.chinadaily.com .cn/bizchina/2007-04/28/content_862645.htm.

42. Xinhua, 2007. China's installed power capacity exceeds 600 GWs in 2006. *Xinhua*, January 12.

43. See, for example, Agence France Presse, 2006; Ng, L., and Wing-Gar, C., 2005. Bracing for record coal prices. *International Herald Tribune*, January 4. Available online at www.iht.com/ articles/2005/01/03/bloomberg/sxpower.php.

44. For more on developments in the Chinese energy sector and the challenges faced by the central government in meeting its stated goals for the sector, see Cummings, M. B., 2006. Helping the dragon leapfrog: A survey of Chinese energy policy and U.S. energy diplomacy at the crossroads. *Environmental Law Reporter* 36 (July): 10526–52.

45. An, F., and A. Sauer, 2004. *Comparison of Passenger Vehicle Fuel Economy and GHG Emission Standards around the World*. Prepared for the Pew Center on Global Climate Change. Available online at www.pewclimate.org/global-warming-in-depth/all_reports/fuel_economy/index.cfm. This comparison does not include the future fuel-economy standards passed with the signing of H.R. 6, the Energy Independence and Security Act of 2007, signed by President Bush on December 19, 2007.

46. See, for example, Sinton, J., and D. Fridley, 2000. *Status Report on Energy Efficiency Policy and Programs in China: Recent and Related Developments*. June. Available online at http://china.lbl .gov/publications/status-rpt-eepolicy-6-00.pdf.

47. China National Development and Reform Commission, 2006. Available online at http:// hzs.ndrc.gov.cn/newzwxx/t20060414_66220.htm; China to require swap of old coal plants for new, 2007. PlanetArk.com, February 1. Available online at http://www.planetark.com/dailynewsstory.cfm/ newsid/40107/newsDate/1-Feb-2007/story.htm; China orders small cement plants to be closed, 2007. Reuters/PlanetArk.com, March 2. Available online at http://www.planetark.com/dailynewsstory .cfm/newsid/40623/story.htm.

48. See, for example, International Energy Agency (IEA), 2006. *China's Power Sector Reforms: Where to Next? Setting Near-Term Priorities to Build a Market-Based Sector*. July, IEA, Paris; Iyengar, J., 2004. China power crisis dims production. September 24. Available online at www.atimes .com/atimes/China/FI24Ad06.html.

49. Lieberthal, K., and M. Herberg, 2006. China's search for energy security and implications for U.S. policy. *NBR Analyis* 17 (1), April 2006. Available online at www.nbr.org/publications/ analysis/pdf/vol17no1.pdf.

50. IEA, *World Energy Outlook 2008*, p. 531. The IEA estimates that coal-fired capacity in 2006 was 449 GW, while earlier projections from the EIA's International Energy Outlook 2006 indicated 477 GW.

51. Calculated using assumptions presented by D. G. Hawkins, NRDC at the Workshop on Power Market Reforms and Global Climate Change, Stanford University, January 28, 2005. Presentation available online at http://iis-db.stanford.edu/evnts/4060/stanfordJan28_05_-_ hawkins.pdf.

52. Background information on the economics of carbon dioxide capture and geological storage can be found in Dooley J., R. Dahowski, C. Davidson, M. Wise, N. Gupta, S. Kim, and E. Malone, 2006. *Carbon Dioxide Capture and Geologic Storage: A Core Element of a Global Energy Technology Strategy to Address Climate Change*. PNNL, Richland, WA. PNWD-3602.

53. MIT Study on the Future of Coal, 2007. Executive summary. Cambridge: MIT Press; xiii–xiv.

54. European Commission, 2005. EU-China summit: Joint statement. September 5. Available online at http://ec.europa.eu/comm/external_ relations/china/summit_0905/index.htm.

55. UK Department of Environmental, Food, and Rural Affairs, 2005. Available online at www .defra.gov.uk/environment/climatechange/internat/ devcountry/china.htm.

56. 2008. Carbon capture milestone in China. *Science Daily* August 4. Available online at www.sciencedaily.com/releases/2008/07/0807311 35924.htm.

57. Official web site of the FutureGen alliance, www.futuregenalliance.org/alliance/members .stm.

58. Smith, R., and S. Power, 2008. After Washington pulls plug on FutureGen, clean coal hopes flicker. *Wall Street Journal*, February 2.

59. Shisen, X. 2006. Green coal-based power generation for tomorrow's power. Thermal Power Research Institute, presentation to the APEC Energy Working Group: Expert Group on Clean Fossil Energy, Lampang, Thailand, February 24.

60. See, for example, U.S. Department of State, 2006. *Work Plan for the Asia-Pacific Partnership on Clean Development and Climate*. January 12. Available online at www.state.gov/g/oes/rls/or/2006/59161.htm.

61. Shourie, D., 2005. Asia-Pacific pact not a substitute for Kyoto protocol: UNEP. *Hindustan Times*, July 29. Available online at www.hindustantimes.com/news/181_1445315,000400 11.htm; 2006. Asia-Pacific partnership no substitute for Kyoto Protocol. Greenpeace, Canada, May 18. Little, A. G., 2005. Pact or fiction? New Asia-Pacific climate pact is long on PR, short on substance. August 4. Available online at www.grist.org/news/muck/2005/08/04/little-pact/.

62. Renewable Energy Law of The People's Republic of China. Adopted at the 14th meeting of the Standing Committee of the 10th National People's Congress on February 28, 2005. Available online at www.resource-solutions.org/lib/librarypdfs/RE_law_english_version.doc.

63. Lewis, J. I., 2005. *From Technology Transfer to Local Manufacturing: China's Emergence in the Global Wind Power Industry*, PhD Thesis, Energy and Resources Group, University of California–Berkeley, August; Yundong, W., 2004. Manufacturing technology of wind turbines (grid connected) in China. *Proceedings of the World Wind Energy Congress*, Beijing, China, October 31–November 4. Reuters, 2005. China seen world leader in clean energy. September 29. Available online at www.chinadaily.com.cn/english/doc/2005-09/29/content_481754.htm.

64. U.S. Department of Treasury, 2008. *Eco-Partnerships: China and the U.S. Partnering for Success*. Available online at www.ecopartnerships.gov/.

Chapter 30

Climate Change and the New China

Paul G. Harris

Since its opening to the world in the late 1970s, China has become intimately connected to the world economy, and Chinese society has been transformed. China has rapidly become the new epicenter of environmental destruction, with most of its waterways polluted, its cities choking in smog, and its people suffering the effects of severe pollution of all kinds.[1] Indeed, only the United States surpasses China in its impact on the health of the Earth; China's economy and changing lifestyles draw in huge quantities of raw materials and commodities, and release vast quantities of pollutants and greenhouse gases.[2] The most profound environmental consequence of the new China is its contribution to climate change.

China's Energy Consumption and Contribution to Climate Change

China's energy use has doubled since the 1980s and will double again by about 2025.[3] Its use of energy is very inefficient by international standards. The ratio of energy consumption to gross domestic product is twice that of India, more than double that of the United States, and five times that of Japan.[4] Two-thirds of China's energy comes from coal, the dirtiest fossil fuel and the primary source of anthropogenic carbon emissions.[5] Indeed, China is the world's largest consumer and producer of coal with coal use increasing 13 percent per year on average.[6]

Although its per capita emissions are still relatively low, China is now the largest source of greenhouse gases (GHGs), surpassing the United States, and its overall emissions will inevitably increase as China's burgeoning middle class intensifies its energy use.[7] Worryingly, China's carbon dioxide emissions over the coming quarter century will likely be twice those of all developed countries put together.[8]

China's Burgeoning Consumer Culture

Between 2004 and 2013, the number of urban households in China with the ability to make "discretionary consumer purchases beyond meeting basic needs" is expected to increase

317

from 31 million (17 percent of households) to 212 million (90.6 percent of households).[9] The advent of discretionary income is concomitant with increases in resource and energy use, waste, pollution, and environmental destruction. The biggest environmental impact of rising per capita and household incomes is the rapid increase in demand for private automobiles. China is now the largest market for cars and appliances.[10] The number of passenger cars in China doubled every thirty months during the 1990s, and official estimates predict that the total number will reach 140 million by 2020.[11] In major cities bikes have been banned on major roads to make way for private automobiles. The cumulative effect of these consumption choices places an even greater and growing burden on the environment.

Environmental Policy in China

The Chinese government recognized some environmental problems and began addressing them as early as the 1950s, but into the 1970s it argued that as a socialist state China did not have environmental problems.[12] However, increasingly obvious damage to China's natural environment and associated impacts on economic development caused the government to be more concerned in the 1980s. In 1982 the Chinese Constitution was rewritten with the government pledging to "protect the environment and natural resources by controlling pollution and its societal impact, ensuring the sensible use of natural resources, and safeguarding rare animals and plants."[13] The following year, environmental preservation was declared one of China's basic national policies, and by the end of the decade China started its first major campaign to combat pollution.[14] During the 1980s the government also instituted new environmental protection laws in the areas of solid waste, noise, air, and water pollution, and in 1989 the government strengthened the Environmental Protection Law.[15]

By the mid 1990s the government was becoming more serious about environmental issues, even closing some polluting factories. In 1994 the Chinese State Council adopted "China's Agenda for the Twenty-First Century." Its main themes were economic development (first) and environmental protection (second). A new environmental awareness at the national level was evidenced in pronouncements of national leaders. For example, in 1995 Premier Li Peng told the National People's Congress (NPC) to follow the national policy of environmental protection, and in 1997 President Jiang Zemin reported to the Chinese Communist Party's (CCP) National Congress that pressures on the environment caused by overpopulation and economic development were harming the country.[16] The agency tasked with promoting environmental protection was elevated to ministerial status, becoming the State Environmental Protection Administration (SEPA). By the late 1990s the central government was allocating substantial (if grossly inadequate) funds to environmental and resource protection, and it reportedly shut down tens of thousands of polluting enterprises and implemented emissions fees and clean technologies.[17]

The central government has for some time required industry to become more energy efficient, although it seldom enforces these requirements for major violators. While efficiency has improved in recent years, China's growing economy is more than outpacing it. Given the country's growing reliance on imported energy, and the horrendous air pollution that has come from the burning of coal for electricity and industrial production and from the use of petroleum-derived fuels for transport, the government has passed new legislation to improve energy efficiency. It has also enacted new taxes on transport fuels and larger automobile engines, and its 2006 eleventh five-year plan set out new requirements to limit energy use. Its automobile fuel-efficiency standards are now ahead of those of most countries, notably the United States.

However, the environmental benefits of these actions have often been limited. Implementation of environmental laws is hindered by lack of money, corruption, the refusal of local authorities to take the laws seriously, and the inability or unwillingness of higher officials to force them to do so.[18] Beijing often has limited control over the vast bureaucracy, particularly outside Beijing, and the institutional structure of China's environmental management system is extraordinarily complex. Underlying the inability to implement environmental protections is a strong nationwide fixation on economic growth.[19] Wealth creation in the short term usually trumps environmental protection. Consequently, despite concern at the highest level, and in spite of many private-sector initiatives (e.g., commercial wind farms and widespread domestic use in some areas of solar hot water heaters), the trend toward increasing greenhouse gas emissions has been mitigated only minimally.

China's Climate Diplomacy

China has been actively involved in international negotiations on climate change. During early negotiations, it sought to protect its sovereignty by opposing any legal requirements for it to limit its GHG emissions and by asserting its freedom to pursue economic growth.[20] It opposed every effort to require GHG limits by developing countries—even those calling for *voluntary* commitments to restrict future emissions *increases*—and it has consistently demanded that developed countries provide assistance to developing countries to help them cope with climate change. It has usually resisted any links between financial and technical assistance from developed countries in the context of the climate change regime and emissions limitations by developing countries. Instead, it has demanded transfers of funds on noncommercial and preferential terms, and until quite recently rejected many of the market-based international mechanisms for emissions reductions advocated by developed countries and their industries.[21] According to one observer, "only when outsiders (e.g., the GEF [Global Environment Facility]) have paid the incremental costs has China been willing to implement global warming projects."[22] However, once it was established that developing countries would not be required to take on any obligatory commitments under the climate convention, it became clear to senior policy makers that joining the Kyoto Protocol could be in China's economic, environmental, and diplomatic interests: the Clean Development Mechanism (CDM) would provide new funds to aid economic development, domestic pollution would be mitigated by the resulting new technologies, and accession to the treaty would show the world that China was leading developing countries in addressing an important global issue. China has already benefited from investment via the protocol's CDM; in 2006 alone, the CDM brought investment valued at nearly $3 billion into China, far exceeding predictions.[23] Furthermore, the International Energy Agency expects international carbon trading to bring China more than $1 billion per year by 2010.[24]

China's climate change policies are now established at a very high level within the government, with the powerful new National Development Reform Commission acting as coordinator. China's current international position on climate change was revealed by its diplomacy in the 2006 working group meetings of the UN's Intergovernmental Panel on Climate Change, which was preparing its fourth assessment report.[25] During those negotiations, Chinese diplomats tried to push through language that would play down the scientific certainty about the causes and consequences of climate change, and it sought to water down wording that would set stringent global standards for limiting concentrations of carbon dioxide in the atmosphere. Chinese diplomats also advocated wording that would increase the amount of global warming that the world could accept. These efforts, albeit mostly rebuffed by other delegations, were clearly attempts to delay the date when China itself would be called upon by the world to accept binding limitations, and eventually cuts, in its own greenhouse gas emissions.

Climate Justice and the New China

Although developed countries are the ones who have historically caused most of the climate change problem, it is not fair, nor environmentally prudent, for the many affluent Chinese and rich elites there and in other developing countries to be absolved of duties regarding climate change.

China is now a major source of GHGs, and this will only become truer with time. In-

deed, affluent Chinese may have more obligation to limit their GHG emissions than do many or most affluent Canadians, for example, because the latter are saddled with infrastructure and long-standing habits that were created *before* they knew that climate change was a problem. Because scientific knowledge about climate change, and associated high-profile international diplomacy, has coincided with China's economic rise, educated Chinese people (and their government, media, and so on) should be aware that by adopting Western ways of living the affluent in China will contribute to climate change.[26] Consequently, it is possible that history will judge affluent Chinese even more harshly than many people in North America, Australia, and Europe because the former had (and still have) a choice about whether to jump on the consumption bandwagon. The Chinese government is responsible and complicit.

Conclusion

Whether climate disaster is averted in the long term will in very large part be determined by whether China and its people are willing and able to change course and find a more climate-friendly future. One thing is now evident: without China's full participation, global efforts to mitigate and reverse climate change will surely fail. The developed world must give China all the help and encouragement it can, but there is much that China can do on its own.[27] Alas, the new China is headed down the same environmentally unsustainable development path of the West. The temptations of modernity are very powerful, and the desire for affluence is extraordinary in China. It is unlikely that the Chinese will be any better at resisting these temptations than have been people in the West.

Notes

1. See Elizabeth C. Economy, *The River Runs Black: The Environmental Challenge to China's Future* (Ithaca, NY: Cornell University Press, 2004).

2. The exception is petroleum, for which the United States remains the largest consumer. Christopher Flavin and Gary Gardner, China, India, and the new world order, in Linda Starke, ed., *State of the World 2006* (New York: W.W. Norton, 2006), p. 6.

3. P. Andrews-Speed, Liao Xuanli, and R. Dannreuther, "Searching for energy security: The political ramifications of China's international energy policy." *China Environment Series* 5 (Washington, DC: Woodrow Wilson Center, 2002), pp. 13–14, cited in Cynthia W. Cann, Michael C. Cann, and Gao Shangquan, "China's road to sustainable development," in Kristen A. Day, ed., *China's Environment and the Challenge of Sustainable Development* (London: M.S. Sharpe, 2005), p. 8.

4. Michael T. Hatch, "Chinese politics, energy policy, and the international climate change negotiations," in Paul G. Harris, ed., *Global Warming and East Asia* (London: Routledge, 2003), p. 46.

5. Paul G. Harris and Chihiro Udagawa, "Defusing the bombshell? Agenda 21 and economic development in China." *Review of International Political Economy* 11 (3) (August 2004), p. 622.

6. Hatch, op. cit., p. 45. International Energy Agency, *World Energy Outlook 2006* (Washington, DC: International Energy Agency, 2006).

7. See Paul G. Harris, ed., *Confronting Environmental Change in East and Southeast Asia* (New York: United Nations University Press/Earthscan, 2005), p. 25.

8. Lawrence Brahm, "State of destruction." *South China Morning Post,* May 15, 2007.

9. J. Garner, *The Rise of the Chinese Consumer: Theory and Evidence.* (Hoboken, NJ: Wiley, 2005), p. 73. This level of "significant discretionary consumer spending" is set by Garner at $5,000.

10. David Wilson, "Designs on sustainable development." *South China Morning Post,* February 14, 2006.

11. Kelly Sims Gallagher, *China Shifts Gears: Automakers, Oil, Pollution, and Development* (Cambridge, MA: MIT Press, 2006). China to have 140 million cars by 2020. *China Daily,* September 4, 2004, www.chinadaily.com.cn/english/doc/2004-09/04/content_371641.htm. This is roughly the same number of cars as in the United States.

12. This section and the next update material in Paul G. Harris, "Environmental politics and foreign policy in East Asia," in Harris, ed., *Confronting Environmental Change in East and Southeast Asia,* pp. 21–26.

13. Edward Tseng, "The environment and the People's Republic of China," in Dennis Soden and Brent Steel, eds., *Handbook of Global Environmental Policy and Administration* (Basel: Marcel Dekker, 1999), p. 383.

14. Ibid.

15. United Nations Environment Program (UNEP), *Global Environment Outlook* (London: Earthscan, 1999), pp. 241–42.

16. Tseng, op. cit. p. 383.

17. UNEP, op. cit. p. 246.

18. See Elizabeth Economy, "Environmental enforcement in China," in Day, op. cit., pp. 66–101.

19. See Paul G. Harris, "Getting rich is glorious": Environmental values in the People's Republic of China." *Environmental Values* 13 (2) (2004), pp. 145–65.

20. Zhihong Zhang, The forces behind China's climate change policy: Interests, sovereignty, and prestige, in Harris, ed., *Global Warming and East Asia,* pp. 66–85.

21. Joanne Linnerooth-Bayer, "Climate change and multiple views of fairness," in Farenc L. Toth, ed., *Fair Weather? Equity Concerns in Climate Change* (London: Earthscan, 1999), p. 59.

22. David Victor, "The regulation of greenhouse gases: Does fairness matter?" in Toth, *Fair Weather? Equity Concerns in Climate Change,* p. 203.

23. See Axel Michealowa, et al., The clean development mechanism and China's energy sector, in Harris, ed., 2003, op. cit., pp. 109–31. This was more than 60 percent of carbon credits awarded through the CDM in 2006. Keith Bradsher, "China knows which way the wind blows." *International Herald Tribune,* May 10, 2007, p. 1. See Antoaneta Beziova, "China sends smoke signals on Kyoto Protocol." *Inter Press Service News Agency,*

January 20, 2006, http://ipsnews.net/news.asp?id news=31842.

24. Ibid.

25. See Intergovernmental Panel on Climate Change (IPCC) Working Group (WG) I, *Climate Change 2007: The Physical Science Basis* (Cambridge: Cambridge University Press, 2006); IPCC WG II, *Climate Change 2007: Impacts, Adaptation and Vulnerability* (Cambridge: Cambridge Uni-versity Press, 2006); and IPCC WG III, *Climate Change 2007: Mitigation of Climate Change* (Cambridge: Cambridge University Press, 2006).

26. China's shift to capitalism began in earnest about 1980, which is about the time the climate science started to become prominent.

27. See also Harris, 2005 op. cit., and Harris, 2003, op. cit.

Chapter 31

India

ASHOK GADGIL AND SHARACHCHANDRA LÉLÉ

India is an important actor in the climate change debate for several reasons. First, India's monsoonal climate, long coastline, and large rural population with substantial poverty make it highly vulnerable to climate change impacts. Second, the Indian economy, because of its size and recent rapid growth, is becoming a significant contributor to greenhouse gas (GHG) emissions. Third, Indian analysts have played important roles in framing the climate debate, and India and China will play important roles in climate negotiations as leaders of the developing world.

Background

India is a large and complex country: the world's most populous democracy, perhaps the most eco-socially diverse region of the world, a complex mixture of the traditional and the modern, and a rapidly changing economy. India's population crossed the 1 billion mark in 2000, and the growth rate, although slowing down, is still just below 2 percent. About two-thirds of this population still depends on agriculture, forests, and fisheries for

its livelihood.[1] Politically, India has been a relatively stable multi-party democracy with a multi-tiered system of governance.

Since independence in 1947, India has made impressive progress on many economic and social indicators, including achieving self-sufficiency in food production and substantially reducing poverty, illiteracy, and fertility. It has established a diverse and strong industrial base, and the Indian economy has grown rapidly (at more than 6 percent annually) since 1994.

The fruits of this rapid economic growth, however, are unevenly shared within Indian society. The top 10 percent of India's population accounts for 34 percent of the GDP, while the bottom 10 percent accounts for about 3 percent, and at least 26 percent of the population lives below the poverty line.[2] In 2008, UNDP ranked India 132nd on the Human Development Index.[3] But economic growth continues to be the main focus of the political and economic leadership in the country today, with poverty reduction believed to be achievable through trickle-down and environmental concerns largely receiving lip service.

GHG Emissions from India

The choice of measure used to characterize a country's contribution—current absolute emissions, aggregate historical emissions, current per capita emissions, and so on—is inextricably linked with one's value position about how responsibility for the climate change problem and its mitigation should be assigned. We provide estimates and projections in different terms and discuss their implications.

Current Situation

In absolute terms, India's annual GHG emissions were 1,228 million tonnes of CO_2 equivalent (tCO_2e) in 1994 and are estimated to have reached about 1,750 million tCO_2e by 2005.[4] This amounts to only about 4 percent to 5 percent of the global GHG emissions rate in this period.[5] In per capita terms, India's GHG emissions were 1.3 tCO_2e in 1994, rising to 1.9 tCO_2e in 2004 (still using 1994 population), and so India is ranked 146th among all countries. In contrast, per capita emissions of the United States were around 23 tCO_2e in 2004.[6]

CO_2 of course constitutes the major share of GHGs emitted from India (65 percent in 1994), but methane (primarily from the livestock and agricultural sector) contributes significantly also (31 percent in 1994).[7] Methane emissions are, however, relatively stagnant, whereas CO_2 emissions are rapidly increasing.[8] The main sectors that accounted for CO_2 emissions from fuel combustion in 1994 were energy transformation including power generation (52 percent), industry (22 percent), and transport (12 percent). Coal, of which India has large stocks, is the main source of energy for power generation and meeting industrial energy requirements, accounting for nearly two-thirds of India's CO_2 emissions. The industrial sector accounts for around 50 percent of total commercial energy consumption, with fertilizers, iron and steel, aluminum, cement, and paper and pulp industries collectively accounting for about two-thirds of total industrial energy consumption. Not much information is available on variations in emissions by region or class.

Future Scenarios

Predictions of future emissions from India based on coarse-scale models of economic growth and sector-wise intensities all suggest substantial increases in CO_2 emissions regardless of the policy scenario, but there are significant differences in the estimated values and scope for mitigation. Shukla and colleagues developed four scenarios for India's development trajectory from 2000 to 2030, ranging from "high economic growth" to "self-reliance."[9] Per these scenarios, energy consumption grows from about 500 million tonnes of oil equivalent (mtoe) to values ranging from 800 mtoe (self-reliance) to 1,300 (high growth) by the year 2030. Coal continues to dominate the energy mix in all these scenarios. Natural gas shows tremendous growth in all scenarios from its relatively low base in 2000. Renewable electric capacity increases by a factor between ten and fifteen, but its absolute contribution remains small. CO_2 emissions increase from about 990 million tonnes to values ranging from 2,020 to 3,120 million tonnes, and per capita emissions grow from about 0.99 tonne to values ranging from 1.47 to 2.38 tCO_2. TERI's more recent predictions suggest much higher emissions in 2031 (around 5 tCO_2 per capita) under business-as-usual, but much greater scope for reductions (down to 1.2 tCO_2/capita) with an ambitious mitigation policy requiring 41 percent of commercial needs to be met by renewables.[10]

In summary, India has so far contributed little in absolute terms and much less in per capita terms to GHG emissions. Its absolute contribution will definitely rise rapidly over the next three decades, but per capita emissions will continue to be below even the current global average and far below the current western per capita emissions. On the other hand, depending upon what is considered to be a global per capita sustainable emissions level for 2030, India may come close to or even cross that level if it continues business as usual.

Climate Change Impacts on India

It is becoming clearer in the Indian subcontinent that significant climate changes are imminent. Global climate models (GCMs) now predict a 3- to 6-degree-Celsius increase in temperature and 15 to 40 percent increase in rainfall by end of this century, and higher variability in temperature and rainfall extremes.[11] All GCMs predict more pronounced warming during winter and in the post-monsoon season—a feature conspicuously observed in current Indian temperature trends. Even if average temperature rise is modest, extreme events will increase significantly. In particular, since the monsoon is a critical phenomenon in the Indian climate system, increased variability of the monsoon is likely to be the major form in which climate change affects India.[12] Similarly, sea-level rise is likely to be less than a meter during this century, but the intensity of tropical cyclones and high surges is already increasing

and will increase further.[13] Although regional models continue to have limitations, several weather-related incidents in recent years have coincided to make even policy makers and laypeople take the idea of climate change seriously.

Given that the majority of the Indian population still depends upon agriculture, forests, and fisheries for their livelihoods, the social impacts of such physical changes and increased unpredictability could be very serious. We try to summarize below what is known and unknown about the different impacts of climate change on India.

Water Resources

Effects of climate change on river flows are likely to vary across the region. For snowfed rivers the retreat of glaciers is a major concern, because initially such melting will cause flooding and later on it may increase intra-annual and inter-annual variability due to reduced buffering. Some models also predict reduced summer flows in such rivers.[14] In purely rainfed areas, changes in annual runoff could be either positive or negative and up to 20 percent in magnitude—but most current analyses are based on scant data and inadequate or unsatisfactory models.[15] In a country that is not only facing increasing physical scarcity of water but also a crisis of governance of water resources and poor systems of hydrological monitoring, climate change–induced floods and droughts are both bound to aggravate the situation.

Agriculture

Studies that are largely based on earlier climate models (increased temperature and increased precipitation) seem to agree that there will generally be a decline in physical and economic productivity of Indian agriculture of 5 percent to 20 percent even after CO_2 fertilization effects and farmer adaptation are factored in.[16] Other predictions differ in magnitude but not in direction. Increased variability of the monsoon, possibly more important than changes in mean precipitation, will also adversely affect agricultural production. Obviously, food security will be threatened. Since rainfed cultivation is 65 percent of the total, much of it for subsistence, the bigger problem is that the impacts will be disproportionately felt by the poorer farmers.

Natural Hazards

Fast glacier melting causes unexpected or sudden glacial lake outbursts, an increasing threat in the upper Himalayas.[17] Increased frequency of cyclonic storms could cause severe loss not just to coastal agriculture and business but also to human life, habitation, and settlements, given the high density of the coastal population.

Biodiversity

Indian forests and other habitats were originally very rich in biodiversity but are currently greatly fragmented and under serious multiple pressures. There has not been much progress in our ability to predict the impact of climate change on forest vegetation, so the predictions of changes in productivity and shifts in species composition must be treated cautiously, given that forests react much more slowly to any environmental changes than annual crops.[18]

Human Health

Most of the attention has been focused on possible increases in malaria. Any predictions about the behavior of this disease are fraught with uncertainty, given the complex interaction between climatic factors, poverty, irrigation practices, waste management, and other socioeconomic factors. But there is room to believe that areas not previously infested may now face outbreaks due to favorable climatic conditions.[19]

On the whole, there are clear signs of multi-dimensional and adverse impacts of unpredictable magnitude already taking place and becoming more severe in the long run. Within India, the impacts vary significantly in space and tend to be disproportionately felt by the poorer economic classes.

Indian Society's Response to Climate Change

There is broad agreement among Indian analysts on four points. First, India and developing countries in general are not historically responsible for the problem of climate change. Most of the historical increase in CO_2 in the atmosphere has come from the developed countries.[20] Second, given that India is just beginning to make the transition (in average terms) from a poor to a middle-income country, India's capacity to make significant contributions to absolute reductions is small, compared to the total reductions required to contain climate change under 2 degrees Celsius. Third, a broadly equal per capita allocation of emissions rights, even if it is in terms of "frozen"1990 population levels and even giving some margin to historical and geographical variations, is the only morally acceptable approach.[21] Fourth, much of the human cost of climate change will be borne by developing countries due to their higher vulnerability, sensitive location, and limited adaptive capacities. There is, however, considerable debate regarding how to translate these into international negotiating positions and domestic policies.

Responding Internationally

India and other developing countries have forged and stuck to the principle of "common but differentiated responsibility" based upon: (1) considerations outlined above, (2) the argument that poverty eradication and meeting minimum development needs cannot be compromised, and (3) developed countries needing to provide substantial additional assistance to help contain future emissions.[22]

Post-Kyoto, the major internal debate was on whether India should sign on to the clean development mechanism (CDM). Critics of CDM noted that selling certified emissions reductions (CERs) to developed countries permits them to meet their small Kyoto commitments cheaply and to postpone or not address the major changes required in their domestic economies and lifestyles. Second, selling off the cheapest GHG reduction options now to the industrial countries amounts to giving them the low-hanging-fruit, whereas when the time inevitably came for India to accept some reductions in its own GHG emissions, India would have to undertake the remaining (more expensive) options to meet its own obligations.[23] Third, CDM-related payments will end up replacing, not supplementing, untied development aid. Supporters of CDM have provided arguments of realpolitik, namely, that developed countries would not

have accepted even the mild commitments under Kyoto if CDM had not been ratified, that CDM is essentially free money, providing funds for efficiency improvements that one might have made anyway, and that the scope for such improvements is so vast that participating in CDM projects will not significantly reduce the potential for future savings.

After intense lobbying by the United States, the Indian government adopted a pro-CDM approach in 2001.[24] India's National CDM Authority came into existence in 2003. In 2005 and 2006, India was the leading developing country in terms of number of CDM projects in the pipeline and CERs issued. However, the results are climatically ineffective; 80 percent of the CERs result from the destruction of HFC-23, a by-product of the manufacture of HCFC-22, that has to be phased out anyway under the Montreal Protocol.[25] The companies involved are finding it much more lucrative to destroy HFC-23 than to actually produce HCFC-22—an example of the distortions that can easily creep into CDM projects.[26] Many other projects in the renewable energy sector seem to be those that would have been profitable anyway. Thus, the achievements under CDM to date are not truly contributing to climate change mitigation (see chapters 27 and 28 for more on CDM).

The inclusion of land-use change and forestry in CDM starting in July 2001 was seen as providing another avenue for CDM gains for countries like India that have undergone historical deforestation. Several analysts have claimed that the 50 to 70 million hectares of wastelands in India can be used for reforestation programs involving local communities, thereby simultaneously providing livelihood enhancement, local environmental benefits, and climate change mitigation.[27] However, this approach is fraught with pitfalls. The wastelands are actually significant sources of firewood, grazing material, and other subsistence products gathered by the rural poor. The major cause of wasteland degradation is not the absence of adequate economic incentives, but the absence of adequate, secure, and enforceable property rights for local users over these lands. Providing payments for carbon sequestration without, among other things, resolving forest rights issues will lead to plantation forestry that will benefit the village elite (who already have non-forest sources of income) and/or the state agencies who control common lands, at the cost of the needs of poorer or forest-dependent households.[28] This has often happened under donor-funded joint forest management (JFM) programs.[29] On the other hand, India does not stand to benefit from the new Reducing Emissions from Deforestation and Degradation (REDD) and so has opposed it, although the problems with REDD may not be fundamentally different from those with CDM-afforestation.

At Bali and afterward, even as the developed world's actions reflect a greater urgency, India and the developing countries in general have come under greater pressure to take on some commitments. While China and some others have shown willingness to consider sustainable development policies and measures (SD-PAMs), India has stuck to its hard line of no commitments. Ironically, this position may have inadvertently aligned India with the United States as the spoilers at Bali.[30] There is room to believe that Indian actions before and at Bali were driven more by geopolitics rather than high principles.

For instance, after ratifying Kyoto, India joined the Asia-Pacific Partnership, an alliance floated by the United States in 2005 as an alternative to Kyoto that allows member countries to set individual reduction targets with no enforcement mechanism. India also has negotiated technology deals or partnerships related to nuclear energy and clean coal technology with the United States. Indian climate negotiators have sought to (mis-)characterize the Indo-U.S. nuclear deal as solving all mitigation concerns.

Addressing Mitigation Nationally

Should India have a proactive emissions reduction policy at home? The economic growth lobby would rather not see reduction measures, except if they come in the form of the free money from CDM. Others, including the government, seem to subscribe to a "development first with climate-co-benefits" approach, whereby emission reductions will happen as a by-product of sound development policies, including local-level pollution control.[31] The 2005 Integrated Energy Policy document of the Planning Commission seemed to reflect such thinking.[32] The focus was entirely on ensuring energy security given the anticipated manyfold increases in energy demand. Climate change was (literally) at the bottom of the agenda. With growth equated to development, almost all expert discussions in India focus on supply-side solutions (e.g., nuclear energy), efficiency improvements, and perhaps slowing population growth.[33]

Not surprisingly, the National Action Plan for Climate Change released in mid-2008 drew attention to the seriousness of impending climate change impacts on India, but has set no concrete targets even for a few sectors. It lays out an ambitious plan for energy efficiency improvements in building, but otherwise reiterates strategies that have already proven to be problematic, such as the use of JFM for greening India or reform of electricity subsidies to farmers, and has no vision in terms of mass transit or other shifts.[34]

But it may be necessary to go beyond "development first" for several reasons. First, to be consistent, the principle of equity espoused internationally has to be applied in the internal allocation of emission rights and of compensation funds received, if any. Already the top 10 percent of the population in India today is probably consuming five times the national average, or as much per capita as the average person of a mid-level developed country.[35] If everyone else in India also aspires to that lifestyle, efficiency improvements alone will not be enough. And long before even milder aspirations are met, the climate space available to the Indian poor will have been usurped by the Indian rich. Second, if some reduction obligations are inevitable in the long run, then major long-term decisions being made currently in the energy infrastructure (such as more coal-fired plants), transportation infrastructure (such as more highways), and industries (more steel, cement, and bauxite plants) need to be reconsidered before they lock the country into a high-emission trajectory for several decades. It may also be argued that Indian business can leapfrog and specialize in low-carbon technologies and processes.[36] Third, even if developmental problems are a high priority, they are likely to be dramatically exacerbated by impending climate change. It may therefore be in India's narrow self-interest to negotiate more flexibly if that can lead to early mitigation actions by developed countries. Embracing the concept of SD-PAMs aggressively might shift the focus back toward the deep emissions reductions required of the developed countries.

Adapting to Climate Change

The second dimension of the internal response to climate change is adaptation. Here again, an approach similar to "development first" would argue that having a greater capacity to adapt to what could be devastating changes in climate and ecosystem functioning requires a high level of development (akin to the West). High rates of economic growth, it is argued, are required to build the necessary human, institutional, and technological capital required to face up to possibly increased water scarcity, the costs of ocean-level rise and extreme storm events, increased disease burden, and so on. The measures contemplated in this approach are more irrigation infrastructure to protect against drought, more infrastructure to protect against floods or cyclones, more investment in early-warning systems and in breeding drought-resistant and pest-resistant crops, and so on.

An alternative developmental perspective might suggest that many of the means through which conventional economic development is achieved, such as increased market integration or mono-cropping in agriculture, are those that reduce the capacity to respond to catastrophic events, especially for the poor. Given that the natural resource-dependent poor are also the most vulnerable, reducing vulnerability may require a flexibility and people-centric dimension that only a highly decentralized, democratic, and environment-friendly system of governance can ensure.

Conclusion

The response of Indian society (as against that of the Indian government) to the threat of climate change will depend on its perception of the immediacy and seriousness of impacts from climate change, and the larger question of the extent to which Indian society and decision makers weigh concerns for environmental sustainability and social justice in their development thinking. Currently, public awareness of the risks to India from climate change is relatively low. Unless there is an overall shift from a simplistic focus on growth (and implicit trickle-down) to an explicit focus on basic needs, sustainable lifestyles, and environmental justice as an integral part of the development process, climate change will not be taken seriously. Such a focus will then call for not just efficiency-improvement programs but also for more attention to renewables and (more important) for major changes in the way the energy sector is governed, and indeed in the mapping of the development trajectory itself, an approach in which climate change–sensitive and climate change–agnostic policies may converge.[37]

Notes

The authors are grateful to M. V. Ramana, Navroz Dubash, Sudhir Chella Rajan, Alisar Aoun, an anonymous referee, and the editors for their useful comments. Core support from the Ford Foundation enabled the second author to contribute to this chapter.

1. GoI, 2003. Agricultural statistics at a glance. Ministry of Agriculture and Cooperation, Government of India, New Delhi.

2. GoI, 2006. Economic survey 2005–2006. Ministry of Finance, Government of India, http://indiabudget.nic.in/es2005-06/general.htm.

3. See http://hdr.undp.org/en/statistics/ visited on January 29, 2009.

4. Throughout this chapter, "tonne" denotes a metric ton, i.e., 1,000 kg, not the U.S. ton that equals 2,000 pounds. MoEF, 2004. India's initial national communication to the United Nations Framework Convention on Climate Change. Ministry of Environment and Forests, Government of India, New Delhi. Garg, A., P. R. Shukla, and M.

Kapshe, 2006. The sectoral trends of multigas emissions inventory of India. *Atmospheric Environment* 40(24): 4608–20.

5. EIA, 2006. International energy annual 2004. Energy Information Administration, Office of Integrated Analysis and Forecasting, U.S. Department of Energy, www.eia.doe.gov/emeu/iea/carbon.html.

6. Anonymous, 2007. List of countries by greenhouse gas emissions per capita. Wikipedia, the free encyclopedia, http://en.wikipedia.org/wiki/List_of_countries_by_greenhouse_gas_emissions_per_capita.

7. MoEF, 2004 op. cit.

8. Garg, Shukla, and Kapshe, 2006 op. cit.

9. Shukla, P. R., R. Nair, M. Kapshe, A. Garg, S. Balasubramaniam, D. Menon, and K. K. Sharma, 2003. Development and climate: An assessment for India. Draft Report, Indian Institute of Management, Ahmedabad.

10. TERI, 2008. Mitigation options for India: The role of the international community, The Energy and Resources Institute, Delhi. Note that the per capita values mentioned in this paragraph are based on predicted populations of 2030/31 under different scenarios, and hence the per capita figure in terms of 1990 reference population will be higher by up to 30%.

11. Kumar, K. R., A. K. Sahai, K. Krishna Kumar, S. K. Patwardhan, P. K. Mishra, J. V. Revadekar, K. Kamala, and G. B. Pant, 2006. High-resolution climate change scenarios for India for the 21st century. *Current Science* 90(3): 334–45.

12. Sulochana Gadgil, Centre for Atmospheric and Oceanic Sciences, Indian Institute of Science, Bangalore, personal communication.

13. Unnikrishnan, A. S., K. R. Kumar, S. E. Fernandes, G. S. Michael, and S. K. Patwardhan, 2006. Sea-level changes along the Indian coast: Observations and projections. *Current Science* 90(3): 362–68.

14. Singh, P., and L. Bengtsson, 2004. Hydrological sensitivity of a large Himalayan basin to climate change. *Hydrological Processes* 18: 2363–85.

15. Gosain, A. K., S. Rao, and D. Basuray, 2006. Climate change impact assessment on hydrology of Indian river basins. *Current Science* 90(3): 346–53.

16. Kumar, K. S. K., and J. Parikh, 2001. Indian agriculture and climate sensitivity. *Global Environmental Change* 11(2): 147–54.

17. Leber, D., 2002. Glacier hazards and risk

response strategies in Himalayan countries (Bhutan, India, Nepal). Discussion Paper, Institute of Geology, University of Vienna, Vienna.

18. Such as Ravindranath, N. H., N. V. Joshi, R. Sukumar, and A. Saxena, 2006. Impact of climate change on forests in India. *Current Science* 90(3): 354–61.

19. Bhattacharya, S., C. Sharma, R. C. Dhiman, and A. P. Mitra, 2006. Climate change and malaria in India. *Current Science* 90(3): 369–75.

20. According to Loske, R., 1996. *Scope of the Report: Setting the Stage: Climate Change and Sustainable Development.* Third Assessment Report of Working Group III, Intergovernmental Panel on Climate Change, Geneva. About 83 percent of the cumulative rise in fossil-fuel–related CO_2 from 1800 onward came from developed countries.

21. As suggested by Byrne, J., Y.-D. Wong, H. Lee, and J.-D. Kim, 1998. An equity- and sustainability-based policy response to global climate change. *Energy Policy* 26(4): 335–43.

22. Soz, S., 1997. Address to the UN Convention on Climate Change by the Union Minister of Environment and Forests, at the 3rd Session of the Conference of the Parties to the Framework Convention on Climate Change at Kyoto, Japan, on December 8, 1997, www.indianembassy.org/policy/Environment/soz.htm.

23. Agarwal, A., S. Narain, and A. Sharma, 1999. *Green Politics* (Centre for Science and Environment, New Delhi). Agarwal, A., 2001. Climate change: A challenge to India's economy. Briefing Paper for Members of Parliament, Centre for Science and Environment, New Delhi.

24. Agarwal, 2001 op. cit. Hausker, K., and K. McGinty, 2001. India's reappraisal of the clean development mechanism. Feature, *Resources For the Future*, Washington, DC.

25. Kumar, V., 2006. CDMs: Destination India. The Energy Research Institute, http://static.teriin.org/teritimes/issue8/news4.html.

26. Balaji, N., 2006. CDM in India and its contribution to compensation for ecosystem services, paper presented at South Asia Regional Workshop on Compensation for Ecosystem Services, organized by World Agroforestry Centre and Institute for Social and Economic Change at Bangalore during May 8–10.

27. Satyanarayana, M., 2004. How forest producers and rural farmers can benefit from the Clean Development Mechanism, in Sim, H. C., S. Appanah, and Y. C. Youn, eds., *Proceedings of the Workshop on Forests for Poverty Reduction: Opportunities with Clean Development Mechanism, Environmental Services and Biodiversity held at Seoul, Korea on 27–29 August 2003,* Food and Agriculture Organization of the United Nations, Regional Office for Asia and the Pacific, Bangkok. Sathaye, J. A., and N. H. Ravindranath, 1998. Climate change mitigation in the energy and forestry sectors of developing countries. *Annual Reviews of Energy and Environment* 23: 387–437.

28. Gundimeda, H., 2005. Can CPRs generate carbon credits without hurting the poor? *Economic and Political Weekly* 40(10): 973–80.

29. Lélé, S., A. K. K. Kumar, and P. Shivashankar, 2005. Joint forest planning and management in the Eastern Plains Region of Karnataka: A rapid assessment. CISED Technical Report, Centre for Interdisciplinary Studies in Environment and Development, Bangalore.

30. Dubash, N., 2007. Inconvenient truths produce hard realities: Notes from Bali. *Economic and Political Weekly* 42(52): 31–35.

31. See, for example, Shukla et al., 2003 op. cit..

32. Anonymous, 2005. Draft Report of the Expert Committee on Integrated Energy Policy. Planning Commission, Government of India, New Delhi.

33. Shukla et al., 2003 op. cit.; Sethi, S. P., 2006. Climate change dialogue: India country presentation. United Nations Framework Convention on Climate Change, Presentation, http://unfccc.int/files/meetings/dialogue/application/vnd.ms-powerpoint/20060516_india_presentation.ppt. Srinivasan, J., 2006,. Hottest decade: Early warning or false alarm? *Current Science* 90(3): 273–74.

34. Anonymous, 2008. Climate change: Not vision, not plan. *Economic and Political Weekly* 43(28): 5–6.

35. Ananthapadmanabhan, G., K. Srinivas, and V. Gopal, 2007. Hiding behind the poor: A report by Greenpeace on climate injustice. Greenpeace India Society, Bangalore.

36. Dubash, 2007 op. cit.

37. Narasimha, D., 2006. Low on fuel: Crisis ahead. *Infochange Agenda,* June, pp. 10–13.

Chapter 32

Australia

CHRIS HOTHAM

Introduction

Late 2006 to early 2007 marked a fundamental shift in climate policy at a national level in Australia. During this time a record drought raised public concern on climate change to new heights, the business community moved toward support for a carbon price within the Australian economy, and mounting political pressure from state governments helped force the debate on climate change into the halls of decision makers in the nation's capital. The result was a significantly altered policy landscape that made climate change, alongside the economy and national security, one of the central issues for the Australian federal election late in 2007.

Ultimately, the perception of federal government inaction on climate change proved to be one of the factors that contributed to the downfall of the incumbents in the 2007 election.

This chapter provides an analysis of Australia's transition leading up to this point—from an early leader on climate policy, to a laggard with its refusal to ratify the Kyoto Pro-tocol—before its recent return as an active participant in the international community. It does this by examining the essential elements: the role of business, public attitudes, the drought, political pressure, and the role of Australian states.

Climate Change Policy in Australia: 1997 to 2006

Initially, Australia played a central role in the development of the Kyoto Protocol and brokered some very favorable outcomes for itself. Through negotiations, Australia was not only successful in arguing special case exemption entitling the country to a target of net increase in emissions on 1990 levels (108 percent), but was also instrumental in the inclusion of Land Use, Land-Use Change, and Forestry (LULCF) as a greenhouse emissions sector.[1]

In 1997 Australia signed the Kyoto Protocol—a precursor to ratification. Australian Prime Minister John Howard hailed the move as a "massive contribution to the world environmental effort to cut greenhouse gas emissions," asserting that Kyoto represented a "win for the environment and a win for Australian jobs."[2]

Australia further established itself as a policy leader on climate change when, in April 1998, it became the first country to establish a dedicated policy agency—the Australian Greenhouse Office (AGO)—to "support a national approach to Australia's greenhouse commitments." The charter of the office included a commitment to explore the "options for a possible domestic emissions trading scheme."[3]

The AGO put in place many of the critical building blocks for a substantive national response. In April 2001, the AGO initiated a major market-pull policy response in the form of the Mandatory Renewable Energy Target (MRET). The target was innovative by global standards at the time and was supported by legislation that required electricity retailers to purchase a specified percentage of their electricity from renewable energy sources.

However, any momentum behind an Australian response to climate change was short-lived. Following the U.S. lead, the Australian government never ratified the Kyoto Protocol and began to criticize it as an ineffective long-term response, citing its failure to involve developing countries and framing it as a threat to Australia's national interest due to its impact on the resources sector.

On the domestic front, the Australian government failed to increase its MRET target, despite the fact that 2002 projections showed the scheme's 2010 target would be met using renewable projects already committed or planned.[4]

Reflecting an intention to find alternatives to Kyoto, the federal government helped establish the Asia Pacific Partnership on Clean Development and Climate (AP6) in January 2006, with Australia, the United States, Japan, South Korea, China, and India as inaugural partners.[5] The intent of the partnership was to "accelerate the development and deployment of clean energy technologies" and emphasize the government's preference for R&D for abatement technologies over the successful market-pull mechanisms of the EU. The partnership evidenced the government's preference for supportive measures that would avoid placing any imposition on business.

By mid-2006, significant nationally focused policy in Australia seemed a long way off. Against this backdrop, the evolution of climate policy in Australia during the last twelve to eighteen months appears extraordinary.

Factor 1: The Role of Business

Due to its crucial importance for Australia's economy, the resource sector has been said to exert a disproportionate influence on climate policy.[6] The last decade has seen strong growth in Australia's energy exports contribute significantly to the national economy. Over this period—driven largely by demand from its largest trading partner, China—Australia has become the world's largest coal exporter, and consequently the notion of a mining boom is high in the national consciousness as the driver for current economic prosperity.[7] The notion of such a strongly resource-focused economy has shaped Australia's argument for special status within the Kyoto process.[8]

Adding to the influence of the resources sector on Australia's climate policy is the country's wealth of uranium deposits. Australia holds more than one-third of the world's uranium, with almost all of the extracted ore exported. The federal government has strongly supported the role of nuclear energy as part of the climate solution in Australia, but with strident opposition from state governments, the role of uranium mining, export, and use in Australia has remained very much a live debate.

A fracturing of the business community had occurred during the last couple of years, with energy companies with significant gas reserves in their portfolio (such as AGL and Origin Energy) perceiving that current government policy was restricting their business opportunities. The CEOs of both companies have subsequently become among the loudest champions of a change in policy in the Australian business community, and new gas peaking plants, as well as baseload facilities, are being planned and built around Australia. Australia's primary producers are also awakening to the opportunities available to them in a shift in policy. Rural interest in biofuel production, biomass for electricity production, and agri-char, forestry, and perennial grasses for carbon sequestration are rising rapidly, and may be a factor in future policies.

As the political climate has shifted, emerging industries have been growing more attractive. Concentrator photovoltaic technologies, such as those manufactured by Solar Systems, have won significant government subsidies and are on track to massively expand their operations. One of the most interesting emerging industries is the geothermal energy sector. Australia has the world's largest proven reserve of hot dry rocks—granites buried at a depth of 4 kilometers in northern South Australia.[9] The energy embedded in these reserves is approximately three times larger than all of Aus-

tralia's combined reserves of natural gas—enough to power the Australian economy for a century. If trials are successful, geothermal energy may become increasingly considered alongside nuclear and clean coal technologies as a future source of baseload power.

A shifting mood within the Australian business community had been signaled early in 2006 when the Australian Business Roundtable on Climate Change (including members of the insurance, energy, and banking sectors) released a report supporting a "long, loud, and legal" framework for the introduction of a carbon price signal into the Australian economy, including the need for a long-term emissions reduction target.[10]

Later that year, the carbon consequences for industry were underlined in a way that few would have predicted. In a ruling with potentially far-reaching consequences for Australia's mining, energy, and manufacturing industries, a New South Wales court found that the coal producer Centennial Coal Company Ltd. should have included the climate impacts in its environmental impacts assessment of the proposed Anvil Hill mine.[11] The court ruled that the developers had failed to adequately account for the greenhouse gas emissions from burning the coal produced at the mine. While the ruling did not invalidate the assessment process for the project, it was another harbinger for a carbon price within the Australian economy.

The inevitability of a carbon price was reflected in moves from Australia's largest resource companies, including Rio Tinto and the "big Australian" BHP Billiton, to support—albeit cautiously—moves toward a national emissions trading scheme.[12] The Business Council of Australia, perhaps the most influential industry body in the country, endorsed "long-term emissions reduction targets."[13]

With the release of the *Stern Review on the Economics of Climate Change* in November 2006, it was clear that the tide had turned.[14] The influence of the resource industries may not have diminished but the sentiment of the business community was certainly changing.

For many Australian businesses, the *Stern Review* increased the pressure to respond to climate change already being felt from action in Europe and the United States. With Stern's assessment of the benefits of early action it was increasingly clear in Australia that the rest of the world would continue to act. The detrimental cost of Australian "unilateral action" was a more difficult case to argue. Indeed the reverse now seemed more likely—Australia faced isolation from new markets and from future trade opportunities if it failed to consider shifts toward a carbon price.

Perhaps the wildcard in the changing attitudes of the business community came in the form of NewsCorp's Rupert Murdoch. As perhaps the most influential global media presence, Murdoch's influence is no less pervasive in the country of his birth, where he owns Australia's highest-selling broadsheet, its highest-selling tabloid, and a controlling share of the market-leading newspapers in most Australian states.

The editorial line of the Murdoch press had previously tended to be a skeptical one on climate change. However, in November 2006, Murdoch called for business to lead the search for solutions and for an international agreement to supersede Kyoto. Murdoch's changing stance was underlined by the announcement in May 2007 that NewsCorp would go carbon-neutral.

Factor 2: Public Attitudes and the Drought

Throughout this time, popular opinion on climate change was evolving dramatically. While it took a number of factors to come together to produce policy change at a national level, more than any other issue, it was public attitudes to the drought that forced the prime minister's hand.

Al Gore's visit to Australia in September 2006 coincided with the start of another summer of drought and economic hardship—the fifth in a row and the worst drought. The majority of Australia's farming land was drought-stricken, estimates of crop yields for the upcoming season were slashed, and livestock losses were widespread. Scientists labeled it as a "one in a thousand year" event.[15]

In the cities, the implications of drought began to bite as reservoir levels dropped and water restrictions were tightened. State governments moved to consider desalinization options, as it was clear that decision makers were beginning to consider water shortages as a future pattern rather than a short-term aberration. The national scientific research agency, the CSIRO, reinforced this picture, suggesting that by 2070 the most populous areas of the southeast could see a 40 percent drop in rainfall under climate change.

Climate change and the drought were on the front page of the national papers nearly every day throughout December 2006 and January 2007. While polar bears on ice caps remained popular in the media as a symbol of climate change, increasingly images showed Australia farmers against a backdrop of barren soils.

Polls stretching back to mid-2003 reflected increasing concern with climate change. In a

February 2007 poll, respondents identified water shortages and climate change as the leading issues of most concern to them, ahead of housing costs and terrorism.[16] A separate poll conducted in early 2007 across seventeen countries (55 percent of the world's population) found Australians were most likely of any country to consider climate change a "serious and pressing problem" (67 percent) and most likely to favor measures to combat global warming (92 percent).[17]

Factor 3: Political Pressure and the Role of States

If there is a lesson for other countries in the Australian example, it is in the role that subnational governments—in this case, states—can play in fostering a national climate change agenda, despite the reluctance of governments at a national level (see also chapters 34 and 36 in this volume).

Australia has a federal system of government with powers divided between a federal national government, six states, and two territories, with states and territories also devolving powers to more than 673 local councils. Constitutionally the federal government has no specific energy or environmental powers, leaving the responsibility for climate policy sitting somewhat uncomfortably between federal and state/territory governments.

For the ten years leading up to the federal election in November 2007, the federal government had been a coalition government led by the center-right Liberal Party, but over this time the center-left Australian Labor Party (ALP) gained power in all states and territories.

Australian states have been active on climate policy for some time. During the period in which the federal government lay largely dormant on effective climate policy, states—particularly New South Wales (NSW), Victoria, and South Australia—were able to establish momentum on climate change action through policies in their own jurisdictions and later, with the support of the federal opposition, collective action toward a national response.

Examples of key policies developed by states include the NSW Greenhouse Gas Abatement Scheme (GGAS, the world's first functional emissions trading scheme) and Victoria's Renewable Energy Target scheme (VRET, a market-pull requirement on electricity suppliers to source 10 percent renewable energy).[18] While these policies realized real emissions reductions, NSW led the way on broader policy directions by becoming the first state to commit to a long-term emissions reduction target of 60 percent by 2050 (a path later followed by South Australia, Victoria, and Western Australia). While state measures attracted some criticism in industry circles for creating a patchwork of policies, the argument was an easy one politically—states were filling the vacuum left by the federal government.

During the last few years, states have continued to up the ante on climate policy, shifting their focus to more ambitious outcomes at a national level. Moves to fill a policy void were exemplified in 2004 by the establishment of a state-based task force to investigate the design features for a national emissions trading scheme in Australia (a body with which the federal government pointedly refused to engage). In April 2007, states endorsed the need for a national emissions trading scheme, effectively surrounding the federal government on the issue. In the same month the federal ALP endorsed a national emissions reduction target of 60 percent by 2050.

The prime minister, whose own task group on emissions trading had been formed in December 2006 to investigate a "global emissions trading scheme in which Australia could participate," was now forced to reconsider a domestic scheme that could be established unilaterally—a significant policy shift. In June 2007, capping this transition, the prime minister endorsed a national emissions trading scheme, remarkably similar to that proposed by the states a year earlier.

Pressure at home on the Australian government was reinforced by international pressure that had been building for some years. Australia had maintained solidarity with the United States in its opposition to the Kyoto Protocol throughout a period where there had been significant shifts in the EU and among U.S. states. When the Democrats secured both houses of Congress in November 2006, it was a development that undoubtedly left the Australian government acutely aware of the precariousness of its position.

A Changing Political Climate

Against the backdrop of a changing business landscape, increasing public concern regarding climate change in the face of severe drought, and the leadership of states in developing a policy response to climate change came the 2007 election season. As electioneering kicked off in earnest, it was clear this would be an issue the opposition would seek to exploit.

The Labour opposition leader Kevin Rudd assumed the mantle of prime-minister-in-waiting as, following a visit from Sir Nicholas Stern, he called a national Climate Change Summit (attracting an impressive array of community and business leaders) and initiated the Garnaut Review, a major review of his own on the economic implications of climate change in Australia.

It was a move that characterized public perceptions: the opposition leader appeared willing to tackle the big ideas while the prime minister was the skeptic with a lot of ground to make up.

Perceptions mattered, particularly when it was difficult to discern the obvious policy distinctions.[19] By the end of the campaign both parties supported an emissions trading scheme, both had announced substantial clean energy or renewable targets, and both had been resolute in their ongoing support for R&D (particularly clean coal). Of course, attitudes toward the Kyoto process remained a point of difference.

Perhaps the last chance for the prime minister to break these negative perceptions and turn around his fortunes was through the Asia-Pacific Economic Cooperation (APEC) in Sydney in September 2007—the forum that had come to represent the government's distaste for the Kyoto Protocol. For months the prime minister had hailed the forum as the most important meeting of its type ever to be held in Australia.

The centerpiece of the APEC convention was intended to be the Sydney Declaration, earmarked by the prime minister as a "very important milestone" toward an international climate change deal. In the end, the declaration was largely symbolic. If the prime minster had believed the declaration would act as a bridge to a successful post-Kyoto agreement at the Bali COP later in the year, then he had seriously underestimated the stakes for the other parties. The resistance of the United States to targets and the Asian countries' (including

China's) strong preference for the UNFCCC process were not to be swayed by this late bout of Australian political opportunism. In the end the declaration lacked any firm commitments, promising instead to "work to achieve a common understanding on a long-term aspirational global emissions reduction goal to pave the way for an effective post-2012 international arrangement."[20]

As one commentator put it, the agreement was effectively "an aspiration to an aspiration."[21] Underwhelming at best, the declaration fell far short of the milestone agreement that the prime minister had hoped for. There was no "new Kyoto" for the prime minister. Instead, when Australians went to the polls, voters remembered the government's legacy of inaction on climate change over ten years.

Conclusion

The rest is history. In the November 2007 election, the Labour opposition swept to power and turned a deficit of sixteen seats into a majority of eighteen. The government's defensive strategy on climate change and inaction for more than a decade typified a government that had lost touch with public concerns, and now with it, lost government.

There is now hope of a new era of effective climate change policy in Australia, including a bipartisan commitment to an emissions trading scheme and an emissions reduction target. Both seemed unthinkable only six months ago. Both are now government policy.

The note of caution is that climate change has long been a partisan issue in Australia. Debate is inherently politicized. In particular, climate policy in Australia risks being reduced to a single outcome—emissions trading, which is particularly problematic when questions remain about the scale of emissions reduction achievable under such a scheme.

So, while a seismic shift in climate policy, especially at the federal level, appears to have occurred, questions linger as to how far down this new path Australia is willing to go. Climate policy in Australia must seek to build on the momentum created during the last year to ensure that responses are both effective and enduring. Important decisions will be made during the next twelve months—among them the nature of domestic emissions targets and an emissions trading scheme. Until then, the adequacy of an Australian response remains in question.

Notes

Thanks to Tim Flannery, who contributed to an earlier version of this chapter.

1. Given that broad-scale land clearance diminished markedly in the mid-1990s, this inclusion essentially allowed Australia a large credit in its 1990 emissions baseline.

2. Why haven't the U.S. and Australia joined Kyoto? *Climate Action Network Australia.* www.cana.net.au/kyoto/.

3. *Output 1: Leading the Agenda.* 1998–1999. Australian Greenhouse Office Annual Report.

4. Australian Greenhouse Office Annual Report 2001–02.

5. See www.asiapacificpartnership.org.

6. How big energy won the climate battle. *The Age,* July 30, 2005.

7. The value of Australia's coal exports rose by 43 percent between 2004/2005 and 2005/2006.

8. Australia is the country with the highest per capita emissions in the world—at 25.6 tons per capita, compared to the United States at 24.5 and China at 3.9. This is due largely to the electricity sector's reliance on coal, which accounts for more than half of Australia's emissions and which has increased by more than 50 percent between 1990 and 2005.

9. Dozens of other companies are now exploring for hot rocks in other regions of Australia.

10. *The Business Case for Early Action*. Australian Business Roundtable, www.businessround table.com.au.

11. *Major Project Assessment: Anvil Hill Coal Project*. NSW Government, Department of Planning, June 2007.

12. Rio Tinto, Submission to the Task Group on Emissions Trading, March 2007. BHP Billiton, Submission to the Task Group on Emissions Trading, March 2007, www.dpmc.gov.au/climate_ change/emissionstrading/submissions.cfm.

13. Setting achievable emissions targets for Australia. Business Council of Australia, www.bca .com.au.

14. Stern, N. H. *The Economics of Climate Change: The Stern Review*. Cambridge University Press, 2007.

15. 2006 Drought Summary. Murray Darling Basin Commission, www.mdbc.gov.au.

16. *Climate of the Nation: Australian Attitudes to Climate Change and Its Solutions*. The Climate Institute, March 2007.

17. Poll finds worldwide agreement that climate change is a threat. Chicago Council on Global Affairs, www.worldpublicopinion.org

18. See www.greenhousegas.nsw.gov.au. See also www.esc.vic.gov.au/public/VRET/.

19. APEC leaders sign climate change pact. www.abc.net.au/news/stories/2007/09/08/2027608 .htm.

20. *2007 Leaders' Declaration*. Sydney APEC Leaders' Declaration on Climate Change, Energy Security and Clean Development. September 9, 2007.

21. MacCallum, M. *Poll Dancing: The Story of the 2007 Election*. Black Inc., 2007.

United States

Chapter 33

National Policy

ARMIN ROSENCRANZ AND RUSSELL CONKLIN

> We need an energy bill that encourages consumption.
>
> —George W. Bush[1]

The past sixteen years of climate policy may best be described as an era of missed opportunities. Rather than embrace the evolving scientific consensus on anthropogenic sources of global warming by mandating proactive action, such as an emissions cap to mitigate its effects, the administrations of Bill Clinton and George W. Bush promoted more limited policies of continued research and development (R&D), tax incentives for cleaner technologies, and international technological cooperation. In addition, the Bush administration, an ally of fossil-fuel interests, sought to discredit the scientific consensus on climate change through censorship of government scientists and official reports.

President Barack Obama campaigned on ending the nation's dependence on fossil fuels, promoting alternative energy and green jobs and addressing climate change with a cap-and-trade system. The 2009 economic stimulus package contained $87 billion for al-ternative energy and green jobs. A central pillar of Obama's 2009 proposed budget is the revenue from auctioning carbon pollution permits in a climate change cap-and-trade system. Most observers believe that Congress will not enact climate change legislation until 2010, and such legislation is likely to be quite different from the Obama administration's vision, especially on the issue of auctioned permits.

In this chapter, we review the climate policies of the George W. Bush presidency, which cast a shadow over the Obama presidency. We conclude with a summary of Obama's public statements on climate and energy, and speculate on how much these statements can be expected to ripen into national policy.

G. W. Bush Administration: Official Policies

President Bush set the tone for his official climate change strategy early in his first term, when he abandoned a campaign promise to regulate carbon dioxide emissions and then, soon after, indicated he would make no attempt to implement or ratify the Kyoto Protocol, effectively ending U.S. participation.[2]

In February 2002, President Bush unveiled his full climate change plan, "A New Approach on Global Climate Change." Bush's domestic policies included transferable credits for voluntary emissions reductions, an improved R&D portfolio, tax incentives for renewables and cogeneration, expanded research on hybrid vehicles and hydrogen-based fuel cells (dubbed "FreedomCar"), incentives for carbon sequestration, and voluntary public-private partnerships.[3] To implement the plan, Bush created a cabinet-level Committee on Climate Change Science and Technology Integration (CCCSTI).[4]

Bush promised that these measures would result in an 18 percent cut in national emissions *intensity*, or emissions produced per unit of GDP, by 2012, not necessarily emissions; economic growth could still lead to increased greenhouse gas (GHG) emissions overall, despite better emissions efficiency.

President Bush's menu of climate change policies and programs expanded considerably during his presidency—although nearly all of the new initiatives center on further R&D, additional voluntary programs, and/or more funding—which had a track record extending back to the Clinton administration of ineffectiveness in reducing overall carbon emissions.

The Bush Approach: Unofficially

Despite these generally positive efforts to harness the power of markets, advanced technologies, and regional alliances to mitigate greenhouse gas emissions, President Bush's commitment to addressing climate change seemed to be severely undercut by the administration's less official efforts to suppress definitive scientific statements and to avoid more rigorous mandatory policies. To this end, the Bush approach was to overstate uncertainties in climate science, censor climate scientists' voices within the federal government, and support increased use of fossil fuels.

Overstating Uncertainties in Climate Science

Despite an almost universal scientific consensus on the anthropogenic causes of global warming, President Bush and administration officials repeatedly overplayed the remaining uncertainties and declared that more research was needed to identify the human contribution to climate change.[5]

For example, in May 2001, the administration requested that the National Academy of Sciences (NAS) identify areas of uncertainty in the science of climate change as well as discrepancies in reports from the Intergovernmental Panel on Climate Change (IPCC), which it then used to justify continued inaction in regulating GHG emissions.[6] The administration cited these uncertainties until well into Bush's second term, in an apparent attempt to undermine public confidence in the scientific consensus.

Suppression of Government Science

Besides overstating the uncertainty of anthropogenic climate change, the Bush administration attempted to suppress the evidence of human impacts on climate. This was shown by congressional testimony revealing a pattern of tampering and censorship that crippled government scientists' ability to accurately report on climate change.[7]

For instance, the administration directed the Environmental Protection Agency (EPA) not only to remove the usual climate change section from its annual air pollution report in 2002, but also to delete references to human causes and global consequences for climate change, and by adding qualifying words such as "potentially" that weakened the report's arguments.[8] The most famous example of this pattern of suppression was when the administration directed NASA's public affairs staff to restrict media access to, and review and censor the work of, James Hansen, director of NASA's Goddard Institute for Space Studies, who had made public statements contradicting official administration policy.

Bush's Energy Policy: No Fossil Fuel Left Behind

In early 2001, Bush directed Vice President Dick Cheney to develop a national energy policy. Cheney's National Energy Policy Development Group met in secret with executives from energy corporations such as Enron, ExxonMobil, Shell, and Peabody Energy (the world's largest coal company), most of which were large contributors to the administration and the Republican Party. Environmental groups complained they were effectively shut out of the process.[9]

While encouraging greater conservation and research in alternative fuels, the resulting plan, released in May 2001, promoted both increased domestic production of fossil fuels and coal-fired electricity generation with reduced government regulation.[10] The plan

recommended $2 billion in federal assistance for continued clean coal research.[11] These were exactly the types of policies that energy corporations had been seeking. In fact, documents obtained from the Department of Energy by the Natural Resources Defense Council in 2002 revealed that many energy industry proposals and recommendations were directly translated into official U.S. policy, sometimes with nearly identical language.[12]

Strange Bedfellows: The Energy Policy Act of 2005

The Energy Policy Act of 2005 (EPACT 2005), a long, complicated law touching on nearly every facet of energy production and use, contains a number of provisions to mitigate greenhouse gas emissions.[13] Among these is a host of subsidies and incentives for renewable energy sources, standards for biofuels, requirements for increased federal purchases of renewable power, provisions for further development and deployment of carbon capture technologies, and the commission of a new NAS study on fuel cells.[14] Moreover, it prohibits new federal or state permits for oil or gas drilling in or under the Great Lakes and does not open the Arctic National Wildlife Refuge to oil and gas exploration. EPACT 2005 also offers a boost to nuclear power, through a number of incentives including a production tax credit and extension of the Price-Anderson liability protections, and creates federal loan guarantees for energy projects that use new, innovative technologies that avoid or sequester emissions of GHGs.[15]

Other provisions of EPACT, however, promote the expansion of fossil-fuel production and use. The law calls for the Department of the Interior to conduct an inventory of oil and gas resources on the outer continental shelf and to open up federal lands for R&D on oil shale; streamlines the permitting process for pipelines and oil and gas development on existing federal sites; and repeals the previous 160-acre limit on coal leases.[16] In addition, the act provides tax incentives for the construction of additional coal-fired power plants, albeit ones using so-called clean coal technologies.[17]

Despite the rhetoric of "clean coal," only 16 percent of proposed new coal plants were expected to use coal gasification technologies that may allow carbon capture and sequestration. Nearly 60 percent were expected to rely on traditional, heavily polluting pulverized coal technology.[18]

President Bush's 2006 and 2007 State of the Union Addresses

In his 2006 State of the Union address, President Bush recognized America's addiction to oil, touting technology as the way to break the addiction, lowering reliance on Middle East petroleum imports, and moving beyond an oil-based economy.[19] He announced that his new Advanced Energy Initiative, primarily an increase in funding for existing clean energy R&D efforts, would develop technologies to provide cleaner, reliable electricity (e.g., "clean coal" again) as well as better batteries and biofuels for cars.

In 2007, President Bush finally uttered the phrase "global climate change" in a State of the Union address.[20] Nevertheless, he offered only a limited mix of policy prescriptions. He emphasized the familiar theme of continued R&D in energy technologies. More specifically, he called for a 20 percent reduction in projected gasoline use during the next ten years, to be achieved through reforms of the Corporate Average Fuel Economy (CAFE) standards for cars and a mandatory fuels standard requiring 35 billion gallons of renewable and alternative fuels (e.g., corn ethanol).[21] Bush's proposals ignored significant sources of greenhouse gases like the electricity-generation and manufacturing sectors.[22] In addition, while promoting greater fuel diversity, Bush called for increased domestic oil production, albeit in "environmentally sensitive ways."[23] After some congressional modification, this list of policies, including the increase in CAFE standards and the renewable fuels standards, was codified into law in the Energy Independence and Security Act of 2007.[24]

Bush Offers His International Proposal

In May 2007, shortly before a Group of Eight (G-8) summit, the Bush administration vehemently rejected a proposal from Germany, which had the support of both Japan and Great Britain, to cut greenhouse gas emissions by 50 percent by 2050.[25] As an alternative, President Bush put forward a proposal for what he called a "new framework" to address climate change. He called on the fifteen nations responsible for 80 percent of world GHG emissions—including the EU, China, India, and Brazil—to hold a series of "Major Economies" meetings to create a framework for a post-Kyoto regime by the end of 2008. A key aspect of the framework would be a long-term, global emissions reduction goal with mid-term targets set on a national basis. Yet, while the discussion of such goals marked a notable shift in Bush administration rhetoric, the proposed emissions reductions would be voluntary rather than mandatory; or, as described by James Connaughton, chairman of the White House Council on Environmental Quality, "aspirational."[26]

The Bali UN Framework Convention on Climate Change (UNFCCC) Meeting: December 2007

In December 2007, at the UNFCCC Conference in Bali, the heated debate centered on the terms of future UNFCCC negotiations. Despite promises that it supported the creation of a road map to a global post-Kyoto agreement, the United States diligently played the role of obstructionist, steadfastly opposing any proposals that might lead to specific emissions reduction targets for developed nations.[27] Initially, the U.S. delegation refused to commit to the measurement and verification of technical and financial assistance to developing countries. The angry audience erupted in a cacophony of boos, followed by a succession of reprimands from developing countries' delegates.[28] Facing the loss of any leadership it might have in future negotiations and with other countries' support for the Major Economies process increasingly at risk, the United States finally relented and joined the consensus, to a round of applause.[29] The next major international climate negotiation will be held in December 2009 in Copenhagen.

In April 2008, President Bush announced that the United States would stop the growth of its GHG emissions by 2025, thus setting a mid-term target consistent with the administration's goals for the Major Economies process and positioning itself for continued negotiations under the Bali Action Plan. In addition, Bush suggested that the United

States might be willing to sign on to a binding international commitment, if the other major economies did likewise.[30] Nevertheless, in his announcement, Bush relied on technology and incentives, and failed to offer a clear vision for a path forward commensurate with the challenge.

Congressional and Judicial Responses

In 1997, at the advent of the Kyoto Protocol, Congress voted 95 to 0 for the Byrd-Hagel "Sense of the Senate" resolution, which declared that the United States should not sign any climate agreement that did not include the full participation of developing countries or that could seriously harm the U.S. economy.[31] Since then, Congress has shown increasing willingness to pass substantive climate change legislation. More than sixty bills were introduced in the 110th Congress (2007–2008). In 2003, Senators Joseph Lieberman (D-CT) and John McCain (R-AZ) submitted their Climate Stewardship Act (2003), which would have capped aggregate emissions from covered industries at 2000 levels starting in 2010, allowed trading of allowances, and included financial penalties for noncompliance. Brought up for a vote in October 2003, it failed 43 to 55, a remarkable change from the Byrd-Hagel result.[32] In June 2005, during negotiations on the Energy Policy Act, the Senate passed a nonbinding resolution submitted by Sen. Jeff Bingaman (D-NM) that recognized that mandatory steps would be needed to cut greenhouse gas emissions.[33]

The 2006 congressional elections reshaped the legislative debate on global warming. Jerry McNerney, a wind energy consultant, defeated Richard Pombo (R-CA), chairperson of the House Resources Committee and proponent of oil drilling in the Arctic National Wildlife Refuge. Barbara Boxer (D-CA) replaced James Inhofe (R-OK), a noted global warming denier, as chair of the Senate Committee on Environment and Public Works, and promised to focus on global warming.[34] Speaker of the House Nancy Pelosi (D-CA) created a Select Committee on Energy Independence and Global Warming, consisting of nine Democrats and six Republicans, to research climate change and recommend legislation.[35] The actual creation of legislation remains with the traditional committees, most notably the House Committee on Energy and Commerce now chaired by Henry Waxman (D-CA), who replaced the automobile industry–favoring former chair, John Dingell (D-MI).[36]

In April 2007, the Supreme Court also weighed in on the global warming debate. In a case brought by twelve states, the District of Columbia, America Samoa, the cities of New York and Baltimore, and a host of NGOs against the EPA, a closely divided Court held that the agency must regulate greenhouse gases under the Clean Air Act (CAA) unless the agency determined that these gases do not contribute to climate change or provided a "reasonable explanation" as to why it could not make such a determination.[37] In a press release of a petition timed to coincide with the oral arguments in this case, representatives of more than 10,000 EPA scientists — or over half of the EPA's workforce — called on Congress to take immediate action on global warming, revealing a clear ideological break with the agency's political leadership.[38]

In October 2007, Sens. Joseph Lieberman (I-CT) and John Warner (R-VA) submitted America's Climate Security Act in the U.S. Senate.[39] The bill covered all U.S. sources that emit carbon dioxide from the combustion of coal, petroleum, and natural gas, using a cap-and-trade system that included the auction of some GHG emission permits.[40] The Lieberman-Warner bill quickly became the consensus climate proposal in the Senate, receiving the key support of the chair of the Senate Committee on Environment and Public Works, Senator Boxer. Nevertheless, the bill failed to secure a majority following a floor debate in June 2008.

The Obama Presidency

In the 2008 presidential campaign, Barack Obama emphasized his support for production tax credits for solar and wind power. Mandatory GHG emissions ceilings combined with energy efficiency and $150 billion in renewable energy infrastructure would, in Obama's view, create new product lines and new jobs. Moreover, Obama's public statements indicated that he understood the need for a strong market price for carbon, and the need also to auction GHG emissions allowances. In the transport sector, Obama pledged to repeal tax breaks for oil companies and use the savings to invest in fuel-efficient cars, help U.S. automakers retool, and have a million plug-in hybrids on the road by 2010. He noted that hybrid electric vehicles are expected to get 150 miles (250 kilometers) per gallon in the near future.

In his inaugural address, President Obama alluded to his plans to double renewable energy supplies in the next three years by "harness[ing] the sun and the winds and the soil to fuel our cars and run our factories." He promised to "restore science to its rightful place," a move that seemed to be an admonition to the Bush administration, which, as explained above, continually suppressed government findings and documents on climate change. He pledged to work with other countries to "roll back the specter of a warming planet."[41]

In the American Recovery and Reinvestment Act of 2009, the $780 billion economic stimulus bill that was enacted by Congress, at President Obama's request, in February 2009, more than 10 percent of the money was dedicated to new energy, energy infrastructure (energy efficiency and a strengthened electrical grid), and energy-related jobs. This energy funding was framed as a response to the financial crisis. Thus, Obama has prioritized energy not only in the context of climate change, but also in the context of the economy and national competitiveness.[42]

The Obama Budget

The Obama budget promises a comprehensive effort to address global warming, slash oil imports, and create a new "green" economy that produces millions of new jobs. The White House estimates that a cap-and-trade program with auctioned permits will produce $645 billion in revenue over ten years beginning in 2012. The revenue from the auctioned permits would finance renewable energy projects and compensate families, communities, and businesses that are hurt by higher energy prices. The revenue would be paid by oil, electric power, and heavy industries that produce the majority of carbon dioxide and other gases blamed for the warming of the planet. Many of these costs would presumably be passed on to consumers.[43]

The White House budget plan for the four years beginning in 2009 includes projected revenue of $79 billion from auctioned carbon permits. The proceeds of the cap-and-trade auction would be used to invest in "clean" energy, help finance Obama's tax credit for workers as well as offset higher energy costs for low- and middle-income people, and clean up costs for small businesses. This move was intended to spur congressional action on climate change, as well as action at the EPA. The EPA's proposed budget is $10.5 billion, an increase of $3 billion from 2008 funding levels and the largest in the agency's history. The EPA's budget includes a $19 million increase for the greenhouse gas emissions inventory and related activities that will provide data for implementing a comprehensive climate change bill.[44]

In March 2009, the Environmental Protection Agency sent a proposal to the White House that would label carbon dioxide a danger to public welfare—a key precursor to regulating greenhouse gas emissions as pollutants. The proposal would permit the agency to begin such regulation. The long-awaited "endangerment" finding stems from the 2007 U.S. Supreme Court decision, *Mass. v. EPA*, in which the EPA was found to have the authority to regulate emissions that contribute to global warming.[45]

Seeking a Comprehensive Energy Policy

As the Obama administration develops its energy plans, it is being lobbied both by oil companies, which ask the president to fulfill his reluctant campaign pledge to support U.S. offshore drilling, and by environmental groups, which demand a renewal of the drilling ban that Congress suspended in September 2008.

In the words of a *New York Times* correspondent:

"President Obama must decide what strategies are most likely to achieve his goals of diversifying the nation's fuel supplies, developing alternative energy sources, reducing oil consumption, and curbing carbon emissions. . . .

"Part of that calculus is what role the administration sees for domestic supplies. Since taking office, it has scrapped rules issued in the final days of the Bush administration that would have opened up vast new areas for offshore drilling well into the next decade. At the same time, the Obama administration has allowed the Interior Department to auction leases in the Gulf of Mexico that includes 4.2 million acres that had been off limits since 1988. . . .

"The United States is the world's top oil consumer, but its domestic output has been falling since 1971. Oil imports make up more than 60 percent of the nation's daily consumption of 19 million barrels. Yet for more than thirty years, drilling off most of the American coastline has been forestalled by opposition from coastal states and environmental groups. About 85 percent of the nation's coasts are now off limits, including most of the Pacific and Atlantic seaboards and the western coast of Florida. . . .

"The Gulf of Mexico, where some drilling is allowed, has seen tremendous growth in domestic oil production since the 1990s, because of . . . technological advances that have enabled drilling in [deep] waters. Estimated reserves in the Gulf of Mexico have grown sevenfold in the last thirty years. . . . The Interior Department estimates that undiscovered oil reserves total 86 billion barrels, four times the nation's official proven reserves. The bulk of that potential oil, nearly 68 billion barrels, is . . . already accessible to drilling in the Gulf of Mexico and Alaska."[46]

American Clean Energy and Security Act (ACES)

On June 26, 2009, by a vote of 219–212, the U.S. House of Representatives passed ACES, also known as the Waxman-Markey bill, HR 2454. The bill's two main sponsors were Rep. Henry Waxman (D-CA) and Rep. Edward Markey (D-MA).[47]

The bill pledges to cut U.S. emissions, especially from power plants, factories, and tailpipes, by 17 percent below 2005 levels by 2020, and 83 percent by 2050. It would establish a "cap-and-trade" system of buying and selling emissions permits, under which 85 percent of such permits would be allocated free. Fifteen percent would be auctioned. Environmentalists and President Obama had wanted the government to auction all permits. Waxman and Markey agreed to give free permits to coal-burning utilities, oil refineries, automakers, and manufacturers competing with China and India (and their low-cost power and labor).

Electricity producers would be required to obtain at least 15 percent of their energy from renewable sources by 2020, with as much as 5 percent more energy saved from new efficiency measures. The two figures must add up to 20 percent. The bill mandates investment in new clean energy technologies and energy efficiency ($90 billion in new investments by 2025), carbon capture and sequestration ($60 billion), electric and other advanced technology vehicles ($20 billion), and basic scientific research and development ($20 billion). It also mandates new energy-saving standards for buildings, appliances, and industry.[48]

Waxman agreed to put the Agriculture Department—rather than the EPA—in charge of the offset program that pays farmers and other landowners to conduct environmentally friendly projects. He also agreed to freely allocate permits to coal and other fossil-fuel–fired power plants. Several industry groups opposed the bill, including the U.S. Chamber of Commerce and the American Petroleum Institute. But a host of companies and utilities supported the bill, including Nike Inc., Starbucks Corp., Exelon Corp., Symantec Corp. and PG&E Corp.—a coalition that House Democrats said was invaluable.[49]

Al Gore, from his Twitter page, said, "Make no mistake, this [Waxman Markey Bill] is the most important environmental vote of this generation. If passed, this legislation will put us on the road to actually solving the climate crisis, in addition to building a green economy."[50]

The U.S. and International Climate Negotiations

Shortly after taking office, President Obama placed the United States at the forefront of the international climate effort and raised hopes for an effective international climate agreement. President Obama's chief climate negotiator, Todd Stern, observed in February 2009 that the United States would be involved in the negotiation of a new treaty—to be signed in Copenhagen in December—"in a robust way."[51] The Copenhagen agreement is expected to provide money and technical assistance to help developing countries cope with climate change.

Conclusion

Despite the present attention at the federal level to global climate change, the last sixteen years of U.S. climate policy may be described as a collection of missed opportunities. By removing itself from the Kyoto process, the United States lost the opportunity to participate in, learn from, and benefit from Clean Development Mechanism projects in the global south and carbon-trading programs in the north. President Bush (and President Clinton before him) failed to use the power of the presidency to convince U.S. corporations and the American public of the need for a significant change from business as usual. U.S. companies fell behind their foreign counterparts in Japan and Europe in designing and marketing low- and non-carbon energy technologies, undoubtedly weakening America's long-term economic competitiveness.[52]

Significantly, the United States ceded the moral high ground it once had on environmental issues, considerably weakening its ability to persuade China, India, and other fast-growing nations to avoid relying on fossil fuels. The Bush administration's mixed strategies—promoting continued research on clean energy technologies on one hand and expanding fossil-fuel exploration and coal-fired electricity generation on the other—wasted valuable time and resources that could have gone to lower U.S. impact on the global climate, and perhaps the global climate's impact on the United States through extreme weather events. Lacking a comprehensive climate strategy that balances continued R&D with effective, mandatory policies, the U.S. government ensured that when the inevitable national transformation occurs, whether to mitigate emissions or adapt to a warming world, it will be more difficult and more painful.

The 2008 elections brought in a Congress that seemed open to addressing global climate change. President Obama has appointed climate progressives to key positions, including Carol Browner as Energy and Climate Coordinator; Steven Chu as Secretary of Energy; and John Holdren as science adviser. But the Obama blueprint for energy and climate change depends on Congress. A watered-down Waxman-Markey cap-and-trade system may pass the Senate during the 111th Congress and be signed into law by President Obama, but its effectiveness in mitigating climate change may be modest.[53] If President Obama and Congress can work effectively together, they may be able eventually to create and enact a stronger climate change policy and rally the nation to the large effort that lies before it.

Notes

1. Bush, George W., "President Bush Calls on Congress to Act on Nation's Priorities," speech at Army National Guard Aviation Support Facility, September 23, 2002, transcript viewed at www.whitehouse.gov/news/releases/2002/09/20020923-2.html, January 9, 2007.

2. Jehl, D., and A. C. Revkin, Bush, in reversal, won't seek cut in emissions of carbon dioxide. *New York Times*, March 14, 2001. U.S. won't follow climate treaty provisions, Whitman says. *New York Times*, March 28, 2001.

3. The FreedomCar and Fuels Partnership, the benefits of which are not likely to be realized for twenty years, replaced Clinton's Partnership for a New Generation of Vehicles, which seemed to be making steady progress on its goal of developing a mid-size car by 2003 that would achieve 80 mpg. The White House, Global Climate Change Policy Book, February 2002, viewed at www.whitehouse.gov/news/releases/2002/02/climatechange.html, December 29, 2006.

4. The White House, 2002 op. cit.

5. See for example, sixteen national academies of science, "The Science of Climate Change," May 17 2001, viewed at www.royalsociety.org/displaypagedoc.asp?id=6206, December 29, 2006. For a list of some of the statements that form this consensus, see Union of Concerned Scientists, "The Scientific Consensus on Climate Change," July 6, 2006, viewed at www.ucsusa.org/ssi/climate_change/climate-consensus.html, January 5, 2007.

6. National Research Council, *Climate Change Science: An Analysis of Some Key Questions, Committee on the Science of Climate Change*, Division on Earth and Life Sciences (Washington, DC: National Academy Press, 2001); see especially appendix A. Abraham, S., The Bush administration's approach to climate change. *Science* 305 (2004): 616–17.

7. Goldenberg, S., Bush administration accused of doctoring scientists' reports on climate change. *The Guardian*, January 31, 2007, viewed at www.guardian.co.uk/usa/story/0,,2002484,00.html, February 4, 2007.

8. Union of Concerned Scientists (UCS), "Scientific Integrity in Policymaking: An Investigation into the Bush Administration's Misuse of Science," March 2004. EPA Internal Memo, "Issue Paper: White House Edits to Climate Change Section of EPA's Report on the Environment," April 29, 2003. This can be found in UCS, 2004 op. cit.; see especially appendix A, 35–38.

9. Pasternak, J., Bush's energy plan bares industry clout. *Los Angeles Times*, August 26, 2001; Milbank, D., and J. Blum, Document says oil chiefs met with Cheney task force. *Washington Post*, November 16, 2005.

10. National Energy Policy Development Group, *Reliable, Affordable, and Environmentally Sound Energy for America's Future: Report of the National Energy Policy Development Group*, National Energy Policy, May 2001, viewed at www.whitehouse.gov/energy, January 4, 2006.

11. Nation Energy Policy Development Group, 2001 op. cit.

12. Van Natta, D., and N. Banerjee, Review shows energy industry's recommendations to Bush ended up being national policy. *New York Times*, March 28, 2002. In return, the oil and gas industries have been good to the administration. In the 2004 electoral cycle, they gave more to President

Bush—more than $2.5 million—than the next eleven presidential and congressional candidates combined. See Center for Responsive Politics (CRP), "Oil & Gas: Top 20 Recipients," viewed at www.opensecrets.org/industries/recips.asp?cycle=2004&ind=E01, January 7, 2006. This is based on Federal Electoral Commission data released on May 16, 2005.

13. U.S. Congress, 109th Congress, *Energy Policy Act of 2005*, P.L. 109-58, 42 USC 15801, August 8, 2005, viewed at http://frwebgate.access.gpo.gov/cgi-bin/getdoc.cgi?dbname=109_cong_public_laws&docid=f:publ058.109, January 1, 2007.

14. Pew Center on Global Climate Change, "Summary of Energy Policy Act of 2005," viewed at www.pewclimate.org/policy_center/analyses/hr_6_summary.cfm, January 5, 2007.

15. U.S. Senate Committee on Energy and Natural Resources, "Conference Report Summary by Fuel," August 1, 2005, viewed at http://energy.senate.gov/public/_files/Conferencereportoverviewexpanded080105.doc. U.S. Department of Energy, "Secretary Bodman Announces $2 Billion Federal Loan Guarantee Program as Part of First Anniversary Celebration of Energy Policy Act," August 7, 2006, viewed at www.lgprogram.energy.gov/press/080706.html, April 15, 2007.

16. U.S. Senate Committee on Energy and Natural Resources, "Conference Report Summary by Title," August 8, 2005, viewed at http://energy.senate.gov/public/_files/PostConferenceBillSummary.doc, January 5, 2007. U.S. Senate Committee on Energy and Natural Resources, "Conference Report Summary by Fuel," 2005, op. cit. McGarvey, J., and M. Murphy, "Summary of the Energy Policy Act of 2005," The National Regulatory Research Institute, September 2005, p. 15, viewed at www.nrri.ohio-state.edu/dspace/bitstream/2068/567/1/Summary+of+Energy+Policy+Act+of+2005.pdf, January 5, 2007.

17. U.S. Senate Committee on Energy and Natural Resources, "Conference Report Summary by Fuel," 2005, op. cit.

18. Madsen, T., and R. Sargent, *Making Sense of the "Coal Rush": The Consequences of Expanding America's Dependence on Coal*, NJPIRG Law & Policy Center, July 2006, viewed at http://njpirg.org/reports/CoalRushNJ.pdf, January 15, 2007.

19. Bush, George, "State of the Union 2006," speech to U.S. Congress, Washington, DC, January 31, 2006, viewed at www.whitehouse.gov/stateoftheunion/2006, January 6, 2007.

20. Bush, George W., "State of the Union 2007," speech to U.S. Congress, January 23, 2007, transcript viewed at www.whitehouse.gov/news/releases/2007/01/20070123-2.html, January 3, 2007.

21. Even if the 20 percent reduction target is met, it is conceivable that overall gasoline consumption could increase if the projected consumption baseline were significantly higher than present consumption. In this way, this target is similar to Bush's energy intensity goal. Bush, 2007 op. cit.

22. Pew Center on Global Climate Change, "Response to 2007 State of the Union," 2007, viewed at www.pewclimate.org/press_room/speech_transcripts/sotu2007.cfm, February 3, 2007.

23. Bush, 2007 op. cit. Interestingly, his proposed budget for fiscal year 2008 would eliminate funding for oil and natural gas R&D. See Simon, R., Energy efficiency, aid to poor suffer in priority shuffle. *Los Angeles Times*, February 6, 2007, viewed at www.latimes.com/news/nationworld/washingtondc/la-na-energy6feb06,1,254497.story, March 20, 2007.

24. U.S. Congress, 110th Congress, *Energy Independence and Security Act of 2007*, P.L. 110-140, 42 USC 17001, December 19, 2007, viewed at http://frwebgate.access.gpo.gov/cgi-bin/getdoc.cgi?dbname=110_cong_public_laws&docid=f:publ140.110, January 31, 2008.

25. Cooper, H., and A. C. Revkin, U.S. rebuffs Germany on greenhouse gas cuts. *New York Times*, May 26, 2007.

26. Fletcher, M. A., and J. Eilperin, Bush proposes talks on warming. *Washington Post*, May 31, 2007. Stolberg, S. G., Bush proposes goals on greenhouse gas emissions. *New York Times*, June 1, 2007, viewed at www.nytimes.com/2007/06/01/Washington/01prexy.html.

27. Pew Center on Global Climate Change, "Thirteenth Session of the Conference of the Parties to the UN Framework Convention on Climate Change and Third Session of the Meeting of the Parties to the Kyoto Protocol." Pew Center COP-13 Summary, viewed at www.pewclimate.org/docUploads/Pew%20Center_COP%2013%20Summary.pdf, March 15, 2008. The United States was not alone is its opposition to specific emissions commitments. At least Russia, Canada, Japan, and India each opposed provisions with such or similar

commitments. Eilperin, J., Bali forum backs climate "road map." *Washington Post*, Deember 16, 2007.

28. Pew Center COP-13 Summary, 2008 op. cit.; Eilperin, 2007 op. cit.

29. Pew Center COP-13 Summary, 2008 op. cit. Thomas and Revkin, 2007 op. cit.

30. Bush, George W., "President Bush Discusses Climate Change," Speech in the Rose Garden, April 16, 2008, transcript viewed at www .whitehouse.gov/news/releases/2008/04/20080416-6.html, April 17, 2008.

31. U.S. Congress, 105th Congress, *A Resolution Expressing the Sense of the Senate Regarding the Conditions for the United States Becoming a Signatory to Any International Agreement on Greenhouse Gas Emissions under the United Nations Framework Convention on Climate Change.* Senate Resolution 98, Report No. 105-54, 1997, viewed at http://thomas.loc.gov/cgi-bin/bdquery/ z?d105:SE00098:, December 26, 2006.

32. U.S. Congress, 108th Congress, *A Bill to Provide for a Program of Scientific Research on Abrupt Climate Change, to Accelerate the Reduction of Greenhouse Gas Emissions in the United States by Establishing a Market-Driven System of Greenhouse Gas Tradeable [sic] Allowances that Could Be Used Interchangably [sic] with Passenger Vehicle Fuel Economy Standard Credits, to Limit Greenhouse Gas Emissions in the United States and Reduce Dependence upon Foreign Oil, and Ensure Benefits to Consumers from the Trading in Such Allowances.* S.139, 2003, viewed at www.thomas.gov/ cgi-bin/bdquery/z?d108:SN00139:@@@S, December 26, 2006.

33. Bingaman, J., et al., Sense of the Senate on climate change. SA 866, *Congressional Record* S7089, 2005, viewed at http://thomas.loc.gov/ cgi-bin/query/F?r109:1:./temp/~r109JWbrGC:e11 9074:, January 7, 2007. The amendment was accepted via voice vote. A vote to table the amendment, sponsored by Sen. Inhofe, failed 44–53.

34. Young, S., Boxer to focus on global warming. *Associated Press*, November 9, 2006.

35. See Associated Press, Lawmakers reach deal on climate committee. *Washington Post*, February 7, 2007, viewed at www.washingtonpost.com/ wp-dyn/content/article/2007/02/06/AR2007020601 604.html, March 20, 2007.

36. For the switch from Dingell to Waxman, see www.politico.com/news/stories/1108/15565 .html. See also www.nytimes.com/2008/11/20/us/ politics/20dingell.html.

37. *Massachusetts, et al. v. Environmental Protection Agency*, 127 S. Ct. 1438; 167 L. Ed.2d 248 (2007).

38. Public Employees for Environmental Responsibility, "EPA Scientists File Mass Petition for Action on Global Warming," November 29, 2006, viewed at www.peer.org/news/news_id.php?row_ id=789, January 10, 2007. The petition can be viewed at www.peer.org/docs/epa/06_29_11_ global_warming_petition.pdf, January 10 2007.

39. U.S. Congress, 110th Congress, *A Bill to Direct the Administrator of the Environmental Protection Agency to Establish a Program to Decrease Emissions of Greenhouse Gases, and for Other Purposes.* S.2191, 2007, viewed at www.thomas.gov/ cgi-bin/bdquery/z?d110:s.02191:, March 15, 2008.

40. Economic analyses of emissions reduction proposals suggest that implementing a cap-and-trade bill may cause U.S. GDP to drop by a mere 0.5 percent to 1.5 percent by 2050. A recent McKinsey & Company analysis of a substantial greenhouse gas pollution reduction target projects that the economic benefits of taking action, chiefly energy savings, could roughly offset the economic costs by 2030. See Creyts, J., A. Derkach, S. Nyquist, K. Ostrowski, and J. Stephenson, *Reducing U.S. Greenhouse Gas Emissions: How Much at What Cost?* U.S. Greenhouse Gas Abatement Initiative Executive Report, December 2007, viewed at www.mckinsey.com/clientservice/ccsi/pdf/US_ ghg_final_report.pdf, March 15, 2008.

41. Transcript of Barack Obama's Inaugural Address, January 20, 2009, www.nytimes.com/ 2009/01/20/us/politics/20text-obama.html.

42. http://www.newenergyworldnetwork.com/ renewable-energy-news/by_technology/energy_ efficiency/president-barack-obama-believes-energy-bill-will-transform-us-economy.html (June 29, 2009)

43. "Remarks by the President on the Fiscal Year 2010 Budget," February 26, 2009, www .whitehouse.gov/the_press_office/Remarks-by-the-President-on-the-Fiscal-Year-2010-Budget/.

44. Walter, L. "Obama Proposes $10.5 Billion EPA Budget for Increased Environmental Protection," March 4, 2009, http://ehstoday.com/ standards/epa/obama-epa-budget-2317.

45. Galbrath, K. EPA proposal calls greenhouse gases a danger to the public. *New York Times*,

March 23, 2009, http://greeninc.blogs.nytimes.com/2009/03/23/epa-proposal-calls-greenhouse-gases-a-danger-to-the-public/?hp.

46. Mouawad, J. Obama tries to draw up an inclusive energy plan. *New York Times*, March 17, 2009. http://www.nytimes.com/2009/03/18/business/energy-environment/18offshore.html?scp=1&sq=business%20energy%20environment%20March%2018%202009&st=cse.

47. http://www.govtrack.us/congress/bill.xpd?bill=h111-2454.

48. http://climateprogress.org/2009/06/02/a-useful-summary-of-the-house-clean-energy-and-climate-bill.

49. Tankersley, J., Climate bill shaped by compromise, http://www.latimes.com/news/nationworld/nation/la-na-energy28-2009jun28,0,7474723.story.

50. http://www.renewableenergyfocus.com/view/2387/the-us-house-approves-the-waxmanmarkey-american-clean-energy-and-security-act-hr-2454-legislation.

51. Rosenthal, E. Obama's backing raises hopes for climate pact. *New York Times*, February 29, 2009, www.nytimes.com/2009/03/01/science/earth/01treaty.html?_r=2&hp.

52. An early sign of this reduced competitiveness may be the Detroit automakers. Overreliant on SUV sales, they have lost market share to Japanese automakers, most notably Toyota, which has developed and heavily promoted fuel-efficient hybrid automobiles. See, for example, Big 3 automakers see sharp decline in sales. *Associated Press*, August 1, 2006, viewed at www.msnbc.msn.com/id/14136797/, March 20, 2007. As we transition to a carbon-constrained regulatory environment, other areas of concern include building materials (e.g., triple-pane glass), appliances, and renewable energy technologies.

53. James Hansen, NASA climate modeler, said he was growing "concerned" that "if they are going to support cap and trade then unfortunately I think that will be another case of greenwash. It's going to take stronger action than that" (see www.guardian.co.uk/science/2009/mar/18/nasa-climate-change-james-hansen).

Chapter 34

Policy in California

Jason Mark and Amy Lynd Luers

The state of California has emerged as the most significant climate policy actor in the United States. Climate action has been motivated by concern over potential climate-related impacts and facilitated by the state's existing framework of energy and environmental polices. A comprehensive suite of climate policies materialized rapidly beginning in 2000 as the national policy vacuum created a political environment favorable to strong state leadership.

California's emerging climate policy action is built upon its legacy of leadership in the energy and environmental policy space. Electricity consumption has long been the focus of energy policies, and the state's air-quality regulations have driven the global introduction of numerous automotive pollution–control technologies. As the state now turns its attention to mitigating climate change, this existing policy basis is being leveraged to yield additional reductions in greenhouse gas emissions. This chapter provides an overview of California's actions to address climate change.

Background

California is often characterized as a "nation state," owing to its size and influence. California is among the top ten largest economies in the world and is the world's twelfth-largest source of carbon dioxide emissions with per capita emissions double the global average.[1] While California is significant because of the size of its population, economy, and aggregate emissions, it is also notable for its low emissions intensity relative to other U.S. states. Efficiency and clean energy policies, combined with its mild climate, have enabled California to rank among the least carbon-intensive states in the country.[2] Transportation is the largest single source of carbon emissions, accounting for 41 percent of gross emissions, followed by the industrial sector (23 percent) and electricity (20 percent) (see figure 34.1). Vehicles, electricity supply, and large energy consumers remain the most significant opportunities for emissions reductions as California seeks to address climate change.

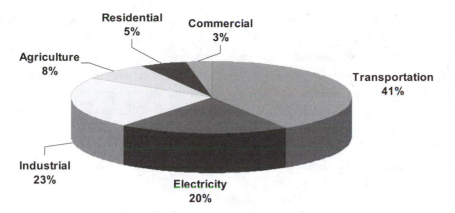

FIGURE 34.1. California greenhouse gas emissions, by sector, 2002.

Assessing Climate Change

California policy makers first turned their attention to climate change nearly twenty years ago, when State Senator Byron Sher spearheaded the adoption of the state Assembly Bill 4420.[3] This 1988 law called for the preparation of the first scientific assessment of the potential impacts of climate change on California and for options for reducing greenhouse gas (GHG) emissions in the state.

In response, two climate reports were prepared—"The Impacts of Global Warming on California" and "Climate Change Potential Impacts and Policy Recommendations"—that provided the first overview of California's vulnerabilities to the changing climate.[4]

In 1999, Field and colleagues produced a comprehensive report on the ecological impacts of continued climate change in the state that helped to educate California policy makers on the risks of climate change.[5] This was followed by the release of the first national assessment, which included a regional assessment of California.[6]

Perhaps one of the most important findings reported in these assessments were those highlighting the potentially severe threats that climate change posed to California's water resources. Maury Roos, California's chief hydrologist in the California Department of Water Resources (DWR), first brought attention to this threat by documenting the declining trend of spring and summer water runoff since records began in the early 1900s.[7] The threats to the state's water supply were further highlighted by studies showing how April 1 snow levels in the Sierra Nevada would be substantially reduced as climate warms.[8]

In the most recent years, a series of reports were prepared that focused on communicating to the California public and to policy makers the implications of divergent emissions scenarios for California's health, economy, and environment. In 2004, a group of eighteen scientists published an assessment in the *Proceedings of the National Academies of Sciences* (PNAS), which contrasted the impacts on California under two possible futures, one under a higher-emissions scenario and another under a lower-emissions scenario.[9] This new analysis highlighted that the severity of climate change impacts depends on whether and how quickly California and the rest of the world reduce emissions. As with Field and colleagues, the science team participated in numerous briefings for state agencies, policy makers, and industry groups on the study's finding, and a nontechnical summary of the report was widely distributed for educational purposes.[10]

The most recent statewide assessment, the 2006 California Climate Scenarios Project, expanded the 2004 *PNAS* study to explore the impacts on more sectors of the state.[11] Most notably, the scientific analysis was produced as part of a comprehensive report that outlined a set of strategies for managing climate change through aggressive mitigation and adaptation approaches.

The combination of the severity of projected climate change impacts and the credibility of the scientists involved has helped raise public concern about the risks of global warming in the state. A 2006 poll by the Public Policy Institute of California showed that 70 percent of likely voters favored actions to address global warming—this was up from 35 percent reported in a 2000 poll.[12]

Climate Policy

Amid growing concern about the potential impacts of climatic change for California's economy, health, and environment, policy makers have enacted an array of progressively significant climate-specific policies. Historic leadership in energy efficiency, renewable energy, and vehicles gave way to climate-specific legislation in the power and transportation sectors starting in 2002. By 2006, the state produced the nation's first comprehensive, economy-wide climate cap.

Historic Basis

California's emerging climate policies are built on decades of experience with energy and air-quality policies that have become the basis for new strategies directed at the climate challenge.

EFFICIENCY

California's efficiency success story is exemplified by the fact that per capita electricity consumption in the state has remained relatively flat during the last three decades while the nation has seen a 45 percent increase. Historic efficiency programs and standards have saved more than 12,000 megawatts (MW) of peak electrical demand, equivalent to more than two dozen large power plants.[13] Standards for appliances and for building design, equipment, and materials account for half of the reductions.[14] The state's efficiency programs are primarily funded through utility resource procurement budgets and surcharges on customer bills. In 2006, California's investor-owned utilities launched the most aggressive energy efficiency program in the history of the utility industry. Over three years, the utilities' $2 billion investment is expected to provide approximately 1,500 MW of savings (equal to three large power plants) and to return nearly $3 billion in net savings to customers.[15]

RENEWABLE ENERGY

California also has a long history of supporting renewable energy generation through incentives and regulations. Today, a portion of the "public goods" surcharge on the bills of investor-owned utility customers is also used for renewable energy incentives. Most recently, the state adopted standards in 2002 requiring retail electricity sellers to increase total renewable electricity sales by at least 1 percent annually, reaching 20 percent by 2017.[16] Municipal utilities, which account for nearly one-quarter of the state's electricity sales, are required to develop and implement renewable electricity standards under their own direction. In 2006, the standards were accelerated to meet the 20 percent level by 2010.[17] California also passed landmark solar energy legislation in 2006 that will provide $3.2 billion in incentives over eleven years for solar electricity.[18]

VEHICLES

For nearly four decades, California has exercised its special authority under Section 209 of the federal Clean Air Act to set its own vehicle emission standards. In 1990, the state created the Zero Emission Vehicle program, establishing a requirement that 10 percent of vehicles sold in the state have no tailpipe emissions by 2003. Although the program's original goals have not been realized, the regulations have guided investments in hybrid, electric, and fuel-cell vehicle technologies that offer medium- and long-term opportunities for reduced greenhouse gas emissions.[19]

Emissions Accounting

State law created the California Climate Action Registry in 2000, a nonprofit public/private organization charged with voluntary reporting of greenhouse gas emissions.[20] The registry provides an opportunity for companies, nonprofits, and government agencies to report emissions as far back as 1990 and have them independently verified. Emissions accounting offers companies insight into their emissions profile while often identifying manufacturing inefficiencies that can yield cost savings. Registering the emissions also creates a basis for garnering credits for early action under future climate regulations.

Transportation Programs

Building on the state's unique authority to set motor vehicle pollution standards, in 2002 California enacted a new state law to develop the nation's first greenhouse gas standards for passenger vehicles.[21] As adopted by the California Air Resources Board in 2004, the regulations require a 22 percent reduction in new-vehicle emissions by 2012 compared to 2002 levels and nearly a 30 percent reduction by 2016.[22] The program is projected to reduce carbon emissions by nearly one-fifth relative to business as usual by 2020.[23] New vehicle costs under the program are anticipated to be offset by operating cost savings for consumers, largely through lower fuel bills. At $2.50 per gallon for gasoline, net consumer savings for vehicles sold under the regulations through 2016 are projected to be more than $14 billion.[24] Legal challenges by the automakers have been rejected in two federal court cases. However, in December 2007, the Bush administration blocked California's program by denying a Clean Air Act waiver needed for implementation. As of the time of writing, the legal debate over the waiver continued, although a new administration could grant the waiver and allow the program to proceed.

In January 2007, Governor Arnold Schwarzenegger signed an executive order creating the world's first carbon-based fuels standard.[25] The low-carbon fuel standard, to be developed through regulation at the California Air Resources Board, would reduce the carbon content of California's passenger vehicle fuels by 10 percent by 2020. The standards will cover the carbon contained in the fuel as well as emissions associated with the production and distribution of fuels (so-called lifecycle emissions). The requirements are expected to further spur investments in low-carbon alternatives to petroleum, such as ethanol, electricity, and hydrogen for transportation.[26]

Electricity Programs

California's long-standing efficiency and renewable energy programs have formed the core of the state's efforts to address greenhouse gas emissions to date. Building on its experience with efficiency and renewable energy programs, California is advancing a series of climate-specific policies. In 2004, the California Public Utilities Commission (CPUC) adopted a new policy requiring investor-owned utilities to account for the financial risk associated with greenhouse gas emissions in evaluating long-term investments. The final rules include a "greenhouse gas adder" to be used for evaluation purposes only (i.e., not for payment) for procurements of at least five years in duration.[27]

In 2006, the CPUC decided to cap retail electricity suppliers' greenhouse gas emissions.[28] This so-called load-based cap will limit the emissions associated with all the resources the retail suppliers use to meet their customers' needs. Since half of the emissions associated with California's electricity consumption comes from imported power, this design is intended to address all the emissions for which customers are accountable.[29]

The CPUC rule making also launched the development of a greenhouse gas performance standard codified by legislation signed into law in September 2006. The legislation requires that new financial commitments of five years or more for baseload electricity be for generation that is as clean or cleaner than a modern natural gas–fired power plant.[30] The performance standard applies to all power providers in the state, not just the retail electricity sellers under the purview of the CPUC. The rule will have dramatic impacts on new investments in coal-fired power, which currently does not meet the standard; however, new advanced technologies that allow carbon dioxide to be captured and stored could meet the standard. In 2005, coal generation (largely imported from outside the state) accounted for 20 percent of the state's electricity supply.[31]

Economy-Wide Emissions Limit

California's flurry of climate policies reached a climax in 2006 as the state enacted the nation's most significant greenhouse gas policy to date, the California Global Warming Solutions Act of 2006.[32] The new law requires the state to develop regulations to restore statewide greenhouse gas emissions to 1990 levels by 2020, estimated to reduce emissions by 29 percent compared to business as usual.[33] The genesis of the new emissions cap was a June 2005 executive order signed by Governor Schwarzenegger that established greenhouse gas emission reduction targets for California, as follows: return to 2000 levels by 2010; return to 1990 levels by 2020; and reduce to 80 percent below 1990 levels by 2050.[34] The subsequent legislation, sponsored by Assembly members Fran Pavley and Fabian Nuñez, focused on 2020 and converted targets into an enforceable cap while creating a process for developing post-2020 emissions limits.

The details of California's new climate regulations will be developed over the span of several years, with reductions to meet the cap beginning in 2012. Policies in place or already under development are estimated to meet nearly 40 percent of the required emissions reductions by 2020.[35] For the remaining emissions reductions, the state must consider a combination of additional sector-specific regulations (e.g., tighter efficiency requirements) as well as market-based mechanisms (e.g., a cap-and-trade program).

Conclusion

California's rapid ascendance as an international climate policy leader has been driven in part by growing levels of public concern over the climate threat. Scientific analysis and outreach—supported by government, academic, and nonprofit initiatives—steadily built the case for action among policy makers, industry leaders, and the public. California also relied on its legacy of highly successful energy and air-quality policies to develop its climate-specific policy agenda. In essence, climate change emerged as yet another reason to continue the pursuit of efficiency and renewable energy strategies, and the policy infrastructure was already in place.

In addition to growing concern about climate change and familiarity with many of the mitigation strategies, political factors played a central role in catalyzing California's climate leadership. In particular, climate-specific policies would likely have been slower to emerge in California had the federal government shown leadership on climate issues. The Bush administration's inaction created a federal policy vacuum, and California's political leaders stepped in—spurred by international appeals, environmental pressure, and acquiescence from some in industry who saw opportunity in early climate action.

Continuing a long tradition of transferring policy firsts to other regions, California's climate policy leadership is already driving change outside the state. For example, twelve states, accounting for more than one-third of the U.S. vehicle market, have already adopted the California vehicle greenhouse gas standards.[36] Europe is pursuing a California-style low-carbon fuel standard. Western and midwestern states have signed regional agreements to set regional greenhouse reduction goals and pursue joint policy action, including a regional cap-and-trade program.[37] If past is prologue, California will continue to be a key force in driving U.S. climate policy during the coming years.

Notes

We would like to thank Devra Wang, John Galloway, and Cliff Chen for helpful comments on an earlier draft of this chapter.

1. Bureau of Economic Analysis, "Western States Led Economic Growth in 2005: Advance 2005 and Revised 1998–2004 Gross State Product (GSP) Estimates," U.S. Department of Commerce,

June 6, 2006, https://bea.gov/bea/regional/gsp.htm, (September 24, 2006); World Bank, "Total GDP 2005," 2006, http://siteresources.worldbank.org/DATASTATISTICS/Resources/GDP.pdf, (September 24, 2006); Bemis, G., and J. Allen, *Inventory of California Greenhouse Gas Emissions and Sinks: 1990 to 2002 Update*, California Energy Commission, CEC-600-2005-025 (Sacramento, CA: California Energy Commission, June 2005); Energy Information Administration, *International Energy Annual 2003* (Washington, DC: Energy Information Administration, June 2005).

2. California per capita emissions calculated from emissions data from the CEC 2005 CA inventory of GHGs (Bemis and Allen, 2005 op. cit.) and population data from the California Department of Finance (State of California, Department of Finance, *Race/Ethnic Population with Age and Sex Detail, 2000–2050*, Sacramento, CA, May 2004). Industrialized nations per capita emissions calculated from IPCC 2000 emissions and population figures (Nakicenovic, N., J. Alcamo, G. Davis, B. de Vries, et al., IPCC, *Special Report on Emissions Scenarios* (New York : Cambridge University Press, 2000).

3. California Legislature. Assembly. 1988, Chapter 1051, 1987–1988 sess. AB 4420, Sher.

4. California Energy Commission, *The Impacts of Global Warming on California*, Interim Report, August 1989. California Energy Commission, *Planning for and Adapting to Climate Change. Global Climate Change: Potential Impacts and Policy Recommendations*, Committee Report, II: 6.1–6.18, 1991.

5. Field, C., G. C. Daily, F. W. Davis, S. Gaines, P. A. Matson, J. Melack, and N. Miller, *Confronting Climate Change in California: Ecological Impacts on the Golden State*, The Union of Concerned Scientists and The Ecological Society of America (Cambridge, MA: UCS Publications, November 1999). Boyd, J., "Statement at the Climate Change Science Meeting" (Sacramento, September 13, 2006).

6. Wilkinson, R. C., *The Potential Consequences of Climate Variability and Change for California. The California Regional Assessment*, Report of the California Regional Assessment Group for the U.S. Global Change Research Program, June 2002.

7. Roos, M. "A Trend of Decreasing Snowmelt Runoff in Northern California." Proc. 59th Western Snow Conf., Juneau, AK, 29–36, 1991.

8. Knowles, N., and D. R. Cayan, 2002. Potential effects of global warming on the Sacramento/San Joaquin watershed and the San Francisco estuary. *Geophysical Research Letters* 29(18): 1891.

9. Hayhoe, K. et al., Emissions pathways, climate change, and impacts on California. *Proceedings of the National Academy of Sciences of the United States of America* 101, no. 34 (2004), www.pnas.org/cgi/content/full/101/34/12422 (June 6, 2007).

10. Union of Concerned Scientists, "Climate Change in California: Choosing Our Future," 2004. This was a series of products based on the NAS study.

11. Cayan, D., A. Luers, M. Hanemann, G. Franco, and B. Croes. *Scenarios of Climate Change in California: An Overview*, report from California Climate Change Center, CEC-500-2005-186-SF, February 2006, www.energy.ca.gov/2005publications/CEC-500-2005-186/CEC-500-2005-186-SF.PDF (March 20, 2007).

12. Baldassare, M., July 2006. *PPIC Statewide Survey: Special Survey on the Environment*, Public Policy Institute of California, www.ppic.org/main/publication.asp?i=699 (March 25, 2007).

13. Bemis and Allen, 2005 op. cit.

14. Brown, S., California Energy Commission, June 2005. *Global Climate Change*, CEC-600-2005-007 www.energy.ca.gov/2005publications/CEC-600-2005-007/CEC-600-2005-007-SF.PDF (June 7, 2007) .

15. California Public Utilities Commission and California Energy Commission (CPUC and CEC), August 2006. *Energy Efficiency: California's Highest-Priority Resource*. ftp://ftp.cpuc.ca.gov/Egy_Efficiency/CalCleanEng-English-Aug2006.pdf (June 7, 2007).

16. California Legislature. Senate. 2002. *Renewable Energy: California's Renewables Portfolio Standard Program*. 2001–2002 sess., SB 1078 Sher. www.energy.ca.gov/portfolio/documents/SB1078.PDF (June 5, 2007).

17. California Legislature. Senate. 2006. *Renewable Energy: Public Interest Energy Research, Demonstration, and Development Program*. 2005–2006 sess., SB 107 Simitian. www.energy.ca.gov/portfolio/documents/SB_107_BILL_20060926_CHAPTERED.PDF (June 5, 2007).

18. California Legislature. Senate. 2006. *Million Solar Roofs*. 2005–2006 sess., SB 1, Murray. www.leginfo.ca.gov/cgi-bin/waisgate?WAISdocID

=08772520425+0+0+0&WAISaction=retrieve (June 5, 2007).

19. California Air Resources Board, *Status Report on the California Air Resources Board's Zero Emission Vehicle Program* (Sacramento, CA: California Air Resources Board, April 20, 2007).

20. California Legislature. Senate. 2000. *Greenhouse Gas Emission Reductions: Climate Change.* 2000–2001 sess. SB 1771, Sher. http://info.sen.ca.gov/pub/99-00/bill/sen/sb_1751-1800/sb_1771_bill_20000930_chaptered.html (June 6, 2007).

21. California Legislature. Assembly. 2002. *California Clean Cars Campaign.* 2001–2002 sess. AB 1493, Pavley. http://198.104.131.213/docs/ABOUTUS/AB1493.pdf (June 6, 2007).

22. California Air Resources Board, December 2004. *Fact Sheet: Climate Change Emission Control Regulations.* www.arb.ca.gov/cc/factsheets/cc_newfs.pdf (October 2, 2006).

23. California Air Resources Board, September 2004. *Addendum Presenting And Describing Revisions To: Initial Statement Of Reasons For Proposed Rulemaking, Public Hearing To Consider Adoption Of Regulations To Control Greenhouse Gas Emissions From Motor Vehicles.*

24. Net present value of savings in 2006 dollars. Author calculation based on CARB, September 2004.

25. Schwarzenegger, A. 2007. *Executive Order S-01-07.* http://gov.ca.gov/executive-order/5172/ (June 6, 2007).

26. Crane, D., and B. Prusneck, "The Role of a Low Carbon Fuel Standard in Reducing Greenhouse Gas Emissions and Protecting Our Economy," January 8, 2007. http://gov.ca.gov/index.php?/fact-sheet/5155/ (May 28, 2007).

27. California Public Utilities Commission, D.04-12-048, in R.04-04-003. *Opinion Adopting Pacific Gas and Electric Company, Southern California Edison Company and San Diego Gas & Electric Company's Long-Term Procurement Plans.*

December 2004. www.cpuc.ca.gov/PUBLISHED/FINAL_DECISION/43224.htm#P180_5449 (June 6, 2007).

28. California Public Utilities Commission, D.06-02-032, in R.04-04-003. *Opinion on Procurement Incentives Framework.* February 2006. http://www.cpuc.ca.gov/Published/Final_decision/53720.htm (June 7, 2007).

29. Bemis and Allen, 2005 op. cit.

30. California Legislature. Senate. 2006. *Electricity: Emissions of Greenhouse Gases.* 2006–2007 sess., SB 1368, Perata. www.energy.ca.gov/ghgstandards/documents/sb_1368_bill_20060929_chaptered.pdf (June 7, 2007).

31. California Energy Commission, April 2006, *Net System Power: A Small Share of California's Power Mix in 2005.* CEC-300-2006-009-F. April 2006. www.energy.ca.gov/2006publications/CEC-300-2006-009/CEC-300-2006-009-F.PDF (June 7, 2007).

32. California Legislature. Assembly. 2007. *Global Warming Solutions Act.* 2006–2007 sess., AB 32, Nuñez/Pavley. www.law.stanford.edu/program/centers/enrlp/pdf/ab_32_bill_20060927_chaptered.pdf (June 6, 2007).

33. Climate Action Team (CAT). March 2006. *Climate Action Team Report to Governor Schwarzenegger and the Legislature* (Sacramento, CA).

34. Schwarzenegger, A. 2005. *Executive Order S-3-05.* www.dot.ca.gov/hq/energy/ExecOrderS-3-05.htm.

35. CAT, March 2006 op. cit.

36. Ward's Automotive Group, "Vehicles in Operation in U.S. by State, 2004," www.wardsauto.com (September 28, 2006).

37. Associated Press, Utah joins pact to reduce gas emissions: Utah joins five other states, British Columbia, in pledge to reduce greenhouse gas emissions. May 21, 2007, viewed at www.cbsnews.com/stories/2007/05/21/ap/tech/main2833474.shtml, May 28, 2007.

Chapter 35

California's Battle for Clean Cars

FRAN PAVLEY

It All Seemed So Simple

After serving four terms as mayor and council member in the small city of Agoura Hills, California, twenty-eight years teaching middle school, and being a member of several environmental organizations, I was elected to the State Assembly in November 2000. In January 2001, after assembling my staff, we put together my first bill package—including AB 1058.

Russell Long of the San Francisco–based environmental group Bluewater Network had made a simple proposition—require a reduction of greenhouse gas emissions from automobiles to address the increasing concerns among scientific and environmental leaders on the causes of global warming. As the world's twelfth-largest contributor of greenhouse gases, and with the sixth biggest economy in the world and 38 million residents, California was posed to make a significant impact on global carbon emissions. Focusing on automobile emissions seemed logical— roughly 58 percent of our greenhouse gas emissions comes from mobile sources. AB 1058 was introduced in January 2001 as a one-page measure. Easier said than done.

Sometimes there is an advantage to being perceived as a mild-mannered and middle-aged schoolteacher and a freshman legislator. I was under the radar of the well-connected lobbyists for some of the most powerful special-interest groups that walk the halls of the State Capitol.

The Process Began

AB 1058 was set for its first hearing in April in the Assembly Transportation Committee. Bluewater Network and the Coalition for Clean Air led by Tim Carmichael were the sponsors. The support list of environmental groups, scientists, California's congressional members, and university professors was growing. Scientific articles and reports accompanied the information distributed to the committee members.

Timing is everything. The same week as our hearing, the April 9, 2001, edition of *Time* magazine had on its cover a huge photograph of an egg frying with a picture of the world silhouetted into the yolk. The headline was "Global Warming," with a subhead question: "Climbing Temperatures. Melting Glaciers.

Rising Seas. All Over the Earth We're Feeling the Heat. Why Isn't Washington?" Just a month prior, President George W. Bush had announced his decision to abandon the Kyoto process.

We armed ourselves for a huge debate on the science of whether or not Earth really was warming due to human activity. Instead, however, the auto dealers and Chamber of Commerce argued mostly for a federal policy instead of just one state taking the lead on a global issue, perhaps hoping for a more favorable forum from an administration that was still questioning whether global warming even existed. Would the automobile manufacturers have to make one "clean car" for California and different models for the other forty-nine states? The roll was taken and the bill was placed "on call," meaning still alive but short by three votes.

I had to somehow convince a few more members. Usually a member of the majority party, in this case the Democrats, is given a little courtesy leeway in a bill's first committee hearing. Assembly Member Joe Simitian crafted an amendment to delay the implementation of the bill by one year in order for the State Legislature to review (not vote on) the proposed Air Resources Board regulations. That appealed to wavering committee members and got the tacit support of Chairman John Dutra. People were shocked that AB 1058 survived this critical committee at all. It was now headed for the Assembly Appropriations Committee to analyze any fiscal impacts.

Obviously we needed to work on the bill, to meet with other members, and to develop a strategy on how to get forty-one votes on the Assembly floor. We were definitely not ready for prime time. I met with then-Speaker Bob Hertzberg and asked him to allow me to get this bill through Appropriations Committee

and I would not bring it to a floor vote that year. I kept my promise.

Throughout the summer and fall of 2001, endless meetings were held to expand our coalition, including meetings with newspaper editors, legislators, and other potential allies. Anne Baker from my staff and V. John White, a lobbyist for the Sierra Club and the Clean Power Campaign, deserve much of the credit for these key organizational strategies. In November, I held a well-attended Select Committee Hearing on the "Impacts of Climate Change on California" in Santa Monica. Based on the testimony of many experts on global warming, we developed a thorough set of findings to add to the legislation as well as to help articulate the need for the bill.

The Coalition Grew

The meetings were beginning to pay off. Our support was growing both in size and diversity. On the local level, several dozen cities and counties joined their elected officials, air districts, and water districts to support the bill. A long list of business leaders, the ski industry, and Environmental Entrepreneurs, or E-2 (a relatively new Silicon Valley–based group of forward-looking venture capitalists, business executives, and environmental entrepreneurs co-founded by Bob Epstein and Nicole Lederer) joined our effort. AB 1058 would be E-2's first effort to lobby legislation.

Churches and religious leaders from a variety of faiths were led by the California Interfaith Council of Power and Light, and spoke to legislators and their congregations about the moral responsibility of taking care of the planet. Physicians and public health organizations became involved because of the health impacts associated with climate change. The

American Lung Association and Physicians for Social Responsibility were early supporters, later joined by the California Medical Association and the California Nurses Association. A long list of environmental groups including the California League of Conservation Voters, Environmental Defense, Union of Concerned Scientists, Planning and Conservation League, Clean Power Campaign, and CALPIRG (the California Public Interest Research Group, which has now renamed their organization Environment California) were added. Natural Resources Defense Council (NRDC) and the Sierra Club joined Bluewater Network and the Coalition for Clean Air as sponsors.

And only in California would we add a long list of entertainment personalities and movie stars to raise the profile of the issue. It wasn't the first time that high-profile celebrities like Robert Redford, Paul Newman, and Leonardo Di Caprio would be involved in important environmental issues, and specifically climate change.

By the end of 2001, the State Assembly had a new Speaker, my seatmate on the Assembly floor, Herb Wesson. I was also able to add more than twenty coauthors to the bill, including the two leaders of the Latino Caucus—Senator Martha Escutia and Assembly Majority Leader Marco Firebaugh. The opposition had no idea how busy we had been and probably thought the bill was dead.

2002 Was Time to Make It Happen

It was January 2002, and we only had until the end of the month to get the bill off the Assembly floor. Newspapers were starting to pay attention and turning the fight into David versus Goliath. Articles and supportive editorials began appearing in papers across the state. The powerful oil and auto industries suddenly became engaged in the debate, with some suggesting that voluntary programs could achieve the same goals. The opposition called the bill poorly crafted and an attempt to end-run around the federal fuel-efficiency standards. However, we were not trying to set a miles-per-gallon standard; we were regulating tailpipe emissions with off-the-shelf available technologies. Assurances were added to the bill language that the standard would allow for "maximum feasible, cost-effective reductions." Final implementation of the bill would be delayed one year, until January 2006. The new Speaker and I were ready to go.

Speaker Wesson provided strong leadership to encourage and provide political cover for many reluctant members who were being heavily lobbied to kill the bill. It passed 42–24 on January 30, 2002. It is noteworthy that two Republicans—Assembly Members Dave Kelly and Tom Harman—broke ranks and voted for the bill.

The opening paragraph of the January 31 edition of the *San Francisco Chronicle* said, "Beating back a lobbying blitzkrieg from carmakers and the oil industry, environmentalists narrowly won Assembly approval of a bill to make California the first state to regulate carbon dioxide emissions from cars." Scores of articles about the passage of AB 1058 appeared in papers from Detroit and New York to England. AB 1058 had gained national and international attention.

The bill then moved to the more environmentally friendly State Senate. Always outspoken and blunt, Senate President Pro Tem John Burton of San Francisco immediately recognized the environmental and political importance of the bill and took personal responsibility to move it through the Senate; AB 1058 would not have had a chance without his leadership. We held many meetings in his office to discuss timing and strategy, and I learned much about the politics of the State Capitol. I also heard more four-letter words in those few months than in my twenty-eight years teaching middle school.

By early spring, Winston Hickox, Secretary of California's Environmental Protection Agency under Governor Gray Davis, began to express interest in AB 1058, which was good news. The bill was heard in Senator Byron Sher's Natural Resources Committee, and it passed on April 1 on a partisan 5–2 vote.

The Opposition Woke Up

After realizing that the bill was making progress, a press conference held by automakers coincided with the publication of full-page newspapers ads, radio spots, and a new, industry-sponsored Web site www.wedrive.org. Automakers warned the public that environmentalists and government bureaucrats should not be allowed to pass a law that would decide what type of car they could drive.

The Automobile Alliance got hold of an unrelated draft report prepared the year before by the California Energy Commission on a wide list of strategies to reduce fuel consumption. They extracted four suggestions from this long list and ran their opposition campaign around them. Using scare tactics that had nothing to do with what the bill actually said or did, the Alliance claimed that AB 1058 would allow the California Air Resources Board (ARB) to increase taxes on gasoline, reduce speed limits, place fees on new vehicles, and even impose a tax on vehicle miles driven; all totally false and impossible under the bill since the ARB has no legal authority to raise taxes or set speed limits. Television commercials featured legendary car salesman Cal Worthington and his dog Spot telling viewers that motorists should fear what was happening in Sacramento.

The Temperature Was Rising

These same scare tactics had been used by the auto industry during the years in its failed resistance to adding catalytic converters and life-saving seat belts and air bags to vehicles. Congressman Henry Waxman, one of the authors of the Federal Clean Air Act, sent me copies of similar statements the automakers used in their failed attempts to undermine the passage of that important clean air public health measure.

California Chamber of Commerce President Allan Zaremberg reported in the Chamber's Alert newsletter that AB 1058 would "limit consumer choice by significantly increasing the cost of larger more powerful vehicles they want to buy and force all drivers to pay much more in the form of gas taxes and mileage fees." The local chambers of commerce, oil companies, and automobile dealers easily brought the state Chamber on board to assist in their efforts to stop the bill. Not once, however, did the Chamber ever talk to me.

Automakers claimed that AB 1058 was a direct attack on California's favorite car, the SUV. It was now an attack on "soccer moms." Conservative radio talk shows, including "John and Ken" from Orange County, fanned the flames. The campaign of misinformation was spreading to motorists spending hours listening to their radios while commuting on Southern California freeways. Talk radio whipped up drivers to place angry, and sometimes threatening, calls to legislators' district offices. Proponents worked hard to help diffuse the well-funded opposition's attacks. CLCV (California League of Conservation Voters) set up phone banks to call legislative offices, many meetings with the media and press conferences were held, volunteer lobbyists descended on the Capitol, and full-page ads with headlines titled "There They Go Again" referred to the automakers repeating the same gloom and doom predictions for the American automotive industry that they had made when trying to defeat California's attempts to pass laws to require unleaded gas, catalytic converters, or even seat belts.

On April 28, the Senate Appropriations Committee passed AB 1058 on an 8–3 vote and sent it to the floor of the Senate. Eron Shosteck, representing the Alliance for Automobile Manufacturers, called this bill "a veiled attack on California's family vehicles" and suggested that SUVs and pickup trucks would no longer be sold in California. John Burton would have none of it and reminded them, "When I was in Congress in the seventies, I had a hearing on air bags and the industry was in there like you were going to steal their firstborn." A visibly angry and frustrated Senator Debra Bowen challenged the carmakers to stop spending money on their lawyers and lobbyists and instead hire engineers to make a cleaner, better car. Peter Welch of the California Motor Car Dealers Association was dismissive that AB 1058 would provide "no climate benefits for California."

Automaker Desperation Was Becoming Clear

I left the state the next day to attend, at my own expense, an international hearing with a series of expert panelists on "Air Pollution and Climate Change." European engineers described their work reducing emissions in automobiles. These newer technologies were available and *already* being used or tested in Europe by some of the same automobile manufacturers that were trying to kill our bill in California. The opponents were against any regulatory approach that would require changes to their vehicles or impact their profit margin. One conversation with an engineer from China brought home the broader implications of creating a market for these new, cleaner technologies. He said that if California had not mandated and required the use of catalytic converters in automobiles, China would not be using them today in their dramatically growing vehicle fleets. California could lead the way, again. AB 1058 wasn't just about one state. It could inspire other states and nations to use these new cars built with currently available technology.

While at the conference, I received a panicked call from my staff. Burton was taking the bill up on the Senate floor! On May 2 on a 22–13 vote, and despite the might of the opposition, AB 1058 passed off the Senate floor and was on its way back to the Assembly for concurrence with Senate amendments.

A lobbying blitz by major automakers and oil companies, as well as the daily hammering from conservative talk radio, helped stall the bill as several lawmakers who originally voted for AB 1058 were having second thoughts. The misinformation and fear tactics were paying off. Radio talk show hosts led their most dogmatic conservative listeners in a protest parade of SUVs up from Southern California to Sacramento, circling the State Capitol, waving banners, honking their horns, and yelling from bullhorns. And we were at least three votes shy of the forty-one needed to get the bill out of the Assembly and to the governor.

To counter and refute the lies that the ARB would be able to limit the miles motorists could drive, raise gas taxes, or deny consumers the choice of buying certain classes of large vehicles such as SUVs, on a quiet Saturday, June 29, the Senate passed a new "gut and amended" second measure—AB 1493—with a 23–6 vote. AB 1493 contained language that listed and *expressly* prohibited the ARB from taking any of these actions. The contents of AB 1058 were placed in the new bill. The Assembly now had two versions to consider.

Race to the Finish Line

Assembly Speaker Herb Wesson took charge of the bill and scheduled it for a full Assembly vote on July 2. Members could now point to the bill language to assure their constituents that AB 1493 could not raise taxes or take away the cars they wanted to drive. The halls of the Capitol were packed with lobbyists, and the relentless attacks by talk radio were still generating up to 200 calls per day to members' offices. Former President Bill Clinton placed a well-timed personal call to Speaker Wesson on the national importance of the bill. Because it had been amended, it went back to the Assembly Transportation Committee, with the chair now opposed. It passed, but by the narrowest of margins.

I presented the bill on the floor of the Assembly, and the Republicans immediately called a caucus. Two of their members left the caucus to be phone-in "guests" on conservative talk radio, but actually to stall for time and whip-up listeners to start calling the Capitol to stop AB 1493. However, it was clear that the bill finally had enough votes. AB 1493 passed the full Assembly by a mere forty-one votes, but that's all it took.

Clean Air Won

AB 1493, the landmark first-in-the-nation bill to reduce greenhouse gas emissions from mobile sources, became law on July 22, 2002. Governor Gray Davis, in front of huge crowds of supporters and the media, signed the bill in two separate ceremonies at the Griffith Park Observatory in Los Angeles and at the Presidio in San Francisco with the dramatic backdrop of the Golden Gate Bridge. Politicians, actors Robert Redford and Rob Reiner, scientists, and others joined in the triumph. Governor Davis summed it up before he signed the law: "Opponents of this bill say the sky is falling. But the sky is not falling. It's just getting a whole lot cleaner."

The California Air Resources Board, under the leadership of Chairman Dr. Alan Lloyd, went to work on the draft regulations, holding multiple hearings and workshops that included all stakeholders. In October 2004, the ARB unanimously adopted regulations to reduce greenhouse gas emissions from automobiles by 30 percent by 2016. Beginning in the model year 2009, using off-the-shelf technologies such as variable valve timing and more efficient air conditioners, along with a two-tier phased-in approach, automobile manufacturers would have to reach the reduction targets. And the regulations were reviewed by the Air Resources Board to insure that they were "cost-effective and feasible" as required by the law. Alternative compliance flexibility was also added to the final regulations. No taxes were raised. Consumers would still have their choice of vehicles. They would just be cleaner.

Automakers Choose to Litigate, Not Innovate

At the end of December 2004, in spite of the fact that they were making the changes in other countries that they claimed were not possible in the United States, automakers filed suit to stop the implementation of the new California regulations. In 2005, the International Automobile Alliance joined with them.

California's Attorney General, Bill Lockyer, with the support of our new governor, Arnold Schwarzenegger, said that California would defend the new law in court. NRDC, the Sierra Club, and Environmental Defense Fund intervened in the defense.

Since then, fourteen other states have adopted California's "Clean Car" regulations. Canada, a signatory to the Kyoto Protocol, obtained a voluntary MOU with these same automakers to reduce mobile source emissions in Canada using nearly identical technology-based reduction strategies instead of fuel-efficiency requirements.

California Will, Again, Lead

In 2005, I introduced AB 32 as an important next step. AB 32 requires mandatory reporting as well as setting reduction targets for both mobile and stationary sources. In order to begin to reduce greenhouse gas emissions, we need first to monitor and track current emissions levels, set a cap, and then begin to schedule reductions, with an enforcement mechanism to ensure compliance. Again, we received Republican support—Assembly Member Shirley Horton voted to support AB 32, and Governor Schwarzenegger signed the "Global Warming Solutions Act of 2006" in September of that year. There is now a cap on California greenhouse gas emissions with a required reduction back to 1990 levels by 2020.

Other states and the world continue to look to California as a leader in the fight for cleaner air and reducing greenhouse gas emissions, and as the home of innovation and new job-creating technologies.

The Public Gets It

Especially given the spiraling cost of gasoline in 2006, the public has been clear in its support for more fuel-efficient vehicles. According to annual polls administered in 2002 through 2006 by the nonpartisan Public Policy Institute of California, air quality, global warming, and their associated public health impacts continue to remain in the forefront of Californians' concerns. We need to continue our investment in the production of cleaner alternative fuels and to reduce our dependence on imported oil from potentially unstable sources of supply.

Chapter 36

U.S. State Climate Action

Joshua Bushinsky

Introduction

In 2004, with evidence of anthropogenic climate change mounting and U.S. climate policy stalled at the federal level, U.S. state policy makers seized the opportunity. U.S. states have taken leadership on climate policy by introducing a variety of initiatives that directly reduce greenhouse gas (GHG) emissions, encourage energy efficiency and renewable energy projects, and prepare for climate impacts. While many of these approaches to climate change are tailored to the opportunities for and vulnerabilities of individual states, policy makers have also formed regional compacts to cooperate in their emission reduction efforts. Such policy experimentation follows the traditional federalist path of environmental policy making in the United States. States have historically functioned as policy laboratories that provide models and data for federal action.[1] Climate change is a global problem that demands global action, including substantial national action by the United States. State and regional climate policies cannot substitute for a coordinated national response, but they provide foundations, models, and impetus for that response.

State Authority and Its Limitations

Although states lack certain powers necessary to implement comprehensive climate policies, they have some jurisdictional powers over key sectors of the economy critical to addressing climate change. States regulate many aspects of the electricity and natural gas sectors, as well as agriculture, infrastructure, and land use. One example is the ability of state public utility commissions to authorize charging ratepayers for the costs of reducing GHG emissions. However, states have limited resources to devote to climate change, and their strict budget requirements can put long-term climate policies in jeopardy, particularly in the face of economic downturn.[2] States are further constrained by federal preemption: the constitutional limitation of state power in the presence of federal policy making.[3]

State Action on Climate Change
Usual and Unusual Suspects

State action on climate change has expanded beyond the usual environmental first movers of California and New England to become a nearly nationwide phenomenon.[4] Twenty states have greenhouse gas emissions targets.[5] Twenty-four states, including states as diverse as Alaska, Montana, and the Carolinas, have active legislative commissions or governor advisory councils charged with developing a climate strategy.[6] Twenty-nine states—including Ohio, Michigan, and Florida—and the District of Columbia have set standards specifying that electric utilities must generate a certain amount of electricity from renewable sources.[7] States are moving on other fronts in the absence of federal action, creating appliance efficiency standards, encouraging renewable fuel sources, improving the carbon footprint of state fleets and government buildings, and developing plans to prepare for climate impacts.

Regional Initiatives

Climate policy provides a productive forum for interstate cooperation because the geographic location of emission reduction is immaterial, and states can combine analytic capacity and reduction opportunities to their mutual benefit.[8] For example, in the context of cap-and-trade programs, a larger number of states and emissions sources provide greater opportunities for low-cost emissions reductions. Increasing the number of participants in a cap-and-trade program also reduces the possibility of emissions increasing in states or sectors not included in the cap. This "leakage" can occur when an emissions cap results in a shift of electricity generation and resulting GHG emissions from capped to uncapped states, thus negating some of the program's actual emissions reductions. Across the country, states are collaborating on regional climate initiatives.

The Regional Greenhouse Gas Initiative (RGGI) established the first mandatory U.S. cap-and-trade program for carbon dioxide and began operation in January 2009. RGGI currently includes ten Northeastern and mid-Atlantic states. The governors of Connecticut, Delaware, Maine, New Hampshire, New Jersey, New York, and Vermont established RGGI in December 2005.[9] Massachusetts, Rhode Island, and Maryland have since joined. The program caps power plant carbon dioxide emissions at current levels and mandates a 10 percent emissions reduction by 2019. The RGGI states collaborated to develop a model rule, and each participating state adopted the rule through legislation or regulation.[10] The member states agreed to set aside the revenue from at least 25 percent of its emissions allowances for public benefit purposes including investments in energy efficiency and renewable energy technologies.[11] One of the major developments in RGGI has been the states' decision to auction a high percentage of their emissions allowances rather than directly allocating those allowances to covered entities, with the revenue going toward public benefit purposes.[12]

In February 2007, the governors of Arizona, California, New Mexico, Oregon, and Washington signed an agreement establishing the Western Climate Initiative (WCI).[13] Since then, Utah, Montana, and the Canadian provinces of British Columbia, Ontario, Manitoba, and Quebec have joined the WCI.[14] The seven states and four provinces jointly set a regional emissions target of 15 percent below 2005 GHG emissions levels and committed to design a regional GHG cap-and-trade market. This effort coincides with the implementation of California's Global Warming Solutions Act (AB 32), and these parallel development processes have informed each other.[15] The WCI released design recommendations for its cap-and-trade program in September 2008, just three months before California adopted its Scoping Plan describing a climate strategy for the state.[16]

The Midwest has provided the most recent forum for state collaboration on climate policy. The Midwest Regional Greenhouse Gas Accord was signed in November 2007 by the governors of Illinois, Iowa, Kansas, Michigan, Minnesota, and Wisconsin, as well as the premier of the Canadian province of Manitoba.[17] Similar to the WCI, the midwestern effort seeks to develop a regional GHG emissions target and a multi-sector cap-and-trade program. The accord is part of a broader Energy Security and Climate Stewardship Platform for the Midwest that intends to promote advanced energy technologies, energy efficiency, and other climate-friendly policies.[18]

Implications for Federal Policy

The existence of state climate policy and the experience of state policy makers have become increasingly important as the federal government considers adopting a national climate policy. Congress and the administration have a wealth of state expertise to draw upon in the course of federal policy design, and states can undertake informed lobbying for policies that reflect their interests. Some state approaches provide building blocks for a federal program: a federal GHG cap-and-trade system, renewable portfolio standard, and GHG emission registry can build directly on existing state models. If federal policy adoption is stalled by an economic downturn, political gridlock, or other causes, states will continue to carry the burden of—and opportunity for—policy leadership.

Compared to other environmental problems, there are many avenues for confronting climate change; the scope of possible solutions equals the scale of the problem. Some approaches at the state level, such as cap and trade and renewable portfolio standards, have counterparts in proposed federal legislation. Other approaches, including low-carbon fuel standards or GHG emission performance standards, have only recently received serious attention at the federal level.

States have already built up a sizable knowledge base that will be valuable to federal policy makers negotiating the passage of climate bills. The states participating in RGGI and the WCI have considered many of the cap-and-trade design decisions relevant to federal policy: inclusion of emission offsets, allocation of emission allowances, and cost-containment mechanisms, among others.[19] Although some aspects of these debates do not bear on federal legislation, many decisions made at the state level could influence federal policy by providing both an institutional model and empirical evidence about policy efficacy. These state-level decisions reflect concrete compromises between stakeholders and policy makers rather than academic exercises, and they may become negotiating positions for states as they engage in the federal policy debate. For example, prior approaches to GHG offsets have included lengthy and uncertain certification processes, exemplified by case-by-case review of projects under the Kyoto Protocol's Clean Development

Mechanism. The RGGI states have developed and adopted new GHG reduction protocols; these projects are as specific as sulfur hexafluoride emissions reductions from electricity transmission equipment.[20] The protocols are designed to provide greater guidance and certainty to project developers, while ensuring that projects result in real emissions reductions. This advance is an example of state-level innovation that could improve federal climate policy.

State climate policy commitments have political implications for business engagement on federal policy. Climate legislation and regulations force companies to engage in state-level policy discussions, educate themselves on climate science, and examine their options for policy compliance.[21] State and regional climate policy adoption results in a "patchwork quilt" of policies across the nation that can duplicate work already accomplished in other states and complicate the regulatory regime for companies doing business across state lines. A coherent federal policy can provide regulatory certainty, which significantly improves corporate decision making, particularly for companies that require long planning cycles for large capital investments like power plants.[22] Federal policy will also expand nationwide opportunities for industries that are both instrumental to climate change mitigation efforts and potential sources of significant economic growth. The combination of an uncertain state patchwork and the potential benefits of federal policy have led in part to increased business support for consistent and certain federal climate policy.[23]

Consideration of federal policy has highlighted two major issues for state leaders: how to provide credit for reductions undertaken before a federal program is implemented and how to structure interactions between state and federal climate policies.[24] States that have made emissions reductions will demand recognition of these early actions. Likewise, companies that have reduced emissions to comply with state mandates will want dispensation under a federal program. Policy makers in some states have voiced their desire to be preempted by a federal policy, whereas other states want to maintain their ability to move beyond any federal requirements.[25] Historically, federal environmental policy has partially preempted state policy by setting a minimum standard and allowing states to choose more stringent standards.[26] Such partial preemption may be untenable for a cap-and-trade program that requires a national cap and consistent emissions reduction standards across all covered entities. The House Committee on Energy and Commerce described one approach in a discussion draft released in October 2008. This proposal would address these challenges by preempting state cap-and-trade programs while setting aside a limited number of emission allowances for entities holding allowances for existing state cap-and-trade programs.[27] Successfully resolving these issues will be a significant hurdle to translating the states' ambitious and varied efforts into a federal response to climate change. These choices will impact the efficacy and efficiency of federal climate policy, particularly for a potential federal cap-and-trade program.[28]

Even under a federal climate regime, states will continue to play a part in climate policy. Federal policy will have to strike a balance between the respective roles of state and federal governments.[29] States are likely to have implementation responsibilities for federal policy, and many of their own policies and programs will continue. Indeed, states have jurisdiction over policy areas where the federal government cannot intervene and where GHG emission-reductions are possible. The role of states as policy innovators will remain important. The first federal climate bill enacted will not be the last bill necessary, and states will continue to provide lessons for future legislation.

Conclusion

State experiments with policy approaches to climate change have begun to demonstrate that strong leadership on climate policy in the United States is politically possible and technically feasible. Continued state innovation on climate policy will inform federal efforts and may expand the constituencies demanding federal leadership. Companies prefer some level of regulatory certainty, clean-tech ventures benefit from a price on carbon, and environmental advocates argue for strong federal action by pointing to states' leadership. State climate efforts present federal policy makers with a strong case for the feasibility of national legislation. While the existence of state climate policy will pose some political and technical hurdles to the adoption of federal policy, state actions provide early lessons and a strong impetus for a robust federal response to climate change.

Notes

1. The description of states as policy laboratories was introduced by Supreme Court Justice Louis Brandeis. See *New State Ice Co. v. Liebmann*, 285 U.S. 262 (1932) (Justice Brandeis, dissenting). Examples of vertical diffusion of environmental policy can be found in Aulisi, A., J. Larsen, J. Pershing, and P. Posner (2007), "Climate Policy in the State Laboratory." World Resources Institute. Online at www.wri.org/publication/climate-policy-in-the-state-laboratory.

2. Forty-nine states require a balanced budget by statute or constitution. See Snell, R. K. (2004), "State Balanced Budget Requirements: Provisions and Practice." National Conference of State Legislatures. Online at www.ncsl.org/programs/fiscal/balbuda.htm.

3. Weiland, P. (2000), Federal and state preemption of environmental law: A critical analysis. *Harv. Envtl. L. Rev.* 24: 237. Litz, F., and K. Zyla, (2008), "Federalism in the Greenhouse: Defining a Role for States in a Federal Cap-and-Trade Program." World Resources Institute. Online at www.wri.org/publication/federalism-in-the-greenhouse.

4. For a more comprehensive treatment of the variety of state actions to address climate change, see Pew Center on Global Climate Change (2008), "Learning from State Action on Climate Change." Online at www.pewclimate.org/policy_center/policy_reports_and_analysis/state.

5. Pew Center on Global Climate Change, "States With Greenhouse Gas Emissions Targets." Online at www.pewclimate.org/what_s_being_done/in_the_states/emissionstargets_map.cfm.

6. Pew Center on Global Climate Change, "States with Active Climate Legislative Commissions and Executive Branch Advisory Groups." Online at www.pewclimate.org/what_s_being_done/in_the_states/climatecomissions.cfm.

7. Pew Center on Global Climate Change, "Renewable Portfolio Standards." Online at www.pewclimate.org/what_s_being_done/in_the_states/rps.cfm.

8. While the location of emissions reductions may not change the climate benefit, recent studies suggest that warming due to greenhouse gas emissions has localized air pollution effects. See Jacobson, M. Z. (2008), On the causal link between carbon dioxide and air pollution mortality. *Geophysical Research Letters* 35, L03809, doi:10.1029/2007GL031101.

9. Regional Greenhouse Gas Initiative, "Multi-State RGGI Agreement." Online at www.rggi.org/agreement.htm.

10. Regional Greenhouse Gas Initiative, "Regional Greenhouse Gas Initiative Model Rule." Online at www.rggi.org/docs/public_review_draft_mr.pdf.

11. Regional Greenhouse Gas Initiative. "Memorandum of Understanding (in Brief)." Online at www.rggi.org/docs/mou_brief_12_20_05.pdf.

12. Regional Greenhouse Gas Initiative, "Date Announced for the Nation's First Auction of Greenhouse Gas Emissions Allowances." Online at www.rggi.org/docs/20080317news_release.pdf.

13. Western Regional Climate Initiative, "Western Regional Climate Initiative." Online at www.governor.wa.gov/news/2007-02-26_Western ClimateAgreementFinal.pdf.

14. Western Regional Climate Initiative, "Western Regional Climate Initiative Update." Online at www.westernclimateinitiative.org/ewebedit pro/items/O104F13074.pdf.

15. California Air Resources Board, "Timeline: California Global Warming Solutions Act of 2006." Online at www.arb.ca.gov/cc/factsheets/ ab32timeline.pdf. California's climate change initiatives are described in detail in chapters 34 and 35 in this volume.

16. Western Regional Climate Initiative, "Design Recommendations for the WCI Regional Cap-and-Trade Program: Complete Report with Appendices." Online at www.westernclimate initiative.org/index.cfm. California Air Resources Board, "AB 32 Climate Change Scoping Plan Document." Online at www.arb.ca.gov/cc/scoping plan/scopingplan.htm.

17. Midwestern Governors Association, "Midwestern Greenhouse Gas Accord 2007." Online at www.midwesterngovernors.org/Publications/ Greenhouse%20gas%20accord_Layout%201.pdf.

18. Indiana, Ohio, and South Dakota are participating as observers in the process.

19. An offset refers to GHG reductions undertaken outside the coverage of an emissions reduction system.

20. Regional Greenhouse Gas Initiative Model Rule, Regional Greenhouse Gas Initiative (RGGI), p. 95. Online at www.rggi.org/docs/public_review_ draft_mr.pdf.

21. For an examination of business engagement with state environmental policy making, see Rabe, B. G. (2005), "Business Influence in State-Level Environmental Policy Formation and Implementation." Paper presented at the annual meeting of the American Political Science Association. Online at www.allacademic.com/meta/p_mla_apa_ research_citation/0/4/1/1/9/p41197_index.html.

22. Lempert, R., S. Popper, S. Resetar, and S. Hart (2002). "Capital Cycles and the Timing of Climate Change Policy." Pew Center on Global Climate Change, Arlington, Virginia.

23. See Barringer, F. (2007), A coalition for firm limit on emissions, New York Times, January 19.

24. Staff of House Committee on Energy and Commerce Climate Change, 110th Congress (2008), "Legislation Design White Paper: Appropriate Roles for Different Levels of Government." Online at http://energycommerce.house .gov/Climate_Change/white%20paper%20st-lcl% 20roles%20final%202-22.pdf.

25. California Senators Barbara Boxer and Dianne Feinstein have argued against proposed federal preemption of California vehicle emissions standards. See Simon, R. (2007), Democrats face off over emissions bill, Los Angeles Times, June 8.

26. Aulisi et al., 2007 op cit.

27. Staff of House Committee on Energy and Commerce, 110th Congress (2008), "Discussion Draft of Climate Change Legislation." Online at http://energycommerce.house.gov/Climate_ Change/index.shtml.

28. The result of these choices will depend on the relative stringency of state and federal programs and the overlap in source coverage. See McGuinness, M., and A. D. Ellerman (2008), "The Effects of Interactions between Federal and State Climate Policies." Massachusetts Institute of Technology. Online at http://tisiphone.mit.edu/RePEc/mee/ wpaper/2008-004.pdf.

29. Litz et al., 2008 op. cit.

Chapter 37

Policies to Stimulate Corporate Action

Eileen Claussen, Vicki Arroyo, and Truman Semans

Addressing the challenge of climate change will require a significant reduction in annual greenhouse gas (GHG) emissions across all sectors of the economy. Unfortunately, the United States lacks a national climate policy that could generate significant efforts to reduce GHG emissions and result in the necessary technology shift to low-carbon energy sources. At the state level, however, climate and energy programs are rapidly emerging and the resulting policy landscape is growing fragmented and complex—with some programs aimed at climate change, some at energy efficiency, others at technology research and development, and still others at renewable energy (see chapters 34 and 36). The lack of regulatory certainty and consistency makes it difficult for companies to move forward.

Business engagement is critical for developing efficient, effective solutions to the climate problem. Not only are GHGs emitted in the production process, they also result from the ongoing energy consumption of manufactured products. The large-scale reductions needed to deal with this issue will require investment in efficiency improvements, widespread use of advanced low-carbon technology, and the use of alternative fuels, in addition to developing more efficient products and climate-related services. The transition to new fuels and technologies will bring not only challenges but also opportunities for increasing energy security, improving general environmental quality and public health, and promoting economic development.

We explore ways that policy can be used to stimulate the type and scale of changes in private-sector behavior to achieve the fundamental transition from the way we currently produce and use energy. In evaluating mechanisms to influence investment decisions, we discuss policies and programs that "push" new technologies through direct investment in research and development, those that "pull" new low-carbon technologies into the market by stimulating demand, and those that seek to overcome other barriers to using low-carbon technology. Finally, we highlight the recent formation of the U.S. Climate Action Partnership (USCAP)—a successful collaboration between important corporations and NGOs to address climate change and outreach to Congress.

Why Industry Engagement Is Critical

Current and forecasted emissions are heavily determined by technologies now in use for electricity generation and transportation, as these sectors account for roughly 60 percent of U.S. emissions.[1] Carbon dioxide generated by the manufacturing sector represents an additional 19 percent.[2] This pattern is difficult to shift quickly because of a relatively fixed reliance on specific processes and types of energy. Energy use and consequent emissions in the U.S. economy are largely a function of the current equipment (or "capital stock") used to extract, produce, convert, and use energy, and this stock can often have a lifetime of up to several decades.[3] Examples include boilers, turbines, buildings and the equipment used to heat and cool them, vehicles, and manufacturing facilities.

Capital stock is generally used until it wears out or breaks down and must be replaced, and this "natural" rate of capital stock replacement is not easily accelerated. It is generally not in a company's financial interest to retire equipment before the end of its useful life or at least before it is fully depreciated. However, requirements and incentives can motivate firms to replace stock with the best available technology when the time comes.[4]

These motivators include putting in place early and consistent incentives that assist in the retirement of old, inefficient capital stock; making certain that policies do not discourage capital retirement; and pursuing policies that shape long-term patterns of capital investment. Even a modest carbon price could stimulate investment in new capital equipment. On the other hand, piecemeal regulatory treatment of pollutants rather than a comprehensive approach could lead to stranded investments in equipment (e.g., if new conventional air pollutant standards are put in place in advance of carbon dioxide controls at power plants). Ultimately, any well-crafted policy to address climate change must consider and harness market factors and policies that drive capital investment patterns.[5]

How and Why Companies Are Responding to Climate Change

Climate change and potential climate policy pose levels of uncertainty and potential market consequences that are large, even by normal business standards. Notably, time horizons associated with climate-related impacts on business are more distant than those typical of planning for most corporations, and there is great uncertainty about policy changes that, in turn, will affect energy supplies, energy prices, potential mitigation costs, broad trends in market demand, public perception, and so on. Some firms view impacts in these areas as unavoidable and choose to proactively implement business strategies to address them, often including efforts to shape future regulations. Other companies choose to be more reactive—for example, responding only when shareholders or customers raise concerns, and only commenting on climate policy as it is proposed.

In the United States, businesses' responses to climate change vary widely. As of this writing, the vast majority of companies do not have specific climate programs underway, although most large corporations do have some form of energy efficiency program in place. According to the latest Carbon Disclosure Project report, more than 135 companies have disclosed data on their energy use, GHG emissions, risks, and opportunities.[6] Numerous companies have undertaken significant emissions reductions, developed risk mitigation measures, and begun to capture new business opportunities.[7] Most of the currently recognized climate leaders have actively integrated climate into business planning for many years, although some climate leaders such as Wal-Mart have recently undertaken significant new efforts on the issue.[8] A growing number are engaging in the policy process, some constructively and some obstructively.

Proactive efforts often stem from a company's desire to attempt to actively change the way business as usual is conducted. The world's largest carpet manufacturer, Interface Inc., for example, was an early adopter in developing its climate change strategy because, as CEO Daniel T. Hendrix noted, "Interface considers mitigating climate change to be of utmost importance in achieving a more sustainable future for everyone."[9] To date, Interface's combined global efforts have reduced absolute carbon dioxide emissions by more than 60 percent from its 1996 baseline.[10] Proactive decision making can open up new markets and opportunities for business. Dowell and colleagues found in 2000 that firms that go beyond the compliance of environmental regulations tend to have a higher adjusted market value than their compliant (or reactive) peers.[11]

Wal-Mart serves as a strong example of how a company's reactive policies can lead to proactive decisions. Starting in 2004, the company devised an ambitious series of climate change–related goals, which were created mainly as a defensive strategy but led to an actual change in corporate philosophy.[12] One of the main forces that brought about this organizational change was the discovery of the potential financial value gained through decreased operating costs, new revenue opportu-

nities, and greater employee satisfaction. Another example is Rio Tinto Energy, which developed a climate program based in part on the risks the company first identified to operations and markets, and from weather-related events across all of its business units.[13]

Regulatory Measures to Address Climate Change

Many investment decisions are made over long time horizons, based on expectations about future obligations and opportunities. Certainty about coming regulation greatly increases the ability of firms to plan investments. A realistic, meaningful schedule of emissions reductions could provide the desired regulatory certainty for purposes of business planning. If this schedule is achieved through flexible mechanisms (such as emissions trading under a cap) and coupled with technology incentives, the challenge of climate change can be met squarely and cost-effectively.

Currently, climate change and potential climate policy pose levels of uncertainty and potential market consequences that are large, even by normal business standards. There is great uncertainty about policy changes that, in turn, will affect energy supplies, energy prices, potential mitigation costs, broad trends in market demand, and a variety of other factors. Clearly defining a federal climate policy allows companies to integrate regulatory expectations into traditional business strategy frameworks in order to develop appropriate climate strategies.

Some companies in the Pew Center's Business Environmental Leadership Council (BELC) report that climate issues are treated similarly to other strategic challenges and that companies seek to define the "business case" around a set of climate strategy elements and targets.[14] In practice, the business cases for corporate climate action are often less quantitatively concrete than some might prefer, due in part to the many policy-related factors affecting markets that are hard to forecast precisely. Nonetheless, the practice of constructing a business case helps provide rigor of assigning values and probabilities where possible and otherwise forcing a logical justification of assumptions.

Corporate Views about the Future

Based upon responses to a Pew Center Survey conducted in Fall 2005, corporations are already integrating expectations about existing and prospective carbon regulations into their business forecasts.[15]

- 90 percent of corporations believe that GHG regulation is imminent, and 67 percent of those expect federal regulations to take effect between 2010 and 2015.
- Respondents gave the highest ranking of importance among "measures of success of climate-related strategies" to the following:
  — Energy efficiency;
  — Operational improvement;
  — Cost savings; and
  — Anticipating and influencing climate change regulation.
- Business units in countries that have ratified the Kyoto Protocol were more than twice as likely to willingly adopt climate-related obligations as those business units in countries that have not ratified.
- Respondents ranked "GHG trading

schemes" highest among the actions that will be important in federal responses to climate change; "voluntary GHG limits" ranked lowest.

The majority of the forty-five companies in the BELC see providing certainty in business planning as among the primary considerations for developing government policy.[16]

A 2005 survey of companies involved in the European Union's Emissions Trading System (ETS) found that 112 of the 167 companies polled wanted the ETS to be extended.[17] Of these companies, 93 percent say the ETS should be extended by ten years or longer. Survey respondents cited as justification the necessary conditions for investment planning and the efficient, undistorted development of carbon markets. In the same survey, 48 percent of respondents currently price the value of CO_2 allowances into daily operations and 71 percent expect to do so in the future.

Policies Can Influence Companies to Act and Drive Technological Innovation
Designing Policy with Business Planning in Mind

The assumptions, drivers, and approaches affecting the development of company strategies to address climate change have several broad implications for the design of policies to encourage climate-friendly behavior in the private sector. Important characteristics include:

- Clarity of financial or compliance signals to firms, in terms of the present value of financial impact.

- Consistency and predictability at least ten years into the future (and longer when targeting sectors with very long-lived capital stock).
- Ensuring that the combined signals sent to firms and investors (through some combination of incentives, disincentives, and other market-based measures) are sufficient to justify private investment in technologies that otherwise could not reach commercialization for many years.[18]
- Targeting, as directly as possible, those firms that have definite control over emissions.
- Where possible, consistency with existing opportunities to enhance profitability and operational efficiency that exist even in the absence of climate considerations.[19]

All of these assume adjustment of policy and program signals to account for the way firms will assess the time value of money and regulatory uncertainty. Various policy mechanisms can be used to motivate business action by providing important certainty regarding regulatory options and financial incentives needed to stimulate the technological change essential to tackling climate change.

Regulatory Options: Technology Demand "Pull" Policies

MARKET CREATION

While both broad (economy-wide) and technology-specific policies are essential, a clear and consistent signal of market value (a price on carbon) is arguably the single most influential driver of private-sector investment in long-term R&D.[20] A cost-effective policy would combine a limit (or "cap") on GHG emissions with market mechanisms, such as trading among different sectors or greenhouse gases. A cap on emissions would send an economy-wide signal favoring emissions reductions, and emissions trading would ensure that reductions are achieved at the lowest possible cost. Furthermore, well-designed emissions trading programs have been shown to achieve environmental goals more quickly and with greater confidence than more costly command-and-control alternatives.[21] Sectors not readily suitable for cap and trade can have programs tailored to address their emissions more directly (such as through standards, discussed in the next section), or can be used as offsets to other sectors' trading programs.

STANDARDS

While less flexible than market-based mechanisms, regulatory standards are an additional option for creating demand for low-carbon technologies. Standards for new appliances and building materials could promote increased efficiency along with policies to promote more efficient energy production through distributed generation or combined heat and power.

Voluntary Incentives: Technology "Push" Policies

Whatever form regulatory approaches take, there is no doubt that to reduce GHGs most cost-effectively, they should be combined with technology "push" policies. Both approaches are essential because two independent market failures exist and require correction.[22] The first failure reflects the presence of environmental externalities associated with fossil-fuel combustion and other activities that generate GHGs. The second is the inability of private actors to benefit fully from their R&D (often called a "spillover" effect since the benefits of private research often accrue to society at large). Different policies are required to address these two market failures. Reliance solely on one approach will be less efficient and result in higher costs than using both.[23]

Traditional technology policy includes: (1) direct government funding for research and development; (2) policies that either directly or indirectly support commercialization and adoption of technology or directly support development; and (3) policies that foster technology diffusion through information and learning.[24] In these cases, a mandatory requirement is not set, so there is no certainty in the outcome. However, consistent policy signals are still very important for guiding corporate behavior and advancing technological innovation.

Direct Government Funding

Calls for more investment in research and technology solutions regarding climate change are widespread across a variety of stakeholders and government officials.[25] For example, investment in energy technology may be justified as insurance against a number of issues, including climate change, oil price shocks, and urban air pollution.[26] Indeed, government investment in certain stages of research, development, and deployment could be critical to the success of a number of technologies presumed to be pivotal in addressing climate change.[27] Private funding alone will not likely deliver the longer-term, more risky, and often more socially beneficial technologies because of the bottom-line and near-term focus of businesses and investors.

One approach is for the government to fund innovation through direct spending on government entities. For example, the Defense Advanced Research Projects Agency (DARPA) has a well-defined mission that promotes innovation for military application.[28] Some have suggested that a similar independent agency with an energy- or climate-focused mission would be desirable, and recent energy legislation has made strides in this direction. Another popular approach for R&D in recent decades is to leverage funding by working through consortia. Public/private partnerships involving industry, government agencies and national labs, and universities can accelerate development and deployment of low-GHG technologies. Such collaboration can draw on expertise from a range of sectors, reduce duplication of effort, exploit economies of scale, provide for knowledge transfer and accelerated commercialization, and realize many other benefits.

Governments can also support demonstration projects for emerging costly technologies where individual firms may be unwilling or unable to take risks. Demonstration projects can support commercialization of potentially viable innovations that are simply too costly or risky for individual private developers to undertake.[29] In the climate change arena, demonstration of carbon capture and storage (CCS) or advanced nuclear technologies are likely to require government support. American Electric Power, for instance, is a participant in the Midwest Regional Carbon Sequestration Partnership (MRCSP), which is a DOE-funded initiative designed to assess and test methods for achieving successful carbon sequestration. The cost of this effort was split among thirty entities and the federal government, allowing for knowledge transfer and also cost mitigation. The MRCSP is now in its second phase of work, conducting multiple sequestration field tests throughout the Midwest.[30]

While increasingly popular, some previous collaborative efforts have not lived up to their promise. Often, there is tension between individual firm interest and the public policy objective of the government sponsor. For example, intellectual property, spillover, and other concerns relating to competitiveness may drive companies to sign up only for consortia that present low risks and therefore offer lower potential rewards.[31] Also, lack of a consistent long-term commitment has hampered successful fruition of some efforts. For example, funding for the Partnership for the New Generation Vehicle was eliminated under the George W. Bush administration and, more recently, federal support for the ambitious advanced coal technology partnership Future-Gen was abruptly halted. One solution may be to focus such efforts on near-term goals and incremental progress rather than long-term, high-risk projects and radical innovation.[32]

Despite the benefits, the government share of U.S. R&D funding overall has declined significantly from its peak in the late 1970s.[33] According to the National Commission on Energy Policy (NECD), public and private spending on energy-related R&D, which includes funding for conservation programs and renewable energy, has been falling.[34] It reports that the energy sector is by far the least R&D-intensive high-technology sector in the U.S. economy.[35] Private-sector investments in energy R&D appear to have fallen from 0.8 percent of sales in 1990 to 0.3 percent of sales in 2004. To put these numbers into perspective, NCEP notes that private-sector investments in R&D are about 12 percent in the pharmaceutical industry and about 15 percent in the aircraft industry. At

the same time, public-sector funding for energy research, development, and deployment (RD&D) has also dropped dramatically. From 1978 to 2004, U.S. government appropriations fell from $6.4 billion to $2.75 billion (constant dollars)—about a 60 percent reduction.[36]

GOVERNMENT SUPPORT FOR PRODUCTION AND COMMERCIALIZATION OF TECHNOLOGY

While government support for R&D in various forms is critical, it is neither sufficient nor the most cost-effective tool to bring about innovation.[37] Incentives such as production tax credits can drive important investment in new energy sources. Properly placed and consistently applied incentives can be used in the climate change arena to incentivize development and diffusion of clean technologies and fuels. The Department of Energy's support for the Integrated Corn Based Biorefineries program allowed DuPont to share risks and costs of its investment in cellulosic ethanol technology. Similarly, incentives can be offered to those who purchase low-emitting vehicles or other products. For example, the government offers tax breaks to those purchasing hybrid or electric vehicles.[38]

Again, certainty and consistency play an important role. In the United States, the wind power industry has experienced a "boom and bust" effect in which production grows and shrinks along with implementation and expiration of the wind production tax credit. In contrast, Germany has seen steady growth in the wind industry as a result of its steady policy support. Policies to push low-carbon technologies must be in place long enough to foster market confidence and sustain market growth over a reasonable period of time.[39]

Enabling Policies

Other policies create an important foundation for innovation and enable the deployment of new technologies. Examples of these are discussed below.

Intellectual Property

Protection of intellectual property such as patents provides powerful incentive for innovation. The lack of protection has at times been a deterrent in domestic development of technology as firms are hesitant to disclose their work in moving beyond bench models to commercialization. Lack of international protection of intellectual property rights also creates challenges for deploying technology internationally—a necessary step in disseminating advanced technologies to the developing world.

While protection of property rights is considered an important stimulus for discovery and invention, it can also limit dissemination of technologies. Care must be taken to achieve the right balance between stimulating innovation and sharing access to information on technological breakthroughs when they occur.

Information Dissemination

Information and transparency are important goals—not just in and of themselves—but also in educating consumers and firms about the options available for cost-effectively reducing GHG emissions. A number of federal programs support outreach and education of firms and consumers regarding best practices in productivity and efficiency.

Imperfect information and the transaction costs associated with getting better information may drive otherwise rational consumers or firms to purchase equipment that is less energy-efficient than they otherwise might have chosen if they had adequate information. Imperfect information has long been recognized as a market failure associated with energy efficiency, and numerous studies have evaluated policy options for overcoming this issue.[40]

For example, the Agricultural Extension Service provides assistance to farmers to enhance their access to the best available science and increased productivity. Efforts to reprogram existing subsidies and outreach in agricultural and other areas to include addressing efficiency and GHG mitigation are being promoted. Notably, Title II of the Farm Bill promotes conservation research and demonstration activities including those related to GHG reduction, carbon sequestration, and energy conservation and management.[41]

Recently, federal support of energy efficiency outreach has declined compared to other sectors.[42] Conservation and efficiency saves consumers and businesses money, puts downward pressure on energy prices, helps decrease reliance on foreign oil, helps to reduce other types of air pollutants, and generally helps the economy overall (since reducing the amount spent on energy will allow it to be invested elsewhere). However, lack of information on these benefits and payback often prevents firms and individuals from taking full advantage of these opportunities and available efficiencies are not fully realized.[43]

U.S. Climate Action Partnership

On January 22, 2007, ten large corporations and four leading nongovernmental organizations publicly recommended to Congress an economy-wide U.S. climate policy including a cap-and-trade system, goals for U.S. global leadership, and mandatory measures and incentives to cut emissions from transportation, buildings, and coal-based energy.

The principal USCAP recommendation was that Congress should enact legislation for an economy-wide program to achieve significant reductions of GHG emissions as soon as possible. Other more specific recommendations from 2007 include the following.[44]

- Aim to stabilize global GHG concentrations over the long term at a carbon dioxide equivalent level between 450 and 550 parts per million.
- Adopt a market-driven, economy-wide approach that includes a cap-and-trade system that ensures emissions reduction targets will be met while generating a price signal for GHGs, thus stimulating investment in necessary technologies.
- Establish short- and medium-term mandatory GHG emissions reduction targets that slow, stop, and reverse the growth of U.S. emissions—100 to 105 percent of today's levels in five years of enactment, 90 to 100 percent in the following five years, and 70 to 90 percent in the third five-year period—and a target zone of 60 to 80 percent reductions from current levels by 2050.
- Adopt additional policies in sectors where the initial price signal under cap-and-trade will not sufficiently reduce emissions and advance new technologies, especially for:
  - Transportation (vehicles, fuels, and behavior)
  - Buildings and efficiency (utility incentives to pursue energy efficiency, extension of federal energy efficiency incentives, better codes and standards, tax policies to advance new high-efficiency technologies, national protocol for measuring and accounting GHGs)
  - Stationary sources including coal-based energy (rapid transition to low- and zero-emission technologies and large-scale geologic sequestration).
- Establish a federal technology research, development, demonstration, and deployment program with stable, long-term financing for low-GHG technologies.

The deeper business rationale for why USCAP companies committed to such a path-breaking initiative is almost as significant as the USCAP recommendations themselves. The USCAP group includes some of the largest direct and indirect GHG emitters in the economy as well as some of the largest current and future producers of climate-friendly fuels and technologies. In some cases, member companies represent both perspectives—facing both major regulatory risks and potential business opportunities. At a high level, they believed that sensibly designed regulation and related programs offer the best path forward for the U.S. economy as a whole. Gaining regulatory certainty will prevent inefficient allocation of money and effort across the whole economy, and robust economic growth is among the most powerful drivers of profitability in many USCAP companies.

The need for regulatory certainty motivated both the companies that face major carbon-sensitive capital investment decisions and those facing critical decisions on research and marketing of low-emission technologies. For companies with relatively high emissions or product footprints, another strategic benefit of USCAP lay in managing risk—regulatory risk, reputational risk, the risk of stranded assets, and so on. Companies also could manage risks associated with uncertain future revenue from customers in sectors that are themselves highly exposed to climate-related regulations.

The impact of USCAP since its inception has been even greater than the group had hoped. Coalition membership has expanded to twenty-five companies and five NGOs; in January 2009 the goup released a more detailed set of recommendations in a document known as the Blueprint for Legislative Action (see www.us-cap.org for more information). The unveiling of USCAP recommendations and intense follow-up work with Congress has elevated visibility of the issue for the public and corporate America, and it has sent a clear message to Congress that support in the business community for federal action is growing rapidly. Notably, many of USCAP's 2009 recommendations on cap-and-trade were incorporated into HR2454, the American Clean Energy & Security Act, which passed the House of Representatives in June 2009.

Conclusion

Policy makers should enact legislation that accurately reflects the real cost of greenhouse gas emissions, and thus appropriately values those technologies and services that reduce or sequester greenhouse gases that contribute to global warming. This is the key to allowing the market to work in a truly efficient manner. A clear, market-based national policy will provide price signals that stimulate private investment in the most economically efficient low-emitting goods and services. Should the modest price signal created initially by a cap-and-trade system be insufficient to drive reductions in key sectors due to market ineffi-

ciencies (such as price inelasticities and barriers to entry), additional complementary policies may be warranted.

Companies that are well attuned to the challenges posed by climate change and related policy realize that a fundamental market transformation is both inevitable and beneficial for society. As a result, firms should work to integrate climate change considerations into their corporate planning and strategies. A company should first assess its carbon footprint, and then begin mitigating those impacts on the climate. Companies need to understand how their market will change, and position themselves strategically for the future. These strategic moves should encompass both risk management and business opportunity. Most companies with a significant carbon footprint will benefit substantially from constructive engagement in the development of the policies that will reshape their markets. Ultimately, many businesses that succeed in bringing all of their key functional strengths to bear on the climate issue—from product development to marketing to government affairs—have the potential to achieve enduring competitive advantages over their peers. If firms move rationally in this direction, and government leaders make the hard choices needed for sound national and international climate policies, the result will be growth that is sustainable both for the economy and the environment.

Notes

1. EPA. 2007. Inventory of U.S. Greenhouse Gas Emissions and Sinks: 1990–2005. U.S. Environmental Protection Agency, Washington, DC.

2. EPA, 2007 op. cit.

3. Lempert, R., S. Popper, S. Resetar, and S. Hart. 2002. *Capital Cycles and the Timing of Cli-*

mate Change Policy. Pew Center on Global Climate Change, Arlington, VA.

4. Ibid. p. 41.

5. Ibid. p. 29.

6. Innovest Strategic Value Advisors. 2005. *Carbon Disclosure Project 2005.* Carbon Disclosure Project, New York.

7. United States Environmental Protection Agency. *Climate Leaders.* Available at www.epa.gov/stateply/index.html.

8. Hoffman, A. 2006. *Corporate Strategies that Address Climate Change.* Pew Center on Global Climate Change, Arlington, VA.

9. http://www.pewclimate.org/companies_leading_the_way_belc/company_profiles/interface_inc/.

10. http://www.pewclimate.org/companies_leading_the_way_belc/company_profiles/interface_inc/.

11. Dowell, G., S. Hart, and B. Yeung. 2000. Do corporate global environmental standards create or destroy market value. *Management Science* 46(8).

12. Gunther, M. 2006. The green machine. *Fortune Magazine,* August 6.

13. Rio Tinto Climate Change Program. 2004. Presentation by Preston Chiaro, Chief Executive Energy, May 7. Available online at www.climatechangecentral.com/resources/presentations/RioTintoPresentationCalgary.pdf. Downloaded July 2006.

14. Pew Center on Global Climate Change. 2006. A BELC membership list, member profiles, and descriptions of their climate strategies can be found at: www.pewclimate.org/companies_leading_the_way_belc.

15. The Pew Center Survey was a 100-question survey of thirty-one companies, including twenty-seven BELC members and four non-BELC members. Findings from the survey were incorporated into the Pew Center's report, *Getting Ahead of the Curve: Corporate Strategies That Address Climate Change.*

16. Based on formal and informal communications between BELC Director Truman Semans and BELC member companies, March 2005 through August 2006.

17. European Commission Directorate General for Environment, McKinsey & Company, and Ecofys, 2005. *Review of EU Emissions Trading Scheme: Survey Highlights.*

18. For example, to rapidly commercialize power generation technologies for coal that allow carbon capture, policy makers probably need to combine near-term but predictably continued subsidies and risk-sharing programs, a CO_2 cap or tax that creates a clear and high price signal, and new barriers to public utility commission acceptance of pulverized coal generation proposals such as changes in definitions of best available control technology.

19. For example, measures to nudge sectors toward energy-efficiency investments that have a positive return but not a high enough return to compete against other internal capital needs.

20. The 10-50 Solution. Proceedings of Pew Center/NCEP "10-50 Solution" workshop. Washington, DC, March 25–26, 2004.

21. Ellerman, D., P. Joskow, and D. Harrison, Jr. 2003. *Emissions Trading in the U.S.: Experience, Lessons, and Considerations for Greenhouse Gases.* Pew Center on Global Climate Change, Arlington, VA, p. iii.

22. Goulder, L. 2004. *Induced Technological Change and Climate Policy.* Pew Center on Global Climate Change, Arlington, VA., p. 6.

23. Ibid. p. 28.

24. Alic, J., D. Mowery, and E. Rubin. 2003. *U.S. Technology and Innovation Policies.* Pew Center on Global Climate Change, Arlington, VA.

25. See, for example, Pew Center's Agenda for Climate Action; National Commission on Energy Policy report entitled "Ending the Energy Stalemate," Washington, DC, 2004; and "Global Climate Change Policy Book," www.whitehouse.gov/news/releases/2002/02/climatechange.html.

26. Schock, R. N., et al. 1999. How much is energy research and development worth as insurance? *Annual Review of Energy and the Environment* 24: 487–512.

27. Electric Power Research Institute (EPRI). 2007. "The Power to Reduce CO_2 Emissions: The Full Portfolio." August. Discussion paper prepared for the EPRI 2007 Summer Seminar Attendees by the EPRI Energy Technology Assessment Center.

28. Alic et. al., 2003 op. cit.

29. Norberg-Bohm, V. 1999. *Creating Incentives for Environmentally Enhancing Technological Change: Lessons from 30 Years of U.S. Energy Technology Policy.* Massachusetts Institute of Technology, Cambridge, MA.

30. Midwest Regional Carbon Sequestration Partnership. Obtained from http://216.109.210.162/ on March 27, 2008.

31. See, for example, discussion of concerns regarding fuel cell research in PNGV process discussed in Alic et al., 2003 op. cit. p. 24.

32. Alic et al., 2003 op. cit.

33. National Commission on Energy Policy (NCEP). 2004. "Ending the Energy Stalemate: A Bipartisan Strategy to Meet America's Energy Challenges." Available online at www.energycommission.org/files/contentFiles/report_noninteractive_44566feaabc5d.pdf. Downloaded July 9, 2006.

34. NCEP, 2004 op. cit.

35. The National Commission on Energy Policy defines R&D intensive as the ratio of investments in R&D divided by the value of sales in the sector.

36. NCEP, 2004 op. cit. p. 101.

37. Goulder, L. *Induced Technological Change*. Pew Center on Global Climate Change, Arlington, VA. Also see Alic et al., 2003 op. cit.

38. U.S. Department of Energy. 2008. "New Energy Tax Credits for Alternative Fuel Vehicles." Available online at www.fueleconomy.gov/feg/tax_afv.shtml.

39. "The 10-50 Solution," 2004 op. cit.

40. Prindle et al. 2007. "Quantifying the Effects of Market Failures in the End-Use of Energy." Report Prepared by the American Council for an Energy-Efficient Economy, for the International Energy Agency (IEA). IEA, Paris. Available at http://www.aceee.org/energy/IEAmarketbarriers.pdf.

41. See Section 1238(A)(d)(2).

42. Alliance to Save Energy. 2006. "Legislative Alert: President's FY 2007 Would Cut Energy Efficiency Funding." Available at www.ase.org/section/_audience/policymakers/fedbudget/. See also www.ase.org/content/article/detail/2894.

43. Nadel, S. 2006. "Energy Efficiency Resource Standards: Experience and Recommendations." American Council for an Energy-Efficient Economy (ACEEE). Available online at http://aceee.org. Downloaded March 2006.

44. These are detailed in *A Call for Action* (see www.us-cap.org for complete text).

Chapter 38

Corporate Initiatives

PAUL DICKINSON, JAMES P. HAWLEY, AND ANDREW T. WILLIAMS

Introduction

This chapter examines the incentives and disincentives facing corporations when confronting issues raised by climate change and greenhouse gas (GHG) emissions. In particular, it focuses on corporate factors that have influenced various measures to reduce GHG emissions and/or to develop carbon-reducing technologies and processes in the absence of U.S. regulatory standards. Thus, it tends to focus on U.S. developments, although not exclusively. Additionally, it examines the role that institutional investors play in responding to climate change.

Several factors, often in combination, tend to pressure companies to respond in some manner to climate change. Among the most important are changes in regulatory requirements and national laws; competitive advantage by developing leading-edge technology or shaping markets and national standards; risk management; and protection or promotion of reputation (hopefully leading to market advantages). The corporate response to climate change has ranged from inaction or outright hostility to any governmental action to reduce GHGs to, in certain exemplary cases, the adoption of a broad range of activities and investments to reduce their carbon footprint.

Why Do Corporations Care about Climate Change?

Generally, firms are concerned about how climate change could raise their costs through, for example, higher insurance rates. They are also concerned about what is called "regulatory risk"—the risk that regulations will be enacted in this area (see chapter 37 in this volume). Some companies may also respond to climate change as a moral or social issue but, primarily, corporations respond to fundamental economic factors; profit margins compel action.

The economic impact of climate change on corporations can be divided into direct and indirect impacts. A direct impact would be the effect of rising sea levels on owners of beachfront property or on a company with a manufacturing facility at sea level. These impacts require a response because their effect on the bottom line is clear and the cost is obvious and quantifiable. Indirect impacts such as rising insurance premiums due to an increase in severe weather events are more difficult for companies to deal with, but may be more important in the long run. Indirect impacts are often buried in a general increase in the cost of doing business.

Companies face further incentives to respond to climate change when competitors adapt more quickly or successfully to a new business environment, when a multinational company standardizes business practices internationally, and, possibly, from an attempt to gain strategic advantage by anticipating stricter regulation in the future. Broad agreements such as the Kyoto Protocol facilitate a response to these pressures by leveling the playing field and by providing positive incentives to respond to problems. However, since the United States has not ratified the Kyoto Protocol, the asymmetry between the United States and much of the rest of the world has led to state-level and regional attempts to address climate change. In this complex political environment, companies that compete across borders and those subject to international competition find it increasingly difficult to deal with multiple regulatory environments (see chapter 37). Sometimes the response is to voluntarily comply with the highest standards regardless of where they may be imposed. In other cases, a race to the bottom has occurred whereby companies try to locate production in less regulated regions and countries such as China or Bangladesh. Alternatively, some companies have realized that significant business opportunities flow from confronting climate change risk.

Institutional investors, such as the California Public Employee Retirement System (CalPERS), own shares across a broad swath of the economy and so have an incentive to discourage the companies in their portfolios from engaging in activities that increase costs to other portfolio companies. In effect, the ability of large, highly diversified "universal owners" to satisfy their fiduciary duties to their beneficiaries depends on the overall performance of their portfolio.[1] Given their broad diversification of investment across the world and over all economic sectors, universal owners have a particular interest in encouraging a uniform response to climate change through mechanisms such as the Kyoto Protocol.

Corporate Responses to Price Incentives

Corporations in Europe and in other countries that have signed on to the Kyoto Protocol face a variety of regulatory and market pressures to reduce or offset carbon emissions. With the initiation of a cap-and-trade program in the EU, more than 40 percent of all GHG emissions from twenty-seven member states face incentives that should lead to reduced emissions.[2] Under the carbon-trading scheme and with EU approval, companies are assigned a certain number of carbon credits by national governments. This should provide economic incentives to reduce overall carbon emissions, but to date results have been disappointing and the EU carbon trading system is being revised to make it more effective.[3]

Nonetheless, a pricing mechanism for carbon emissions, either through caps or taxes, redistributes the costs of doing business. Initially the effect has been most strongly felt through electricity rates. Thus, electric utility companies with low emission sources, such as hydropower and nuclear, have benefited, although only marginally to this point. All utilities have been able to pass the higher cost of carbon dioxide emissions to customers, thus heavy electricity users such as steel plants have seen a considerable increase in their electricity costs.[4] Introducing a price for carbon emissions has the potential to act as a powerful incentive to increase energy efficiency leading to reduced GHG emission, but it may also strain trade relations between countries limited and unlimited by Kyoto.

Corporate Initiatives in Response to Climate Change

At the firm level, some companies are responding to the threat of climate change based on perceived risks and business opportunities. General Electric (GE), for example, has taken a proactive approach to climate change. While its U.S. operations are not carbon constrained, GE is nonetheless a global firm attempting to position itself on the leading edge of innovation. Thus, in 2005 it launched its "Ecoimagination" initiative with a 2012 goal of reducing greenhouse gas emissions 1 percent below their 2004 levels. Without this reduction, emissions were expected to grow by 30 percent during that period.

GE views climate change as a business opportunity, and as a leading manufacturer of utility-scale wind turbines it provided 60 percent of the capacity installed in the United States in 2005. It is also a major supplier of gas turbine combined-cycle technology and by increasing efficiency in this area contributes to reducing greenhouse gas emissions. GE also reports GHG-reducing technological advances in jet engines, locomotives, and integrated gasification combined-cycle technology, which produces more efficient and cleaner electricity from coal. In addition to voluntarily setting a goal to reduce worldwide emissions below current levels, GE operates about a dozen manufacturing plants in Europe that are, of course, subject to the EU Emissions Trading Scheme.[5]

Investors agree with GE's emphasis on the economic case for reducing GHG emissions. Writing in *Business Week*, Todd S. Thomson, CEO of Citigroup Global Wealth Management, stated:

> Forward-thinking businesses that realize environmental standards must tighten will thrive. As political pressure mounts to make reduction of greenhouse gases mandatory, companies with a head start on eco-friendly technology will have the credibility to participate in, or even shape, the debate over how to further reduce emissions. The amount of innovation necessary to stop global warming will be staggering, but it also could be *staggeringly profitable* for global companies. (emphasis added)[6]

However, some industries such as the oil and gas industry, a major source of CO_2 emissions, have resisted a climate policy. Even within that industry, some oil companies are actively responding to climate change risk. British Petroleum (BP), for example, with its "Beyond Petroleum" slogan, is a long-standing leader in this area. In 1997 it was the first company to state publicly that climate change requires precautionary action. The company set a 2010 goal of reducing operational greenhouse gas emissions 10 percent below 1990 levels, which it achieved in 2001 at a reported savings of $650 million.[7] For this accomplishment, *Business Week* ranked BP number two in its list of the top ten companies of the decade for its total reduction of greenhouse gases relative to company revenues.[8] Its current goal is to hold absolute emissions steady through 2012. BP has a major alternative energy business in solar, wind, hydrogen, and combined-cycle power generation, and currently it is one of the world's leading producers of solar panels.[9]

At first, the "green" makeover by Wal-Mart of both its supply chain and its operations were perceived by some commentators as a superficial response to bad publicity. However, after an in-depth cost/benefit analysis, Wal-Mart concluded that green business practices would produce significant cost savings.[10] Thus, in October 2005, CEO Lee Scott committed Wal-Mart to an ambitious program to boost energy efficiency, increase organic food sales, and reduce waste and greenhouse gas emissions. The goal is to cut greenhouse gas emissions 20 percent by 2012, much below the Kyoto target of 7 percent for the United States, while targeting 100 percent renewable energy and zero waste. The company has opened two stores to pilot environmental concepts that will then be adopted system wide.[11] While many are skeptical of Wal-Mart's environmental commitment, high-profile announcements from the CEO coupled with specific goals make it likely that Wal-Mart will take some action to reduce its greenhouse gas emissions. Interestingly, Scott says that while Wal-Mart first came to this position from a risk-management perspective (including reputational risk), the company now embraces it as "the greatest opportunity we have to create value for our customers, cut costs, increase morale, grow responsibly, and do the right thing for the planet."[12]

These examples and others reflect positive corporate responses to climate change. Prominent among motivations are cost savings, regulatory anticipation and/or conformance, reputation enhancement, and taking advantage of business opportunities.

Role of Institutional Investors in Climate Change Initiatives

Institutional investors also play an important role in the movement to mitigate climate change risk. For our purposes, institutional investors active in this area are of three types: socially responsible investors (SRI), such as the Domini and Calvert families of mutual funds in the United States; mainstream investors, especially some U.S. public pension funds but increasingly some investment and other banks; and mainline insurance companies and re-insurance companies. These investors possess a huge potential to impact firms' carbon emissions practices and to encourage the development of low carbon technologies since they either own shares and carry debt issued by firms; lend to firms and serve as their investment advisers; and/or insure them. While SRI investors (far smaller than mainstream investors) have long focused on climate change issues, only recently has this focus moved into the far larger and more influential mainstream investment world. Our discussion focuses on this latter group.

Mainstream investors have sometimes worked with nongovernmental organizations (NGOs) such as Ceres, Greenpeace, Friends of the Earth, and the National Resources Defense Council to engage with firms in order to get them to adopt carbon reduction or mitigation strategies. In the United States and many other highly industrialized countries public and much private equity is directly and indirectly owned by institutional owners; the most important are mutual funds, pension funds, and insurance companies. More than 60 percent of U.S. publicly traded equity is currently owned by U.S. and non-U.S. institutional owners. Increasingly some of these large institutional owners have come to view climate change as a form of long-term risk to their investments while also recognizing potential business opportunities. As a result they are developing programs and investment funds to address carbon risk as part of what they view as their long-term financial and fiduciary interests and obligations.

For large public (and some other) pension funds, these activities have taken two primary forms: corporate governance initiatives and the creation of various "alternative" investment funds to create lower carbon impact portfolios and/or to invest in "clean tech" — innovative and emerging industries and technologies. For example, the largest and third largest U.S. public pension funds, the California Public Employees' Retirement System (CalPERS) and the California State Teachers' Retirement System (CalSTRS), have developed environmental investment initiatives in order "to explore ways in which we can marry the jet stream of finance and the capital markets with public purpose [in order to achieve] . . . positive financial returns, while fostering sustainable growth and sound environmental practices" (quoting CalPERS). The program invests in two areas: a $500 million public equity mandate for a portfolio that screens firms for good environmental practices and a $1.1 billion investment in private equity to develop environmental technologies that are more efficient and less polluting than current ones.[13] While these are relatively small amounts for an institution that in early 2009 had $187 billion in assets, the funds are nonetheless significant in absolute terms. The goal of the investment in public equity is to test the waters with a screened environmental fund by determining its returns during a five-or-so-year time period. In addition to the public and private equity investments, the program seeks a 20 percent reduction in energy use at CalPERS's

large core real estate holdings and the use of corporate governance clout as a shareowner to "improve transparency and timely disclosure of corporations' environmental impacts."[14]

Another example of institutional investor action is the 2006 corporate governance engagement process adopted by a group of New York retirement funds led by the New York City Employees' Retirement System (NYCERS). The group filed shareholder proposals at seven large utilities asking each of them "to disclose cost-effective ways to reduce carbon dioxide and other emissions from their current and proposed power plant operations due to potential risks associated with climate change." Four of the targeted companies quickly agreed to take the actions requested in the shareholder resolutions.[15]

While actions by individual institutional investors or small groups of investors are important, collective action on a broader scale may be even more effective. For example, CalPERS and NYCERS are founding members of the Investor Network on Climate Risk (INCR), a coalition of more than seventy large institutional investors managing more than $7 trillion in assets "dedicated to promoting better understanding of the financial risks and investment opportunities posed by climate change."[16] The INCR was initiated by Ceres and has become a major U.S. and global NGO working closely with firms and institutional investors to reduce climate risk. Additionally, CalPERS and many other institutional investors have signed onto the Carbon Disclosure Project (CDP), whose signatories collectively represent more than $57 trillion in assets. The CDP seeks timely disclosure of climate risk information in order to facilitate a focus on particular firms and industrial sectors (e.g., auto, utilities) to limit carbon emissions and to spotlight activities that would delay limiting emissions, such as the auto industry's campaign against California's law requiring reduced auto fleet carbon content in the next decade.

Some banks and investment managers, such as Bank of America, Bank of New York, and Barclays Global Investors, have lent support to investor climate change coalitions and some have adopted principles of sustainable environmental practices. Articulating the best strategy to date, Goldman Sachs, one of the largest U.S. investment banks, issued a strategy statement in 2005 that argued that green is not merely an SRI issue (which it supports, but not as a fundamental financial strategy) and that environment and climate change, in particular, are a fundamental issue for financial analysis.[17] This strongly suggests that climate change needs to be taken into account when analyzing investment risk and opportunity, along with a number of other factors not related to climate risk.

It should be noted that while financial support, sponsorship, and overall strategic analysis focusing on climate risk are important steps, to our knowledge no mainline investment or commercial bank has taken the next step to make its lending and advising contingent on climate risk mitigation or reduction, or even on monitoring and disclosure. This, however, may be in the process of changing. The International Financial Corporation of the World Bank (a multilateral agency) has set de facto standards with its Equator Principles, which have been signed onto by banks representing about 80 percent of global project finance, mostly in the developing world. The principles call for a full social and environmental impact assessment for all project lending to less developed and developing economies and now includes, for the first time, GHG monitoring as part of the updated principles.[18] This

is a small but potentially significant step toward GHG disclosure as a requirement for banks when lending, advising, and investing funds. As with banks and institutional owners, insurance and re-insurance companies are becoming deeply concerned about climate change. Re-insurance companies, which insure insurance companies, have long been warning about the risks of climate change and have been adjusting their insurance policy terms and rates accordingly. Munich Re and Swiss Re have been important leaders in this area. Another example is a recent report by the Allianz Group in conjunction with the World Wide Fund for Nature on "Climate Change and the Financial Sector," which reviews risks and opportunities, recommends polices for various types of financial institutions, and in particular focuses on the financing of low carbon forms of energy, disclosure of carbon emissions, and changes in governmental and intergovernmental carbon policy.[19] These actions will increasingly pressure mainstream investors to take account of climate risk and other environmental risks in their merger and acquisition, as well as lending, activities.

In sum, using the language of accounting standards, there is increasing recognition that climate change represents a significant material-value component of assets (in the case of emission reduction) or liabilities (in the case of continual emitters, especially large ones) that needs of necessity concern all investors.[20] This growing recognition is the result of increasing pressure from institutional investors, NGOs, and some governments, particularly Norway, the UK, and France. We believe that we are witnessing the beginning of such developments in the financial sector, both among commercial and investment banks, and importantly among some sectors of large institutional shareowners of corporations. The latter are concerned at this stage mostly with various forms of risk to firms they own that are large carbon emitters. Additionally, some see the potentials in the "green is green" investment strategy. As time passes, many large institutional owners will come to see the threat of climate change to many sectors of their entire portfolios, not only to those firms that are significant emitters. Such recognition flows from these large owners being universal owners.[21]

Conclusion

While some U.S. firms, either under pressure from their institutional owners and other stakeholders and/or for competitive reasons (including their global reach), have taken initial actions to confront climate change, these actions are likely to prove inadequate to lead to significant GHG reductions. Thus, it is highly significant that firms such as the electrical utility Duke Energy have called on the U.S. government to establish mandatory regulatory standards for GHG emissions, suggesting their recognition of the collective action nature of the problem.[22] We expect this view among U.S. corporations to grow as the pressures from institutional owners and others grow, influencing both corporate greening and corporate pressure on government for major policy initiatives for carbon reduction. And there is hope that the new administration will promulgate policies that will strengthen trends already underway and provide real incentives to greenhouse gas reduction on the part of firms.

Notes

1. Hawley, J. P., and A. T. Williams, *The Rise of Fiduciary Capitalism: How Institutional Investors Can Make Corporate America More Democratic*. Philadelphia: University of Pennsylvania Press, 2000.

2. Climate Action Network Europe, "Emission Trading in the EU," www.climnet.org/EU energy/ET.html. Accessed January 12, 2009.

3. Kanter, J., EU carbon trading system brings windfalls for some, with little benefit to climate, *International Hearld Tribune*, December 9, 2008. www.iht.com/articles/2008/12/09/business/wind fall.php. Accessed January 13, 2009.

4. "Soaring CO_2 prices have helped drive electricity prices to record levels across much of Europe, handing windfall profits to utilities," in Penson, S. "ANALYSIS: Europe Greenhouse Gas Trade Hots Up as Prices Soar," July 13, 2005, www.planetark.com/dailynewsstory.cfm/newsid/31644/story.htm. Accessed January 12, 2009.

5. Cogan, D. G., *Corporate Governance and Climate Change: Making the Connection*, Ceres, March 2006, pp. 149–51. Available on the Investor Network on Climate Risk Web site www.incr.com.

6. Thomson, T. S., Green is good for business, *Business Week*, May 8, 2006, p. 124.

7. For a description of its greenhouse gas emission program see the BP Web site at www.bp.com/sectiongenericarticle.do?categoryId=900 7646&contentId=7014542.

8. "Top Ten Companies of the Decade," www.businessweek.com/magazine/content/05_50/b39 63415.htm. Accessed January 12, 2009. Dupont was number one on the list, having "reduced energy consumption 7 percent below 1990 levels, saving more than $2 billion—including at least $10 million a year by using renewable sources."

9. Cogan, 2006 op. cit., pp. 209–11.

10. Little, A. G., "Always Low Prices—and Now Eco-Accountability," *Outside Magazine*, December 2006. http://outside.away.com/outside/culture/200612/code-green-wal-mart.html. Accessed January 12, 2009.

11. Bulter, R. A., "Green Wal-Mart? Wal-Mart Embraces Environmental Sustainability," http://news.mongabay.com/2006/0208-walmart.html. Accessed January 12, 2009.

12. "Our CEO on the Environment," press release, http://walmartstores.com/GlobalWMStores Web/navigate.do?catg=664&contId=5597. Accessed April 19, 2006.

13. "CalPERS Environmental Programs Advance: Energy Reduction in Real Estate Sets Pace

of Programs," CalPERS, November 20, 2008, www
.calpers.ca.gov/index.jsp?bc=/about/press/pr-2008/
nov/environmental-programs-advance.xml.   Accessed January 12, 2009.

14. CalPERS, "Environmental Investment
Initiatives," 2006, pamphlet and www.calpers.ca
.gov. Accessed January 12, 2009.

15. "Thompson Calls on Power Companies to
Assess and Mitigate Carbon Dioxide and Other
Emissions," press release, February 21, 2006. www
.comptroller.nyc.gov/press/2006_releases/pr06-02-
023.shtm. Accessed January 12, 2009.

16. Investor Network on Climate Risk, "About
INCR," www.incr.com/Page.aspx?pid=261. Accessed January 12, 2009.

17. Goldman   Sachs,   "Portfolio   Strategy
United States," August 26, 2005. www2.goldman
sachs.com/ideas/environment-and-energy/port-strat-
growing-interest-pdf.pdf. Accessed January 12,
2009.

18. Press Release No.: 06/0035, International
Financial Corporation, February 21, 2006. www
.equator-principles.com/principles.shtml.   Accessed January 12, 2009.

19. www.ren21.net/pdf/Studie_Climate_and_
Finance.pdf. Accessed July 1, 2009.

20. United Nations Environmental Pro-
gramme-Financial Initiative, *The Materiality of Social, Environmental and Corporate Governance Issues to Equity Pricing*, June 2004, Geneva,
Switzerland. See also "A legal framework for the integration of environmental, social and governance
issues into institutional investment," Freshfields
Bruckaus   Deringer,   October   2005.   www
.unepfi.org/fileadmin/documents/freshfields_legal
_resp_20051123.pdf. Accessed July 1, 2009.

21. Hawley and Williams, 2000 op. cit., and
"The Universal Owner's Role in Sustainable Eco-
nomic Development," in R. Sullivan and C.
MacKenzie, *Responsible Investment*. London:
Greenleaf, 2006.

22. "Duke's Energy Position," www.duke-
energy.com/environment/climate-change/duke-
energy-position.asp. Accessed January 12, 2009.

Chapter 39

Carbonundrums: The Role of the Media

MAXWELL T. BOYKOFF

Introduction

Media are key contributors—among a number of others—that shape climate change science and policy discourse as well as action. Previous studies have found that the public garners much of its knowledge about science (and more specifically climate change) from the mass media.[1] For one thing, citizens typically do not start their day with a morning cup of coffee and the latest peer-reviewed journal article. For another, Ungar has asserted, "science is an encoded form of knowledge that requires translation to be understood."[2] Mass media fill these roles.

"Mass media" have been broadly defined as the publishers, editors, journalists, and others who constitute the communications industry and profession, and who produce, translate, interpret, and disseminate information, largely through newspapers, magazines, television, radio, and the Internet. The mass media serve a vital role in communication processes between science, policy, and the public; thus, representations of climate change shape many perceptions and considerations for action. These interactions are dynamic and

highly contested. As discussed in other chapters in this volume, discussions surrounding climate change mitigation and adaptation cut to the heart of our carbon-based societal structures and behaviors.

First, this chapter briefly surveys historical interactions at this "triple-interface" of climate change science, policy, and media. Next, it explores multifaceted external and internal pressures at the triple-interface by touching on the salient challenges of language/translation and uncertainty. Then the chapter looks at these interactions by way of "climate contrarian" influences. The chapter aims to provide another insight into climate science and policy processes, and to particularly complement the accompanying chapters in this volume that address media and climate change.[3]

A Brief History of the Climate Change Science-Policy-Media Triple-Interface

Modern scientific investigations into various aspects of climate change and the development of mass media communications began concurrently in the 1700s and 1800s. Commercialization and increased dissemination in the early 1900s carried conflicting impulses of expanding democratic speech and corporate capitalist pursuits of profit. As these developments continued, the power of mass media became both amplified and more entrenched in society.[4] Concurrently, military interests developed and used media technologies and climate science to achieve strategic goals, as a sort of "military-climate industrial complex." Besides the advantages of intra-military battlefield communications, propaganda campaigns were launched in order to weaken foreign enemies while whipping up domestic patriotism (dubbed "the manufacture of consent").[5]

At the same time, funding for particular military programs also served to contribute significantly to the progress of climate science research, effectively catalyzing many climate science inquiries. For example, in the 1950s Gilbert Plass used military funding to conduct research on atmospheric CO_2 and infrared radiation absorption.[6] While this helped with the study of infrared absorption of heat-seeking missiles, it also added to a growing body of anthropogenic climate science research. Also in the 1950s, funding from the U.S. Navy and the U.S. Atomic Energy Commission supported Hans Seuss and Roger Revelle's research on radiocarbon dating and isotope decay, to both examine fallout from nuclear bomb tests and trace the distinct isotopic signature of anthropogenic carbon emissions into the atmosphere. During this time, Charles David Keeling began studying the interactions of atmospheric CO_2 and temperature. The initial stages of his research were paid for by funds from the U.S. Atomic Energy Commission (after 1963, funding was

continued through the U.S. National Science Foundation).[7] This research—now referred to as the "Keeling Curve"—is considered some of the most valid and reliable evidence regarding anthropogenic climate change.

The two spheres of climate science and mass media first came together in coverage of climate change beginning in the 1930s. In 1932, *New York Times* staff wrote, "The earth must be inevitably changing its aspect and its climate. How the change is slowly taking place and what the result will be has been considered."[8] Media coverage of human contributions to climate change appeared more clearly in the 1950s. For instance, the *Saturday Evening Post* published a story entitled "Is the World Getting Warmer?" that explored links between atmospheric temperature change and agricultural shifts as well as sea-level rise.[9] In 1956, Waldemar Kaempffert wrote in the *New York Times*:

> Today more carbon dioxide is being generated by man's technological processes than by volcanoes, geysers, and hot springs. Every century man is increasing the carbon dioxide content of the atmosphere by 30 percent—that is, at the rate of 1.1 degrees Celsius in a century. It may be a chance coincidence that the average temperature of the world since 1900 has risen by about this rate. But the possibility that man had a hand in the rise cannot be ignored.[10]

In 1957—the International Geophysical Year—Robert Cowen wrote an article in the *Christian Science Monitor* called "Are Men Changing the Earth's Weather?" He began:

> Industrial activity is flooding the air with carbon dioxide gas. This gas acts like the glass in a greenhouse. It is changing the

Earth's heat balance. It could bring anything from an ice age to a tropical epoch. . . . Every time you start a car, light a fire, or turn on a furnace you're joining the greatest weather "experiment" men have ever launched. You are adding your bit to the tons of carbon dioxide sent constantly into the air as coal, oil, and wood are burned at unprecedented rates.[11]

In the subsequent three decades, mass media coverage regarding climate change remained sparse. There was scant newspaper, radio, and television news coverage on topics such as U.S. National Academy of Sciences reports in the 1960s and 1970s that made repeated reference to emergent climate science and links to anthropogenic sources. However, coverage increased again in the 1980s at a time when international and domestic climate policy began to take shape. These media-science-policy spheres collided prominently in 1988, when media coverage of climate change science and policy increased substantially. Many factors contributed to this rise in coverage. Among them was NASA scientist James Hansen's testimony to Congress that summer. Hansen testified that he was "99 percent certain" that warmer temperatures were caused by the burning of fossil fuels and not solely a result of natural variation, and that "it is time to stop waffling so much and say that the evidence is pretty strong that the greenhouse effect is here."[12] This statement served to generate substantial media coverage and became a spectacle that signified solidified scientific concern for anthropogenic climate change. Moreover, that 1988 summer was one marked by extreme drought and high temperatures throughout North America. These concomitant events were thought to sensitize many in the climate science and pol-

icy communities, as well as the media and public, to the issue of climate change. In the science and policy spheres, 1988 was also the year in which the United Nations Environment Program and the World Meteorological Organization created the Intergovernmental Panel on Climate Change (IPCC). These climate change science and policy events and activities were pivotal in shaping media coverage from 1988 forward, during the time when multinational media corporations underwent further and significant consolidation, through various mergers and acquisitions.[13] Since the late 1980s, this triple-interface has become an increasingly politicized arena. Many factors, such as the emergence of a cohesive group of "climate contrarians" (discussed later in this chapter), have fueled an atmosphere of contention and conflict up to the present time.

Contemporary Media Coverage of Climate Change

Mass media have become significantly influential translators between science, policy, and citizen communities. W. Lance Bennett has commented, "Few things are as much a part of our lives as the news . . . it has become a sort of instant historical record of the pace, progress, problems, and hopes of society."[14] Research on multifarious factors involved in the processes of media reporting on climate science and policy has been pursued through a variety of methods and approaches (see chapters 40 and 41 in this volume). Overall, these studies have sought to carefully examine the role of the media by moving beyond anecdotes and platitudes to provide explicit, detailed, and empirical examples of factors shaping these dynamic and contested spaces.

Climate change science and policy have shaped media reporting and public understanding; however, journalism and public concern have also shaped ongoing climate science and policy decisions. Focusing on media, editors and reporters must navigate through many pressures while reporting the news, ranging from political economic to social, cultural, ethical, biophysical, and journalistic norms and values. These permeate multiple scales, from the global to the community to the individual. These challenges are very difficult to disentangle, as many of these pressures are interrelated and cross scales, and are nested within others. These multiscale factors interact, feedback, and re-embed themselves through time. For instance, everyday journalistic practices are made in the context of larger political economic pressures, where journalist's constraints on time-to-deadlines and space exist within a predominantly corporate-controlled media environment.[15] Research has documented that deadlines and space considerations constrain journalists, as do editorial preferences and pressure from publishers.[16] Moreover, economic considerations have led to decreased mass-media budgets for investigative journalism and fewer independent news sources, and quick deadlines can lead to one-source stories.[17] Together, such pressures can be particularly troubling when covering a multifaceted and complex issue like climate change. Furthermore, these multiple pressures shape the ongoing process of media production as well as framing of news on climate change for policy and the public. In this mix of pressures and influences, two challenges to media coverage of climate change science and policy gain salience: issues of language and translation, and dealing with uncertainty.

Language and Translation

Scientists have a tendency to speak in cautious language when describing their research findings and have a propensity to discuss implications of their research in terms of probabilities. For journalists and policy actors, this is difficult to translate smoothly into the crisp, unequivocal commentary often valued in communications and decision making.[18]

In combination, these factors feed into ongoing differences in language use in media coverage. Moreover, in order to compete in the mass-media "attention economy," many journalists feel pressure to keep stories short and simple.[20] Through these factors, media translations of complex climate science inevitably shape perceptions and, in turn, influence policy considerations.

Uncertainty

Uncertainty is an inherent feature of inquiry and action. It crops up in places such as business, marketing, and insurance endeavors; it informs, yet does not prohibit action. All scientific inquiry contains uncertainty by definition, as it operates past the bounds of certainty in examinations, critiques, and analyses of the unknown. At the interface of climate change science, policy, and media, uncertainty often garners a great deal of attention and is a battlefield for meaning. Inaccurate amplification or diminution of uncertainty can cause troubles in communications across this interface and obfuscate or confuse many important aspects of the subject.[19]

Climate scientists often have difficulty placing the uncertainty associated with their research into a familiar context, through an appropriate analogy; in other words, "translating error bars into ordinary language."[21] Unfortunately, admission of various forms of scientific uncertainty can be reframed as scientific incompetence, contention, and confusion in order to "invalidate the overall public concern for global warming as an environmental-social problem."[22]

Contemporary Media Courtesans: Climate Contrarians

In the case of climate change and media influence, a cohesive opposition group emerged in the late 1980s with activities often funded by carbon-based industry interests. These climate contrarians—also dubbed "climate skeptics" or the "carbon club"—gained significant discursive traction through the media and, as a result, have significantly affected policy and public understanding.[23] Research by McCright and Dunlap focused on this movement and examined how climate contrarians developed competing discourses that disempowered top climate science and reframed climate change science and policy issues with greater uncertainty, thus breeding greater public confusion and doubt. The authors also examined links between contrarians and conservative think tanks, antienvironment movements, and carbon-based industry.[24] Climate contrarians include scientists S. Fred Singer, Sallie Baliunas, Robert Balling, Richard Lindzen, David Legates, Sherwood Idso, Frederick Seitz, and Patrick Michaels. This camp of contrarians—while heterogeneous in some ways—has spoken out stridently against dominant views in climate science over time. By taking advantage of media outlets, these dissenting views have been able to signifi-

cantly shape public perception as well as climate policy considerations (also see chapter 40 in this volume).[25]

Organizational entities often housing climate contrarians have also staged deliberate disinformation campaigns through the media. There have been many revelations of such activities over the last two decades in relation of climate change. For instance, in February 2007 *The Guardian* uncovered an effort by the American Enterprise Institute (AEI)—an ExxonMobil-funded think tank—to undermine the recently released IPCC Working Group I Summary for Policy Makers. Letters were sent by AEI to many scientists and economists offering $10,000 to write articles that would emphasize uncertainties and weaknesses in the IPCC report.[26]

It is important to point out that the funding of these individuals and organizations by carbon-based industry itself is not necessarily a problem. As demonstrated above through the historical links between climate science and military funding, there is a complex history of funding streams and their actual effects. However, ethical concerns arise when such funding comes with implicit or explicit demands, and when inductive scientific inquiry is turned on its head so that the desired result (such as emphasizing uncertainties in the recent IPCC report) drives and structures the process. Naomi Oreskes has pointed out, "the issue is that the research is supported by a sponsor who wants a *particular* result . . . and the researchers know in advance what that outcome is, producing an explicit conflict of interest, which undermines the integrity of the research performed."[27]

Previously, Schneider and Kuntz-Duriseti evaluated ways forward amid the perennial challenges of uncertainty in climate change policy. Focusing on analytical frames and methodological approaches to climate change uncertainty (based on assumptions and research decisions), the authors concluded that—among a number of considerations—there is a vital need for "improved communication of uncertainties" between groups of researchers and policy makers.[28] Furthermore, research by Corbett and Durfee examines issues of controversy and context in climate change news reporting. Through an experiment design of four treatment groups given four different articles on the subject of the Antarctic ice sheet, the authors were able to control for particular elements of context and controversy. They conclude, "The media's attraction to controversy, no matter the source or topic, is unlikely to wane. It is heartening, however, that the simple inclusion of scientific context may help mitigate the uncertainty stirred by scientific controversy."[29] These are just a few among many suggestions and opportunities to improve interactions at this triple-interface.

Conclusion

The many factors, pressures, and processes surveyed above help to explain why climate change science and policy have struggled for fair and accurate attention in the media over time. Through many interlocking factors, the mass media have contributed to a complex and dynamic terrain of ongoing environmental, political, and discursive struggle. Through time, research at the climate science–policy–media triple-interface has demonstrated that understanding the role of the media and improving reporting on climate change science and policy are critical to promoting better international environmental governance on cli-

mate policy, better links between climate science and policy, and improved public understanding of climate change science and policy. William Ruckelshaus—first U.S. Environmental Protection Agency administrator—has said, "If the public isn't adequately informed [about climate change], it's difficult for them to make demands on government, even when it's in their own interest."[30] Research has shown that accurate knowledge of the causes of global climate change is the strongest predictor of a person's stated intentions to act.[31]

Overall, the way that climate change is covered in news media can have far-reaching consequences in terms of ongoing climate scientific inquiry as well as policy makers' and the public's perceptions of climate change. These factors all contribute to the perceived range of possibilities for action. Through media representational practices, people's behaviors can range from being galvanized into action to being resigned to passivity, and our collective future rests on these critical choices.

Notes

1. Nelkin, D., *Selling Science: How the Press Covers Science and Technology* (New York: W.H. Freeman, 1987); Wilson, K., Mass media as sources of global warming knowledge. *Mass Communication Review* 22, no. 1&2 (1995): 75–89.

2. Ungar, S., Knowledge, ignorance, and the popular culture: Climate change versus the ozone hole. *Public Understanding of Science* 9 (2000): 308.

3. This chapter focuses attention primarily on Western media in English-speaking countries; therefore, analyses inevitably can only produce a partial reading of this wide-ranging subject.

4. McChesney, R. W., *Rich Media, Poor Democracy: Communication Politics in Dubious*

Times (Urbana and Chicago: University of Illinois Press, 1999).

5. Lippman, W., *Liberty and the News* (New York: Harcourt, Brace and Howe, 1920).

6. Plass worked at Johns Hopkins University and served as a consultant for the Office of Naval Research as well as Lockheed Aircraft Corporation. Weart, S., *The Discovery of Global Warming* (Cambridge, MA: Harvard University Press, 2003).

7. Fleming, J., *Historical Perspectives on Climate Change* (Oxford University Press, Oxford, UK, 1998).

8. Next great deluge forecast by science, *New York Times*, 1932.

9. Abarbanel, A., and T. McClusky, Is the world getting warmer? *Saturday Evening Post*, July 1, 1950: 22–23, 57, 60–63.

10. Kaempffert, W., Science in review: Warmer climate on Earth may be due to more carbon dioxide in the air. *New York Times*, 1956.

11. Cowen, R., Are men changing the Earth's weather? *Christian Science Monitor* (1957): 13.

12. Shabecoff, P., Global warming has begun, expert tells Senate. *New York Times*, June 24, 1988.

13. Bagdikian, B., *The New Media Monopoly* (Boston: Beacon Press, 2004).

14. Bennett, W. L., *News: The Politics of Illusion*, 5th Ed. (New York: Longman, 2002), p. 10.

15. Bagdikian, 2004 op. cit.

16. Schudson, M., *Discovering the News: A Social History of American Newspapers* (New York: Basic Books, 1978); Schoenfeld, A., R. Meier, and R. Griffin, Constructing a social problem: The press and the environment. *Social Problems* 271 (1979): 38–61.

17. McChesney, 1999 op. cit.; Dunwoody, S., The science writing inner club: A communication link between science and the lay public, in *Scientists and Journalists: Reporting Science as News*, eds. S. Friedman, S. Dunwoody, and C. Rogers (New York: The Free Press, 1986), 155–69.

18. Boykoff, M., and J. Boykoff, Climate change and journalistic norms: A case study of U.S. mass-media coverage. *Geoforum* 38, no. 6 (2007): 1190–1204.

19. Scientific uncertainty can take on multiple characteristics, as described by Wynne's four-part taxonomy: *risk* (knowing the odds), *uncertainty* (don't know odds but know the parameters), *ignorance* (unknown unknowns), and *indeterminacy* (causal chains are open, thus defying prediction).

Wynne, B., Uncertainty and environmental learning: Reconceiving science and policy in the preventive paradigm. *Global Environmental Change* 2, no. 2 (1992): 111–27.

20. Ungar, 2000 op. cit.

21. Pollack, H., Can the media help science? *Skeptic* 102 (2003): 77.

22. Williams, J., The phenomenology of global warming: The role of proposed solutions as competitive factors in the public arenas of discourse. *Human Ecology Review* 72 (2000): 70.

23. Leggett, J., *The Carbon War: Global Warming and the End of the Oil Era* (New York: Routledge, 2001). In terms of environmental issues more broadly, William Freudenberg discusses constructions of "*non*-problematicity." Through embedded power and leveraged legitimacy, "if one person or social group is able to obtain privileged access to valued resources without having other persons or groups challenge that privilege—or perhaps even notice it—so much the better." Freudenberg, W., Social construction and social constrictions: Toward analyzing the social construction of "the naturalized" as well as "the natural," in *Environment and Global Modernity*, eds. G. Spaargaren, A. Mol, and F. H. Buttel (London: Sage, 2000), 106.

24. McCright, A., and R. Dunlap, Defeating Kyoto: The conservative movement's impact on U.S. climate change policy. *Social Problems* 503 (2003): 348–73.

25. See Boykoff, M., Lost in translation? United States television news coverage of anthropogenic climate change, 1995–2004. *Climatic Change* 86, no. 1 (2008): 1–11; Boykoff, M., From convergence to contention: United States mass media representations of anthropogenic climate science. *Transactions of the Institute of British Geographers* 32, no. 4 (2007): 477–89; and Boykoff, M., Flogging a dead norm? Media coverage of anthropogenic climate change in United States and United Kingdom, 2003–2006. *Area* 39, no. 4 (2007): 470–81.

26. Sample, I. Scientists offered cash to dispute climate study. *The Guardian*, February 2, 2007.

27. Oreskes, N., Science and public policy: What's proof got to do with it? *Environmental Science and Policy* 7 (2004): 381.

28. Schneider, S. H., and Kuntz-Duriseti, K., Uncertainty and climate change policy, in *Climate*

Change Policy: A Survey, ed. S. H. Schneider, A. Rosencranz, and J. O. Niles (Washington, DC: Island Press, 2002), 66.

29. Corbett, J., and J. Durfee, Testing public uncertainty of science: Media representations of global warming. *Science Communication* 26 (2004): 142.

30. Ruckelshaus, W., Journalists/Scientists Science Communications and the News Workshop, organizers Anthony Socci and Bud Ward, University of Washington, November 8–10, 2004.

31. Bord, R., R. O'Connor, and A. Fisher, In what sense does the public need to understand global climate change? *Public Understanding of Science* 9 (2000): 205–18.

Chapter 40

Newspaper and Television Coverage

AARON M. McCRIGHT AND RACHAEL L. SHWOM

Introduction

During the last two decades, social scientists have conducted an impressive number of studies on U.S. mass media coverage of climate change. In this chapter, we summarize some of the more robust findings from research on the amount and content of this coverage.[1] We limit our review to newspaper and television coverage for two reasons. First, despite rising numbers of people using the Internet, newspapers and television news remain the two most widely used media sources. Second, most scholars to date have focused their research on television news and especially newspapers. We end this chapter with some brief suggestions for increasing the quality of mass media coverage of climate change to increase public awareness and understanding.

The Amount of Climate Change Coverage

Although some climate scientists had been working to focus public attention on climate change throughout the 1970s and 1980s, mass media attention to climate change was minimal prior to 1988.[2] Then, as seen in figure 40.1, climate change experienced a rush of media attention between the middle of 1989 and early 1990 for several reasons.[3] Concerned scientists and environmental activists effectively connected climate change to more popular issues such as nuclear winter and ozone depletion.[4] Also, the extreme drought during the summer of 1988 led many Americans to feel greater vulnerability to climatic forces.[5] Finally, in his dramatic testimony in front of a Senate committee in June 1988, James Hansen attributed the abnormally hot weather plaguing our nation to global warming, confirming some Americans' fears.[6]

Between 1993 and 1996, media attention decreased to levels lower than the peak coverage from 1990 to 1992 but higher than pre-1988 levels.[7] This is consistent with the expectations of the issue-attention cycle model and the public arenas model.[8] Both perspectives predict that after an initial media explosion, later media coverage will experience short bursts of increased attention corresponding with significant social or political events; yet, these short bursts rarely are expected to ex-

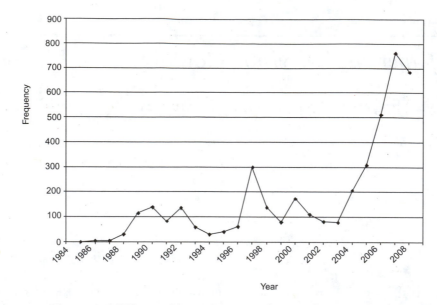

FIGURE 40.1. Frequency of Climate Change News Articles in Top Five Circulating U.S. Newspapers, 1985 to 2006. Data were compiled for the *USA Today, Wall Street Journal, New York Times, Los Angeles Times,* and *Washington Post* for all news articles containing the terms "climate change" or "global warming" in the headline or lead paragraph.

ceed that of the initial peak. However, we see that the case of climate change in major newspapers does not follow this pattern.

The total number of climate change stories for 1997 more than doubled the previous peaks in 1990 and 1992, but has been overshadowed by the unprecedented upsurge in coverage in recent years. We may attribute the 1997 peak almost solely to the events surrounding the December 1997 Kyoto Conference, which established climate change as a major issue in global political arenas. Yet, newspaper coverage of climate change in the six years following Kyoto dropped to levels equivalent to those of the early 1990s. However, beginning in 2004, we have experienced a sizable spike in climate change articles. Unlike the 1997 peak, no one event or story has driven this increased media attention. Indeed, newspaper editors have published climate change news articles in all sections of these major newspapers, on topics including sci-

ence, entertainment, sports, health, economics, and lifestyles. Arguably, climate change appears to have become routinized, whereby editors perceive it as a wide-ranging, newsworthy topic deserving regular attention in and of itself, independent of any new event or new finding.

Some Existing Questions for Future Research

Reporters have been writing news stories about a large range of social, political, economic, and cultural topics related to climate change. It appears as though climate change coverage is becoming more ubiquitous, not just in newspapers but also in popular magazines, radio news, and television news. This raises important questions. What is fueling this trend of increasing media coverage? How long will it continue? Might there eventually be some degree of saturation or burnout among news audiences? Is all media coverage equal, and what impact will this media coverage have on American public opinion and policy making?

What fuels media coverage of climate change continues to be an important question, and existing theoretical perspectives often fall short of a satisfactory explanation. The relationship between meteorological phenomena and climate change coverage in newspapers and television news is far from clear. For instance, an investigation of the relationship between local temperature in New York City and the District of Columbia and the coverage of climate change in the *New York Times* and *Washington Post*, respectively, found no relationship between temperature and coverage in Washington, but did find a modest positive relationship in New York City.[9] Weather was "undoubtedly *not* the most important determinant of attention to climate issues."[10] Instead, political events and the release of scientific reports are more likely to influence the amount of attention climate change receives.[11] Yielding similar results for television media, a study covering the period 1968 to 1996 found no association between U.S. network television coverage of extreme weather events and the amount of climate change coverage.[12] These initial findings about a possible relationship between meteorological events and climate change coverage in newspapers and television news should motivate scholars to systematically investigate the extent to which recent extreme meteorological phenomena (e.g., the 2004 tsunami and the 2004 and 2005 Atlantic hurricane seasons) have had an influence on climate change coverage in the mass media.

Two related factors conspire against a clear explanation of the amount of climate change media coverage since the mid-1990s. First,

the two main theoretical perspectives guiding much of this research (the issue-attention cycle and the public arenas model) say little about the amount of media attention an issue garners after its initial media explosion. Indeed, the sheer magnitude of climate change news articles in 1997 and since 2004 contradicts the expectations of both perspectives, thus demanding substantial theoretical revision.

Second, social scientists have not yet analyzed mass media coverage since the late 1990s as systematically and in the same detail as they did for coverage a decade earlier. This gap is significant since two major IPCC reports have been published in this time period.[13] Greater analysis of mass media coverage during this time will help us examine the reasonable assertion that the publication of scientific reports significantly affects climate change news coverage.[14] Furthermore, the large increase in climate change news coverage in 2006 and early 2007, while Al Gore's *An Inconvenient Truth* was making millions of dollars in theaters and then winning Academy Awards, calls for an analysis of the roles of political elites and/or social celebrities for driving media attention.

Thus, we put a premium upon examining the long-term trends in U.S. mass media coverage of climate change. One promising example of such research indicates that U.S. coverage is characterized by a cyclicity that is tied to specific journalistic cultural practices, a topic to which we turn in the next section.[15]

The Content of Climate Change Coverage

Early news stories on climate change relied heavily on conventional climate scientists as sources. Over time, however, political actors and corporate representatives edged out scientific experts as the dominant sources in these news stories.[16] With this shift in sources around 1991 to 1992, the news media altered its focus from stories about climate science to stories about policy responses.[17] At the same time, opposition to mainstream climate science began to emerge with the growing concern over the economic costs of binding action and the ascent of the George H. W. Bush administration.[18] In general, support for mainstream climate science knowledge claims was greater in news stories than in opinion-editorial articles, where the ideas of a few climate change contrarians flourished.[19]

Important Influences on Media Content

In general, U.S. mass media coverage of climate change since the early 1990s has focused disproportionately upon the uncertainty of climate science knowledge claims, scientific controversy, and the economic costs of binding international action.[20] For example, the New York Times emphasized conflicts between scientists and politicians and potential negative impacts of climate change policy significantly more than did the French newspaper Le Monde.[21] Likewise, the New York Times and Washington Post reported uncertainty about global warming theory in 57 percent and 58 percent of their articles, respectively, while Finland's Hesinging Sanomat and the New Zealand Herald each only highlighted such uncertainty in 9 percent of their climate change articles.[22] In general, a pro-corporate bias often arises in newspaper coverage of climate change in the Christian Science Monitor, New York Times, San Francisco Chronicle, and Washington Post.[23]

Many scholars have tried to explain these trends by highlighting (1) the political economy of American mass media; (2) prevailing journalistic norms in America; (3) the occupational culture within American journalism; and (4) the rise of organized interests attempting to exploit these factors for their own gain.

In explaining the pro-corporate biases and persistence of uncertainty in U.S. news coverage of climate change, some researchers have focused on the broader structure of the media and corporate power in the United States.[24] Since the U.S. economy is much more dependent upon fossil fuels than are the economies of Finland and New Zealand, economic realities influence trends in media coverage: "There is a vested interest on the part of the petrochemical industries to extend the debate and to sow uncertainty regarding the overwhelming scientific consensus regarding global warming. Without such a vested interest, New Zealand and Finland have media that generally follow scientific consensus on the matter."[25]

Second, some scholars have argued that prevailing American journalistic norms facilitate the perpetuation of dominant ideologies and the status quo.[26] Central to this blueprint is the media's "balancing norm," or the equation of "objectivity" with presenting "both sides of the story." Thus, news stories on controversial topics follow a pro-and-con model, where extreme views are contrasted and the reporter concludes by claiming the issue is unresolved—allowing the dramatic narrative to continue but also instilling confusion and passivity in the general public.[27]

Several scholars have expressed concern about how the media's balancing norm in science reporting produces what we refer to above as the "dueling scientists scenario."[28] Reporters solicit statements from scientists holding the most extreme views regarding a scientific issue, even when most scientists hold moderate positions between the extremes and may tend toward a consensus position. This false dichotomy confounds what is widely accepted knowledge, what is a highly speculative claim, and what is a value judgment.[29]

Third, other scholars also have claimed that the occupational dynamics and culture of American journalism further facilitate recent trends in climate change coverage in the U.S. mass media. A few studies have begun to engage news reporters directly to better understand what they know and how they make decisions in the contexts of power and culture described above. Indicating the difficulty in separating out science and politics in climate change, weather forecasters, a group with expectedly high science literacy, had widespread misconceptions about basic climate science, which were connected to the forecaster's values and beliefs about climate change.[30] Those television and newspaper reporters who use scientists as sources and spend the most time reporting environmental issues had the greatest amount of knowledge about climate science and areas of consensus.[31] More than a third of surveyed reporters identified newspapers, which often reported "duels," as the main source of their climate change information; only 20 percent of reporters identified scientists as their primary information source, and only 15 percent relied on science journals.

The practice of using other newspaper reports as a source of information is a troubling one as "food chain" journalism is likely to decrease the accuracy of the story. Wire services play a significant role in this; an explosion of misinformation ripples through the mass media when wire services or news service providers get significant amounts of information in their news stories from climate change contrarians with known ties to the fossil fuels industry.[32]

Fourth, anti-environmental groups, such as the American conservative movement and the fossil fuels industry, have mobilized since the early 1990s to challenge the legitimacy of climate science knowledge claims supporting the assertion that global warming is a real problem.[33] These vested interest groups were successful in capitalizing on the pro-business bias and existing journalistic norms to promote the voices of climate change contrarians.[34] Trends in the use of two groups of scientific sources (five elite climate scientists and five climate change contrarians) in all climate change news articles from 1990 to 1997 in seven of the top circulating U.S. newspapers reveal that the five contrarians achieved approximate parity in citations with some of the most renowned experts in the field. The 1994 Republican takeover of Congress and the concomitant rightward shift in national political culture created opportunities for anti-environmental groups to aggressively manipulate the journalistic balancing norm that produces the dueling scientists scenario.

While some anti-environmental groups have accepted the international scientific consensus on climate change, a surprising number of organizations in the American conservative movement (e.g., Competitive Enterprise Institute) and in the fossil fuels industry (e.g., ExxonMobil) continue to challenge climate science in order to prevent the likely regulation of carbon dioxide emissions.[35] More recently, the increasing amount of mass media coverage devoted to the politicization of climate science, such as NASA's efforts to keep climate scientists from telling the truth about their findings, may increase the public's awareness of how political actors manipulate science to promote their own agenda: "Complaints about the Bush administration's interference with communication of climate science have led to a 'public accountability' frame that has helped move the issue away from uncertainty to political wrongdoing."[36]

Increasing the Quality of Climate Change Coverage

Given that most Americans get their information about climate change from mass media and since most Americans misunderstand climate change, we argue for increasing the quality of mass media coverage of climate change.[37] To this effect, we end with four suggestions for increasing public awareness and understanding of climate change.[38]

Newspapers and television impose great limitations on scientific communication. Television news stories are often less than one minute. Newspaper articles are written for a fifth-grade level of comprehension. At a minimum, all parties responsible for communicating climate change to the general public (henceforth "communicators") should be aware of the nature and extent of these limitations. We may increase this awareness by better educating scientists about journalistic norms and journalists about scientific norms. Moving beyond awareness to providing scientists with the tools they need to best communicate within these constraints, such as how to prepare for an interview, is also an important step.

Second, given that stories about climate change are steeped in scientific details, communicators should convey the scientific consensus and limitations to current knowledge according more to scientific norms of evidence rather than to journalistic norms of "balance." Whenever possible, communicators should help increase the scientific literacy of their mass audience by explaining how scientists become more confident about knowledge claims, especially regarding the use of probability statements. Furthermore, communicators should clarify that just because the implications of scientific findings may be controversial with some groups in society, this does not mean that the actual scientific theories, methods, and bodies of evidence are controversial within the scientific community.

Third, communicators should consistently expose the motivations, strategies, and goals of the climate change contrarians who lend pseudoscientific legitimacy to attempts at obfuscating scientific communication for the narrow material and ideological interests of fossil fuels organizations and conservative think tanks.[39] In September 2006, Britain's leading scientific association, the Royal Society, disclosed its findings that ExxonMobil had funded thirty-nine organizations that misrepresent the consensus on climate change and asked ExxonMobil to stop these practices.[40] In January 2007, the Union of Concerned Scientists in the United States followed suit by blowing the whistle on ExxonMobil for the $16 million it provided between 1998 and 2005 to forty-three ideological and advocacy organizations to mislead the U.S. public by discrediting the science behind global warming.[41]

Finally, communicators should acknowledge that the crux of the climate change debate at this time is a conflict over values, not science. By putting the conflicting values directly in the public eye, communicators may more honestly discuss the larger political, cultural, social, and economic contexts of climate change. Along these lines, communicators should highlight what different individuals and groups (e.g., organizations, cities, and states) are doing in response to climate change. Communicators could use varied cases embodying different values to promote an insightful national discussion of the political, cultural, social, and economic contexts of climate change.

Notes

The sections on "The Amount of Climate Change Coverage" and "The Content of Climate Change Coverage" draw upon a previous work by one of the authors: McCright, A. M., and R. E. Dunlap. 2000. Challenging global warming as a social problem: An analysis of the conservative movement's counter claims. *Social Problems* 47, 499–522.

1. In addition to the studies we cite throughout this chapter, the following are also valuable works on the amount and content of media coverage: Bell, A. 1994a. Climate of opinion: Public and media discourse on the global environment. *Discourse and Society* 5 (1), 33–64; Bell, A. 1994b. Media (mis)communication on the science of climate change. *Public Understanding of Science* 3, 259–75; Boykoff, M. T., and J. M. Boykoff. 2004. Balance as bias. *Global Environmental Change* 14, 125–36; Boykoff, M. T., and J. M. Boykoff. 2007. Climate change and journalistic norms: A case study of U.S. mass media coverage. *Geoforum*, forthcoming; Hansen, A. 1991. The media and the social construction of the environment. *Media, Culture, and Society* 13, 443–58; Liftin, K. T. 1995. Framing science: Precautionary discourse and the ozone treaties. *Millennium: Journal of International Studies* 24 (2), 251–77; Mazur, A. 1998. Global environmental change in the news: 1987–1990 versus 1992–1996. *International Sociology* 13, 457–72; Newell, P. 2000. *Climate for Change: Non-State Actors and the Global Politics of the Greenhouse*. Cambridge: Cambridge University Press; Ungar, S. 1992. The rise and (relative) decline of global warming as a social problem. *The Sociological Quarterly* 33, 483–501; Ungar, S. 1995. Social scares and global warming: Beyond the Rio convention. *Society and Natural Resources* 8, 443–56; Ungar, S. 1998. Bringing the issue back in: Comparing the marketability of the ozone hole

and global warming. *Social Problems* 45, 510–27; Wilkins, L., and P. Patterson. 1991. Science as symbol: The media chills the greenhouse effect. In L. Wilkins and P. Patterson, eds., *Risky Business: Communicating Issues of Science, Risk, and Public Policy* (pp. 159–76). Westport, CT: Greenwood.

2. Miller, M., J. Boone, and D. Fowler. 1990. The emergence of the greenhouse effect on the issue agenda: A news stream analysis. *News Computing Journal* 7, 25–38; Mazur, A., and J. Lee. 1993. Sounding the global alarm: Environmental issues in the U.S. national news. *Social Studies of Science* 23, 681–720.

3. Trumbo, C. 1995. Longitudinal modeling of public issues: An application of the agenda-setting process to the issue of global warming. *Journalism and Mass Communication Monographs*. J. Soloski, ed., no. 152; Williams, J., and R. S. Frey. 1997. The changing status of global warming as a social problem: Competing factors in two public arenas. *Research in Community Sociology* 7, 279–99; McComas, K., and J. Shanahan. 1999. Telling stories about global climate change: Measuring the impact of narratives on issue cycles. *Communication Research* 26, 30–57.

4. Mazur and Lee, 1993 op. cit.; Williams and Frey, 1997 op. cit.

5. Ungar, 1992 op. cit.; Mazur and Lee, 1993 op. cit.

6. Miller, Boone, and Fowler, 1990 op. cit.; Mazur and Lee, 1993 op. cit.; Trumbo, 1995 op. cit.

7. Ungar, 1992 op. cit.; Williams and Frey, 1997 op. cit.

8. Downs, A. 1972. Up and down with ecology: The "issue-attention" cycle. *Public Interest* 28, 38–50; Hilgartner, S., and C. Bosk. 1988. The rise and fall of social problems: A public arenas model. *American Journal of Sociology* 94, 53–78.

9. Shanahan, J., and J. Good. 2000. Heat and hot air: Influence of local temperature on journalists' coverage of global warming. *Public Understanding of Science* 9, 285–95.

10. Ibid, p. 293.

11. Ibid.

12. Ungar, S. 1999. Is strange weather in the air? A study of U.S. national network news coverage of extreme weather events. *Climatic Change* 41, 133–50.

13. Intergovernmental Panel on Climate Change. 2001. *IPCC Third Assessment Report:*

Climatic Change 2001—Synthesis Report. Geneva: IPCC; Intergovernmental Panel on Climate Change. 2007. *IPCC Fourth Assessment Report: Climatic Change 2007—Synthesis Report.* Geneva: IPCC.

14. Shanahan and Good, 2000 op. cit.

15. Brossard, D., J. Shanahan, and K. McComas. 2004. Are issue-cycles culturally constructed? A comparison of French and American coverage of global climate change. *Mass Communication and Society* 7, 359–77.

16. Miller, Boone, Fowler, 1990 op. cit.; Lichter, S. R., and L. S. Lichter. 1992. The great greenhouse debate: Media coverage and expert opinion on global warming. *Media Monitor* 6 (10), 1–6; Wilkins, L. 1993. Between facts and values: Print media coverage of the greenhouse effect, 1987–1990. *Public Understanding of Science* 2, 71–84; Trumbo, C. 1996. Constructing climate change: Claims and frames in U.S. news coverage of an environmental issue. *Public Understanding of Science* 5, 269–83.

17. Lichter and Lichter, 1992 op. cit.; Trumbo, 1995 op. cit.

18. Mazur and Lee, 1993 op. cit.; Williams and Frey, 1997 op. cit.

19. Wilkins, 1993 op. cit.

20. Zehr, S. C. 2000. Public representations of scientific uncertainty about global climate change. *Public Understanding of Science* 9, 85–103.

21. Brossard, Shanahan, and McComas, 2004 op. cit.

22. Dispensa, J. M., and R. J. Brulle. 2003. Media's social construction of environmental issues: Focus on global warming—A comparative study. *The International Journal of Sociology and Social Policy* 23, 74–105.

23. Nissani, M. 1999. Media coverage of the greenhouse effect. *Population and Environment* 21, 27–43.

24. Dispensa and Brulle, 2003 op. cit.; Nissani, 1999 op. cit.

25. Dispensa and Brulle, 2003 op. cit., p. 98.

26. Gitlin, T. 1980. *The Whole World Is Watching.* Berkeley: University of California Press; McCright, A. M., and R. E. Dunlap. 2003. Defeating Kyoto: The conservative movement's impact on U.S. climate change policy. *Social Problems* 50, 348–73.

27. Epstein, E. J. 1973. *News from Nowhere: Television and the News.* New York: Random House.

28. Freudenburg, W. R., and F. H. Buttel. 1999. Expert and popular opinion regarding climate change in the United States. In R. Coppock, E. McGarraugh, S. Rushton, and R. D. Tuch, eds., *Climate Change Policy in Germany and the United States* (pp. 49–57). Washington, DC: German-American Academic Council Foundation; McCright and Dunlap, 2003 op. cit.; Schneider, S. H. 1993. Degrees of certainty. *Research and Exploration* 9 (2), 173–90.

29. Ibid.

30. Wilson, K. M. 2002. Forecasting the future: How television weathercaster's attitudes and beliefs about climate change affect their cognitive knowledge of the science. *Science Communication* 24, 246–68.

31. Wilson, K. M. 2000. Drought, debate, and uncertainty: Measuring reporters' knowledge and ignorance about climate change. *Public Understanding of Science* 9, 1–13.

32. Antilla, L. 2005. Climate of scepticism: U.S. newspaper coverage of the science of climate change. *Global Environmental Change* 15, 338–52.

33. McCright, A. M., and R. E. Dunlap. 2000. Challenging global warming as a social problem: An analysis of the conservative movement's counter claims. *Social Problems* 47, 499–522; McCright and Dunlap, 2003 op. cit.; Gelbspan, R. 1997. *The Heat Is On.* Reading, MA: Addison-Wesley Publishing.

34. McCright and Dunlap, 2003 op. cit.

35. Levy, D. L., and D. Egan. 1998. Capital contests: National and transnational channels of corporate influence on the climate change negotiations. *Politics and Society* 26, 337–61; Newell, 2000 op. cit.

36. Nisbet, M. C., and C. Mooney. 2007. Framing science. *Science* 316, 56.

37. Wilson, K. M. 1995. Mass media as sources of global warming knowledge. *Mass Communication Review* 22, 75–89; Sussman, G., B. W. Daynes, and J. P. West. 2002. *American Politics and the Environment.* New York: Addison Wesley Longman; Bostrom, A., M. G. Morgan, B. Fischhoff, and D. Read. 1994. What do people know about global climate change? 1. Mental models. *Risk Analysis* 14, 959–70; Dunlap, R. E. 1998. Lay perceptions of global risk: Public views of global warming in cross-national context. *International Sociology* 13, 474–93; Kempton, W. 1991. Public understanding of global warming. *Society and Nat-*

ural Resources 4, 331–35; Read, D., A. Bostrom, M. G. Morgan, B. Fischoff, and T. Smuts. 1994. What do people know about global climate change? 2. Survey studies of educated laypeople. *Risk Analysis* 14, 971–82.

38. Space limitations prevent us from elaborating much on these suggestions. Interested readers may find the following sources full of helpful suggestions for improving the communication between scientists and the general public (often through the mass media): McCright, A. M. 2006. Dealing with climate change contrarians. In S. Moser and L. Dilling, eds., *Creating a Climate for Change: Communicating Climate Change—Facilitating Social Change* (pp. 200–212). New York: Cambridge University Press; Montgomery, S. T. 2003. *The Chicago Guide to Communicating Science.* Chicago: University of Chicago Press; Moser, S. C., and L. Dilling. 2004. Making climate hot: Communicating the urgency and challenge of global climate change. *Environment* 46, 32–43; Moser, S. C., and L. Dilling, eds. 2006. *Creating a Climate for Change: Communicating Climate*

Change—Facilitating Social Change (New York: Cambridge University Press).

39. This is important since Krosnick and colleagues (2006) recently found that exposure to climate change contrarians tended to decrease acceptance that global warming exists, at least among the lesser educated. Krosnick, J. A., A. L. Holbrook, L. Lowe, and P. S. Visser. 2006. The origins and consequences of democratic citizens' policy agendas: A study of popular concern about global warming. *Climatic Change* 77, 1, 7–43.

40. The Royal Society. 2006. Letter from Bob Ward, senior manager for policy communication at the Royal Society, to Nick Thomas, director of corporate affairs in UK public affairs at ExxonMobil. September 4. Available online at www.royalsoc.ac .uk/document.asp?tip=1andid=5851. Last accessed on April 22, 2007.

41. Union of Concerned Scientists. 2007. *Smoke, Mirrors, and Hot Air: How ExxonMobil Uses Big Tobacco's Tactics to Manufacture Uncertainty on Climate Science.* Cambridge, MA: Union of Concerned Scientists.

Chapter 41

Media and Public Education

DALE WILLMAN

It was a typical March Madness kickoff party with friends of friends gathering at a local bar. I found myself among a group of well-educated managers with jobs that allowed them to escape the office on a mid-week afternoon to drink beer and watch college basketball. Seated next to me was someone I didn't know, and when he discovered I was an environmental journalist he quickly launched into something about which he had obviously given a great deal of thought. He declared that the theory about how human contributions to greenhouse gases were adding to global warming was all wrong. "Oceans produce more carbon dioxide than humans do anyway," he said. "When the water evaporates, it breaks up into carbon and oxygen, and that's where the problem really comes from."

It was a long afternoon.

This is just one of a long list of anecdotes that show a lack of understanding on the part of Americans when it comes to climate change and most other topics involving science.

One indication of this comes from an annual survey conducted for several years by the National Environmental Education and Training Foundation (NEETF). A sampling of Americans was questioned on their knowledge of basic environmental issues. In the seventh such poll, 2,000 people were asked ten questions in an effort to determine their level of environmental knowledge.[1] The respondents averaged just 2.2 correct answers. Random guessing would have produced an average of 2.5. Even those respondents with college degrees averaged just 3.1 correct answers. These results were not an aberration. The findings were similar to those from six previous surveys. In fact, the results from year to year were so consistent that NEETF stopped conducting that portion of the survey on an annual basis.

Many other surveys over the years have found similar results. When it comes to science, most Americans are poorly educated.

The Failure of the Media in Covering Climate Change

After leaving school, the public turns to traditional mass media for most of its knowledge about science, and that knowledge comes primarily from broadcast television or radio.[2] Yet scientific topics receive scant media coverage. Content analysis during the last fifteen years shows that most broadcast television news outlets devote less than 2 percent of their total news coverage to science issues.[3]

When journalists do cover science issues the stories are overwhelmingly poorly produced. Both print and broadcast media generally provide just crisis coverage—simplistic, superficial reporting looking at the more sensational aspects of news where a great deal of conflict exists but where there is little substance.[4] Evidence of this is found in a study comparing coverage of climate issues by U.S. and French media outlets. It concluded that "coverage in the United States emphasized conflicts between scientists and conflicts between politicians and tended to focus on domestic politics."[5]

The lack of media attention for science-related topics and the level of crisis reporting given to those few science topics that do receive coverage do not just expose a journalism industry in trouble. More importantly, this can seriously hurt society. Arguably the best current example of how is seen in the coverage of climate change.

Despite the historic portrayal of climate science in the media, there has been clear scientific consensus on this issue for a number of years. Yet even today, after more than twenty years of coverage on the anthropogenic causes of climate change, most journalists still lack an understanding of the issue and fall back on what most think is "balanced" reporting. But this kind of lazy drive-by journalism diminishes the preponderance of scientific evidence and hurts public understanding of an issue critical to our future. Award-winning journalist Ross Gelbspan says, "If the public relations specialists of the oil and coal industries are criminals against humanity (for their

efforts in muddying this issue) the U.S. press has basically played the role of unwitting accomplice by consistently minimizing this story, if not burying it from public view altogether."[6]

It would seem that journalists ought to be able to examine their reporting practices in an effort to change the systemic problems that lead to such coverage in the first place. But that is not likely to happen.

The coverage of tobacco provides a case in point. Scientific studies began documenting the role of tobacco in certain cancers, as well as other health issues related to smoking, as early as 1950.[7] By the mid-1950s the tobacco industry had hired a public relations firm and created the Tobacco Industry Research Council. The council began a campaign to create "doubt about the health charge without actually denying it, and advocating the public's right to smoke, without actually urging them to take up the practice."[8] That campaign was used to create confusion in the public's mind about health concerns from smoking. That confusion remained for some people for decades and was exacerbated by media coverage of the issue.

A desire for "balance" led reporters to regularly quote from the questionable research, even when there might have been reason to doubt its veracity. The need for that "opposite" viewpoint was much stronger than a desire for clarity.

Similar stories can be told about media coverage of other issues, from Rachel Carson's book *Silent Spring* in the 1960s to endocrine disruption in the 1990s, where the same techniques were used with similar outcomes.[9] Or, as one scientist puts it, "The vilification of threatening research as 'junk science' and the corresponding sanctification of industry-commissioned research as 'sound science' has become nothing less than standard operating procedure in some parts of corporate America."[10]

So why have the media (with a few notable exceptions) been fooled again? Journalists seem unable to learn from the past.

One obvious reason for this pattern is that the convention of balance seems to actually work with other subjects, according to Seth Borenstein, who covers science for the Associated Press.[11] "The pro and con balance works in politics, and government stories and policy stories." As an example he cites the Bush administration position on climate science. "There is nothing inherently wrong with the administration's decision to do nothing. That's a legitimate decision that you may or may not agree with." However, saying climate change isn't real, he says, is a different story. And it's here, in the coverage of complex science issues, that the use of balance falls down.

So why is balance used to cover science? One reason is the reality that journalists and editors are little different from the general public when it comes to science knowledge. In my own research, I surveyed television news gatekeepers—producers and news directors who determine what is reported—to ascertain their knowledge of environmental issues.[12] When asked basic knowledge questions about science and the environment, the gatekeepers scored little better than the general public. A large minority even answered incorrectly when asked what is known in surveying circles as the Flintstone Question: "Did humans and dinosaurs coexist?"

Borenstein says a lack of understanding when it comes to complex science issues makes it too easy for a journalist to be spun by just about anyone promoting a slanted point of view. "You have reporters who are not expert in the field who are trained for the ulti-

mate in balance," and without the knowledge needed to adequately cover an issue like climate change, he says, those reporters fall back on their training and rely heavily on the technique of balance. "But when it comes to science like this [climate] it's a false dichotomy here, it's just a false sense of reality." In other words, a lack of science knowledge drives many journalists to rely on a false balance, assuming that by giving two sides to the issue they are providing adequate coverage.

There are, of course, exceptions. A few journalists are well-versed enough to cover climate issues with some authority, such as Andrew Revkin of the *New York Times*. Considered the dean of climate reporting by many of his peers, Revkin is in an enviable position. He works for a prestigious outlet that has an immense news hole for science coverage, at least compared to other media. Yet even Revkin has difficulty getting front page coverage for climate issues.[13]

In the case of the *Times*, Revkin says a major impediment is simply "the disconnect between the shape of the climate story and the traditional notion of what is a news story." He says the culture of a newsroom is built around a twentieth-century template that involves a very specific language, "which is the language and vocabulary of politics, commerce, human events, not of global climatologists, ice physics, things like that." And that, he says, makes climate stories a tough sell. "Outside of the Science section it's always a struggle."

That language and vocabulary of journalism come in large part from how people are promoted at media outlets. Revkin says that "easily 90 percent [of executive and managing editors] cut their personal teeth as political reporters or foreign correspondents." Rarely have these editors dealt with science or the environment on their way up the corporate ladder. That makes it difficult for them to understand whether the reporter has done an adequate job of communicating a complex issue. And it makes it even more difficult for them to understand the value of such a story in the first place.

The complexity of science issues may be the biggest problem of all. As Revkin points out, "Essentially the big stories of the twenty-first century . . . are laden with uncertainty; they have this long time scale." Issues such as the loss of biodiversity, the depletion of the oceans, climate change—these issues are driving a lot of instability around the planet, yet because of their subtlety and complexity it is hard to get the interest of editors. Environmental journalists are fond of saying that environmental stories don't break, they ooze. Perhaps that description best fits this problem.

The situation for broadcasters is similar. For reporter Jeff Burnside of television station WTVJ, the NBC affiliate in Miami, climate is "a very difficult story to sell, perhaps one of the most difficult of all time."[14] One reason, he says, is its overwhelming scale. "The challenge of the reporter is convincing managers that there are tangible, local examples of climate change." Television is obviously a visual medium, and many television managers feel climate stories are often a tough story to get on camera. "How do you visualize climate change?" asks Burnside.

Today Burnside has bosses who let him cover climate issues, but he says it took him three years of convincing to let him do climate stories. The shift came only once the issue began showing up on the front cover of several national magazines. "Editors are less willing to be at the vanguard of any given story, and more likely to follow the lead of national print publications." That, says Burnside, is a prob-

lem. "I think there should be constant coverage of this issue. It's not just a global story, it's a hyper local story and good reporters can make it happen. But the big challenge is convincing others" that the story is significant.

According to a number of scientists and reporters asked about this issue, perhaps the greatest failure in climate coverage has come from public radio. Many people believe that public radio covers science, and news in general, better than any other U.S. medium. But when it comes to climate, until recently National Public Radio (NPR), the major news provider in public radio, hasn't appeared to have been doing a good job. While researching this chapter, I conducted a search of NPR's Web site using the phrase "climate change" and turned up twenty-two stories in a one-month period. Of those twenty-two, most were either an analysis of government policy or reports on reports—stories about journal studies. Only one of the twenty-two stories actually involved a reporter out in the field producing a sound-rich, contextual story about a climate issue. Yet it is the more fully produced pieces that are most capable of getting at the nuances of this topic, by including additional voices and perspectives, thus allowing a fuller examination of the topic.

Since that analysis, NPR has received a large amount of money for climate stories, and its coverage has shown a marked increase in both quantity and quality. But it took an infiltration of cash and a partnership with *National Geographic* magazine to push NPR's climate coverage beyond the realm of journal articles.

More extensive examinations conducted by others have found results similar to mine. While not a scientific sampling of NPR stories, these examinations point to a major problem with NPR's historic climate coverage, according to one freelance reporter. Because she continues to do some work for NPR, she asks that her name not be used, so she'll be called Carol.[15] Carol says you'll find much of NPR's science coverage based on publications released during the previous week, primarily the journals *Science* and *Nature*. "The problem is," she says, "that's not very compelling." More important, because so much of the climate coverage is focused on competing study results, it becomes a disservice to the listener because "it's kind of like the tennis match of science, and that can be very confusing."

Journal studies are what *New York Times* reporter Andy Revkin calls "the period at the end of the sentence." While media sources should inform the public of the results of important studies, focusing solely on study results means much of the work's context will be missing—the rest of the sentence is being ignored. And that means rather than being able to see the broader view of scientific research, the listener bounces from one latest result to the next, like a spectator watching the ball go back and forth at a tennis match. So while study results should be part of a news story, they should not be the story's entire focus.

Changing News Climate

Today, the landscape for climate coverage is drastically different. After twenty years of poor coverage, Americans are suddenly receiving a steady diet of climate stories. What has changed? The answer can be expressed in two words—Al Gore. Gore and what he refers to as his "little slideshow," the documentary *An Inconvenient Truth*, have done more to affect change in climate coverage than twenty years of research and public hearings ever could.[16] Attention to Gore's documentary has led to

extended coverage of climate change not only in the mainstream media, but also in other arenas. For example, climate has made it to the cover of *Vanity Fair* and even *Sports Illustrated*, with each magazine, neither of which is noted for its science coverage, providing its unique take on the topic. Even Fox News has produced a major news special on climate change.

There is a theory among many social scientists known as "agenda setting."[17] This theory states that the media do not tell the public what to think, but by deciding what topics receive coverage they tell the public what to think about. Now that climate issues are receiving more media attention, the general public is indeed talking much more about climate than ever before, and this increased coverage has begun to make a difference in public awareness of climate issues.

Awareness, however, is not the same as understanding. Awareness can occur relatively quickly, but understanding takes much more time, especially when such a complex issue is involved. And understanding requires not just more coverage, but more insightful, intelligent coverage as well.

What Can Be Done

While the impediments to accurate reporting on climate change issues remain great, scientists and policy makers can still play a role in making certain these issues are covered with depth and context. My own research offers some insights on what that role might be. Based on the answers of those television news gatekeepers I surveyed, I developed a multidimensional construct combining both a qualitative and a quantitative measure of the scientific knowledge of those gatekeepers. I then compared the construct to the gatekeeper's view of his or her own station's coverage of environmental news and found a clear relationship. Simply put, the higher the general scientific knowledge of the gatekeeper, the greater the environmental coverage on his or her station.

I also made a second discovery. The survey showed that those gatekeepers who perceived a high public interest in environmental news were more likely to assign a reporter to cover environmental stories. This results in more such stories being broadcast weekly than on those stations whose gatekeepers did not perceive such a high interest. More specifically, having a reporter assigned to cover environmental issues was positively correlated with the number of such stories aired each week. So there is a mediating connection between a gatekeeper's awareness of public interest and the number of stories aired, run through the assigning of an environmental reporter.

These results focused on environmental issues. However, it is likely that similar results might be found with other science topics. This means that the amount of coverage given to science issues, as well as the way in which those issues are covered, might be increased in two ways—first, by increasing the knowledge of news gatekeepers, and second, by increasing the perception of gatekeepers that the public is interested in these topics. Policy makers and scientists can help to make both of these things happen by engaging the media.

Several years ago, journalists and scientists met in a series of workshops sponsored by the National Science Foundation. They discussed how to better communicate science information to the general public. At each of these meetings at least some of the scientists admitted they have never called a reporter or editor, even when they read or heard something egregiously wrong. Scientists have spent years gathering specialized knowledge that can be used to educate gatekeepers. With that knowledge comes a responsibility to share it with the public.

It's just as important for policy makers to engage the media. It's the policy makers who are on the front lines of climate change. Their communities and states are seeing the effects of climate change on their access to sufficient water supplies, their land-use policies, and even regional health care issues. Without an informed constituency it will be difficult for policy makers to make the hard decisions needed to ensure that their region is prepared for what comes tomorrow.

What scientists and policy makers can do to engage the media:

1. Call reporters and editors when you find a factual error.

2. Congratulate reporters and editors on exemplary coverage.

3. Cultivate relationships with gatekeepers and reporters and become a trusted source of information.

4. Contact as many news sources as possible, especially regional and local outlets.

5. Repeat your message to achieve high public awareness.

6. Request greater coverage from news gatekeepers.

7. Identify local connections to climate change.

8. Encourage your colleagues to engage the media.

Andy Revkin of the *New York Times* talks about one example of how media engagement can work. Several years ago, the executive editor of his newspaper was approached by the dean of the forestry school at Yale, who tutored the editor on why climate coverage is important. That contact, says Revkin, led to the commissioning of a multipart series on climate change, "which shows . . . that if [scientists] can get in close to gatekeepers in the media, I think that can make a difference."

Until that happens in earnest, I continue to worry about the state of science journalism. On good days, I'm afraid nothing will change. In my more cynical moments, I'm certain nothing will. But then I bump into someone in a bar, as I did during that March Madness week. After listening to his "theory" on natural forcing of climate change by evaporating seawater, I patiently discuss the science behind what he is saying. By the time I leave, his mind is open to the possibility of human contributions to greenhouse gases being a significant problem.

Now suppose that person is a reporter, or perhaps a television news producer. Only then can we begin to understand the ability of both policy makers and scientists to make a difference in how the general public perceives science issues in general, and climate change in particular.

Notes

1. NEETF and Roper Starch Worldwide, *The National Report Card on Environmental Knowledge, Attitudes, and Behaviors*, December 1998.

2. Nelkin, D., *Selling Science: How the Press Covers Science and Technology*, New York: W.H. Freeman, 1987; Radio and Television News Director's Foundation, *Americans Rely on Local Television News, Rate It Highly, Consider It Fair*, 1998; National Science Board, *Science and Engineering Indicators 2006*. Arlington, VA: National Science Foundation.

3. Lichter, R., *Assessing Local Television News Coverage of Health Issues*. Washington, DC: Cen-

ter for Media and Public Affairs, 1998; Tyndall, A., *The Tyndall Report*, New York, 1997.

4. Klite, P., *1998 National Survey: Not in the Public Interest*, Rocky Mountain Media Watch, 1998; Lichter, 1998 op. cit.; Project for Excellence in Journalism, *Changing Definitions of News: A Look at the Mainstream Press over 20 Years*, March 6, 1998.

5. Brossard, D., Shanahan, J., and McComas, K., Are issue-cycles culturally constructed? A comparison of French and American coverage of global climate change. *Mass Communication and Society* 7, no. 3 (2004): 359–77.

6. Gelbspan, R. *Boiling Point: How Politicians, Big Oil and Coal, Journalists, and Activists Have Fueled the Climate Crisis—And What We Can Do to Avert Disaster*. New York: Basic Books, 2004.

7. Doll, R., and Hill, A. B., Smoking and carcinoma of the lung. *British Medical Journal* 30 (1950): 739; Levin, M., Goldstein, H., and Gerhardt, P. Cancer and tobacco smoking: A preliminary report. *Journal of the American Medical Association* 143 (1950): 336–38; Wynder, E. L., and Graham, E. A., Tobacco smoking as a possible etiologic factor in bronchiogeneic carcinoma. *Journal of the American Medical Association* 143 (1950): 329–36.

8. Stuart, E., *Captains of Consciousness: Advertising and the Social Roots of Consumer Culture*. New York: McGraw-Hill, 1976, p. 50.

9. Carson, R. *Silent Spring*. Boston: Houghton Mifflin, 1962.

10. Michaels, D., Doubt is their product. *Scientific American* 292, no. 6 (June 2005).

11. Author's interview, November 8, 2006.

12. Willman, D. "Commercial Television News Gatekeepers and Their Attitudes Toward Environmental News: A Survey." Antioch University, 2001.

13. Author's interview, October 28, 2006.

14. Author's interview, November 16, 2006.

15. Author's interview, November 17, 2006.

16. See the official Web site for *An Inconvenient Truth* at www.climatecrisis.net.

17. Cohen, B. C., *The Press and Foreign Policy*, Princeton: Princeton University Press, 1963, p. 13.

Mitigation Options to Reduce
Carbon Emissions

Chapter 42

The Road Forward

Robert T. Watson and André R. Aquino

Introduction

Climate change is a complex global problem directly involving sectors that are the backbone of the economy, including energy, agriculture, transportation, industry, and construction. Mitigating climate change will take a concerted global effort to design and implement new policies and to develop and adopt new technologies aimed at reducing greenhouse gas (GHG) emissions so that the costly or even irreversible consequences for human well-being may be avoided. Adapting to climate change is also essential to human welfare. The capacity to adapt, however, varies substantially across countries. Those countries likely to suffer most from the consequences of climate change will be least able to adapt, due to their limited technical and financial capacity.

Mitigation of GHG Emissions

Mitigation of GHG emissions must be seen as prudent investment. They are costs incurred now to avoid future adverse consequences.

Significant reductions in net GHG emissions are technically feasible due to an existing array of technologies and practices in energy supply and demand, waste and land management, and industrial sectors. All economic sectors have to be involved in climate change mitigation since GHG emissions are pervasive across economic activities. Realizing this technical potential, however, will involve the design and implementation of enabling policies to overcome barriers to the diffusion of these technologies into the marketplace; increased funding for research and development (R&D); and effective technology transfer across countries.

Technologies

Reducing GHG emissions will entail the adoption of a broad portfolio of technologies by various economic sectors, such as the energy production and supply sector; transportation and building; and land management. In the energy sector, technologies aim at (1) increasing energy efficiency; (2) further developing and adopting renewable energies; and

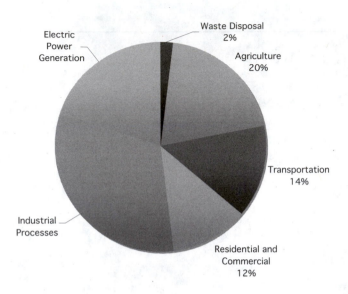

FIGURE 42.1. Sources of anthropogenic GHG emissions.

(3) capturing and storing carbon. Each of these technologies will be explored in detail in the following chapters.

ENERGY PRODUCTION AND SUPPLY

1. *Energy Efficiency.* Improved efficiency in energy use offers one of the greatest opportunities to address energy security, price, and environmental concerns. Nevertheless, if energy efficiency measures are not coupled with other policies to reduce energy consumption, such as higher energy prices, they may work as a perverse incentive by decreasing energy prices and increasing consumption. Among the energy efficiency supply technologies are:

—Reduced energy transmission and distribution losses; and

—Increased power plant efficiency, such as subcritical to supercritical thermal power plants.

2. *Renewable Energy.* This still must overcome a series of technological, economic, and political obstacles before being a large-scale substitute for fossil fuel sources of energy. Nevertheless, it has increasingly drawn the attention of public and private investors in previous years: the annual production of solar cells jumped 45 percent in 2005, while the construction of new wind farms, especially in Europe, has boosted the worldwide generating capacity of wind power tenfold in the past decade.[1] Technologies include:

—Biomass, solar, wind, and large hydropower; and

—Nuclear power.[2]

3. *Carbon Capture and Storage (CCS).* CCS technologies aim to prevent CO_2 from reaching the atmosphere by separating out much of the CO_2 that is generated during coal conversion and transporting it to sites where it

can be stored deep underground, mainly in depleted oil or gas fields or in saline formations. CCS brings two important risks: sudden escape of CO_2 and gradual leakage. Combining the newer integrated gasification combined-cycle (IGCC) plants, in which coal is not burned but oxidized, with CCS could substantially decrease carbon emissions into the atmosphere.

Transportation and Building

To complement low-carbon energy supply technologies, there are numerous technologies that can improve the efficient end-use of energy in the transportation, buildings, and industrial sectors, and in urban management.

- *Transportation*. Efficient gasoline/diesel engines, alternative fuels such as biofuels, urban planning (smart growth), urban mass transport systems, modal shifts to inter- and intra-city rail, and water transport;
- *Buildings*. Insulation, advanced windows, new lighting technology, passive solar, efficient space cooling and heating, water heating, refrigeration, and other appliances;
- *Industry*. Cogeneration, waste heat recovery, pre-heating, new efficient process technologies, efficient motors/drives and improved control systems, incineration of waste gases; and
- *Municipalities/Urban Local Bodies*. District heating systems, cogeneration, efficient street lighting, and efficient water pumping and sewage systems.

Land Management

Improved land use may not only mitigate GHG emissions but also contribute to carbon storage. In agriculture, efficient irrigation pumps and land management techniques, such as no-till agriculture, may prevent carbon from reaching the atmosphere. In forestry, afforestation, reforestation, and agroforestry provide a wide range of opportunities to increase carbon uptake, while deforestation releases GHGs and reduces carbon sinks. Land use, land-use change, and forestry activities have the potential to sequester up to 1 to 2 gigatons of carbon per year during the next fifty years, which is equivalent to 10 to 20 percent of projected fossil fuel emissions over the same period.

Policies

Many energy-efficient and renewable energy technologies have not been adopted widely for a variety of reasons, including insufficient information, poor pricing policies, transaction costs, and institutional constraints, both regulatory and legal. Significant restructuring of the energy system will require energy-sector reform, including both command and control and market-based policies. While the former entails mostly standards and regulations, the latter are schemes that try to make the price of GHG-intensive fuels reflect their true social costs. In addition to these, funding for R&D and education and awareness-raising policies must be undertaken.

Regulations and Standards

These kinds of policies entail setting a standard, such as the maximum level of permissible emissions, and then monitoring and enforcing the standard. Emissions standards can be performance-based standards or technology-based standards. Performance-based standards stipulate emissions limits that each firm is expected to obey. Technology-based standards specify not only emissions limits, but also the "best" technology that must be used. The Corporate Average Fuel Economy (CAFE) regulations in the United States, aimed at improving the average fuel economy of cars and light trucks in the country, is an example of such a policy.

Establishing quantifiable and enforceable restrictions on total GHG emissions should be coupled with market-based mechanisms (discussed in the next subsection) to achieve reductions in a cost-efficient way. Other kinds of regulation include renewable portfolio standards, which require electricity providers to obtain a minimum percentage of their power from renewable energy resources by a certain date. Germany, for instance, is planning to generate 20 percent of its electricity with renewable sources by 2020, while Sweden intends to give up fossil fuels entirely. A similar approach could be used to promote CCS: electricity producers should be expected to include a growing fraction of decarbonized coal power in their supply portfolios. Appliances and industry energy efficiency and fuel-economy standards follow the same logic.

Regulatory policies may also mandate producers to provide information on the energy consumption of their products so that consumers can make more informed decisions. Labeling, such as the U.S. Energy Star program, is an example of this kind of policy. Industries may also be required to disclose their total energy consumption, and the government may establish mandatory energy audits.

Finally, governments need to establish credible legal and regulatory frameworks that provide stability for rules and prices that will induce investments into financially viable products. Governments should also make an effort to avoid the creation of different regulations at different levels of government, as these create transaction costs for private companies. Currently, that is the case in the United States, where states and municipalities are adopting local regulations in the face of the federal government's slow response to climate change (see chapters 34, 36, and 38).

MARKET-BASED POLICIES

Market-based policies use economic variables, such as subsidies, penalties, or permits, to provide incentives for polluters to reduce harmful emissions or adopt preferred energy technologies. They may also create and change property rights. Regulation and market-based policies feed back into each other — for example, legal and institutional responses give rise to economic incentives that in turn will push technological initiatives such as renewable energy and energy efficiency. The following are market-based policies that could be used to mitigate GHG emissions.

- Carbon taxes on fossil fuel–intensive energy sources.
- Removal of subsidies to reflect the true cost of energy supply, including subsidies to oil extraction and refining.
- Provision of subsidies for clean energy technologies.
- Establishment of emissions trading, after setting maximum emission amounts (cap-and-trade system).
- Risk mitigation instruments.
- "Soft" policies, such as voluntary programs, education, awareness-raising, and training.

RESEARCH AND DEVELOPMENT

The current level of investments in energy technology R&D in the public and private sectors is significantly less in real terms than it has been historically. Dwindling R&D hampers the large-scale dissemination of clean technologies needed to mitigate climate change. In the near term, these technologies include CCS, second-generation biofuels, and end-use efficiency options (zero-emission vehicles, for instance, and more efficient buildings). Over the long term, technologies that use hydrogen as an energy carrier, such as fuel cells and nuclear fusion, may be deployed.

Finally, these policies will not be effective in mitigating climate change if not coupled with attempts to change consumption habits and total energy consumption. In the end, humans will have to change some of their behaviors to mitigate climate change.

Adaptation to Climate Change

Given that the Earth's climate has already changed and that further climate change is inevitable, human societies and ecosystems will need to adapt to new conditions. Some of the adverse effects of climate change may be ameliorated with current coping systems, while others may need radically new behaviors. Anticipating and adapting to these effects to minimize both social and environmental costs is a significant challenge for all nations.

Climate change needs to be factored into national and sector-wide development plans. This task is particularly daunting for developing countries, which lack the necessary financial and technical resources to adapt to climate change. Climate variability is already a major impediment to reducing poverty and will become increasingly so as the Earth's climate warms. Most of the steps needed to adapt to a future climate are compatible with those necessary to reduce vulnerabilities to current climates; it is just a matter of degree. However, there are two important differences. First, the question of sharing the burden for adaptation activities is quite contentious, since climate change is an anthropogenic phenomenon while climate vulnerability is natural. Second, climate change is leading to extremely rapid changes. Thus, historic climate records are far less reliable to guide adaptation to climate change, which makes adaptation to climate change a highly uncertain activity.[3]

Adaptation Policies

Successful adaptation will require efforts at different levels. The international community should provide increased levels of aid for vulnerable countries to cope with climate change, national governments should include adaptation in sector and national development planning, and local communities need to devise new behaviors to better cope with changing conditions. Strategic responses to adaptation must encompass economic, agricultural, trade, and transportation policies, among others. Adaptation activities include economic measures, such as insurance for extreme events, capacity-building for alternative crop cultivation, and management of the impacts of sea-level rise; and investments in infrastructure for water storage, groundwater recharge, storm protection, flood mitigation, shoreline stabilization, and erosion control (see box 42.1).

Small island states and low-lying coastal areas need immediate assistance from the international community to develop the necessary infrastructure and capacity to deal with sea-level rise and storms. For most developing countries, the longer-term challenge is in the key sectors related to agriculture and associated water resource management. Climate change will pose a particular challenge to African countries, since variability in rainfall and other climatic extremes are expected to increase in the continent, while at the same time the ability of African institutions and people to adapt to anticipated climate change impacts during the next twenty to thirty years is limited by widespread poverty, fragile ecosystems, weak institutions, and ineffective governance.

Adaptation will require the transfer of existing technologies, the development of new technologies, and the revision of planning standards and systems. Priority funding is needed to develop typologies of country cases to better understand options and costs, to establish better planning and screening tools (especially for hydrological and biological resource management), and to "climate proof"

agriculture through a new generation of drought- and water-resistant seeds and breeds. Much of the technology and knowledge needed for adaptation is either currently available or can be developed at relatively low cost. Given the probability of more extreme weather events, there is also an urgent need to upscale emergency response mechanisms.

Financing Needs and Sources
Mitigation

The incremental costs of mitigating GHG emissions is estimated to range from less than US$10 billion a year to more than US$200 billion (in 2005 dollars), depending on the stabilization target, the pathway to stabilization, and the underlying development path-

---

BOX 42.1. POLICIES FOR CLIMATE CHANGE ADAPTATION

Information—Effective strategies must rest on the best available data of the nature and severity of likely impacts over different time frames in given locales, and of the cost and efficacy of possible response measures.

Capacity—An overriding priority is strengthening capacities in the technical and planning disciplines most relevant to understanding potential climate impacts and devising response strategies.

Financial Resources—Poorer countries will require resources to improve capacity, undertake specific adaptation measures, and cope with impacts as they occur.

Institutions—While adaptation must be integrated across existing institutions, focal points are needed at the national and international levels to garner expertise, develop and coordinate comprehensive strategies, and advocate for broad-based planning and action.

Technology—As in climate mitigation, adaptation success depends in part on access to, and in some areas development of, technologies suited to the specific needs and circumstances of different countries.

Burton, I., et al. *Adaptation to Climate Change: International Policy Options*. November 2006. Pew Center on Climate Change.

ways of developing countries. The central estimate for stabilizing carbon dioxide at 550 parts per million (ppm), for instance, is about US$60 billion per year. The *Stern Review on the Economics of Climate Change* projected that the cost of emissions reductions consistent with a trajectory leading to stabilization at 550ppm CO_2 equivalent is likely to be around 1 percent of gross domestic product (GDP) by 2050.[4]

Climate change mitigation is to a large extent an energy problem. There are major upfront costs to install new energy-saving or clean technologies, making it hard for developing countries to make the transition. However, the decarbonization of developing countries is essential, particularly in the face of the rapid coal-based industrialization process some countries are undergoing. Nevertheless, the financing sources for mitigation in these countries are still limited.

So far, the main sources of funding for mitigating GHG emissions in developing countries have been international grants and emissions and project-based trading. The Global Environmental Facility (GEF), the official financial mechanism to fund projects seeking to achieve the goals of the UNFCCC, is an important financing agent for climate change mitigation projects, but its funds fall well short, by a factor of 10 to 1,000, of the needed investments. The GEF provides approximately $250 million annually in grants for developing countries to cover the incremental costs of reducing or avoiding GHG emissions in the areas of energy efficiency, renewable energy, and sustainable transportation. Most GEF funds go to soft areas, such as removal of market barriers through capacity development and institutional strengthening, rather than infrastructure investment.

Although international grants are impor-

tant, emissions trading is likely to confer the largest flow of funds to developing countries—between $20 billion and $120 billion per year. However, this would require a long-term equitable global regulatory mechanism that includes all major emitters. In 2006, for instance, Kyoto carbon trading funds to help finance clean energy projects reached around $3 billion. The World Bank is also proposing a "clean energy financing vehicle" to fund efforts by developing countries to decarbonize.[5] The fund would lend money on soft terms to help developing countries cut their emissions, in return receiving payments plus carbon credits equivalent to the amount of the emissions reduction, which the fund could then sell.

From the perspective of developed countries, emissions trading is also attractive, since it may reduce the costs of mitigation by allowing countries to invest in projects that generate the greatest GHG emissions reductions at the lowest marginal cost. Emissions trading, however, is strongly criticized by some analysts. Critics argue that it will be easier to buy credits than to reduce emissions, resulting in the purchase of a license to pollute rather than efforts to mitigate real emissions at home. Another criticism is that, instead of important technology transfer to developing countries, the free-trade mechanisms will lead to further dependency. Finally, by treating emissions as commodities, the structural inequity between north and south in commodity trading in general may continue.[6]

Another cost-efficient way to mitigate climate change is by providing credits to countries that avoid deforestation. Currently, deforestation is responsible for approximately 20 percent of global emissions. The Kyoto Protocol, however, does not allow avoided deforestation to receive credits in the clean development mechanism. This way, developing countries do not have financial incentives from the international community to halt deforestation in their territories. Providing these incentives should be seen as a priority to avoid the emission of billions of GHGs into the atmosphere (see chapters 9 and 48 in this volume on the value of tropical forests, which discuss this issue at length).

Developed countries can generate resources to mitigate climate change in a number of ways. A carbon tax or the auctioning of emission allowances after establishing a cap-and-trade mechanism for GHG emissions could generate revenues to be employed in cutting distortionary taxes on labor and capital, in funding R&D, and also in increasing international aid to developing countries, including technology transfer. To deal with the scale of investment needed in climate change, countries also have to establish a long-term, stable, and predictable regulatory system, possibly encompassing common policies, energy efficiency improvement goals, and technology standards or targets. Ideally, a framework should be established that reaches out to 2050 to produce market certainty, stimulate R&D, and allow time for appropriate policies to be enacted. Even with an improved regulatory environment and the use of political risk mitigation instruments, the challenge of financing incremental costs and reducing technology risks will be significant.

Adaptation

Currently, most of the funding for adaptation activities in developing countries comes from the GEF. It has provided approximately $170 million for the preparation of national communications, which address both mitigation and adaptation. The GEF is also administering three funds for adaptation: the Least Developed Countries Fund, the Special Climate Change Fund, and the Adaptation Fund. The first two funds are supported by voluntary contributions from donor countries; the third by a levy of 2 percent on the proceeds of the Kyoto Protocol's clean development mechanism. This is criticized by developing countries since it does not mean additional resources for adaptation, but rather the employment of their own resources from emissions trading to adaptation activities (see chapter 21 in this volume). The funds pledged by developed countries, however, have largely not yet been made available to developing countries, postponing adaptation actions in these countries.

The GEF has also financed the preparation of National Adaptation Programmes of Action (NAPAs) in more than forty of the least developed countries. The NAPAs are meant to draw on existing information and community-level input to assess vulnerability to current climate variability and areas where risks will be heightened by climate change, and to identify priority actions. The GEF recently approved the first allocations for implementation projects through a $50 million Strategic Priority on Adaptation (SPA) initiative.[7] Although the international effort to date has delivered some soft support to developing countries (information, resources, and capacity building), much needs to be done in terms of on-the-ground implementation of adaptation infrastructure and technology development or access.

The Pew Center on Global Climate Change report on "Adaptation to Climate Change: International Policy Options" identifies three options at the global level to generate the resources for climate change adaptation:

1. *Increased Funding through the UNFCCC.* To the degree possible, assistance provided for planning or implementation should serve simultaneously to build or strengthen national capacities so that, over time, countries are better able to adapt on their own. Also critical to long-term success would be adequate, predictable, and sustained funding. This would require supplementing or replacing the present system of pledging-plus-CDM levy with a stronger, dedicated source such as a wider levy on the emissions market or funding commitments under an agreed formula.

2. *Development Support.* Integrate adaptation considerations across the full range of development support through measures such as mandatory climate risk assessments for projects financed with bilateral or multilateral support.

3. *Insurance Approach.* Two possibilities are described here: (1) an international response fund, where donor countries would commit to regular contributions to a multilateral fund to assist countries suffering extreme and/or long-term climate impacts; and (2) an insurance "backstop," where donor countries support the introduction or expansion of insurance-type instruments in vulnerable countries by committing funds to subsidize premiums or to reinsure governments or primary insurers.

There are many obstacles to generating the funds necessary to allow developing countries to adapt to climate change. Developed countries may not be willing to commit substantial long-term funding, especially in the face of the local nature of adaptation (this is different from mitigation, which is essentially a global problem). Developing countries, on the other hand, may resist bearing the costs of addressing a problem largely created by developed countries. Creative solutions, such as the insurance approach noted above, will be required to break this stalemate.

Conclusion

This chapter presents an introduction to the problems of mitigation and adaptation to climate change and the financial resources needed to tackle them. While mitigation is related to preventive measures taken in the present to avoid future damages to climate change, adaptation should be understood as remedial measures to avoid damages from GHGs emitted in the past as well as projected future changes. There is a broad range of technologies and policies that could lead to GHG emissions mitigation, and all countries should adopt those immediately to avoid more costly or irreversible consequences in the future. Funds for adaptation are a particular concern. Countries that contributed the least to climate change and that are least able to cope with its consequences (in terms of technical and financial capacity) will be affected the most by its consequences. Therefore, there is a moral imperative for developed countries to provide assistance so that vulnerable countries can adapt to climate change.

Notes

1. Kammen, D. The rise of renewable energy. *Scientific American* 295, no. 3 (September 2006).

2. Natural gas should also be used as a bridging fuel in a transition period until renewable energy technologies become commercially available.

3. Kammen, 2006 op. cit.

4. Stern, N. *The Economics of Climate Change: The Stern Review*. Cambridge University Press, Cambridge, 2007.

5. Interview with Robert Watson to Reuters on February 2, 2007 ("World Bank needs $10 bln climate fund").

6. Shaw, A. "Climate Change Flexibility Mechanisms." Available at www.globalissues.org/EnvIssues/GlobalWarming/Mechanisms.asp#EmissionsTrading.

7. Ibid.

Chapter 43

Energy Efficiency

Audrey B. Chang, Arthur H. Rosenfeld,
and Patrick K. McAuliffe

Introduction

The production of energy is among the largest sources of greenhouse gas (GHG) emissions that contribute to global climate change. Fortunately, several strategies can reduce emissions associated with energy use: the use of less energy to provide the same or better service; increased use of alternative energy technologies such as renewable resources that have zero emissions; improved technology for burning the fossil fuels (such as coal, natural gas, and petroleum) that are conventionally used to produce electricity and run our cars; and permanent and safe disposal of the emissions we are not able to reduce. The first of these—a concept commonly known as end-use energy efficiency—is the primary focus of this chapter. Making more efficient use of energy in order to provide the services we need and desire—such as lighting, hot showers, or cold drinks—is the fastest, cleanest, cheapest, and most effective near-term technological option to reduce GHG emissions.

In this chapter, we examine the United States and California in particular as case studies of energy efficiency. We first discuss ef- ficiency gains in the United States and then turn to a more detailed discussion of California's experience with energy efficiency, primarily in the electricity sector.

Surprising Energy Efficiency Gains in the United States

The United States has made energy efficiency improvements that have benefited its economy and reduced its GHG emissions. A common measure of energy efficiency is energy intensity, defined as the quantity of primary energy, not just electricity, consumed per unit of gross domestic product (E/GDP). Energy intensity in the United States has declined at five times the historical rate since the 1973–1974 oil crisis raised the price of energy, awareness of energy consumption, and the profile of energy efficiency. If, instead of the actual 2.1 percent decline per year experienced since 1973, U.S. energy intensity had decreased by only the business-as-usual pre-1973 rate of 0.4 percent per year, energy use in the country would have risen by approximately an additional 70 quadrillion Btus

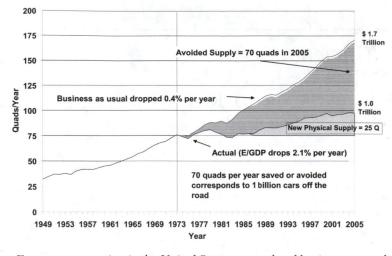

FIGURE 43.1. Energy consumption in the United States at actual and business-as-usual rates of declining energy intensity. The vertical line marks the 1973 oil embargo. Source: Energy Information Administration, *Monthly Energy Review (MER)*, available online at: www.eia.doe.gov/emeu/mer/contents.html; calculations by authors.

(quads) in 2005 (see figure 43.1). Even with this improvement, primary energy use still climbed by 25 quads during these three decades.

The U.S. government's own energy policy reports from 2001 have confirmed that the decreasing energy intensity of the country's economy is due to gains in energy efficiency:

> Had energy use kept pace with economic growth, the nation would have consumed 171 quadrillion British thermal units (Btus) last year instead of 99 quadrillion Btus. About a third to a half of these savings resulted from shifts in the economy. The other half to two-thirds resulted from greater energy efficiency.[1]

This avoided supply of energy, the majority of which was displaced by improved energy efficiency, corresponds to approximately $700 billion of annual savings (avoided expenditures that are available to be invested in growing other parts of the economy). We estimate

that the 70 quads per year of avoided energy supply in 2005 is equivalent to avoided emissions of 4.2 billion metric tons of carbon dioxide (CO_2), or the emissions reductions comparable to taking nearly one billion cars off the road (consider that there are currently only 600 million cars in the world). The actual U.S. emissions in that year were nearly 6 billion tons.[2] In addition, the 70 quads per year avoided is equivalent to an oil flow of approximately 33 million barrels a day, two-fifths of the world's current oil production of 84 million barrels a day.[3]

Such gains are laudable and important from both an economic and an environmental perspective. Prior to 1973, primary energy use, gross domestic product (GDP), and CO_2 emissions from combustion increased nearly in lockstep—energy intensity changed little. After 1973, these relationships dramatically changed: GDP and energy use were uncoupled, and energy intensity improved rapidly.[4] Interestingly, from 1979 to 1983 CO_2 emissions actually declined, caused by individual state adoption of energy efficiency standards for buildings and appliances, increased fuel prices, and federal corporate average fuel economy (CAFE) standards in the transportation sector. However, CAFE standards had until recently remained virtually unchanged since 1985, and many other energy-saving policies also lapsed or failed to become more aggressive with time. As a result, CO_2 emissions in the United States since 1984 have continued to increase, at a rate of 1.3 percent per year.[5]

Thus, additional gains in energy efficiency will be essential to enable the United States to curb its rising emissions. A recent McKinsey report that developed supply curves for greenhouse gas abatement in the United States identified 710 million tons of CO_2-equivalent of emissions reduction potential available from increased building and appliance energy efficiency in its mid-range case for annual emissions reductions by 2030. These energy efficiency improvements could decrease forecasted electricity demand by 24 percent. Importantly, the vast majority of these efficiency savings are available at negative cost, potentially avoiding $300 billion (in 2005 dollars) of projected power generation capital investments.[6]

California: The Role of Energy Efficiency Policy in Reducing GHG Emissions

We use the experience of California as a case study of the potential to reduce GHG emissions through end-use energy efficiency. Since the mid-1970s, California has established itself as a leader in promoting energy efficiency. Recently, the state has also identified

energy efficiency as its top priority energy resource and a primary tool for reaching its GHG emissions reduction goals. We focus primarily on electric and natural gas end-use efficiency in our examination of California, because states control regulation of utilities in these sectors.[7]

Figure 43.2 shows the relative contribution to GHG emissions by end-use consumption in California in 2004. The transportation sector is the largest source of emissions (37 percent). Electricity consumed in California (including electricity imported from outside of the state) is the second largest source of emissions (25 percent). Natural gas use in buildings and industry, excluding electricity generation, is another substantial contributor to state emissions (14 percent).[8] Significant opportunities exist to improve the efficiency of any of these end uses and thus reduce GHG emissions.

To fully capture the emissions impact of California's electricity sector, it is essential to consider the emissions associated with all of

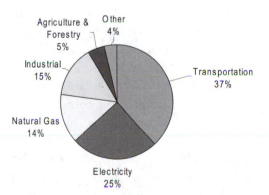

Figure 43.2. 2006 California GHG Emissions by Sector, including electricity imports. Total emissions in 2004 were 480 million metric tons of CO_2-equivalent (MmtCO2eq). Source: California Air Resources Board, *Greenhouse Gas Inventory Data — 1990 to 2004*, November 2007, www.arb.ca.gov/cc/inventory/data/data.htm.

the electricity the state consumes, rather than solely emissions from in-state electricity generation. California imports 22 to 32 percent of its electricity from out-of-state generating units, many of which are high-emitting coal-fired facilities.[9] On the whole, out-of-state electricity consumed in California is more than double the carbon intensity of in-state generation. Although imported power represents less than one-third of the total electricity consumed in the state, it is responsible for more than half of the GHG emissions associated with electricity use in California.[10]

Energy Efficiency: California's Foundation for Reducing Its GHG Emissions

In 2006, California's Global Warming Solutions Act (Assembly Bill 32) established the first mandatory economy-wide limit on GHG emissions in the United States. AB 32 requires emission reductions to 1990 levels by 2020, representing a 30 percent cut from business-as-usual emissions.[11] An additional aggressive long-term target was previously established by Governor Schwarzenegger to reduce GHG emissions to 80 percent below 1990 levels by 2050.[12] In the state's "scoping plan" for implementing AB 32, adopted in December 2008, energy-efficiency strategies figure prominently. The combination of continued and expanded efficiency programs and building and appliance standards are expected to avoid 20 million metric tons of CO_2-equivalent (MmtCO$_2$eq) of emissions in 2020. In all, electricity and natural gas end-use efficiency will contribute at least 12 percent of the total 169 MmtCO$_2$eq of emissions reductions anticipated to be needed to meet the state's 2020 limit (see figure 43.3).[13]

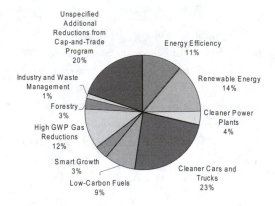

FIGURE 43.3. Strategies for Meeting California's 2020 GHG Reduction Goals. Source: California Air Resources Board, *Climate Change Scoping Plan*, December 2008. Graphing by authors.

Energy efficiency more broadly is an important foundation to help California meet its emission reduction goals. Reducing emissions from power plants ("cleaner power plants") will also involve deploying more efficient technologies, such as combined heat and power applications. Outside of the electricity and natural gas sectors, a large portion of the estimated reductions from the transportation sector are expected to come from technological improvements to enhance that sector's end-use energy efficiency ("cleaner cars and trucks"). Specifically, vehicle tailpipe GHG emissions standards established by AB 1493, and their extension through 2020. These standards have since spurred the United States to announce it will establish similar nation-wide vehicle GHG standards.[14] In total, energy efficiency strategies across all sectors, including end-use electricity and natural gas, electricity generation, and vehicle efficiency, will likely contribute about half of California's total targeted GHG emission reductions by 2020. Considerable effort will be needed to realize all these savings, but efficiency policies in the electricity and natural gas sectors provide clear and convincing evidence that such efforts can be cost-effective and environmentally sound.

Historical Energy Efficiency Accomplishments in California

California has long pursued strong energy efficiency programs and policies, beginning with the establishment of the state's appliance (Title 20) and new-building (Title 24) standards in 1976 and 1978, respectively, and concurrent investments in energy efficiency programs across the state. Figure 43.4 shows that California's historical energy efficiency policies have enabled the state to hold per capita electricity sales essentially constant, while in the United States as a whole, per capita electricity use increased by about 50 percent since the mid-1970s.[15]

Differences in energy policy between California and the rest of the United States partially explain these divergent paths in per capita electricity consumption. Although California's relatively low per capita consumption is partly due to a milder climate, the state's gradual transition over time from a manufacturing-based economy to a service-based economy, and the demand-dampening effect of higher electricity prices, a substantial portion of the difference in per capita electricity use, as compared to the rest of the United States, is due to policies aimed at more efficient use of electricity. If California's per

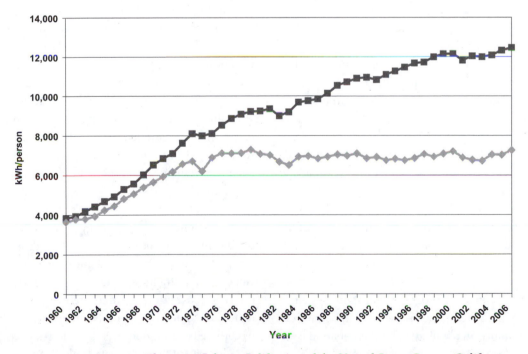

FIGURE 43.4. Per Capita Electricity Sales in California and the United States. Source: California Energy Commission, *2007 Integrated Energy Policy Report*, CEC-100-2007-008-CMF, November 2007, p. 3.

capita consumption had grown at the same rate as the rest of the country since 1975, the state would have needed approximately fifty additional 500 megawatt power plants.

California's success in energy efficiency policy rests on an integrated three-pronged approach:

1. Research and development activities generate new energy-efficiency technologies and strategies. California's Public Interest Energy Research budget is about $80 million per year.[16]

2. As these new technologies are introduced commercially, the state's utilities administer rebate and education programs to accelerate their penetration into the marketplace. The state's investor-owned utilities' (IOUs) energy-efficiency investments in 2008 totalled $935 million.[17]

3. Finally, as these energy-efficient technologies become more commonplace, their higher level of performance is incorporated into building and appliance standards. The total cost of developing these standards is in the range of $10 million to $20 million per year.[18]

Roughly half of California's policy-driven energy savings have come from building and appliance standards that have been progressively strengthened every few years. The other half of these savings has resulted from utility programs that promote deployment of energy-efficient technologies. Figure 43.5 shows the annual energy savings from California's energy-efficiency programs and standards since 1975. Through 2003, these policies have cumulated in about 40,000 GWh of annual energy savings and have avoided 12,000 MW of demand (MW data not shown in graph)—the

same as twenty-four 500-MW power plants.[19] These savings will only continue to grow.

When summed together, the three decades of energy-efficiency programs and standards have resulted in annual efficiency savings today equivalent to approximately 15 percent of California's annual electricity consumption, as shown in Figure 43.5.[20] These savings have reduced CO_2 emissions from the electricity generation sector by nearly 20 percent compared to what otherwise might have happened without these programs and standards.[21] This equates to an avoidance of CO_2 emissions in the state as a whole of about 4 percent due to historical energy efficiency programs and standards.[22]

California was the first state in the nation to adopt efficiency standards for appliances in 1976. Other states soon followed, eventually leading to federal standards in the National Appliance Energy Conservation Act of 1987. This pattern continues today; building and appliance efficiency standards first adopted by California are frequently adopted by other states, the federal government, and other countries around the world, including Russia and China. These standards are regularly updated and strengthened every few years, ensuring that California's buildings and appliances will remain the most energy efficient in the nation.

A key element of California's success in utility efficiency program energy savings is the establishment of a policy to remove the disincentive for utility investments in energy efficiency. Under traditional utility regulation, a utility's recovery of its infrastructure investment costs is tied to how much energy it sells. According to this model, energy efficiency results in lower-than-anticipated sales and decreased revenues, thus preventing utilities

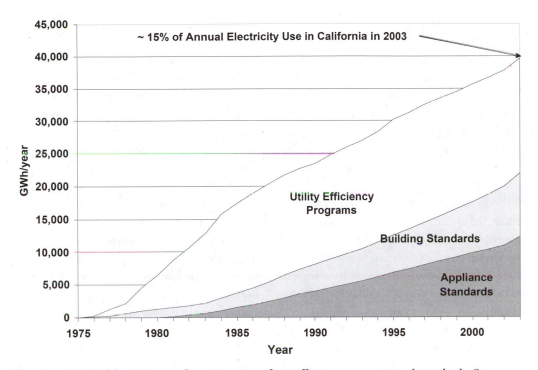

FIGURE 43.5. California's annual energy savings from efficiency programs and standards. Source: California Energy Commission, *Implementing California's Loading Order for Electricity Resources*, Staff Report, CEC-400-2005-043, July 2005, Figure E-1, p. E-5.

from fully recovering their fixed costs. As a result, traditional regulation deters utilities from investing in energy efficiency. However, since 1982 (with a brief hiatus in the mid-1990s, when market restructuring took resource-planning responsibilities away from the utilities), California law has required the state's investor-owned utilities to use modest regular adjustments to electric and gas rates to sever the link between the utilities' financial health and the amount of electricity and natural gas they sell.[23] This concept, known as "decoupling," removes significant regulatory and financial barriers to utility investments in cost-effective energy efficiency improvements and helps align the interests of utilities and customers. Regulators in several other states in

the United States have recently followed California's lead by adopting decoupling mechanisms for electric or natural gas utilities (including Idaho, Ohio, Oregon, Maryland, New York, North Carolina, and Utah), and others are considering proposals to do so (including Iowa, Montana, and New Mexico).[24]

While California's energy efficiency standards and programs have helped reduce the state's GHG emissions, they have also delivered substantial net economic benefits to California. The state's efficiency standards, designed to be cost-effective, accelerate energy savings across the state. The cost of efficiency programs from the utility perspective has averaged two to three cents per kWh saved, which is less than half the cost of the baseload generation, the type of energy resource most often displaced by energy efficiency.[25] During the last decade alone, these efficiency programs have provided net economic benefits of more than $8 billion to California's customers from reduced energy expenditures.[26] Though California is often maligned for its high electricity retail *rates* compared to the rest of the United States, the state's energy efficiency policies have reduced overall energy *bills* for its residents and businesses. Average monthly residential electricity bills in California are 15 percent lower than in the rest of the country and almost half that of Texas.[27]

Not only is energy efficiency the cheapest and cleanest resource, it also has the shortest lead-time. Whereas a new power plant takes a minimum of several years to plan, permit, and build, energy-efficiency measures can be installed in a matter of months or weeks. For example, in a period of fifteen months from 2000 to 2001, inefficient incandescent traffic lights with energy-intensive colored lenses were replaced with light-emitting diodes (LEDs) across the state. This relatively simple retrofit saved 186 GWh annually and 29 MW in reduced peak electricity demand.[28] Because of its short lead-time and ability to be incrementally deployed, energy efficiency can also be used to more closely follow customer load than the building of power plants, which are typically limited to providing new capacity in relatively large portions.[29]

California's New Era of Energy Efficiency

Despite California's historical success, significant cost-effective energy efficiency potential still remains, and state regulators have developed a policy framework to improve on California's record. Beginning in 2003, energy efficiency programs in California have been guided by a formal state policy that places cost-effective energy efficiency above all other energy resources. The Energy Action Plan, adopted by the state's energy agencies and endorsed by Governor Schwarzenegger, established a "loading order" of preferred energy resources—(1) all cost-effective energy efficiency and demand response; (2) renewable energy generation; and (3) cleaner and more efficient fossil-fueled generation.[30] Since 2005, energy efficiency has been codified as the state's top priority procurement resource.[31] The loading order now guides all of the state's energy policies and aligns with California's goal of reducing its GHG emissions in the most cost-effective manner.

In 2005, California regulators adopted a new administrative structure for the delivery of energy-efficiency programs that charges the state's regulated utilities with fully integrating energy efficiency into their resource procurement process.[32] Utilities are now required to invest in energy efficiency whenever it is cheaper than building new power plants. Rigorous evaluation of these program savings will be essential to ensure that can be relied upon for resource planning purposes, as well as for the state's GHG emission reduction goals.[33]

Based on the potential for cost-effective achievable energy efficiency improvements in the state, California regulators established the most aggressive long-term energy savings goals in the nation, since updated to extend through 2020 to align with the AB 32 timeline.[34] While energy efficiency efforts across the United States deliver annual savings of 0.21 percent of total electricity sales, the annual California savings now exceed 1 percent of electricity sales.[35] Figure 43.6 illustrates the annual savings targets, which significantly surpass histor-

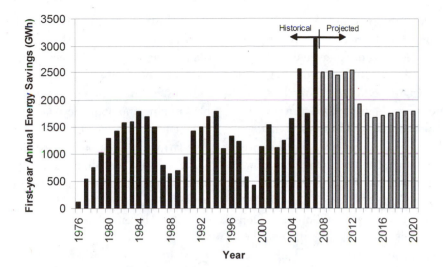

FIGURE 43.6. Historical and Projected Electricity Savings for California Investor-Owned Utilities. Source: 1976 to 1997 data from California Energy Commission, personal communication with Sylvia Bender, 2005; Pacific Gas & Electric Company, Southern California Edison Company, San Diego Gas and Electric Company, and Southern California Gas Company, *Annual Earning Assessment Reports*, 1998–2005; California Public Utilities Commission, Energy Efficiency Groupware Application (EEGA), http://eega2006.cpuc.ca.gov/, compilation of IOUs' Energy Efficiency 2006–2008 program reports; California Public Utilities Commission, Decision 08-07-047, "Decision Adopting Interim Energy Efficiency Savings Goals for 2012 through 2020, and Defining Energy Efficiency Savings Goals for 2009 through 2011," July 31, 2008.

ical reductions. In a few years' time, California's per capita electricity consumption, while it has remained steady during the last three decades, should begin to decline. The long-term energy savings goals will avoid the construction of a new 500-MW power plant every year. Customers will also obtain some relief from rising natural gas bills through the tripling of annual gas savings by 2013 compared to the years prior to the establishment of long-term savings goals in 2004.[36]

The historical data in figure 43.6 also serve to illustrate the importance of consistent and effective policies to encourage energy efficiency. It is worth noting that energy savings during 1998 and 1999 were the lowest of any year since 1976. This period coincides with the restructuring of California's utility indus-

try and the introduction of retail competition, since suspended. Voluntary investments by utilities in energy efficiency investments fell by the wayside and were protected only by a mandated system benefits charge. In restructured electricity markets all over the country, the use of system benefits charges have provided a minimum level of energy efficiency funding in the absence of utility-funded programs. Although energy efficiency programs in California continue to be funded in part through the system benefits charge (approximately $303 million in 2006 for the investor-owned utilities), California's utilities now also more than double this amount with additional investments from their supply-side resource procurement budgets ($310 million in 2006 for the IOUs).[37]

In 2006, California's utilities launched aggressive programs to execute their energy savings goals. The utilities budgeted $2 billion to deliver their 2006 through 2008 energy efficiency program portfolios.[38] This three-year investment is anticipated to return $2.7 billion in net benefits to California's economy through reduced energy bills and the saved costs from avoided construction of new power plants. Moreover, by 2008, these programs are expected to reduce the state's annual GHG emissions by more than 3 million metric tons of CO_2, which is equivalent to the emissions produced by about 650,000 cars.[39]

In 2007, California regulators adopted a performance-based energy efficiency incentive mechanism for the state's IOUs. This policy recognized that the removal of financial disincentives was not sufficient to encourage aggressive investments in energy efficiency at the high levels expected of California's utilities. Providing performance-based incentives is intended to focus the attention of the utilities' management on accomplishing these savings, which will produce both economic benefits for customers and reduced GHG emissions. If the utilities exceed energy savings thresholds, then they are rewarded with a small share of the net benefits the programs provide to customers, up to a maximum of $150 million statewide each year, reflecting less than 1 percent of consumers' total annual cost for electricity and natural gas. These potential rewards are balanced by symmetrical penalties that are assessed for poor performance.[40] The incentive mechanism will spur sustained utility interest in successfully accomplishing the long-term energy savings goals that will help meet AB 32's emissions limit.

It is clear that cost-effective electric and natural gas end-use efficiency has a critical role in combating climate change while providing for the energy needs of the state. California has had a successful history of implementing energy efficiency programs and standards, but further energy-saving potential remains. Future improvements to the California building and appliance standards, ensuring the investor-owned utilities meet their energy savings goals through 2020 and beyond, and efficiency contributions from the state's publicly owned utilities are all necessary components of California's ambitious plans to reduce its GHG emissions.

Conclusion

Policies to encourage energy efficiency are a realistic and negative-cost strategy to reduce the growth of energy demand, thereby lowering GHG emissions. The experience of the United States and California are examples of how energy efficiency can accelerate the decline of energy intensity. California's sustained electric and natural gas energy efficiency policy efforts illustrate the need for continuity of energy efficiency policies to achieve reductions in energy demand growth and GHG emissions. California has far outperformed the United States in energy efficiency, but improvements in the United States have also yielded decisive economic and environmental benefits.

Notes

1. Cheney, D., et al., *National Energy Policy: Report of the National Energy Policy Development Group*, May 2001, p. 1–4.

2. To convert the 70 quads/year of fuel avoided to tons of CO_2, we use the current CO_2 emission rate of 60 million metric tons of CO_2 per quad of primary energy. Data source: Energy Information Administration, *International Energy Annual*, Table H.1, "Carbon Dioxide Emissions from the

Consumption and Flaring of Fossil Fuels, 1980–2004," July 2006, www.eia.doe.gov/pub/international/iealf/tableh1CO2.xls; and Energy Information Administration, *International Energy Annual*, Table E.1, "World Primary Energy Consumption (Quadrillion Btu), 1980–2004," July 2006, www.eia.doe.gov/pub/international/iealf/tablee1.xls.

3. U.S. Department of Energy, Energy Information Administration, Chapter 3: World oil markets, in *International Energy Outlook 2006*, Report # DOE/EIA-0484(2006), June 2006.

4. Data source: Energy Information Administration, *Monthly Energy Review (MER)*, available online at www.eia.doe.gov/emeu/mer/contents.html. Calculations by authors.

5. Data source: Energy Information Administration, *International Energy Annual*, Table H.1, "Carbon Dioxide Emissions from the Consumption and Flaring of Fossil Fuels, 1980–2004," July 2006, www.eia.doe.gov/pub/international/iealf/tableh1CO2.xls. Calculations by authors.

6. McKinsey & Company and The Conference Board, *Reducing U.S. Greenhouse Gas Emissions: How Much at What Cost?* December 2007, pp. 34, 28, and 29.

7. For the remainder of this chapter, we use "energy efficiency" to refer to electric and natural gas end-use efficiency, unless otherwise stated.

8. California Air Resources Board, *Greenhouse Gas Inventory Data—1990 to 2004*, November 19, 2007, www.arb.ca.gov/cc/inventory/data/data.htm. Authors recombined emissions to depict by end-use sector. Data include emissions associated with electricity generated outside of California but consumed within the state.

9. California Energy Commission, *2007 Integrated Energy Policy Report*, CEC-100-2007-008-CMF, November 2007, p. 19.

10. California Air Resources Board, *Climate Change Scoping Plan*, December 2008, p. 12.

11. For the full text of Assembly Bill 32, see www.arb.ca.gov/cc/docs/ab32text.pdf.

12. Executive Order S-3-05, June 2005, www.dot.ca.gov/hq/energy/ExecOrderS-3-05.htm.

13. California Air Resources Board, *Climate Change Scoping Plan*, December 2008, author's calculations.

14. Pavley, F. "Vehicle Emissions: Greenhouse Gases," California Assembly Bill 1493, 2002. See also chapter 35 in this volume.

15. California Energy Commission, *2007 Integrated Energy Policy Report*, CEC-100-2007-008-CMF, November 2007, p. 3.

16. California Energy Commission, *2007 Public Interest Energy Research (PIER) Annual Report*, CEC-500-2008-026-CMF, April 2008, p. 1.

17. California Public Utilities Commission, Energy Efficiency Groupware Application (EEGA), http://eega2006.cpuc.ca.gov/, compilation of IOUs' Energy Efficiency 2006–2008 program reports.

18. California Energy Commission staff estimates of the costs of promulgating and supporting Title 20 and 24 regulations.

19. California Energy Commission, *2005 Integrated Energy Policy Report*, CEC-100-2005-007, November 2005, p. 70.

20. In reality, the actual statewide savings have likely been even greater, since the utility efficiency programs shown here include only those savings reported by the regulated investor-owned utilities in the state and do not include efficiency program savings from the municipal utilities, which account for about a quarter of the state's electricity sales.

21. This calculation is based on a marginal CO_2 emissions rate of nearly 0.5 ton per MWh for a natural gas–fired powerplant with a marginal heat rate of just over 9,000 Btu/kWh.

22. This calculation is the product of 25 percent of the state's CO_2 emissions from electricity consumption (figure 43.2) and the reduction of 15 percent in electricity use due to standards and programs (figure 43.5).

23. California Public Utilities Code Section 739.10 states: "The commission shall ensure that errors in estimates of demand elasticity or sales do not result in material over or under-collections of the electrical corporations." These rate adjustments typically are made on an annual basis and consist of an adjustment, up or down, of less than two percent of a residential customer's energy bill.

24. Kushler, M., D. York, and P. Witte. *Aligning Utility Interests with Energy Efficiency Objectives: A Review of Recent Efforts at Decoupling and Performance Incentives*, Report Number U061, October 2006. Washington, D.C.: American Council for an Energy-Efficient Economy. More recent developments are also included, based on summaries compiled by the NRDC.

25. The cost over the lifetime of energy efficiency initiatives undertaken during 2001 will be an average of 3¢/kWh (Global Energy Partners, *California Summary Study of 2001*, for the

California Measurement Advisory Council, Report ID# 02-1099, March 2003.) The average cost of saved energy of PGC funded efficiency from 1990 to 1998 was about 2.5¢/kWh (Sheryl Carter, *Investments in the Public Interest: California's Public Benefit Programs under Assembly Bill 1890*, Natural Resources Defense Council, January 2000). The average cost of 2000 to 2004 efficiency programs was 2.9¢/kWh. California Energy Commission, *Funding and Energy Savings from Investor-Owned Utility Energy Efficiency Programs In California for Program Years 2000 through 2004*, CEC-400-2005-042-REV, August 2005. The programs continue to average 3¢/kWh.

26. Pacific Gas and Electric Company, Southern California Edison Company, San Diego Gas & Electric Company, and Southern California Gas Company Energy Efficiency Annual Reports, May 1999–2006, California Public Utilities Commission, Energy Efficiency Groupware Application [EEGA], http://eega2006.cpuc.ca.gov/, compilation of IOUs' Energy Efficiency 2006 through 2008 program reports.

27. Next 10, *California Green Innovation Index: 2008 Inaugural Issue*, 2007, p. 21.

28. York, D., and M. Kushler, *America's Best: Profiles of America's Leading Energy Efficiency Programs*, Report Number U032, March 2003. Washington, DC: American Council for an Energy-Efficient Economy.

29. Although California is now able, due to its sustained energy efficiency efforts during the last thirty years, to quickly deploy energy efficiency programs and implement new standards, we note that this market acceptance of energy efficiency does not necessarily exist outside of California. California's aggressive pursuit of energy efficiency has enabled an infrastructure and market environment that is conducive to these improvements. However, this infrastructure may not exist elsewhere, but other states and countries can look forward to these reduced barriers if they put forward a similar sustained effort and commitment to pursuing energy efficiency.

30. California Consumer Power and Conservation Financing Authority (CPA), California Energy Resources Conservation and Development Commission (CEC), and California Public Utilities Commission (CPUC), *Energy Action Plan*, 2003. Available online at www.energy.ca.gov/energy_action_plan/2003-05-08_ACTION_PLAN

.PDF. Letter from Governor Schwarzenegger to CPUC President Peevey, April 28, 2004. CEC and CPUC, *Energy Action Plan II*, September 21, 2005. Available online at http://www.energy.ca.gov/energy_action_plan/2005-09-21_EAP2_FINAL.PDF.

31. California Public Utilities Code Sections 454.5(b)(9)(C) and 454.56(b) require that all electric and natural gas utilities meet their unmet resource needs first through "all available . . . efficiency and demand reduction resources that are cost effective, reliable, and feasible."

32. California Public Utilities Commission, Decision 05-01-055, "Interim Opinion on the Administrative Structure for Energy Efficiency: Threshold Issues," January 27, 2005.

33. The independent evaluation, measurement, and verification of the energy savings achieved by the utility programs is so important that an average of 6 percent of the utilities' total 2006 through 2008 energy efficiency budgets was set aside for this purpose. (California Public Utilities Commission, Decision 05-11-011, "Interim Opinion: Evaluation, Measurement, and Verification Funding for the 2006–2008 Program Cycle and Related Issues," November 18, 2005; and California Public Utilities Commission, Decision 05-09-043, "Interim Opinion: Energy Efficiency Portfolio Plans and Program Funding Levels for 2006–2008 — Phase 1 Issues," September 22, 2005.)

34. California Public Utilities Commission, Decision 08-07-047, "Decision Adopting Interim Energy Efficiency Savings Goals for 2012 through 2020, and Defining Energy Efficiency Savings Goals for 2009 through 2011," July 31, 2008.

35. Eldridge, M., M. Neubauer, D. York, S. Vaidyanathan, A. Chittum, and S. Nadel, *The 2008 State Energy Efficiency Scorecard*, Report Number E086, October 2008, p. 10. Washington, DC: American Council for an Energy-Efficient Economy. Kushler, M., et al., Meeting Aggressive New State Goals for Utility-Sector Energy Efficiency: Examining Key Factors Associated with High Savings, Report Number E091, March 2009, Washington, DC: American Council for an Energy-Efficient Economy.

36. Pacific Gas & Electric Company, Southern California Edison Company, San Diego Gas and Electric Company, and Southern California Gas Company, *Annual Earning Assessment Re-*

ports, 1998–2005. California Public Utilities Commission, Energy Efficiency Groupware Application (EEGA), http://eega2006.cpuc.ca.gov/ (compilation of IOUs' Energy Efficiency 2006 through 2008 program reports). California Public Utilities Commission, Decision 08-07-047, "Decision Adopting Interim Energy Efficiency Savings Goals for 2012 through 2020, and Defining Energy Efficiency Savings Goals for 2009 through 2011," July 31, 2008.

37. California Public Utilities Commission, Decision 05-09-043, "Interim Opinion: Energy Efficiency Portfolio Plans and Program Funding Levels for 2006–2008—Phase 1 Issues," September 22, 2005, attachments.

38. These programs include rebate programs for energy efficient technologies in buildings and industry, as well as general education about energy efficiency.

39. California Public Utilities Commission, Decision 05-09-043, "Interim Opinion: Energy Efficiency Portfolio Plans and Program Funding Levels for 2006–2008—Phase 1 Issues," September 22, 2005, p. 3.

40. California Public Utilities Commission, Decision 07-09-043, "Interim Opinion on Phase 1 Issues: Shareholder Risk/Reward Incentive Mechanism for Energy Efficiency Programs," September 20, 2007.

Chapter 44

Renewable Energy

Daniel M. Kammen

Renewable energy science and technology have undergone dramatic advances in technical performance and economic effectiveness across a wide range of specific energy generation systems during the last several decades.[1] Far from being a set of appealing but "boutique" technologies as they were even a decade ago, renewable energy and energy efficiency have now come of age to the point where cities, states, and nations plan and implement low- and even near zero-carbon energy paths. California has committed through AB 32 (Global Warming Solutions Act) and an executive order (E 3-05) to about 25 percent cuts in greenhouse gas emissions by 2020 and 80 percent cuts by 2050.[2] The United Kingdom has committed to 60 percent emissions reductions by 2050, Sweden has passed an environmental mandate to get off fossil fuels entirely by 2030, and Austria has committed to 80 percent reductions by 2020. Other nations, including Norway, New Zealand, Japan, and Germany, are developing and enacting a number of similarly aggressive measures.

At present the U.S. economy is decarbonizing—or reducing the greenhouse gas emissions per unit of GNP—at just over 1 percent per year, largely due to advances in energy efficiency and industrial technology and process. Despite a history of successes in energy efficiency—efficient lighting, refrigeration, heating and cooling, and appliance standards—increased demands for electricity as well as energy for transportation continue to drive energy needs markedly upward. In this context, low- and no-carbon sources of energy are critically needed along with carbon sequestration and long-term dedication to continual advances in energy efficiency. Only with a major commitment to low-carbon supply-side technologies can we realistically achieve an 80 percent or greater reduction in carbon emissions by roughly mid-century. This level of reductions has been identified by the Intergovernmental Panel on Climate Change (IPCC), which shared the 2007 Nobel Peace Prize and for which I have served as an author, as necessary to significantly address global warming.[3] This level of emissions reduction, while great, is now achievable but will require scientific, technical, political, and economic commitment. We are now in an era where the economics and political opportunities for renewable energy sources are unprece-

446

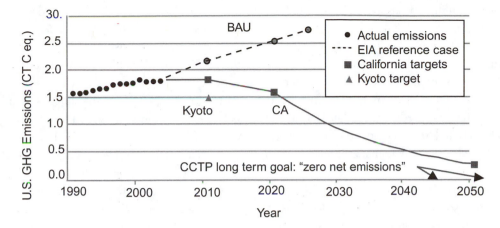

FIGURE 44.1. Actual U.S. greenhouse gases (GHG) emissions from 1990 through 2003 (EPA 2005) in gigatons of carbon equivalent and an extrapolation based on California's state-level targets. Four paths for future U.S. emissions are shown; circles show the business as usual (BAU), or "reference case," as calculated by the Energy Information Agency (EIA). The diamond shows the administration's GHG intensity target for 2012 of 18 percent below 2002 level in tons of carbon per unit of GDP, or a 3.6 percent reduction in emissions from BAU. The squares show U.S. emissions if the nation were to meet the percentage reductions that have been announced in California for 2010, 2020, and 2050 (California Executive Order 3-05, and California AB 32, the "Pavley-Nuñez Bill"). The triangle shows the U.S. target for 2010 under the Kyoto Protocol. Arrows indicate the levels required to meet the CCTP's long-term goal of "levels that are low or near zero" (p. 2-2).

dented, making this a moment of choice and opportunity to dramatically advance clean power for decades to come.

Renewable Energy Technologies

The global solar cell (photovoltaic) industry has grown by more than 20 percent per year each year for the past decade, and solar cells, on a percentage basis, are the world's fastest-growing supply-side sources of energy. In 2005 the global photovoltaic market grew by a remarkable 50 percent, reaching a total production volume of more than 1,100 MW and 200 MW in the U.S. Solar cells can now be made from a range of materials—from the classic multi-crystalline silicon wafers, which still dominate the market, to thin-film cells as well as plastic and organic compounds. Thin-film photovoltaics, generally manufactured in simple reel-to-reel lamination processes or by printing, not only cost less than conventional crystalline cells, but also involve fewer materials overall.

Solar Photovoltaic and Thermal Energy

Solar cell efficiencies today are highest from crystalline silicone solar cells, at 30 percent or more in laboratory tests, while commercially available cells are today in the 15 to 20 percent range. Solar photovoltics (PV) represent a particularly easy-to-use form of renewable energy—with direction installation on rooftops, on buildings, and in arrays possible. The state of California has joined Japan and Germany in leading global push for solar installations, with a new "Million Solar Roof" commitment that will bring more than 10,000 MW of new solar photovoltaics. The potential for this technology is virtually unlimited. My research group predicts that solar PV installed in the United States could grow from the current level of 0.5 GW to more than 700 GW in only twenty years.[4]

PV remains is a relatively expensive renewable source, with costs of 20 to 25 cents per kWh for crystalline cells. For comparison, coal-fired electricity is 4 to 6 cents per kWh, natural gas is 5 to 7 cents per kWh, biomass-fired power costs 6 to 9 cents per kWh, and nuclear, where the most disagreement over costs exists, is between 3 and 12 cents per kWh, depending on what is included and what is excluded from the analysis.[5] That said, significant technical advances (e.g., thin-film solar cells), significant advances in power electronics (e.g., the development of "micro" inverters so that the output of each cell can be converted efficiently to AC power), and significant financing innovations (e.g., solar installations paid for over the lifetime of the panels without up-front costs through the use of regional assessment districts) are changing, and potentially revolutionizing, the economics of solar energy.

Large photovoltaic installations, such as the 675,000 watt system on the Moscone Center in San Francisco, CA, have installed costs as low as $5 per watt. Costs have been falling consistently during the last decade, and aggressive programs in Japan (where 290 MW of solar was installed last year) and in California (where 50 MW was installed) have each observed significant year to year cost reductions, of more than 8 percent per year in Japan and 5 percent per year in California.

Thin-film and amorphous silicon solar cells are also in commercial use today, with Kenya, surprisingly, the global leader in solar systems (not watts) installed per year. More than 30,000 very small (12- to 30-watt solar panels) systems are sold in Kenya each year, at costs—for the panels, wiring, often a simple battery, lights, and outlets for a few bulbs, radio, or TV—of as little as $100 per system. More Kenyans make new electricity connections through solar each year than through new connections to the grid. The efficiency of amorphous solar cells are lower than crystalline, by a factor of two or more, but the costs are less by a factor of at least four, making these more affordable and useful for the 2 billion people on Earth without current access to electricity.

Photovoltaics may be the most prominent form of solar power, but solar thermal systems are also undergoing a resurgence. Solar thermal systems, combining focusing mirrors on a working fluid such as oil, are highly efficient and very low cost central-station power plants. In the fall of 2005, Stirling Energy Systems (SES) signed two long-term electricity contracts with California utilities that will enable it to build two large solar-dish power plants in southern California. AUSRA and other companies are now building manufacturing facilities for large-scale solar thermal deployment, with densely packed 1 square mile arrays capable of producing around 180 MW (peak) of output, at costs of roughly 10 cents per kWh.

Installed system costs are a subject of considerable debate and interest, but estimated levelized electricity costs for current concentrating solar power technologies are 7 to 9 cents per kWh for towers, 7 to 11 cents per kWh for focusing troughs, and 9 to 13 cents per kWh for dishes. Projected costs, assuming reasonably attainable cost reductions in the next ten years are 4 to 6 cents per kWh for towers, 6 to 9 cents per kWh for troughs, and 7 to

10 cents per kWh for dishes. With their very large-scale and long-term contracts, the two SES projects are likely to have costs that are closer to the 7 to 10 cents per kWh projected estimates than the 9 to 13 cents per kWh estimate for current technologies.

Wind Energy

Growth in the wind industry has been nothing short of explosive. In 1994 there was a mere 1,600 MW of wind installed in Europe, but it grew to reach 44,000 MW in 2007. An aggressive construction and installation program in Germany has resulted in more than 21,000 MW installed. The north German state of Schleswig-Holstein currently meets more than 25 percent of annual electricity demand with more than 2,400 wind turbines. Schleswig-Holstein has met more than 50 percent of demand for selected months during each of the past four years. In addition to Germany's progress, there are 10,000 MW in Spain, 3,000 MW in Denmark, and more than 1,000 MW in Great Britain, the Netherlands, and Portugal.

Led by the United States, China, and Spain, more than 20,000 MW of new wind power was installed in 2007, bringing worldwide installed capacity to 94,112 MW; that figure that climbed to more than 100,000 early in 2008. This is an increase of 31 percent compared to the 2006 market and represents an overall increase in global installed capacity of about 27 percent.

Accounting for about 30 percent of the country's new power-producing capacity in 2007, the United States installed 5,244 MW in 2007, more than double the 2006 figure. Overall U.S. wind power generating capacity grew 45 percent in 2007, with total installed capacity at the end of the year at 16.8 GW. By

some forecasts the United States will overtake Germany as the leader on wind energy by the end of 2009.

The top five countries in terms of installed capacity are Germany (22.3 GW), the United States (16.8 GW), Spain (15.1 GW), India (8 GW), and China (6.1 GW). In terms of economic value, the global wind market in 2007 was worth about $36 billion in new generating equipment.[6]

The United States has a tremendous land-based wind resource with about 11 trillion kWh of achievable production, or more than three times the total of all electricity produced in the United States last year. To access these resources, the global wind industry has also been producing increasingly large and efficient turbines, with 4-, 5-, and 6-MW turbines now in production for land and ocean-based applications. In terms of resources available and utilized, it is striking that North Dakota, for example, where only 350 MW is installed today, has a larger wind resource than Germany. Wind-generated electricity is also, in many locations, the cheapest form of new power, with costs generally around 4 to 7cents per kWh range (see figure 44.2). While the wind energy industry is strongly dependent on tax incentives (the production tax credit) for continued growth, due to the barriers a new power-provider must overcome in many current market settings, wind energy is now directly competitive, and in some cases lower cost than fossil-fuel competitors.

Biofuels

Biofuels, too, offer a diverse range of promising alternatives both as liquid fuels for transportation and for use in stationary power plants. Today, ethanol constitutes 99 percent of all biofuels in the United States, with the

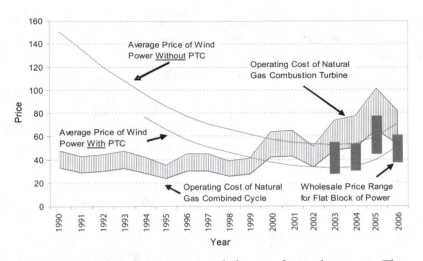

FIGURE 44.2. Comparison of wind power prices with the cost of natural gas power. The production tax credit (PTC) is currently 1.9 cents/kWh and is a constant source of debate politically over the extension of this credit. Source: Lawrence Berkeley National Laboratory (2007); Wiser, R., Bolinger, M., et al. (2007), *Annual Report on U.S. Windpower Installation, Cost, and Performance Trends 2006*. U.S. Department of Energy, Office of Energy Efficiency and Renewable Energy. http://eetd.lbl.gov/ea/ems/reports/ann-rpt-wind-06.pdf.

3.4 billion gallons of ethanol blended into gasoline in 2004 amounting to about 2 percent of all gasoline sold by volume and 1.3 percent (2.5×10^{17} joules) of its energy content. Interest in ethanol from corn, sugarcane, and cellulosic (woody materials containing lignin) sources has reached a fever pitch, with scientific advances moving almost as fast as in the investor interest in a sector that some analysts forecast could eventually provide a quarter or more of our total energy needs.

In a recent study my colleagues and I created a model to compare the energy and climate impacts of biofuel use versus that of gasoline. The energy "balance" of ethanol has been long debated, but we found that the studies that correctly account for the total life-cycle input to ethanol production yield a consistent picture of a positive net energy of about 4MJ/L to 9MJ/L for corn-based ethanol.[7] We conclude that use of ethanol can displace gasoline dramatically, by roughly 95 percent. The greenhouse gas impacts of using corn-based ethanol versus gasoline are more uncertain, but no significant savings appears to exist, with the likely point estimate of net greenhouse gases for corn ethanol at 18 percent below conventional gasoline, very close to our initially reported value, but with a wide uncertainty range of −36 percent to +29 percent.

Thus, corn-based ethanol may offer modest energy security and likely financial benefits, but not an opportunity for decarbonization. Cellulosic ethanol, however—meaning ethanol derived from woody plants—still holds considerable promise for further development. A major issue, however, has recently emerged that complicates the biofuel industry. A set of recent studies have shown that the greenhouse gas benefit of all biofuels grown on agricultural land, or on land cleared, could

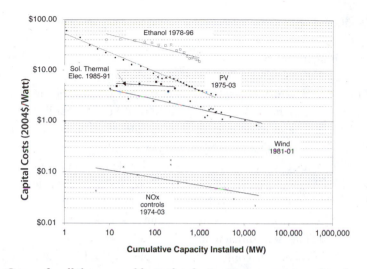

FIGURE 44.3. Curves for all the renewables technologies. Costs per unit produced generally decline by about 20 percent for each doubling of total production Source: Duke, R. D., and Kammen, D. M. (1999). The economics of energy market transformation initiatives. *The Energy Journal* 20(4), 15–64.

incur a significant carbon debt due to the indirect land use effect.[8] By these assessments, the benefits of biofuel production may be swamped by the loss of carbon storage and uptake in lands elsewhere that are converted to food production through the dynamics of market response to reductions in agricultural commodity production. As the biofuel industry evolves, it is now increasingly clear that a global perspective, and appropriate metrics to judge impacts on not only food availability, but also on water demand, soil, watershed impact, and cultural impact, will all be needed. We have termed this need for a sustainable fuel standard.[9]

Each of these technologies are now at or near what is often called the "tipping point": technological and economic stages in their evolution where investment and innovation, as well as market access, could move these attractive but generally marginal contributors to playing major roles in the regional and global energy supply.

Linking Transportation and Stationary Power

Plug-in hybrid electric vehicles (PHEV) provide a means to address: (1) the geopolitical consequences of imported oil; (2) the looming specter of climate change; and (3) the competitiveness of the American economy in the age of globalization.[10]

PHEVs are cars and trucks running fully or partially on batteries that can be charged by plugging into an electrical socket. They also have conventional engines that run on ordinary liquid fuels such as gasoline or ethanol, which removes any limitation on the distance they can be driven. PHEVs are not exotic vehicles of the distant future, but real cars that are in production today and have yet to take advantage of the dramatic improvements in battery technology that are beginning to occur, and which could be accelerated dramatically with a commitment to research, development, and deployment.

While conventional sedans have a maximum fuel economy of about 30 miles per gallon (mpg) and non-plug-in hybrids such as the Toyota Prius average about 50 mpg, PHEVs get an equivalent of 80 to 160 mpg of gasoline, and even more as gasoline in the PHEV's tank is replaced by biofuels. The implications for our national oil addiction are profound: if the current U.S. vehicle fleet were replaced overnight with PHEVs, oil consumption would decrease by 70 to 90 percent, completely eliminating the need for oil imports and leaving the United States self-sufficient in petroleum supply for many years to come.

A switch to PHEVs has equally profound implications for protecting Earth's fragile climate, not to mention the elimination of smog. By providing most of the energy for cars from power plants instead of fuel tanks, the environmental impacts of driving are concentrated upstream in a few thousand central station generating plants, instead of 100 million cars and all the individual consumer decisions associated with them. This focuses the problem of climate protection squarely on the issue of reducing the greenhouse gas emissions from electricity generation. While making electric power greener is itself no small undertaking, the outlines of a solution are at least visible, whether one prefers renewable energy sources such as wind and solar, or nuclear power, or the capture and burial of the carbon dioxide from coal.

Why would the electricity industry want to offer such rates? For the same reason they were proposed in 1906: it could solve the industry's fundamental economic problem since the time of Edison—the enormous difference between on-peak and off-peak demand. Electricity soars to high levels at certain times of the day and year, such as summer afternoons when air conditioning is in use. Utilities must spend money on generating units to meet peak demand, but lose money on their investment when these units sit unused the rest of the time. By charging electrical vehicles while people sleep, electricity demand could be made much more constant, allowing it to be met with inexpensive baseload units rather than expensive peakers. In California, for example, the replacement of 10 million conventional cars, or about 60 percent of the state fleet of 17 million vehicles, with PHEVs that were charged overnight would increase electricity demand to nearly the same level as daytime demand. In addition, electric vehicles not in use during the day could be configured to supply electricity to local distribution networks at times when the grid was under strain. These changes could transform the economics of the utility industry and ameliorate the financial impacts of a transformation to greener power generation.[11] As with most green technologies, we could encourage rapid and large-scale PHEV adoption by incorporating the real costs of carbon dioxide emissions into the price of energy.

Investing in Innovation: Retreating Instead of Advancing

The federal government and private industry are both reducing their investments in energy R&D at a time when geopolitics, environmental concerns, and economic competitiveness are all increasing the need for a major expansion in our capacity to innovate in this sector. Our ability to respond to the challenges of climate change or to the economic vulnerability of the nation to disruptions in

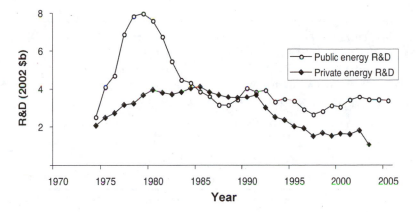

FIGURE 44.4. Patent/funding relationships for renewables. Source: Kammen, D. M., and Nemet, G. (2005). Reversing the incredible shrinking energy R&D budget. *Issues in Science & Technology,* Fall, 84–88.

our energy supply has been significantly weakened by the lack of attention to long-term energy planning. Calls for major new commitments to energy R&D have become common—the White House PCAST study of 1997 and the 2004 bipartisan NCEP recommend doubling federal energy R&D, and calls for energy "Manhattan Projects" have become frequent. Is the expansion of the U.S. energy research portfolio to much higher levels feasible? What is the "insurance value" of energy R&D against risks such as oil price shocks, electricity supply disruptions, local air pollution, and climate change?

We reviewed spending patterns of the six previous major federal R&D initiatives since 1940 and compared them to scenarios of increasing energy R&D by factors of five and ten (see table 44.1). Based on IPCC assessments of the cost to stabilize atmospheric CO_2 at 550 ppmv and other studies that estimate the probable success of energy R&D programs and the resulting savings from the technologies that would emerge, $15 to 30 billion per year in the United States would be sufficient. We find that the fiscal magnitude of a large energy research program—dramatically increasing our meager $3 billion annual national investment in energy research—is well within the range of programs in other sectors, each of which have produced demonstrable economic benefits as well as meeting their direct program objectives. U.S. energy companies could also increase their R&D spending by a factor of ten and still be below average relative to the R&D intensity of U.S. industry overall.

We recommend a sustained increase in funding—by at least a factor of five—to meet the energy challenges of the twenty-first century. The economic benefit of such a bold but long-overdue move would repay the country in job creation and global economic leadership—as well as being the scale of investment seen as needed to transform our economy into what at one time seemed impossible: a vibrant, environmentally sustainable engine of growth.

TABLE 44.1

Comparison of energy R&D scenarios and major federal government R&D initiatives. Sources:
Kammen, D. M., and Nemet, G. (2005). Reversing the incredible shrinking energy R&D budget.
Issues in Science & Technology, *Fall, 84–88; Nemet, G. F., and Kammen, D. M. (2007). U.S. energy*
research and development: Declining investment, increasing need, and the feasibility of expansion.
Energy Policy *35(1): 746–55.*

Program	Sector	Years	PEAK YEAR (2002$ Billions)		PROGRAM DURATION (2002$ Billions)		
			Spending	Increase	Spending	Extra Spending	Factor Increase
Manhattan Project	Defense	1940–1945	$10.0	$10.0	$25.0	$25.0	n/a
Apollo Program	Space	1963–1972	$23.8	$19.8	$184.6	$127.4	3.2
Project Independence for new energy technologies	Energy	1975–1982	$7.8	$5.3	$49.9	$25.6	2.1
Reagan defense expansion	Defense	1981–1989	$58.4	$27.6	$445.1	$100.3	1.3
Doubling NIH budget	Health	1999–2004	$28.4	$13.3	$138.3	$32.6	1.3
Post 9/11 War on Terror	Defense	2002–2004	$67.7	$19.5	$187.1	$29.6	1.2
5x energy scenario	Energy	2005–2015	$17.1	$13.7	$96.8	$47.9	2.0
10x energy scenario	Energy	2005–2015	$34.0	$30.6	$154.3	$105.4	3.2

Conclusion

Make Energy and the Environment a Core Area of Education in the United States

We must develop in both K–12 and college education a core of instruction in the linkages between energy and both our social and natural environment.

Establish a Set of Energy Challenges to Develop a Low-Carbon Economy

- Buildings that cleanly generate significant portions of their own energy needs (zero-energy buildings);
- Commercial production of 200-mile-per-gallon vehicles;
- Zero-energy appliances (appliances that generate their own power); and
- Clean power generated at residences, businesses, and industries.

Invest in Clean Energy Commensurate with Its Importance

Clean energy production—through investments in energy efficiency and renewable energy generation—has been shown to be a winner in terms of spurring innovation and job creation. This should be reflected in federal economic assessments of energy and infrastructure investment. Grants to states, particularly those taking the lead on clean energy systems, should be at heart of the federal role in fostering a new wave of "clean-tech" innovation in the energy sector.

Make the Nation the Driver of Clean Vehicle Deployment

Purchase for state transportation needs only vehicles meeting a high-energy efficiency target, such as 40 miles per gallon for sedans and 30 miles per gallon for utility vehicles. These standards are now possible thanks to improvements in vehicle efficiencies and the wider range of hybrids (including SUV models) now available. A key aspect of such a policy is to announce from the outset that the standards will rise over time, and to issue a challenge to industry that a partnership to meet these targets will benefit their bottom line and our nation.

Expand International Collaborations that Benefit Developing Nations at a Carbon Benefit

In many cases, tremendous opportunities exist not only to offset future greenhouse gas emissions and to protect local ecosystems both at very low cost, but also to directly address critical development needs such as sustainable fuel sources, the provision of affordable electricity, health, and clean water.

Notes

For additional information on efforts to promote renewable energy worldwide, please visit

www.REN21.net; see also REN21 Renewable Energy Policy Network (2005). *Renewables 2005 Global Status Report* (Worldwatch Institute: Washington, DC).

1. Farrell A. E., Plevin, R. J., Turner, B. T., Jones, A. D., O'Hare, M., and Kammen, D. M. (2006). Ethanol can contribute to energy and environmental goals. *Science* 311, 506–8; Kammen, D. M., Farrell, A. F., Delucchi, M. A., Plevin, R., Jones, A., and Nemet, G. (2007). "Energy and Greenhouse Gas Impacts of Biofuels: A Framework for Analysis," OECD Position Paper (Paris, France).

2. Kammen, D. M. September 27, 2006: A day to remember. *San Francisco Chronicle*, September 27, 2006.

3. IPCC Fourth Assessment Report, www.ipcc.ch.

4. See the Web site of the Renewable and Appropriate Energy Laboratory (http://rael.berkeley.edu).

5. Hultman, N., Koomey, J. G., and Kammen, D. M. (2007). What can history teach us about costs of future nuclear power? *Environmental Science & Technology* 40, 2088–93.

6. American Wind Energy Association, www.awea.org.

7. This model is freely available at http://rael.berkleley.edu/EBAMM.

8. Searchinger, T., Heimlich, R., Houghton, R. A., Dong, F., Elobeid, A., Fabiosa, J., Tokgoz, S., Hayes, D., and Yu, T.-H. (2008). *Science* 319, 1238–40; Fargione, F., Hill, J., Tilman, T., Polasky, S., and Hawthorne, P. (2008). Clearing and the biofuel carbon debt. *Science* 319: 1235–38.

9. Kammen, D. M. (2008).Reducing emissions in transportation fuels. *Bulletin of the Atomic Scientists*, March. Available online at www.thebulletin.org/columns/daniel-kammen/20080314.html.

10. See the April 2006 issue of *Scientific American*.

11. Lemoine, D., Kammen, D. M., and Farrell, A. E. (2008). An innovation and policy agenda for commercially competitive plug-in hybrid electric vehicles. *Environmental Research Letters* 3, 1–8.

Chapter 45

Designing Energy Supply Chains Based on Hydrogen

WHITNEY G. COLELLA

One potential stabilization strategy to limit the growth of atmospheric greenhouse gases is to change the energy technologies we use to ones that consume hydrogen fuel (H_2).[1] H_2 fuel can be consumed in energy conversion devices to power vehicles, to provide electricity, to heat spaces, or to cool them. Using currently available technologies, it is possible to design a future economy based on H_2 such that total emissions of greenhouse gases are reduced to zero. Hydrogen is one of the only fuels, or "energy carriers," that could enable this transition to zero emissions (another being electricity).[2] To build an H_2 economy that reduces emissions, policy makers and engineers must pay careful attention to the detailed design of each of the energy conversion devices, processes, and linkages within this new H_2 supply chain. We will learn supply chain analysis, the building blocks with which to design a future H_2 supply chain, and a methodology for selecting an H_2 supply chain design based on a society's relative weighting of its social values.

H_2 is a fuel with a high energy content that contains no carbon (C). By contrast, almost all other commercial fuels such as gasoline, natural gas, coal, and biofuels contain carbon. For these fuels, figure 45.1 shows their carbon content in terms of the mass of carbon in the fuel per unit of energy in the fuel shown in kilograms (kg) per gigajoule (GJ).[3] All other things being equal, fuels with low carbon content emit low carbon dioxide (CO_2) emissions when consumed.

Not all energy supply chains based on H_2 would achieve the same environmental benefits. The term "energy supply chain" refers to the multiple processes connecting the initial sources of the energy extraction with the final end use of that energy. An energy supply chain includes a series of processes from raw material extraction, through processing and energy production, and finally to end use of that energy and waste management. Although an energy supply chain may contain many processes, the most benefit is gained by focusing design efforts on the chain's bottleneck processes.[4] An environmental bottleneck is a process where the most environmental damage or the greatest energy loss occurs. Environmental bottlenecks can be identified through back-of-the-envelope energy consumption calculations and analysis of emis-

Fuel	Chemical Formula	Carbon Content: Mass of Carbon Per Unit Fuel Energy (kg of carbon/kJ)
Coal	$C_nH_{0.93n}N_{0.02n}O_{0.14n}S_{0.01n}$ *(s)*	29.0
Gasoline	$C_nH_{1.87n}$ *(l)*	19.6
Ethanol (a biofuel)	C_2H_6O *(l)*	19.4
Methanol	CH_4O *(l)*	18.7
Natural Gas	$C_nH_{3.8n}N_{0.1n}$ *(g)*	15.5
Methane	CH_4 *(g)*	15.0
Hydrogen	H_2 *(g)*	0.0

FIGURE 45.1. Carbon content: Mass of carbon per unit fuel energy.

sions data. An energy supply chain based on H_2 often has at least three primary bottleneck processes: (1) H_2 manufacture; (2) H_2 storage; and (3) H_2 end-use consumption (for example, in vehicles or power plants).

H_2 Can Be Manufactured in "Dirty" or "Clean" Ways

H_2 at room temperature is a gas that occurs naturally on earth only in minute quantities. As a result, H_2 must be manufactured from other chemicals that contain atomic hydrogen (H). For example, H_2 can be manufactured from pure H_2O, biomass, fossil fuels, or alcohols originally derived from biomass (ethanol) or fossil fuels (methanol), such those shown in figure 45.1. In 2005, the United States produced 8 MT/yr-H_2 (million metric tons of H_2 per year) for industrial use, approximately 14 percent of the 57 MT/yr-H_2 the United States would need to run its on-road vehicle fleet on H_2 fuel cells. A large portion of this H_2 is added to heavy petroleum to upgrade it to gasoline fuel. In 2005, the world produced about 50 MT/yr-H_2.[5]

Depending on the source of the H_2 and the processes involved, H_2 can be manufac-

tured in either "dirty" or "clean" ways. Dirty processes typically use fossil fuels, convert energy inefficiently, and produce high levels of emissions and solid waste.[6] Clean processes typically use renewable energy, convert energy efficiently, and produce low emissions and little solid waste. Of the numerous options for making H_2, each has different impacts on the environment and economic and political goals.

The three most promising H_2 production methods are (1) coal gasification; (2) steam reforming of natural gas; and (3) electrolysis powered by renewable sources, roughly in order of dirtiest to cleanest. Figure 45.2 compares these three main H_2 production processes against one another, relative to environmental, political, and economic goals.

Coal Gasification

Figure 45.1 also shows the carbon content of fuels per unit of atomic hydrogen (H). Coal has the highest C/H ratio. As a result, using coal alone as a source to make H_2 could release high levels of CO_2 compared to other fuels. The process of coal gasification combines coal at high temperatures under pressure with

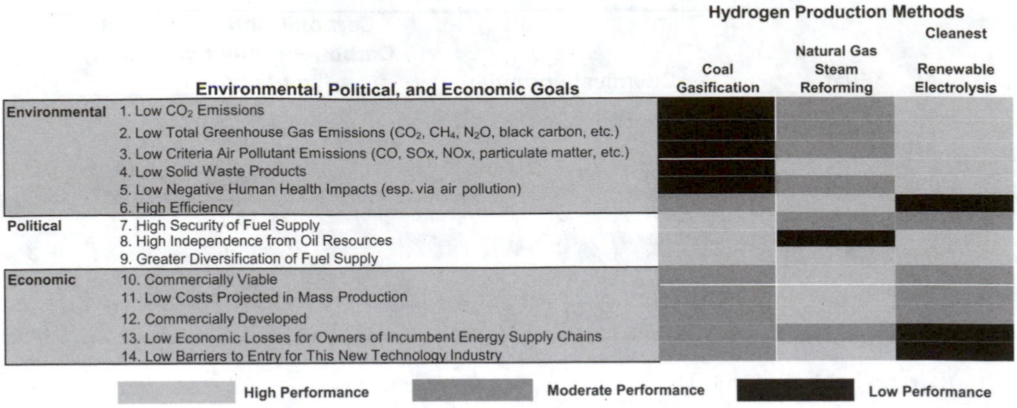

Environmental, Political, and Economic Goals	Coal Gasification	Natural Gas Steam Reforming	Renewable Electrolysis (Cleanest)
Environmental 1. Low CO_2 Emissions			
2. Low Total Greenhouse Gas Emissions (CO_2, CH_4, N_2O, black carbon, etc.)			
3. Low Criteria Air Pollutant Emissions (CO, SOx, NOx, particulate matter, etc.)			
4. Low Solid Waste Products			
5. Low Negative Human Health Impacts (esp. via air pollution)			
6. High Efficiency			
Political 7. High Security of Fuel Supply			
8. High Independence from Oil Resources			
9. Greater Diversification of Fuel Supply			
Economic 10. Commercially Viable			
11. Low Costs Projected in Mass Production			
12. Commercially Developed			
13. Low Economic Losses for Owners of Incumbent Energy Supply Chains			
14. Low Barriers to Entry for This New Technology Industry			

High Performance   Moderate Performance   Low Performance

FIGURE 45.2. Hydrogen production methods.

H_2O and oxygen (O_2) to produce H_2 and carbon monoxide (CO) (and other gases). The energy required to break the bonds in H_2O and in O_2 is provided by the fuel energy in coal. Additional H_2O is then added to the CO to convert these to CO_2 and more H_2, a process known as the water gas shift reaction.

Producing H_2 via coal gasification has benefits and drawbacks, summarized in figure 45.2. Certain countries could gain political benefits from coal-derived H_2, including (1) greater security of fuel supply; (2) a higher level of independence from oil resources; and (3) a greater diversification of transportation fuel feedstocks. One benefit for the United States is its large coal reserves within its borders compared to other fuels, which could enhance the "security of supply" of H_2 fuel to the United States. In part for this reason, the U.S. Department of Energy (DOE) has investigated the idea of phasing in coal gasification for H_2 production, so as to potentially reach more than one-fourth of total H_2 production in 2050.[7] Also, if the United States gradually replaced gasoline as its transportation fuel with H_2 derived from coal, the United States would become more independent from petroleum oil resources, which are situated in politically unstable parts of the world such as Venezuela and the Middle East, and would also diversify the types of fuel resources upon which it depends.

Compared to other H_2 production processes, a few drawbacks of coal-derived H_2 include that (1) it produces the highest CO_2 emissions; (2) it probably emits the highest air pollutant emissions of the three production methods described; (3) it has the potential to produce appreciable levels of unburned solid waste by-products; and (4) although the technology is established, it is not currently commercial.[8] CO_2 emissions from coal gasification can be reduced with carbon sequestration, the long-term storage of carbon in the ground in geological repositories, the oceans, or biological surroundings.

Steam Reforming of Natural Gas

While still consuming fossil fuels, using natural gas to produce H_2 can achieve significant reductions in greenhouse gases.[9] Currently, most H_2 is produced from natural gas through a process known as steam reforming. The steam reforming reaction requires energy input. This energy can come from an additional one-third mole of natural gas being oxidized to provide heat for the reaction.[10]

Producing H_2 through steam reforming has benefits and drawbacks, summarized in figure 45.2. A few benefits include that, compared with other H_2 production processes from fossil fuels, (1) it releases the lowest CO_2 and total greenhouse gas equivalent emissions (please refer to figure 45.1); (2) it emits the lowest air pollutant emissions; (3) it creates the least negative human health impacts stemming from air pollution; and (4) it can operate at relatively high efficiency.[11] It would also produce relatively low levels of solid waste products (mostly spent catalysts) and a greater diversification of the transportation fuel supply if used in conjunction with H_2 vehicles, and it is currently commercial.

For the United States, a significant concern regarding natural gas–derived H_2 is the availability and ownership of natural gas fuel. Within the United States, natural gas reserves are modest, offering little additional security of supply. However, Canada holds extensive natural gas reserves, and these are considered to contribute to U.S. fuel security in the event of security threats or political instability in other countries. Even considering Canada's reserves, a significant drawback of natural gas–derived H_2 is that it is not a resource that is financially and physically independent of oil; it is usually extracted from wells as a by-product of oil production. Natural gas is often extracted from oil fields because it collects as a gaseous layer above the liquid oil within the underground gas field. As a result, oil extraction companies are often natural gas production companies. If these companies perceive that they are more profitable selling oil than

selling H_2, they may perceive H_2 sales as cannibalizing demand for a more profitable product—oil—and may have an incentive to create barriers to switching to H_2. Two drawbacks to natural gas–derived H_2 are the potential economic losses to owners of the incumbent energy supply chain (oil companies) and the consequential financial barriers to entry for this technology. At the same time, some traditional energy companies are choosing to develop prototype H_2 generators for refueling stations, often in conjunction with government sponsorship.[12] To overcome these competing incentives, one option for public policy makers is to encourage the financial separation of the oil and natural gas industries.

Electrolysis of Water with Renewable Sources

One of the most environmentally benign ways to produce H_2 is through the electrolysis of water powered by renewable electricity. By passing an electric current through liquid H_2O, the individual H_2O molecules can dissociate into the separate molecules of H_2 and O_2. The electrolysis of one mole of H_2O produces one mole of H_2 gas and half a mole of O_2 gas. The electricity needed for this reaction could be provided by renewable energy devices such as hydroelectric power plants, geothermal power plants, wind turbines, solar photovoltaic arrays, solar thermal-electric power plants, wave power devices, biomass plants, and tidal power devices.[13]

H_2 production through renewable electrolysis has several benefits, summarized in figure 45.2. For example, it has very low negative environmental impacts. It emits either no or extremely low (1) CO_2 emissions; (2) greenhouse gases; and (3) criteria air pollutants. For example, if the entire U.S. on-road vehicle fleet were switched to fuel cell vehicles, this switch would result in an annual reduction in total U.S. anthropogenic CO_2 emissions of approximately 23 percent if the H_2 were produced with renewable electrolysis as defined above.[14] Consequently, it is understood to have low negative human health impacts, such as from air pollution. Also, renewably generated H_2 produces very low levels of solid waste. Two political benefits of H_2 production through renewable electrolysis include a greater independence from petroleum oil resources and a greater diversification of the fuel supply for vehicles, power plants, heating, and cooling. (Please note that H_2 production from biomass or biofuels also may be considered renewable, but these processes can result in high levels of solid waste and air pollutant emissions.)

H_2 production through renewable electrolysis has several limitations, also summarized in figure 45.2. Two of its largest impediments are the large economic losses it would inflict on powerful technology incumbents, such as incumbent fuel and energy supply companies. These types of companies typically have neither large vested interests nor a technical expertise in renewable energy technologies. As a result, in response to these types of threats, incumbents with large resources have successfully destroyed their competitors. For example, in the early twentieth century, General Motors (GM) organized a company, National City Lines, "to engineer the demise of forty-five electrical mass-transit systems in sixteen states," first by buying all of the electric trolley and transit systems in the United States and then by replacing their services with that of GM buses using internal combustion engines (ICEs).[15] These actions eliminated a transportation substitute for the ICE vehicles

GM sold.[16] In a similar way, traditional energy companies are likely to see H_2 production through renewable electrolysis as a threat to their profit streams and, therefore, can be expected to engage in destructive competition, often not prohibited by law. To reduce the destruction of new environmentally friendly energy technology companies, public policy makers could introduce legislation to make illegal this takeover-and-shutdown business strategy, especially in those industries aimed at improving the environment.

An additional concern of H_2 production via renewable electrolysis is its financial viability. Renewable-H_2 generated by hydroelectric, geothermal, or wind could compete economically per mile with gasoline fuel if consumed in a high-efficiency H_2 fuel cell vehicle.[17] Renewable-H_2 is less economical when generated from more expensive solar photovoltaic, solar thermal, wave, and tidal power devices. Although, for the same source of renewable electricity, fuel cell vehicles are more expensive than battery-electric vehicles as of 2008, batteries suffer from low energy density, which limits their vehicles' range and therefore their commercial viability.

H_2 Can Be Transported, Stored, and Delivered in "Dirty" or "Clean" Ways

H_2 can be transported, stored, and delivered as either a gas or a liquid. Regarding the gaseous method, in order to store H_2 it must be compressed at high pressures to achieve practical energy densities. This compression process consumes at least 10 percent of the energy value of the fuel. Currently, gaseous H_2 is most often compressed onsite prior to delivery. In the future, it could be piped from manufacturing centers to H_2 refueling stations, similar to the way that natural gas is piped to homes and businesses today. Because existing carbon steel pipelines would degrade too severely in the presence of H_2, a parallel stainless steel pipeline infrastructure would need to be built to carry H_2, at significant cost. Regarding the liquid method, in order to create liquid H_2 it must be cooled and slightly compressed. This process consumes energy, equivalent to approximately 30 percent of the energy value of the fuel. Liquid H_2 is usually trucked from manufacturing centers to refueling stations or end users. When H_2 is not produced and consumed onsite at a chemical plant, it almost always is delivered this way. Despite its greater energy consumption requirements and costs, liquid storage is often used in place of gaseous storage when high energy density is important, such as when storing H_2 onboard vehicles. Liquid storage can hold several times the amount of hydrogen as practical gaseous storage (depending on the pressure) per unit volume.

Figure 45.3 compares gaseous and liquid H_2 storage against environmental and economic goals. Because the compression of gaseous H_2 consumes less energy than liquefying H_2, it is generally considered the more environmentally friendly and the more economical of the two options. All other variables remaining constant, and assuming the same electrical power source for compression as for liquefaction, the creation of gaseous H_2 can be expected to produce less CO_2 emissions, greenhouse gases, air pollution, solid waste, and negative health impacts from air pollution created during power generation. Both gaseous and liquid H_2 are technically developed to the point of being commercially viable in different applications, although compression is generally cheaper due to lower energy costs. The use of both would threaten incumbent energy supply chain owners, but the barriers to entry for either technology are only moderate because both already have been commercialized for other applications.

H_2 Can Be Consumed in "Efficient" or "Inefficient" Ways

H_2 can be consumed either efficiently or inefficiently in different types of energy conversion devices, including (1) the ICE; (2) the gas turbine engine; (3) the fuel cell; and (4) the gas burner. Of these, (1) and (2) convert the chemical energy in H_2 to heat and then to mechanical work; (3) converts the chemical energy directly to electricity and heat; and (4)

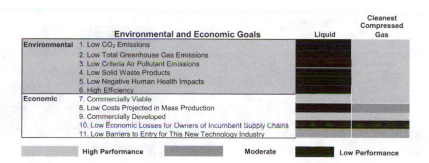

FIGURE 45.3. Hydrogen storage methods.

converts the chemical energy solely into heat. If these energy conversion devices consumed renewable H_2, they would achieve significant reductions in CO_2 and other emissions compared to devices consuming hydrocarbon fuels.

Fuel cells have the potential to achieve extremely high efficiencies in converting a fuel's chemical energy into electricity, over and above the efficiencies of engines.[18] Engineers define efficiency as the ratio of useful work done (such as the production of electricity) to the energy supplied (for example, by a fuel), usually expressed as a percentage. In theory, the maximum efficiency of an H_2 fuel cell is higher than that of an H_2 ICE at low temperatures.[19] In practice, the measured efficiencies of fuel cells are between 45 percent and 70 percent, compared to the measured efficiencies of ICEs around 35 percent. When electricity is desired over mechanical work, fuel cells have an added advantage over engines or turbines, because they transfer fuel energy directly to electricity and skip the additional process of transforming mechanical work into electricity, which results in even lower engine system efficiencies.

Figure 45.4 compares engines, turbines, and fuel cells against environmental and economic goals. Fuel cells and turbines can achieve efficiencies that are higher than ICEs. As a result, their environmental impacts are lower. For example, if coal-derived H_2 is used to fuel all three, the fuel cell is likely to emit the lowest air pollution and to produce the least negative human health impacts from air pollution. H_2 engines will produce nitrogen oxides (NO_x) within certain operating regions.[20] Also, unlike batteries, all three technologies are likely to produce low levels of toxic or hazardous solid waste.

Of the three technologies, fuel cells are the least economically competitive and technically developed. Although projected to be economical in mass production, fuel cells are currently economical only in niche markets.[21] Although manufacturers have built fuel cells that are compact and light enough for vehicular applications, the cells have only limited lifetimes, about 20 percent of their long-term target goal for lifetime (1,000 hours of operation versus the 5,000-hour long-term goal).[22] An additional drawback of fuel cells includes the high economic losses their adoption potentially could have on the owners of the incumbent energy supply chain, such as ICE

FIGURE 45.4. Hydrogen consumption methods.

manufacturers and all of their upstream and downstream partners, including engine companies, engine researchers, and oil companies, as well as traditional power plant manufacturers and traditional utility operators.

Example Comparison of Hydrogen Vehicle Supply Chains

Permutations for a future H_2 economy can be evaluated by linking together different combinations of the processes in the energy supply chain. One option from each of the three figures (figures 45.2, 45.3, and 45.4) can be chosen and linked together to build a possible future H_2 supply chain. We juxtapose two possible H_2 supply chains, one environmentally detrimental and the other beneficial. The dichotomy in results they produce underscores the importance of paying close attention to the detailed engineering design of all of the H_2 supply chain processes to ensure that a change in technologies produces the desired environmental, economical, and political goals (shown in the first column of figures 45.2, 45.3, and 45.4).

An H_2 supply chain could be designed that is more detrimental than our current energy supply chains: (1) coal fuel is gasified at 60 percent efficiency into H_2, coal having the highest carbon content per unit energy and the highest carbon-to-hydrogen ratio of fossil fuels (figure 45.1); (2) this H_2 is liquefied at 70 percent efficiency (as previously discussed) and trucked to refueling stations; and (3) the liquid H_2 then refuels a fleet of ICE vehicles that are 28 percent efficient from tank-to-wheel.[23] In total, from "well-to-wheel," this coal-liquid H_2-ICE supply chain would be approximately 12 percent efficient (60 percent × 70 percent × 28 percent). For comparison, for

a gasoline ICE vehicle supply chain, the "well-to-tank" efficiency of crude oil production, chemical processing, and gasoline transport is about 88 percent efficient, and the "tank-to-wheel" efficiency of the most advanced gasoline ICE vehicles is about 22 percent, for a total "well-to-wheel" efficiency of about 19 percent (88 percent × 22 percent).[24] Therefore, the proposed coal-liquid H_2-ICE supply chain would have a well-to-wheel efficiency that is lower by about 7 percentage points, which is approximately 40 percent worse, compared with a gasoline ICE chain. These results are summarized in figure 45.5. It would also increase CO_2 emissions significantly, due to the higher carbon content of coal compared with gasoline (figure 45.1). Skeptics of the H_2 economy often point to such detrimental H_2 supply chains to undermine serious consideration of a switch.

An H_2 supply chain also can be designed to be beneficial: (1) natural gas is piped to refueling locations where it is chemically converted through steam reforming and solar thermal heating to H_2 gas at an efficiency of 80 percent, natural gas having the lowest carbon content per unit energy (see figure 45.1) and the lowest carbon-to-hydrogen ratio (figure 45.3) of fossil fuels (figure 45.1); (2) this H_2 is compressed at 90 percent efficiency (as previously discussed); and (3) the H_2 then refuels a fleet of fuel cell vehicles that are 53 percent efficient from tank-to-wheel.[25] In total, this natural gas-gaseous H_2-fuel cell vehicle supply chain would be approximately 38 percent efficient (80 percent × 90 percent × 53 percent) from "well-to-wheel." It would be about twice as efficient as a gasoline ICE chain (around 19 percentage points). These results are summarized in figure 45.5. It would also decrease CO_2 emissions significantly, due to the lower carbon content of

	Conventional supply chain: gasoline-ICE	Dirtiest supply chain: coal-liquid H_2-ICE	Cleanest supply chain: natural gas-gaseous H_2–fuel cell
Well-to-tank efficiency	88%	42%	72%
Tank-to-wheel efficiency	22%	28%	53%
Total well-to-wheel efficiency	19%	12% (40% worse)*	38% (100% better)*

()* relative to the conventional supply chain of gasoline ICE

FIGURE 45.5. Supply chains.

natural gas compared with gasoline, and decrease air pollution emissions and the human health effects associated with them.[26]

Conclusion

These two examples present a striking dichotomy of the potential impacts of a future H_2 economy. H_2 supply chains can be designed to be either energy inefficient or energy efficient. A future H_2 chain could either increase or decrease current greenhouse gas emissions, depending on the choice of processes in that future chain, the availability of natural resource inputs to that chain, and the relative weighting of other environmental, economic, and political values. Although one greenhouse gas stabilization wedge could be based on H_2, implementing this future H_2 economy would require the concerted efforts of engineers and policy makers alike, designing the individual pieces of this chain to meet society's goal.

Notes

1. Colella, W. G., et al. Switching to a U.S. hydrogen fuel cell vehicle fleet: The resultant change in emissions, energy use, and global warming gases. Journal of Power Sources 150 (2005), 150–81.

2. Some engineers prefer to refer to hydrogen not as a "fuel" but as an "energy carrier," because they regard "fuels" as typically being mined or unearthed, whereas "energy carriers" such as hydrogen must be manufactured from another substance and contain a significant quantity of energy that can be converted into work at a later stage. The most well-known "energy carrier" is electricity.

3. Derived from: for all liquid fuels: Heywood, J. B. Internal Combustion Engine Fundamentals (New York: McGraw-Hill, 1988), especially table D.4, "Data on Fuel Properties," p. 915; for coal: Starkman, E. S. Combustion-Generated Air Pollution (New York-London: Plenum Press, 1971), via the Ohio Supercomputer Center (OSC) Web site www.osc.edu/research/pcrm/emissions/coal.shtml. Calculations based on Lower Heating Values (LHV).

4. Goldratt, E. M., and Cox, J. The Goal (New York, NY: North River Press, 1992).

5. Raman, V. "Hydrogen Production and Supply Infrastructure for Transportation," Workshop Proceedings: The 10-50 Solution: Technologies and Policies for a Low-Carbon Future, The Pew Center on Global Climate Change and the National Commission on Energy Policy, 2004.

6. Kuhn, I., Thomas, S., Lomax, F., James, B., and Colella, W. Fuel Processing Systems for Fuel Cell Vehicles, report for the U.S. Department of Energy, 1997. James, B., Lomax, F., Thomas, S., and Colella, W. PEM Fuel Cell Power System Cost Estimates: Sulfur-Free Gasoline Partial Oxidation and Compressed Direct Hydrogen, report for the U.S. Department of Energy, 1997.

7. Post-Presentation Oral Discussion with Margaret Singh, Argonne National Laboratory, "Regional Hydrogen Demand, Production and Cost Estimates," UC Davis ITS Hydrogen Case Studies Workshop, Slide 11, June 28th, 2005.

8. Colella, W. G., et. al., 2005 op. cit., especially table 10; for example, Rizeq, G., Project Leader, Fuel Conversion Lab, GE Global Research, 18 Mason, Irvine, CA 92618, e-mail communication, October 16, 2006; Kulkarni, P. P.,

Subia, R., Wei, W., Cui, Z., Zamansky, V., Shisler, R., McNulty, T., Rizeq, G., and Gillette, G. *Advanced Unmixed Combustion/Gasification: Potential Long Term Technology For Production of H_2 and Electricity from Coal with CO_2 Capture*, Oral Presentation for 23rd International Pittsburgh Coal Conference, 2006. For example, Rizeq, G., Project Leader, Fuel Conversion Lab, GE Global Research, 18 Mason, Irvine, CA 92618, telephone interviewed by Whitney Colella, August 19, 2004; Colella, W. G., et al., 2005 op cit. See also www.tfhrc.gov/hnr20/recycle/waste/cbabs1.htm; and *Gasification Plant Cost and Performance Optimization* (Washington, DC: U.S. Department of Energy, 2003), DE-AC26-99FT40342, chapter V and appendix H, Subtask 1.7 "Coal to Hydrogen Plant."

9. Colella, W. G., et al., 2005 op. cit.

10. O'Hayre, R., Cha, S., Colella, W., and Prinz, P. *Fuel Cell Fundamentals* (New York: Wiley, 2006), example problem 10.6, pp. 301–3.

11. *Fuel Processor Reactor Data Sheet: FP05 & FP06 Fuel Processor*, Johnson Matthey Fuel Cells, Blount's Court, Sonning Common, Reading RG4 9HN, United Kingdom, 2005; Colella, W. G., et. al., 2005 op. cit. p. 18; compiled from (1) *PC 25 Model C Fuel Cell Power Plant Design and Application Guide*, Revision B, November 2001, FCR-15389B, UTC Fuel Cells, 2001; and (2) e-mail communication with Joe Staniunas, Engineer, United Technologies Fuel Cells Inc., February 15, 2005.

12. For example, please see three presentations from the *2006 U.S. Department of Energy Hydrogen Program Review Meeting*, May 16–19, 2006: (1) Keenan, Greg. Air Products and Chemicals Inc., "Validation of an Integrated Hydrogen Energy Station"; (2) Guro, David. Air Products and Chemicals Inc., "Development of a Turnkey H_2 Refueling Station"; and (3) Liss, William E., "Development of a Natural Gas-to-Hydrogen Fueling System."

13. For example, please see The Geysers Geothermal Electric Facilities, operated by Calpine Company, Mayacamas Mountains, CA: www.geysers.com/. Ogden and Nitsch, "Solar Hydrogen," in ed. Johansson, T. B., *Renewable Energy: Sources for Fuel and Electricity* (Island Press, Washington, DC, 1993), p. 958. For another example, please see Solar Electric Generating Systems, operated by KJC Company, Kramer Junction, CA:

www.solel.com/products/pgeneration/ls2/kramer junction/; Limpet, built by Wavegen, Inverness, Scotland: www.wavegen.co.uk/; and Marine Current Turbines, Bristol, UK: www.marineturbines.com/home.htm.

14. Colella, W. G., et al., 2005 op. cit. This analysis assumes approximately the same energy and material consumption requirements for the production and maintenance of fuel cell systems as for current internal combustion engines. This assumption is reasonable for the long term, as progress in research and development increases the durability and lifetime of fuel cells.

15. Bradford Snell, testimony to Senate Subcommittee on Antitrust and Monopoly, Industrial Reorganization Act Hearings, Part 3, 93rd Congress, 2nd session, 1974, p. 1810.

16. Adams, W., and Brock, J. *The Bigness Complex* (New York: Pantheon Books, 1986), p. 68.

17. Jacobson, M. Z., Colella, W. G., and Golden, D. M. Effects on air quality and health of converting the U.S. fleet of on road vehicles to hydrogen fuel cell vehicles. *Science* 308 (2005), 1901–5.

18. Ostwald, W. Z. *Elektrochem.* 1 (1894), 122.

19. Colella, W. *Combined Heat and Power Fuel Cell Systems*, Doctoral Thesis, Department of Engineering Sciences, the University of Oxford, 2003. Carnot, S. N. *Réflexions sur la puissance motrice du feu et sur les machines propres à développer cette puissance* (Reflections on the Motive Power of Fire) (Paris: 1824). Kartha, S., and Grimes, P. Fuel cells: Energy conversion for the next century. *Physics Today*, November 1994, pp. 54–61.

20. Freymann, R. Director, BMW Group Research and Technology in Munich, Germany. "Hydrogen Research at the BMW Group: Our Vision of a Sustainable Mobility," Stanford University Mechanical Engineering Seminar Series, April 25, 2005. White, C. "A Technical Review of Hydrogen Internal Combustion Engines," California Air Resources Board: ZEV Technology Symposium, September 25, 2006, slide 17.

21. Lomax, F. D., James, B. D., Baum, G. N., and Thomas, C. E. *Detailed Manufacturing Cost Estimates for Polymer Electrolyte Membrane (PEM) Fuel Cells for Light Duty Vehicles*, Directed Technologies, Inc., Arlington, VA, October 1997.

Colella, W., Niemoth, C., Lim, C., and Hein, A. "Evaluation of the Financial and Environmental Feasibility of a Network of Distributed 200 kWe Cogenerative Fuel Cell Systems on the Stanford University Campus," *Fuel Cells: From Fundamentals to Systems*, 2005.

22. Wipke, K. "Controlled Hydrogen Fleet and Infrastructure Analysis," National Renewable Energy Laboratory, 2006 U.S. Department of Energy Hydrogen Program Review Meeting, May 19, 2006, slide 5; slides 6, 31, and 40; slide 21. Wipke, K., Welch, C., Thomas, H., Sprik, S., Gronich, S., Garbak, and Hooker, D., "Controlled Hydrogen Fleet and Infrastructure Demonstration and Validation Project: Project Overview and Fall 2006 Results," California Air Resources Board: ZEV Technology Symposium, September 25, 2006; slide 30; slide 17. Wipke, K., "Controlled Hydrogen Fleet and Infrastructure Analysis," 2007 U.S. Department of Energy Hydrogen, Fuel Cells & Infrastructure Technologies Program Review, May 17, 2007, slide 5; slide 17.

23. This efficiency is based on (1) the efficiency of one of Honda's most advanced 2005 gasoline ICE vehicles tested against the U.S. Environmental Protection Agency's City Driving Cycle being 22 percent; (2) an H_2-ICE efficiency of 38 percent and a gasoline ICE efficiency of 30 percent at part-load; and (3) the assumption of the same mechanical drivetrain efficiency between vehicles brings the ICE vehicle efficiency to equal 22 percent × 38 percent / 30 percent = 28 percent. E. Villanueva, Senior Engineer, Honda R&D Americas Inc., Presentation at Honda R&D, Torrance, CA, March 22, 2005. T. Kawanabe, Managing Director, Honda R&D Co. Ltd., Presentation at EVS-20, Long Beach, November 19, 2003. Berger, E. BMW Group. "BMW Hydrogen Near Zero Emission Vehicle Development," CARB ZEV Technology Symposium, September 2006, slide 12.

24. Meier, P. J., and Kulcinski, G. L. *Life-Cycle Energy Cost and Greenhouse Gas Emissions for Gas Turbine Power*, Fusion Technology Institute, University of Wisconsin, December 2000; Colella, W. G., et. al., 2005 op. cit., p. 64.

25. Kawanabe, 2003 op. cit. Meier and Kulcinski, 2000 op cit. This fuel cell vehicle efficiency is the same efficiency as Honda's most advanced 2007 hybrid fuel cell vehicles tested against the U.S. EPA's City Driving Cycle. It is consistent with 2006 DOE dynamometer test results for the best performing prototype fuel cell passenger cars, which report a gasoline-gallon-equivalent mileage of approximately 74 miles per gallon on the combined City & Highway Driving Cycle and measured on-board fuel cell system efficiencies ranging between 52 percent and 58 percent for all fuel cell vehicles. By contrast, the U.S. fleet average is 21 miles per gallon. E-mail communications between W. Colella and B. Knight, Vice President, Honda R&D Americas Inc., March 25, 2005 and April 11, 2005. http://world.honda.com/FuelCell/.

26. Colella, W. G., et al., 2005 op. cit.

Chapter 46

Nuclear Energy

BURTON RICHTER

Why Nuclear?

Nuclear energy is undergoing a renaissance, driven by two very loosely coupled needs: the first, to provide more energy to support economic growth worldwide; the second, to mitigate global warming driven by the emission of greenhouse gases from fossil fuel. Many forecasts of energy demand in the twenty-first century have been made and all give roughly the same answer: barring some unforeseen circumstance, world energy demand will track world economic growth with the largest increases taking place in the developing world. The International Institute of Applied Systems Analysis, for example, predicts in its midgrowth scenario that primary energy demand will increase by a factor of two by mid-century and by nearly another factor of two by the end of this century.[1] By the year 2015 the developing countries taken together are projected to pass the industrialized ones in primary energy use. China has already passed the United States as the largest energy consumer.[2] It is worth noting that economic growth in China and India is currently higher than assumed in that scenario.

Supply constraints on oil and natural gas are already evident. The only fossil fuel in abundant supply is coal. However, coal has serious pollution problems. Economics and the environment are driving the need for new large-scale energy sources, especially for carbon-free energy sources. Nuclear energy is one such source. While it cannot be the sole solution to energy supply problems, nuclear energy has the capacity to address economic and climate change concerns, provided the public can be assured that it is safe, that nuclear waste can be disposed of safely, and that the risk of weapons proliferation is contained.

For example, nuclear power supplies 78 percent of electricity in France, which contributes to its relatively low carbon intensity at half the world average (0.56 kg CO_2 per \$GDP versus 0.28 in France). If global carbon intensity matched that of France, carbon emissions would be reduced by about 3.5 billion tons per year, and global warming would be greatly slowed. The global-warming issue has caused prominent environmentalists to rethink their opposition to nuclear power. The question to be confronted is which devil

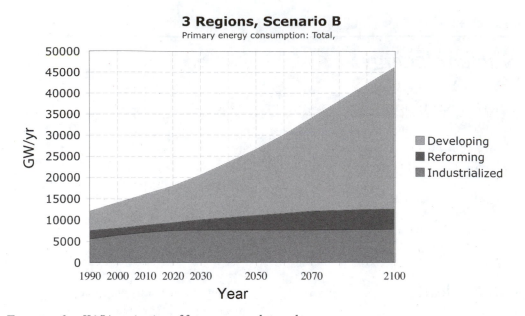

FIGURE 46.1. IIASA projection of future energy demand.

TABLE 46.1

CO$_2$ intensity Source: International Energy Agency, "Key World Statistics 2003."

Area	GDP (ppp) (Billions of U.S. Dollars)	CO$_2$/GDP Kg/$(ppp^a)
World	42,400	0.56
France	1,390	0.28

^{a}Exchange rates are at purchasing power parity.

would they rather live with, global warming or nuclear energy?

Nuclear Power Growth Potential

At present there are about 440 reactors worldwide supplying 16 percent of world electricity and 20 percent of U.S. electricity. Projections for growth in nuclear power keep increas-ing. The International Atomic Energy Agency (IAEA) projections as of July 2004 for the year 2030 ranged from a high of 592 GWe (gigawatt-electrical) to a low of 423 GWe. This represents a net growth of between 16 percent and 60 percent during the next twenty-five years. A recent MIT study, "The Future of Nuclear Power," projected about 1,000 GWe by 2050, while a recent Electricité de France projection was for about 1,300 GWe by 2050.[3] The most recent tabulation as of December 2007 lists thirty-four reactors under construction, ninety-four in the advanced planning stage, and 222 more under discussion.[4]

Energy security is probably the main driver in this expansion. The costs of nuclear energy will probably not be a limiting factor. Recent presentations by Westinghouse, General Electric, and AREVA to a DOE special committee on the future of nuclear energy in the United States claimed that the cost of electricity from a new nuclear plant in the United States would be comparable to a coal

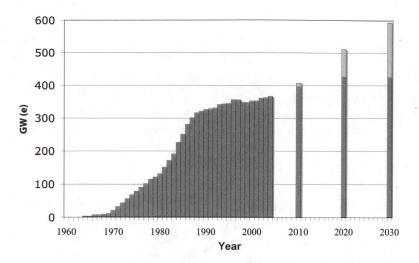

FIGURE 46.2. Nuclear power projection to 2030.

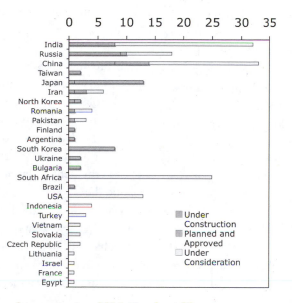

FIGURE 46.3. World nuclear expansion, 2006: Number of future reactors.

plant after first-of-a-kind engineering costs have been recovered and after manufacturing experience is gained with five or so new plants. Even so, projections like those above represent the expenditure of $1 to $2 trillion on nuclear plants in the next fifty years. It is not clear that we will have the personnel trained in nuclear science and engineering for the construction, operation, and regulatory needs of a system that large.

Safety

The new generation of light-water reactors has been designed to be simpler to operate and maintain than the old generation. They have more passive safety systems—for example, emergency cooling systems that rely only on gravity feed rather than pumps. Some designs are claimed to be passively safe in any kind of emergency.

Nevertheless, only with a strong regulation and inspection system can the safety of nuclear systems be assured. Without one, the risks grow. No industry can be trusted to regulate itself, especially when the consequences of a failure extend beyond the bounds of damage to that industry alone. Recent examples of corrosion problems in a U.S. reactor and in several Japanese reactors show again the need for rigorous inspections. In the Davis-Besse reactor in the United States and in several reactors of the Tokyo Electric Company corrosion of the reactor vessel at penetration points had proceeded for much longer than should have been the case with the expected regular inspection programs. The procedures have since been tightened up by the regulators.

Spent Fuel Treatment

There are three main elements of spent reactor fuel: uranium (95 percent), fission frag- ments (4 percent), and other long-lived components (1 percent). In principle, there is no real difficulty with the uranium that makes up the bulk of the spent fuel. It is not particularly radioactive and usually contains more residual U-235 than natural ore. It could be used as input for enrichment or put back in the mines from which it came. While fission fragments are highly radioactive, the vast majority of them have to be stored for only a few hundred years for radioactive decay to reduce the hazard to negligible levels. Robust containment is simple to build to last the requisite time. The problem comes mainly from the last 1 percent of the spent fuel, which is composed of plutonium (Pu) and the minor actinides: neptunium (Np), americium (Am), and curium (Cm), all four collectively known as the transuranics or TRU. For some of the components of this mix, the toxicities are high and the lifetimes are very long. There are two general ways to protect the public from this material: isolation from the biosphere for hundreds of thousands of years or transmutation by neutron bombardment to change them into shorter-lived fission fragments.

Permanent isolation is the principle behind the "once-through" system as advocated by the United States from the late 1970s until recently as a weapons-proliferation control mechanism. In this system all components of the spent fuel are kept together. The radiation from the fission fragments is a lethal barrier to

TABLE 46.2

Components of spent reactor fuel

Component	Fission Fragments	Uranium	Long-Lived Component
Percent of Total	4	95	1
Radioactivity	Intense	Negligible	Medium
Untreated Required Isolation Time (years)	200	0	300,000

theft, and the intact spent fuel is isolated in a geological repository. The plutonium in the spent fuel is not separated from the rest of the material and, thus, cannot be used in a nuclear weapon. In a world with a greatly expanded nuclear power program, the once-through system may not be workable, due to a combination of public suspicion that the material cannot remain isolated from the biosphere for hundreds of thousands of years and technical issues that would require a very large number of repositories.

To use the United States as an example, if nuclear energy were to remain at the projected 20 percent fraction of U.S. electricity needs through the end of the century, the spent fuel in a once-through scenario would need nine repositories of the capacity of the permanent waste repository under development in Nevada at Yucca Mountain.[5] If the number of reactors in the United States increases by mid-century to the 300 GWe projected in the MIT study, the United States would have to open a new Yucca Mountain every six or seven years. This would be quite a challenge since the United States has not been able to open the first one.

The alternative to once-through is a reprocessing system that separates the major components and treats each component appropriately. France has the most developed reprocessing system. In France, spent fuel is first separated into its three main components: uranium, fission fragments, and the TRU, which is further split into Pu and the three minor actinides. All of the TRU is produced by the capture of some of the fission neutrons by the uranium in the fuel. Some of it, particularly the Pu can itself, be used as fuel in a reactor increasing the total amount of energy coming from the original uranium fuel by about one-third. The Pu is said to be usable in weapons even though it is not the nearly pure plutonium (Pu-239) that weapons builders prefer.

Next, mixed oxide fuel, MOX, is made by mixing the Pu-oxide with an appropriate fraction of the uranium-oxide in the spent fuel. The extra uranium can now be safely disposed of. The fission fragments and minor actinides are embedded in a special glass (vitrified) for eventual emplacement in a repository. The glass used in vitrification appears to have a lifetime of many hundreds of thousands of years in the clay of the proposed French repository.

The French plan keeps the spent MOX fuel unreprocessed until fast-spectrum reactors are deployed commercially.[6] These fast-spectrum reactors have higher average neutron energy than the LWRs now in use and can use a mix of plutonium and uranium-238 as fuel and, in principle, fission all of the minor actinides as well. It is possible to create a continuous recycling program where the plutonium from the spent MOX fuel is used to start the fast-spectrum system, the spent fuel from the fast-spectrum system is reprocessed, all the plutonium and minor actinides go back into new fuel, and so forth. In this system nothing but fission fragments go to a repository, and as mentioned before, these would only need to be stored for several hundred years.

Although this is plan sounds good, there is still much work to be done before putting it into practice. Currently, the only fast-spectrum reactors are the sodium-cooled fast reactors (SFRs), which have been built and operated in France, Japan, Russia, and the United States. However, only plutonium-uranium fuel is qualified for the SFR; fuel containing minor actinides is not. Moreover, facilities to test and qualify the new fuels are in short supply. The United States has foolishly killed off its Fast Flux Test Facility at Hanford, Washington. France plans to shut down the

SFR in 2009. Unless new ones are built, the only facilities that will be left are in Japan and Russia. Clearly a coherent international program is needed to support and to use these remaining facilities in an international R&D program.

Until recently, the United States has opposed reprocessing on the grounds that it produced separated plutonium and there was an increased risk that this material would find its way into nuclear weapons. President G. W. Bush in January 2006 announced a change in policy and initiated what is called the Global Nuclear Energy Partnership (GNEP). The purpose of GNEP is to develop the technology for continuous reprocessing in fast-spectrum reactors in partnership with other interested nations. The goal is to advance the technology for a system to make a fuel for the fast reactors that does not separate pure plutonium. When it is all worked out, the only materials that go to a repository are fission fragments and the long-lived actinides that leak into the fission-fragment waste stream because of small inefficiencies in the separation process. If these can be held to about 1 percent or less, which has already been demonstrated possible on a laboratory scale, the required isolation time is on the order of 1,000 years, which can be assured with very high confidence. However, it will take about twenty years to develop and test the GNEP technology.

Proliferation Prevention

Preventing the proliferation of nuclear weapons continues to be an important objective of the international community, but becomes more complex in a world with a much expanded nuclear-energy program involving more countries. The science behind nuclear weapons is well known and the technology seems not that hard to master through internal development or illicit acquisition. It should be clear to all that the only way to limit proliferation by nation-states is through binding international agreements.

One issue that is being revisited is the relative proliferation resistance of the "once-through" fuel cycle compared to those of various reprocessing strategies. An analysis had been done in 2004 by an international group of experts for the U.S. Department of Energy.[7] The methodology created in this analysis was to give a risk score for every phase of the nuclear fuel cycle and then sum the risks over time. Surprisingly, the once-through and the variants of reprocessing have about the same score.

In this model, the increased risk of diversion during the phase where plutonium is available in reprocessing scenarios is balanced by the decreased risk of diversion during enrichment because use of MOX fuel decreases the amount of enrichment required by about 30 percent. In addition, the spent MOX fuel has less of the plutonium isotopes of interest to weapons makers and more of those that make building a weapon more difficult. Note that these scores should not be read as precision measurements. All they really say is that, to sensible people, once-through is not

TABLE 46.3

Relative proliferation resistance score (higher is better)

Cycle	Total Nuclear Security Measure
Once-Through PWR Cycle	0.657
LWR MOX w/ PUREX	0.641
LWR MOX w/ UREX	0.644

that different from reprocessing as far as proliferation potential is concerned.

In 2005, the IAEA director general Dr. El-Baradei and U.S. president George W. Bush separately proposed that internationalization of the nuclear fuel cycle begin to be studied seriously. In an internationalization scenario, there would be countries where enrichment and reprocessing occur. These are the supplier countries. The rest are user countries. Supplier countries make the nuclear fuel and take back spent fuel for reprocessing, separating the components into those that are to be disposed of and those that go back into new fuel.

If such a scheme were to be satisfactorily implemented, there would be enormous benefits to the user countries, particularly the smaller ones. They would not have to build enrichment facilities nor would they have to treat or dispose of spent fuel, neither of which is economical on small scales. Moreover, repository sites for 100,000-year storage may not be available with the proper geology in many small countries. In return for these benefits, user countries would give up potential access to weapons-usable material from both the front and back ends of the fuel cycle.

If this is to work, an international regime has to be created that will give the user nations guaranteed access to the fuel that they require. This is not going to be easy. If a country agrees to be a user country, it will get its enriched fuel from someplace. If the supplier is another country, how can the user be sure that it will not be cut off from its needed fuel for political reasons? For example, Europe gets a large fraction of its natural gas supply from Russia through a pipeline that runs through the Ukraine. In 2006, in a dispute with Ukraine, Russia turned off the gas. It only lasted a short time, but Ukraine had to agree to Russia's terms, and Europe's confidence in the reliability of supply was badly shaken. In the energy area, to be very heavily dependent on a single source of supply is economically and politically dangerous. We have been through this with oil supply in the 1970s. The Arab members of OPEC cut off oil supplies to the West to bring pressure to bear on Middle Eastern problems with Israel. It was disruptive, but we did get through it. The response was to diversify suppliers and to build reserve storage capacity. An analogous system will have to be worked out for nuclear fuel. One option that has been discussed is to create an enriched fuel bank under the control of the IAEA. The bank might, for example, contain a five-year supply of reactor fuel for each user country. There are other options, but an internationally agreed-upon system would have to be worked out.

Reducing the proliferation risk from the back end of the fuel cycle will be at least as complex as from the front end. It is essential to do so because, as we have seen with North Korea, a country can quickly "break out" from an international agreement and develop weapons if the material is available. North Korea withdrew from the Nuclear Non-Proliferation Treaty at short notice, expelled the IAEA inspectors, and reprocessed the spent fuel from its Yongbyon reactor, thus in a very short time acquiring the plutonium needed for bomb fabrication.

However, the supplier countries that should take back the spent fuel for treatment are not likely to do so without a solution to the waste-disposal problem. In a world with a greatly expanded nuclear power program, there will be a huge amount of spent fuel generated worldwide. The projections mentioned earlier predict more than a terawatt (electric) of nuclear capacity producing more than 20,000 tons of spent fuel per year. This spent fuel contains about 200 tons of plutonium and minor actinides and 800 tons of fission fragments. The once-through fuel cycle cannot handle it without requiring a new repository on the scale of Yucca Mountain every two or three years.

Reprocessing with continuous recycle in fast reactors can handle this scenario since only the fission fragments have to go to a repository and that repository need only contain them for a few hundred years rather than a few hundreds of thousands of years. Thus, the supplier-user scenario might develop as follows. First, everyone uses LWRs with all of the enrichment performed by the supplier countries. Then, the supplier countries begin to install fast-spectrum systems as burners. These would be used to supply their electricity needs as well as to burn down the actinides. The GNEP program, mentioned above, is aimed at developing the technology necessary to realize this vision.

It is doubtful that all nations will subscribe to a scheme that limits their potential to acquire nuclear weapons if they feel a great need for them. However, if a large fraction of nations that come to use nuclear energy will subscribe to an enhanced nonproliferation regime, the world will be a safer place.

Conclusion

Nuclear energy can be an important component of a strategy to give the world the energy resources it needs for economic development while reducing consumption of fossil fuels with their greenhouse gas emissions. If this is to happen on a large scale, advances in both science and technology and international political relations will be required.

The sciences and technology can produce better and safer reactors, better ways to dispose of spent fuel, and better safeguards technology. This can best be done in an international context to spread the cost and to create an international technical consensus on what should be done. Countries will be more comfortable with what comes out of such developments if they are part of them.

While the sciences and technology development can best be done in an international context, the political actions to create better mechanisms for proliferation control can only be done internationally. The IAEA seems to be the best place to start and, indeed, the first steps may already be in progress. However, it will be difficult for an organization as large as the IAEA to create a framework for a new international nuclear enterprise if too many voices are involved at the start. It might be better if a broadly based but compact subgroup does the initial work. The minimum membership of such a group would include Canada, China, France, India, Japan, Russia, South Korea, the United States, and representatives of the larger potential user states (e.g., Brazil and Indonesia). In the end, the political advances will be more difficult to achieve than the technical.

Notes

1. Global Energy Perspectives, International Institute of Applied Systems Analysis and World Energy Council, Cambridge University Press, 1999. An interactive version of the energy projection can be found at www.iiasa.ac.at/cgi-bin/ecs/book_dyn/bookcut.py.

2. Each year the International Energy Agency publishes a new version of its "World Energy Outlook" series. Many of the numbers used here come from the year 2006 version. The most recent version can be found at www.worldenergyoutlook.org.

3. The Future of Nuclear Power, MIT, July 2004 (web.mit.edu/nuclearpower).

4. There are two good Web sites with information on all topics related to nuclear energy. They are those of the World Nuclear Association (www.world-nuclear.org) and the Nuclear Energy Agency (www.nea.fr). The NEA has a particularly good general overview called "Nuclear Energy Today."

5. The capacity of Yucca Mountain is now limited by federal legislation to 70,000 tons of spent fuel. The physical capacity of the site has been estimated to be two to four times this limit.

6. Refer to the Web site in note 3.

7. "An Evaluation of Proliferation Resistant Characteristics of Light Water Reactor Fuels," November 2004, is available on the DOE's Web site (www.nuclear.gov) under Advisory Committee Reports.

Chapter 47

Coal Capture and Storage

DAVID HAWKINS

Introduction

Coal is the most abundant and widely distributed of all fossil fuel resources. Due to its ubiquity and low production costs, coal has been the mainstay of most industrializing economies for the past several centuries.

Because of coal's abundance and high carbon content per unit of energy, even if oil and gas use were eliminated, full exploitation of the world's coal endowment would result in atmospheric concentrations of the largest anthropogenic greenhouse gas (GHG), carbon dioxide (CO_2), reaching levels about six times higher than preindustrial levels unless nearly all of coal's carbon content is captured and disposed of in permanently isolated repositories. While energy efficiency and use of renewable energy resources and lower carbon content fuels can slow the rate at which coal is consumed (even eliminate coal use in theory), the low delivered price of coal makes it unlikely that substantial shifts away from coal will occur absent adoption of strong policies to curb CO_2 emissions. Nearly all energy forecasters predict that coal will continue to be consumed in ever-increasing amounts during the next century or more.

Technologies exist today to capture CO_2 from large industrial sources for disposal in geologic formations. However, in the absence of CO_2-control policies, such systems will be deployed only in niche situations. Policies now being adopted and considered in the United States and elsewhere could change this situation.

Global Coal Resources

Total world coal resources are estimated at about 3,700 billion metric tons of carbon (GtC).[1] By comparison, estimated world resources of oil and gas, conventional and unconventional, are much smaller, with oil resources estimated at 700 GtC and gas resources at 550 GtC.[2] Full exploitation of all oil and gas resources is estimated to approximately double preindustrial CO_2 concentrations, resulting in global temperature increases from preindustrial levels between 1.5 and 4.5 degrees Celsius. In contrast, global coal resources, if fully exploited, would by themselves produce estimated CO_2 concentrations about six times preindustrial levels. Using the same climate sensitivity range of 1.5 to 4.5 degrees Celsius for a doubling of CO_2, the estimated global temperature increase from exploiting all fossil resources rises to between 3.5 to 10.5 degrees Celsius.[3]

If one believes climate sensitivity will turn out to be at the lower end of the range, it is possible to conclude that all the world's oil and gas could be consumed without producing extreme increases in global temperatures. However, when coal is added to the picture it becomes clear that its carbon cannot be released to the atmosphere if extreme global temperature increases are to be avoided.

Distribution and Current Production and Consumption of Coal

Coal is distributed broadly across all major regions of the world. Total world proved reserves at the end of 2005 are estimated at approximately 900 billion tons.[4] More than 75 percent of the world's proved coal reserves are found in five countries: the United States (27.1 percent); Russia (17.3 percent); China (12.6 percent); India (10.2 percent); and Australia (8.6 percent). They accounted for more than 73 percent of global coal production of 5,800 million tons in 2005: China (37.4 percent); the United States (17.6 percent); India (7.3 percent); Australia (6.3 percent); and Russia (5.1 percent). The top five coal-consuming countries accounted for 72 percent of global coal consumption in 2005: China (36.9 percent); the United States (19.6 percent); India (7.3 percent); Japan (4.1 percent); and Russia

(3.8 percent). It is worth noting that of the top coal-producing or consuming countries, only one, Japan, has agreed to limits under the Kyoto Protocol that require reductions in emissions from current levels. The European Union (EU-25), which accounted for 10.2 percent of global coal consumption in 2005, also has agreed to such limits.

CO_2 Emissions from Historic and Forecasted Coal Use

Estimated cumulative carbon releases from the use of coal from 1751 to 2003 are about 150 GtC—about half of total carbon releases from all fossil fuel use during this period.[5] While coal's annual share of total carbon emissions has declined as oil use has surged, the total quantities of carbon emissions from coal use continue to grow. More than 37 percent of the total carbon emissions from coal in the past 250 years occurred in the past twenty-five years. Yet these historic emissions of carbon from coal use are just the tip of the iceberg if current forecasts come to pass.

Forecasts by the International Energy Agency (IEA) indicate very rapid growth in coal use and its carbon emissions in the next twenty-five years, absent changes in policy and practice. The IEA's latest World Energy Outlook report estimates cumulative carbon emissions from global coal use from 2004 to 2030 will total more than 100 billion tons—an emission rate nearly double that of the previous twenty-five years.[6] Beyond 2030, scenario analysts forecast ever greater carbon emissions from coal use. Scenarios recently completed by three modeling teams for the U.S. Climate Change Science Program estimate in the reference case that global coal consumption in the year 2100 will be four to eight times higher than in the year 2000, indicating annual carbon emissions between 8.7 and 17.5 GtC from coal use alone in the year 2100.[7]

Coal and Long-Lived Capital Investments

Today most coal is used for electric power generation and heat supply. The IEA reports that such facilities accounted for 68 percent of global coal use in 2004.[8] An even larger share of forecasted increases in coal consumption will be for power and heat uses: the IEA forecasts that 81 percent of the growth in global coal consumption will be dedicated to meeting increased demand for power and heat. This trend presents a crucial challenge to GHG management because of the long operating lifetimes of power and heat plants. The U.S. Energy Information Administration predicts that nearly all the U.S. coal capacity built since the mid-1950s will have operating lives greater than sixty years.[9] In 2030, according to the EIA's forecasts, nearly half of U.S. coal-fired power plants will be fifty years or older and more than 40 percent of U.S. carbon emissions from coal-fired electricity generation will come from those plants.

The IEA predicts an explosion of new coal power investments in the next twenty-five years. The IEA's reference case forecast estimates that 1800 GW of new coal capacity will come on line from 2003 to 2030 globally. Assuming that this new capacity operates for sixty years at an average capacity factor of 75 percent and vents its CO_2, the lifetime emissions from this single tranche of energy investment will equal 188 GtC, an amount that is 125 percent of carbon emissions from all prior human use of coal.

Thus, to prevent enormous growth in global carbon emissions, if significant use of coal continues, as seems inevitable for at least the next several decades, it is essential to apply technologies to prevent CO_2 from coal use from being released to the atmosphere.

CO_2 Capture, Transport, and Disposal (CCD) Systems
CO_2 Capture

In broadest strokes, prevention of CO_2 emissions from coal use involves capture of CO_2 prior to release and disposal in locations where it cannot make its way to the atmosphere. Regarding CO_2 capture, there are several approaches at various stages of technical maturity that will be described in this section. Regarding disposal, the approach that appears most viable relies on injection into stable geologic formations under land masses or under the ocean floor.

In the power sector, the approach to CO_2 capture will differ fundamentally according to the method used to generate power from coal. In a conventional coal-fired power plant, coal and air are burned in a boiler to raise steam. The only other approach to electricity generation from coal now in commercial use involves gasification of coal in oxygen rather than combustion in air. The conventional process produces an exhaust gas containing CO_2 while the gasification process produces a synthesis gas containing carbon monoxide (CO) and hydrogen. In the gasification process, the synthesis gas exiting the gasifier is then burned in a combustion turbine.

Conventional coal combustion is the most prevalent technology used for power generation with all but a handful of the world's existing plants based on these designs. Most of the announced new capacity is also based on the conventional combustion approach. Coal gasification is commercially demonstrated and is used primarily in the chemical industry. In the power sector about twenty-eight integrated gasification combined cycle (IGCC) plants are operating worldwide, with seven of them using coal as the feedstock (the others use petroleum coke).

CO_2 capture approaches from industrial sources are divided into three broad categories: pre-combustion capture; post-combustion capture; and oxyfuel combustion.[10] Pre-combustion capture is commercially demonstrated at gasification plants. Capturing CO_2 from a gasification plant first requires the synthesis gas to be converted into a mixture of hydrogen and CO_2. The most common form of pre-combustion capture involves use of chemicals, such as amines, to absorb the CO_2 from the converted synthesis gas. These capture systems are in use at commercial plants that use coal or petroleum coke to make fertilizer, hydrogen, or other industrial chemicals. Once the CO_2 is captured, it is either vented to the air or compressed for use in enhanced oil recovery operations or in the food and beverage industry.

Post-combustion capture involves the use of the same types of chemicals to absorb CO_2 from the exhaust gas of a power plant after combustion. While technically feasible, due to high costs and large energy penalties, such systems are currently used by only a few power plants at small scale to produce CO_2 for sale to the food and beverage industry. Because air consists of 78 percent nitrogen, the exhaust gas from combustion-based plants has much greater volume and CO_2 is present in lower concentrations and at lower pressures than in the gas stream of a coal gasification plant.

Novel concepts for post-combustion capture have begun to be researched in the past few years although none had progressed beyond the laboratory bench-scale as of early 2008. A pilot scale (5 MW) test of a new chilled-ammonia process began in 2008 at a plant in Wisconsin.[11] Other concepts are also being announced for small scale-field testing. While development of a technically feasible and economically viable post-combustion process could in principle allow existing combustion-based power plants to be retrofit for CO_2, today there is no adequate basis to conclude whether any of these concepts will be successful.

Oxyfuel combustion is another approach that involves combustion of coal in oxygen rather than air. In this process, CO_2 would be recycled into the combustion chamber to produce an exhaust gas with high concentrations of CO_2. This concept has been researched in the laboratory and a pilot-scale test (10MW) has been announced by Vattenfall at a plant in Germany. The concept also has been considered for use at a larger project in Canada.

As of today, pre-combustion capture of CO_2 has the most commercial experience (albeit due to the lack of an economic driver, not in the power sector) and is applied at the largest scale. For example, the Dakota Gasification plant in North Dakota is capturing about 1.5 million tons of CO_2 per year from its lignite gasification chemical and fuels plant there.[12] The captured CO_2 is pipelined 300 kilometers to the Weyburn oil field in Saskatchewan.

CO_2 Transport and Disposal

As indicated by the Dakota Gasification example, the second step in the carbon capture and disposal (CCD) system is to transport the captured CO_2 from the emission source to a disposal site. CO_2 pipelines are commercially demonstrated for this purpose and have been in use in the United States and other countries for several decades to move CO_2 from sources (primarily natural CO_2 deposits) to oil fields where it is injected for enhanced oil recovery (EOR).[13]

The final step in the CCD system is to inject the compressed CO_2 into a geologic formation where natural barriers prevent its escape to the atmosphere. Compressed CO_2 and other buoyant fluids such as oil can be retained in geologic formations for millions of years as is evidenced by natural accumulations of these substances that persist today.

The ideal formation for permanent disposal of CO_2 consists of a highly permeable geologic formation 800 to 1,000 meters or more below the surface with a thick, highly impermeable cap rock formation above it. Oil and gas fields are an example of such formations although their total capacity and proximity to major emission source clusters will operate to limit the amount they could be exploited. Deep saline formations are more widespread and upper limit estimates of their capacity far exceed that of oil and gas fields. The IPCC has estimated it is likely that the global technical potential capacity of geologic formations is at least 2 trillion tons of CO_2.[14] Given current CO_2 annual emissions from all fossil power generation (coal, oil, and gas) of about 10 Gt CO_2, there appears to be adequate capacity to accommodate CO_2 from a substantial fraction of fossil generation for a century or more.

Core questions that are asked about geologic CO_2 disposal include whether it can be injected without local environmental, health, and safety risks, whether it can be retained permanently underground, and whether such injection would pose risks to other resources such as drinking water supplies. Because injected CO_2 is initially buoyant compared to the surrounding formation fluids, it could migrate upward through pathways like faults, fractures, or wells with degraded cement. There are two types of "leakage" concerns: acute, large releases that could pose a threat to vegetation, wildlife, and human health (CO_2 in high concentrations is an asphyxiant); and slow, chronic releases that would compromise the efficacy of CCD as a GHG mitigation technique.

The IPCC in its 2005 *Special Report on Carbon Dioxide Capture and Storage* concluded that given an adequate regulatory regime the "local health, safety, and environment risks of geological storage would be comparable to the risks of current activities such as natural gas storage, EOR and deep underground disposal of acid gas."[15] With respect to long-term slow leakage, the IPCC concluded that the fraction of injected CO_2 into appropriately selected geological reservoirs "is very likely to exceed 99 percent over 100 years and is likely to exceed 99 percent over 1,000 years."[16] The IPCC reached these conclusions based on the existence of natural analogues (oil and gas fields and natural CO_2 domes) where retention for millions of years has been demonstrated and several decades of industrial experience with injection of CO_2 in EOR operations, limited duration large scale CO_2 injection projects, acid gas injections, and operation of natural gas storage facilities.

In the United States, injection of CO_2 for EOR operations has been underway since the 1970s and currently more than 34 million tons of CO_2 are injected annually. Since most of these fields are not subject to monitoring, there is no instrumental verification of industry assertions that no leakage has occurred. Such verification is available for two large projects that have a number of years of operational experience. The Sleipner project involves the injection of about 1 million tons per year of CO_2 into a sub–sea bed geologic reservoir off the coast of Norway. The project has been underway since 1996 and published monitoring information indicates no leakage has occurred. The Weyburn EOR project in Canada is also subject to a comprehensive monitoring regime. It has been injecting about 1 million tons per year of CO_2 since 2000 and no leakage has been observed. A third comprehensively monitored large-scale project in Algeria has been underway since 2004 with no reports of leakage.

By definition, empirical proof of long-term, high-retention capability for large-scale CO_2 injection projects cannot be obtained in advance. The IPCC concluded that risks of leakage diminish over time because of transformation of injected CO_2 after injection. On the scale of decades to a century or more after injection the buoyant CO_2 enters into solution with the formation fluids and loses its buoyancy. On a time scale of hundreds of years, the CO_2 in solution gradually becomes mineralized. Additional large-scale projects will add to the knowledge base about the behavior of CO_2 in large-scale injections, but in the interim decisions whether to require CO_2 to be captured from power plants will have to be based on the relative risks of large-scale CCD compared to continued venting of the CO_2 from the plants' stacks directly to the atmosphere. An initial analysis of this type, using extreme assumptions about hypothetical leak rates from early vintages of geologic repositories, found that total releases would be about an order of magnitude less under an early CCD deployment scenario compared to delayed deployment cases.[17]

Policy Implications

No electric power plant in the world today employs CCD to prevent emissions of CO_2 for the simple reason that CCD necessarily increases costs compared to the free venting of CO_2, and the limited policies in place today to limit GHG emissions are too lax to create an incentive to capture and dispose of CO_2. Cost estimates based on today's demonstrated capture technologies indicate that CCD from coal power plants would not begin until the cost of emitting CO_2 exceeds \$25 to \$30 per ton of CO_2.[18]

Meanwhile, scores of billions of dollars are being invested each year in new coal plant capacity that may never be economically feasible to retrofit with CO_2 capture systems should viable systems be developed. Promoters of new conventional coal plants are beginning to describe them as "capture ready," though in reality this amounts to nothing more than leaving space at the plant for as yet undeveloped methods to capture CO_2 on a post-combustion basis.

As discussed above, the carbon loadings during the long lifetime of a new coal plant are very large. Policies that result in rapid deployment of CCD for new coal plants could avoid this carbon "lock-in" problem. Such policies are beginning to be considered and even adopted in a few jurisdictions. In September 2006, the State of California enacted a CO_2 emissions performance standard for new power supply investments by the state's electricity providers. The standard is set at the level achievable by a modern natural gas combined-cycle plant—an emission rate about half that from an efficient coal plant. Accordingly, any new contracts to supply power to California from coal plants (the state gets about 16 percent of its power from out-of-state coal plants) would require the application of CCD to meet the standard.[19]

Variants of CO_2 performance standards have been proposed but not yet adopted by the U.S. Congress. A bill introduced by Senators Sanders and Boxer in the 110th Congress, 2007, includes a performance standard for CO_2 (also set initially to equal the emission rate of a new natural gas combined-cycle plant) that would apply to new plants coming online after 2011 and would apply to all plants by 2030.[20] In addition, the bill establishes a requirement for an increasing fraction of power supplied by coal-based power plants to come from units that meet a CO_2 emission standard of 250 pounds per megawatt-hour (equivalent to about 87 percent capture from a coal unit).

In Europe, a task force of government, industry, researchers, and other nongovernmental organizations has formed the European Technology Platform on Zero-Emission Fossil Fuel Power Plants and recommended the adoption of a set of policies and measures intended to deploy CCD starting early in the next decade with the aim of applying CCD to all new fossil power plants coming online after 2020.[21]

However, new coal plants in the developing world will make up two-thirds of the 1,800 GW of new construction forecast by the IEA in the next twenty-five years. Some of this capacity can be avoided or deferred by aggressive promotion of efficiency and renewable energy resources but a large amount of new coal capacity is still likely to be constructed in this period.[22] Speeding deployment of CCD in the developing world will likely require a partnership between the wealthier industrial countries and the developing countries where these new plants are being built. For example, an international fund to cover the incremental costs of applying CCD to new coal plants in the developing world could be established as a joint initiative to prevent the potential lock-in of more than 100 billion tons of carbon emissions from these new coal plants. The cost of such an initiative would be billions of dollars per year but it would amount to only several percent of current electricity expenditures by OECD countries.[23]

The high carbon emissions from coal plants and the pace of global construction of new coal plants place a premium on prompt action to avoid emission consequences that could foreclose options to stabilize GHG concentrations at protective levels. Efficiency, renewable resources, and lower-emitting power sources likely all will play important roles but it appears that substantial use of CO_2 capture and disposal will need to be a part of the overall program.

Notes

1. Metz, B., et al., eds. *Climate Change 2001: Mitigation* (Cambridge Univ. Press, New York, 2001), p. 6.

2. Ibid.

3. Caldeira, K., and M. E. Wickett. Ocean model predictions of chemistry changes from carbon dioxide emissions to the atmosphere and ocean. *Journal of Geophysical Research* (Oceans) 110, C09S04, doi:10.1029/2004JC002671, 2005.

4. British Petroleum, *BP Statistical Review of World Energy* (June 2007), p. 32.

5. Marland, G., et al. *Global, Regional, and National Fossil Fuel CO_2 Emissions* (2006), online at http://cdiac.ornl.gov/trends/emis/em_cont.htm.

6. IEA, *World Energy Outlook 2006* (Paris, 2006).

7. U.S. Climate Change Science Program,

Scenarios of Greenhouse Gas Emissions and Atmospheric Concentrations (Washington, DC, 2007).

8. IEA, 2006 op. cit.

9. U.S. Energy Information Administration, *Annual Energy Outlook 2007* (Washington, DC, 2007).

10. Metz, B., et al. *Special Report on Carbon Dioxide Capture and Storage* (Cambridge Univ. Press, New York, 2005).

11. Alstom, Inc. *Alstom to Build Pilot Plant in the U.S. to Demonstrate Its Unique CO_2 Capture Process* (press release, Oct 2, 2006), online at www.power.alstom.com/pr_power/2006/october/27531.EN.php?languageId=EN&dir=/pr_power/2006/october/&idRubriqueCourante=3981.

12. Metz, op. cit., p. 99.

13. Ibid., p. 29.

14. Ibid., p. 12.

15. Ibid.

16. Ibid., p. 14.

17. Hawkins, D., and S. Bachu. Deployment of Large-Scale CO_2 Geological Storage: Do We Know Enough to Start Now? paper presented at 8th International Conference on Greenhouse Gas Technologies, June 2006, online at https://events.adm.ntnu.no/ei/viewpdf.esp?id=24&file=d%3A%5CAmlink%5CEVENTWIN%5Cdocs%5Cpdf%5CC950Final00299%2Epdf.

18. Hawkins, D., et al. *Scientific American* 295, 68 (2006).

19. California Energy Commission. *2006 Gross System Electricity Production*, online at www.energy.ca.gov/electricity/gross_system_power.html.

20. S. 309, *Global Warming Pollution Reduction Act*, online at http://thomas.loc.gov.

21. In March 2007, the Council of the European Union adopted conclusions endorsing this program. See *Presidency Conclusions of the Brussels European Council* (8/9 March 2007), p. 22.

22. IEA, 2006 op. cit.

23. Hawkins, D. *Taking Down the Wall of Denial*, presentation to Princeton Science, Technology, and Environmental Policy Seminar (November 13, 2006).

Chapter 48

Tropical Forests

PHILIP M. FEARNSIDE

Introduction

Deforestation contributes significantly to global warming, which means that actions to reduce deforestation have a valid role as part of strategies to mitigate climate change—especially since avoided deforestation is a relatively cost-effective mitigation strategy.

This chapter discusses policy debates over the use of avoided deforestation as an option for mitigating global warming. The value of forests as a carbon sink is reduced (but not eliminated) by the lack of permanence of holding carbon out of the atmosphere in the case of individual forest tracts and by the greater uncertainty associated with forests as compared to fossil carbon. However, these effects can be offset by avoiding substantially larger amounts of carbon emission than the amount of any carbon credit that is granted. Permanence of rainforest carbon can also be addressed by a sequence of temporary credits (as is currently done for silvicultural plantations under the Kyoto Protocol). In addition to avoiding greenhouse gas emissions, avoided deforestation generates climatic benefits by maintaining evapotranspiration and water cy-

cling, in addition to its fundamental role in maintaining biodiversity.

Brazilian Amazonia is the focus of the present chapter. High rates of deforestation in the region are a major source of emissions: the net committed emission for 1990 from deforestation in Brazilian Amazonia was 218.1 to 227.8×10^6 Mg of CO_2-equivalent carbon per year for biomass emissions only, and 230.0 to 239.7×10^6 Mg of CO_2-equivalent carbon per year including soils and other sources, updated based on revised wood density estimates.[1] Deforestation in 1990 (the standard base year for national inventories under the United Nations Framework Convention on Climate Change) was 13.8×10^3 square kilometers (in primary forest only, not counting clearing of savannas or re-clearing of secondary forests). The deforestation rate in 2004 was 27.4×10^3 square kilometers per year, which corresponds to a net committed emission of $456.7–475.9 \times 10^6$ Mg of CO_2-equivalent carbon per year. This is almost six times Brazil's approximately 80×10^6 Mg of CO_2-equivalent carbon annual emission from fossil fuels and cement. Thus, the case of Amazonia is both illustrative and substantive.

Avoided Deforestation as a Mitigation Option

Proposals to avoid tropical deforestation as a means of mitigating global warming have been the source of considerable controversy. As a matter of disclosure, my role as the originator of such proposals in the early days of this discussion, and my participation as a combatant in the debates over the succeeding decades, makes me clearly partial to using this option to the fullest extent possible.[2] The threat to tropical forests posed by climate change has been a key part of this debate. Opponents of granting credit for avoided deforestation claim that the eventual demise of the forests will release stored carbon in any case and that credit should not be given for temporary carbon storage. Some background on the controversy surrounding carbon credit for avoided deforestation is needed.

Prior to the December 1997 Kyoto Protocol, slowing tropical deforestation to avoid greenhouse gas emissions was regarded as a top priority by European governments and environmental nongovernmental organizations (NGOs) headquartered in Europe.[3] With the advent of the protocol, these governments and NGOs suddenly reversed their positions due to the concern that avoided deforestation would be a temporary and uncertain mitigation strategy.[4] Ruling out avoided deforestation as a mitigation strategy would level the playing field between Europe and North America by forcing the United States to raise fuel prices to reduce emissions instead of relying on the substantial credit that would come from avoiding deforestation. Environmental NGOs headquartered in other parts of the world outside Europe virtually all continued to support credit for avoided deforestation.[5] Grassroots organizations in Brazilian Amazonia overwhelmingly supported credit for avoided deforestation.

On the other hand, the Brazilian foreign ministry opposed credit from avoided deforestation based on the belief that Brazil's sovereignty over Amazonia is under permanent threat and that the major economic interests represented by carbon credit could lead to international pressures that might jeopardize the country's control over the region. Although belief in a threat of "internationalization" of Amazonia is widespread in Brazil, the view that carbon credit for avoided deforestation poses a danger in this regard is not shared by most sectors of Brazilian society outside of the foreign ministry. Brazil's Ministry of the Environment has long favored carbon credit for avoided deforestation.[6] The nine state governments in Brazil's Amazon region have all favored carbon credit for avoided deforestation and one has even attempted to sell it on international commodity exchanges.

At the Conference of Parties (COP) in Bonn in July 2001, the countries that remained in the Kyoto Protocol (after U.S. president George W. Bush withdrew the United States) agreed to exclude avoided deforestation from crediting in the 2008 to 2012 first commitment period. While the Bonn agreement excluded tropical deforestation, it allowed credit for plantations of trees such as *Eucalyptus*. The only country that wanted credit for plantations but not for avoided deforestation was Brazil, which has one of the world's largest plantation industries.[7]

Following the Bonn agreement, the European governments and NGOs have since reverted to their original positions of support for including avoided deforestation in the measures for credit in the second commitment period (2013 to 2017). The geopolitical situation surrounding the current negotiations for the second commitment period is very different from the one that applied to the first commitment period during the 3.5-year-long battle over this issue between the signing of the Kyoto Protocol and the Bonn agreement. For the second commitment period the emissions quotas (assigned amounts) and the rules for crediting (for example, for avoided deforestation) will be negotiated simultaneously, thereby eliminating parallel advantages that countries can get for themselves by excluding avoided deforestation; this negates any argument for a climatic advantage to be achieved by allowing only the minimum possible amount of mitigation in the forest sector. If tropical forests are excluded from credit, then the industrialized countries that would have purchased the credit will simply agree to more modest cuts in their national emissions.

In the current negotiation for including tropical forests in the second commitment period, it is important that countries (or other actors) must take both the benefit and the onus of commitments to reduce deforestation. It is not enough to take credit when deforestation goes down and incur no penalty when deforestation goes up. Proposals advanced in this regard essentially treat avoided deforestation as speculating on the stock market, where the objective is to "buy low and sell high." In other words, the natural oscillations in annual deforestation rates would generate credit even without any change in the behavior of deforesters. Several proposals are under consideration. One is that of the fifteen-country Coalition for Rainforest Nations led by Papua New Guinea and Costa Rica, which presented a proposal in Montreal in December 2005 to grant carbon credit that could be sold and used to meet emissions reduction commitments made under Kyoto Protocol.[8] Brazil submitted a competing proposal in Nairobi in

December 2006 for a voluntary fund for financing deforestation reduction that would not produce credit toward achieving targets for reducing use of fossil fuels.[9] Whatever solution is adopted, the important role that tropical forests play in global warming means that sooner or later measures to reduce deforestation are likely to be funded as mitigation measures.

Underlying Issues in Counting Mitigation Benefits
The Value of Time

In order to reflect a preference to receive benefits as soon as possible and delay incurring costs as long as possible, economists and entrepreneurs apply a discount rate to all future income and expenses. An annual discount rate is a percentage by which future quantities are devalued for each year between the present and the expected credit or debit (after adjustment for any inflation). Financial decisions are often based on annual discount rates on the order of 10 or 12 percent, and are essentially based on the rate at which money can be made from alternative investments in the economy. Policy decisions intended to address different social concerns use other (generally lower) discount rates.

The Kyoto Protocol has adopted a formulation for calculating the equivalence between greenhouse gases with widely differing atmospheric lifetimes based on global warming potentials, or GWPs, that are based on a 100-year time horizon with no discounting during the course of the time horizon.[10] This formulation gives a value to time that is equivalent to an annual discount rate of approximately 1 percent.[11]

By giving any value to time greater than

zero the value of delaying global warming is recognized. If warming by a given amount is delayed by, for example, fifty years, all of the impacts that otherwise would have occurred during those fifty years represent a permanent benefit with real value. Temporary storage of carbon, for example in trees, delays global warming and therefore has a value. While the value of temporary storage is less than that of permanent storage, it is not zero. Even if Amazonian forest is in fact destroyed by climate change in eighty years (as the Hadley Center model indicates under a business-as-usual scenario; see chapter 9 in this volume), those eighty years have value that must be compensated if deforestation is avoided.

Various formulations have been proposed to account for time based on the "ton-years" that the carbon remains out of the atmosphere.[12] The weak point of such formulations is that they require a negotiated agreement on a discount rate or other alternative time-preference weighting. A means of avoiding an explicit negotiation was found by relying on market mechanisms as embodied in the "Colombian Proposal," which creates temporary carbon credits that have to be renewed at defined intervals, either by purchasing another temporary credit or by making a permanent reduction through avoided fossil-fuel emission.[13]

The question of time preference has come to the fore with the recent discovery that living terrestrial vegetation, including tropical forest, may be emitting methane to the atmosphere.[14] How should the small amount of methane a forest emits per hectare per year be weighed against the large immediate impact of cutting down a hectare of tropical forest? Each hectare of deforestation in Brazilian Amazonia releases net committed emissions totaling 170 Mg CO_2-equivalent carbon with altered adjustments for hollow trees and form factor and wood density.[15] Therefore, with no discounting, it would take 665 years for the methane emission from a hectare of standing Amazonian forest to offset the impact of deforesting that hectare, with the range of uncertainty extending from 423 to 1,566 years. Even the low end of this range should make clear the tenuous nature of arguments that would sacrifice the benefits of forests over the next several centuries in the interests of climatic gains that will only begin to accrue several hundred years in the future. If any discounting or other form of adjustment is made for the value of time, keeping the forest becomes the best choice regardless of the time horizon. Any discount rate above a mere 0.15 percent annually would negate forever the benefit of sacrificing tropical forest to avoid its natural emissions of methane.

The Role of Uncertainty

The global-warming impacts of tropical deforestation, and the benefits of any measures taken to reduce it, are inherently more uncertain than are comparable emissions and reductions in fossil-fuel combustion. At each stage of the process—from the planning of a mitigation measure or activity to the execution of the plan to the later evaluation and monitoring—a forest-sector measure will invariably be more uncertain than an energy-sector one.

Uncertainty (the variation in outcomes due to lack of knowledge) and risk (the variation due to known causes) are everyday considerations in financial decisions of all sorts (see chapter 15 in this volume). These concerns are incorporated into the sum of costs and benefits by means of the expected monetary value (EMV), with appropriate adjustments for factors such as risk aversion. EMV represents the sum of the products of the value of each possible outcome times its respective probability of occurrence. For example, if one is betting in the lottery, one may get 1 million dollars if one wins, but the probability of winning will be, say, one in 10 million, making the EMV of a $1 lottery ticket only 10 cents. In the case of carbon from avoided deforestation, the reward may only have a modest probability of being achieved, but its EMV is still considerable because of the large jackpot if avoided deforestation is indeed successful.[16]

The best way to ensure that the climate is not stuck with the losses from overly optimistic expectations of mitigation benefits is to insist on a "pay-as-you-go" policy. This also avoids sovereignty issues that are sometimes raised as objections to avoided deforestation, especially in Amazonia. Any advances of funds on the basis of future expected carbon benefits would have to come from normal financial markets, not from governments or international guarantees.

Unfortunately, uncertainty has often been raised as an objection to using avoided deforestation as a global-warming mitigation option. Representatives of the Association of Small-Island States (AOSIS) insisted that avoided deforestation is too uncertain and that fossil fuels should be the exclusive focus of mitigation efforts. I argue that restricting mitigation to fossil fuels is not in the best interests of those who, like small-island residents, are most at risk from global warming because the expected benefit is substantially higher from avoiding deforestation than it is from the same investment in reducing fossil-fuel emissions. The carbon benefits are simi-

lar to the EMV of financial decision making. The device of insisting on complete or nearly complete certainty has the result of ruling out forests as mitigation options.

Additional Benefits of Avoided Deforestation

Biodiversity

Climate change and biodiversity conservation are intimately linked in various ways. These are explained in more detail in chapters 3 and 4 in this volume. Both the climatic and biodiversity functions of tropical forests are vulnerable in the face of catastrophic impacts that have been predicted for Amazonia. The recent finding of multiple extinctions of Costa Rican frog species due to pathogens whose spread was aggravated by climate change underlines the widespread and poorly understood nature of these effects.[17] An analysis of the Hadley Center results under a business-as-usual emissions scenario indicates that 43 percent of a representative sample of sixty-nine angiosperm plant species would become unviable by 2095 due to shifts in climatic zones.[18] Both climate and biodiversity concerns are also linked by the benefit of avoiding deforestation: saving a hectare of forest from deforestation both mitigates climate change and preserves biodiversity. In addition, Amazonian forests recycle a tremendous amount of water, supplying water vapor to the atmosphere that sustains rainfall in the Amazon Basin that is necessary to maintain the forest itself.[19] This water also maintains rainfall in heavily populated parts of Brazil such as São Paulo.[20]

There is a natural alliance of interests between those who want to conserve the Amazonian forest for its biodiversity benefits and those who want to conserve it for its climatic benefits. However, this general alignment can break down when it comes to identifying which pieces of forest should receive top priority.[21] Often, biodiversity conservation focuses on the very long-term future, or what "will be left" after deforestation has presumably run its course for many years and left a landscape of remnants, mostly in protected areas (therefore favoring investments in large reserves far from the present deforestation frontier). Global warming mitigation benefits, on the other hand, are generally judged in terms of "additionality" over the span of a five-year Kyoto commitment period, and would favor reserves near the deforestation frontier if credit were granted.[22]

Biodiversity and climate considerations lead to sharp differences in priorities for the tropical forest locations that are most important to protect. Biodiversity is often discussed in terms of "hotspots" where many endemic and endangered species occur.[23] These include the Yungas region along the eastern foothills of the Andes, the Atlantic Forest on the southeastern coast of Brazil, Central America, and Madagascar. With the partial exception of the Yungas, all of these areas represent the last remaining fragments of forests that have suffered centuries of degradation. From the point of view of climate, these forests have lower priority than the vast expanses of remaining forest in Amazonia because any change that might be achieved in public policy to reduce future deforestation in these last remaining forest remnants would affect a minimal area of forest and stock of carbon, whereas even a slight change in deforestation rates in Amazonia affects an incomparably larger stock of carbon.[24]

Reducing emissions globally will require using every existing mitigation option, among which reducing tropical deforestation is one of the most cost-effective.[25] The IPCC has identified deforestation as the dominant form of potential mitigation in the tropics.[26] In other words, keeping forests standing for their carbon value helps avert the climate change that threatens biodiversity and, through its feedback to emissions, provokes still more climate change.

Equity

Equity issues are intimately linked to climate change impacts, mitigation, and the future of tropical forests. (These issues are covered in more detail in chapters 24 and 25 in this volume.) The tropical forest areas of the world are economically poor as compared to the industrialized areas that are responsible for most of the world's release of greenhouse gases. The misconception is common that avoiding tropical deforestation means preventing poor farmers from feeding themselves with slash-and-burn agriculture in order to let rich Americans drive luxury cars. However, in Brazilian Amazonia (in contrast to some other parts of the tropics) deforestation is mostly done by the rich.[27] This presents an opportunity through what this author refers to as the "Robin Hood strategy," or taking from the rich to give to the poor by halting the deforestation by wealthy ranchers and land speculators and using the value of the environmental services from this as a means of sustaining Amazonia's poor rural population.[28]

A recurrent question is how to compensate for the environmental services of standing forest without rewarding *grileiros* (land thieves) and those who have been victorious in the often-bloody struggle for land. Compensation for carbon makes it even more profitable to enter into Amazonian land grabs as an economic activity. Social impacts represent grounds for concern in a wide variety of potential mitigation projects in tropical forests.[29] While many policy and legal safeguards (and social struggles) will be needed to ensure that disadvantaged segments of the region's population benefit from mitigation activities, the first step in any plan to tap the value of the forest's environmental services must be creation of that value in the first place. The existence of equity concerns indicates the need for social changes, not that the carbon value of tropical forests should be denied.

The place of indigenous people in maintaining Amazonian forest is a crucial part of the debate on the role of avoided deforestation in mitigating global warming. Indigenous areas are a primary bulwark against deforestation and account for much more forest than do conservation units.[30] The notion that those concerned with global warming can just pocket the environmental contributions of indigenous peoples for free is gravely mistaken and is likely to lead to erosion and loss of the protection these forest guardians now provide.[31]

Conclusion

Because tropical forests contain large stocks of carbon that are released as greenhouse gases when the forests are cleared, substantial benefit for climate is achieved if deforestation is avoided. Advances in creating international mechanisms that grant carbon credit for avoided deforestation are needed. Other environmental services, such as maintenance of biodiversity and of water cycling, also have value that could be tapped to help slow deforestation.

Incorporating tropical forests into efforts to mitigate global warming has proved to be a complex and controversial task. Establishing an equivalence between permanent avoidance of fossil-fuel emissions and nonpermanent storage of carbon in forests requires assigning a value to time, either explicitly or by indirect market mechanisms. It also requires acknowledging the greater uncertainty of forest carbon. However, the greatest impediments have been political rather than technical, and the current prospects of reaching agreement are much better than at any time since the 1997 Kyoto Protocol. Keeping tropical forests standing not only avoids emissions but also benefits biodiversity, traditional peoples, and social equity for the region's population as a whole.

Notes

The Conselho Nacional de Desenvolvimento Científico e Tecnológico (CNPq: Proc. 470765/01-1, 306031/2004-3, 557152/2005-4, 420199/2005-5) and the Instituto Nacional de Pesquisas da Amazônia (INPA: PPI 1-1005, PRJ05.57) provided financial support. I thank the editors, Stanford student volunteer "Group 2" (Andre, Christine, Julia, and Meg), and an anonymous reviewer for helpful suggestions.

1. Fearnside, P. M. 2007. Uso da terra na Amazônia e as mudanças climáticas globais. *Brazilian Journal of Ecology* 10: 83–100; Nogueira, E. M., P. M. Fearnside, B. W. Nelson, and M. B. França. 2007. Wood density in forests of Brazil's "arc of deforestation": Implications for biomass and flux of carbon from land-use change in Amazonia. *Forest Ecology and Management* 248(3): 119–35.

2. Fearnside, P. M. 1985. Brazil's Amazon for-

est and the global carbon problem. *Interciencia* 10(4): 179–86; Fearnside, P. M. 1989. Forest management in Amazonia: The need for new criteria in evaluating development options. *Forest Ecology and Management* 27(1): 61–79.

3. Deutscher Bundestag. 1990. *Protecting the Tropical Forests: A High-Priority International Task.* Referat Öffentlichkeitsarbeit, Deutscher Bundestag, Bonn, Germany, 968 pp; Leggett, J., ed. 1990. *Global Warming: The Greenpeace Report.* Oxford University Press, Oxford, U.K., 554 pp.; Myers, N. 1989. *Deforestation Rates in Tropical Forests and Their Climatic Implications.* Friends of the Earth, London, U.K., 116 pp.

4. WWF Climate Change Campaign. 2000. "Make-or-Break the Kyoto Protocol." World Wildlife Fund-US, Washington, DC (available at www.panda.org/climate); Greenpeace International. 2000. *Should Forests and Other Land Use Change Activities Be in the CDM?* Greenpeace International, Amsterdam, the Netherlands, 24 pp.

5. Fearnside, P. M. 2001. Saving tropical forests as a global warming countermeasure: An issue that divides the environmental movement. *Ecological Economics* 39(2): 167–84.

6. Fearnside, P. M. 2006. Mitigation of climatic change in the Amazon, pp. 353–75 in W. F. Laurance and C. A. Peres, eds., *Emerging Threats to Tropical Forests.* University of Chicago Press, Chicago, Illinois, 563 pp.

7. Fearnside, P. M. 1998. Plantation forestry in Brazil: Projections to 2050. *Biomass and Bioenergy* 15(6): 437–50.

8. Papua New Guinea and Costa Rica. 2005. Submission by the Governments of Papua New Guinea and Costa Rica: Reducing Emissions from Deforestation in Developing Countries: Approaches to Stimulate Action. Eleventh Conference of the Parties of the UNFCCC, Agenda Item No. 6. United Nations Framework Convention on Climate Change (UNFCCC), Bonn, Germany, 23 pp. (www.rainforestcoalition.org/documents/COP-11AgendaItem6-Misc.Doc.FINAL.pdf); Laurance, W. F. 2007. A new initiative to use carbon trading for tropical forest conservation. *Biotropica* 39(1): 20–24.

9. Brazil. 2006. The 12th Conference of the Parties of the UNFCCC, Nairobi, Kenya. Positive incentives for voluntary action in developing countries to address climate change: Brazilian perspective on reducing emissions from deforestation.

United Nations Framework Convention on Climate Change (UNFCCC), Bonn, Germany. 4 pp. (http://unfccc.int/files/meetings/dialogue/application/pdf/wp_21_braz.pdf).

10. Schimel, D., et al. 1996. Radiative forcing of climate change, pp. 65–131 in J. T. Houghton, L. G. Meira Filho, B. A. Callander, N. Harris, A. Kattenberg, and K. Maskell, eds., *Climate Change 1995: The Science of Climate Change.* Cambridge University Press, Cambridge, U.K., 572 pp.

11. Fearnside, P. M. 2002. Why a 100-year time horizon should be used for global warming mitigation calculations. *Mitigation and Adaptation Strategies for Global Change* 7(1): 19–30.

12. Fearnside, P. M., D. A. Lashof, and P. Moura-Costa. 2000. Accounting for time in mitigating global warming through land-use change and forestry. *Mitigation and Adaptation Strategies for Global Change* 5(3): 239–70.

13. Blanco, J. T., and C. Forner. 2000. *Expiring CERs: A Proposal to Addressing the Permanence Issue for LUCF Projects in the CDM.* Unpublished manuscript, Economic and Financial Analysis Group, Ministry of the Environment, Bogotá, Colombia. FCCC/SB/2000/MISC.4/Add.2/Rev.1, 14 September 2000. (available at www.unfccc.de).

14. Keppler, F., J. T. G. Hamilton, M. Brass, and T. Röckmann. 2006. Methane emissions from terrestrial plants under aerobic conditions. *Nature* 439: 187–91; Hopkin, M. 2007. Missing gas saps plant theory. *Nature* 447: 11.

15. Fearnside, P. M. 1997. Greenhouse gases from deforestation in Brazilian Amazonia: Net committed emissions. *Climatic Change* 35(3): 321–60; Fearnside, P. M. 2000. Greenhouse gas emissions from land-use change in Brazil's Amazon region, pp. 231–49 in R. Lal, J. M. Kimble, and B. A. Stewart, eds., *Global Climate Change and Tropical Ecosystems.* Advances in Soil Science. CRC Press, Boca Raton, Florida, 438 pp.; Nogueira, E. M., P. M. Fearnside, B. W. Nelson, and M. B. França. 2007. Wood density in forests of Brazil's "arc of deforestation": Implications for biomass and flux of carbon from land-use change in Amazonia. *Forest Ecology and Management* 248(3): 119–35; Nogueira, E. M., B. W. Nelson, and P. M. Fearnside. 2005. Wood density in dense forest in central Amazonia, Brazil. *Forest Ecology and Management* 208(1–3): 261–86; Nogueira, E. M., B. W. Nelson, and P. M. Fearnside. 2006. Volume and biomass of trees in central Amazonia:

Influence of irregularly shaped and hollow trunks. *Forest Ecology and Management* 227(1–2): 14–21; Nogueira, E. M., P. M. Fearnside, B. W. Nelson, R. I. Barbosa, and E. W. H. Keizer. 2008. Estimates of forest biomass in the Brazilian Amazon: New allometric equations and adjustments to biomass from wood-volume inventories. *Forest Ecology and Management* 256: 1853–57.

16. Fearnside, P. M. 2000. Uncertainty in land-use change and forestry sector mitigation options for global warming: Plantation silviculture versus avoided deforestation. *Biomass and Bioenergy* 18(6): 457–68.

17. Pounds, J. A., M. R. Bustamante, L. A. Coloma, J. A. Consuegra, M. P. L. Fogden, P. N. Foster, E. La Marca, K. L. Masters, A. Marino-Viteri, R. Pushcendorf, S. R. Ron, G. A. Sánchez-Azofeifa, C. J. Still, and B. E. Young. 2006. Widespread amphibian extinctions from epidemic disease driven by global warming. *Nature* 439: 161–67.

18. Miles, L., A. Grainger, and O. Phillips. 2004. The impact of global climate change on tropical biodiversity in Amazonia. *Global Ecology and Biogeography* 13(6): 553–65.

19. Lean, J., C. B. Bunton, C. A. Nobre, and P. R. Rowntree. 1996. The simulated impact of Amazonian deforestation on climate using measured ABRACOS vegetation characteristics, pp. 549–576 in J. H. C. Gash, C. A. Nobre, J. M. Roberts, and R. L. Victoria, eds., *Amazonian Deforestation and Climate*. Wiley, Chichester, U.K., 611 pp.; Foley, J. A., G. P. Asner, M. H. Costa, M. T. Coe, R. DeFries, H. K. Gibbs, E. A. Howard, S. Olson, J. Patz, N. Ramankutty, and P. Snyder. 2007. Amazonia revealed: Forest degradation and loss of ecosystem goods and services in the Amazon Basin. *Frontiers in Ecology and the Environment* 5(1): 25–32.

20. Fearnside, P. M. 2004. A água de São Paulo e a floresta amazônica. *Ciência Hoje* 34(203): 63–65.

21. Fearnside, P. M. 2003. Conservation policy in Brazilian Amazonia: Understanding the dilemmas. *World Development* 31(5): 757–79.

22. Credit under Kyoto's Clean Development Mechanism is not a possibility until after 2012.

23. Myers, N., C. G. Mittermeier, R. A. Mittermeier, G. A. B. da Fonseca, and J. Kent. 2000. Biodiversity hotspots for conservation priorities. *Nature* 403: 853–58.

24. Fearnside, P. M. 2001. The potential of Brazil's forest sector for mitigating global warming under the Kyoto Protocol. *Mitigation and Adaptation Strategies for Global Change* 6(3–4): 355–72.

25. Fearnside, P. M. 2003. Deforestation control in Mato Grosso: A new model for slowing the loss of Brazil's Amazon forest. *Ambio* 32(5): 343–45; Fearnside, P. M., and R. I. Barbosa. 2003. Avoided deforestation in Amazonia as a global warming mitigation measure: The case of Mato Grosso. *World Resource Review* 15(3): 352–61; Moutinho, P., and S. Schwartzman, eds. 2005. *Tropical Deforestation and Climate Change*. Instituto de Pesquisa Ambiental da Amazônia (IPAM), Belém, Pará, Brazil & Environmental Defense (EDF), Washington, DC, 131 pp.; Santilli, M., P. Moutinho, S. Schwartzman, D. Nepstad, L. Curran, and C. Nobre. 2005. Tropical deforestation and the Kyoto Protocol. *Climatic Change* 71: 267–76.

26. Nabuurs, G. J., O. Masera, K. Andrasko, P. Benitez-Ponce, R. Boer, M. Dutschke, E. Elsiddig, J. Ford-Robertson, P. Frumhoff, T. Karjalainen, O. Krankina, W. Kurz, M. Matsumoto, W. Oyhantcabal, N. H. Ravindranath, M. J. S. Sanchez, and X. Zhang. 2007. Forestry. Chapter 9 in B. Metz and O. Davidson, eds., *Climate Change 2007: Mitigation of Climate Change*. Contribution of Working Group III to the Fourth Assessment Report of the Intergovernmental Panel on Climate Change. Cambridge University Press, Cambridge, U.K.

27. Fearnside, P. M. 1993. Deforestation in Brazilian Amazonia: The effect of population and land tenure. *Ambio* 22(8): 537–45; Fearnside, P. M. 2005. Deforestation in Brazilian Amazonia: History, rates, and consequences. *Conservation Biology* 19(3): 680–88.

28. Fearnside, P. M. 1997. Environmental services as a strategy for sustainable development in rural Amazonia. *Ecological Economics* 20(1): 53–70.

29. Fearnside, P. M. 1996. Socio-economic factors in the management of tropical forests for carbon, pp. 349–61 in M. J. Apps and D. T. Price, eds., *Forest Ecosystems, Forest Management and the Global Carbon Cycle*, NATO ASI Series, Subseries I "Global Environmental Change," vol. 40. Springer-Verlag, Heidelberg, Germany, 452 pp.

30. Nepstad, D. C., S. Schwartzman, B. Bamberger, M. Santilli, D. Ray, P. Schlesinger, P. Lefebvre, A. Alencar, E. Prinz, G. Fiske, and A.

Rolla. 2006. Inhibition of Amazon deforestation and fire by parks and indigenous lands. *Conservation Biology* 20(1): 65–73.

31. Fearnside, P. M. 2005. Indigenous peoples as providers of environmental services in Amazonia: Warning signs from Mato Grosso, pp. 187–98 in A. Hall, ed., *Global Impact, Local Action: New Environmental Policy in Latin America*, University of London, School of Advanced Studies, Institute for the Study of the Americas, London, U.K., 321 pp.

Chapter 49

Engineering the Planet

DAVID W. KEITH

While the scope of human environmental impact is now global, we have yet to make a deliberate attempt to transform nature on a planetary scale. I call such transformation geoengineering.[1] More precisely, I define geoengineering as intentional, large-scale manipulation of the environment. Both scale and intent are important. For an action to be geoengineering, environmental change must be the goal rather than a side effect, and the intent and effect of the manipulation must be large in scale. Two examples demonstrate the roles of scale and intent. First, intent without scale: Ornamental gardening is the intentional manipulation of the environment to suit human desires, yet it is not geoengineering because neither the intended nor realized effect is large-scale. Second, scale without intent: Climate change due to increasing carbon dioxide (CO_2) has a global effect, yet it is not geoengineering because it is a side effect of the combustion of fossil fuels to provide energy. Pollution, even pollution that alters the planet, is not engineering. It's just making a mess.[2]

Manipulations need not be aimed at changing the environment, but rather may aim to maintain a desired environment against perturbations—either natural or anthropogenic. In the context of climate change, geoengineering entails the application of countervailing measure, one that uses additional technology to counteract unwanted side effects without eliminating their root cause, a "technical fix."

Sun Shades

If we decreased the amount of sunlight absorbed by the Earth we might engineer a cooling effect sufficient to counterbalance the warming caused by CO_2. Cooling might be achieved by adding aerosols, fine particles suspended in air, to the atmosphere, where they would scatter sunlight back into space and might also increase the lifetime and reflectivity of clouds.[3] Alternatively, it might be possible to engineer giant shields in space to scatter sunlight away from the planet.[4] These are the oldest and best-known geoengineering proposals so I will discuss them in some detail.

Like many other tools for geoengineering, the use of aerosols imitates nature. Sulfate

494

aerosols injected into the stratosphere by large volcanoes can cause rapid global cooling. The eruption of Mount Tambora in present-day Indonesia, for example, was thought to have produced the "year without a summer" in 1816. Likewise, the 1991 eruption of Mount Pinatubo in the Philippines caused a rapid decline in global temperatures that persisted over several years. In fact, "artificial volcanoes" have been proposed to deliberately inject sulfate aerosols into the stratosphere.[5]

As well as imitating natural processes, proposals for geoengineering often mimic existing human impacts: combustion of coal already creates great quantities of aerosols that offset part of the warming caused by CO_2. Geoengineering might therefore be seen as adding one pollutant—aerosols—to counteract the effect of another—CO_2. Like any technology, geoengineering entails risks and side effects. Sulfate aerosols injected into the stratosphere will, for example, generate impacts such as ozone loss. But, geoengineering is not pollution. Intent matters. The political implications of geoengineering, the institutional coordination required to implement it, and the moral implications of so doing all differ radically from the aerosol pollution that arises as a by-product of fuel combustion. Geoengineering may generate pollution as a side effect, but it is not simply a continuation of our long history of polluting the planet. Deliberate planetary engineering would open a new chapter in humanity's relationship with the Earth.

There is a surprisingly rich history of proposals to engineer the climate. As early as the 1960s, when modern knowledge of the CO_2-climate problem was in its infancy, there were suggestions that climate control using aerosols be used to offset the effects of rising CO_2 concentrations. Consider, for example, "Restoring the Quality of Our Environment," a report delivered to U.S. president Lyndon Johnson in 1965 by the Presidential Science Advisory Committee, which was the first high-level government policy document to draw attention to the threat of CO_2-driven climate change. While the report's discussion of climate science is consistent with that found in similar reports today, the sole suggested response to the CO_2-climate problem is geoengineering, which reflects extreme confidence in human technological prowess: "The possibilities of deliberately bringing about countervailing climatic changes therefore need to be thoroughly explored." The report suggests dispersing of buoyant, reflective particles on the sea surface, concluding that "a 1 percent change in reflectivity might be brought about for about $500 million a year. . . . Considering the extraordinary economic and human importance of climate, costs of this magnitude do not seem excessive."[6] The report does not mention the possibility of reducing fossil fuel use; this surprising fact illustrates that our thinking about the appropriate tools for managing the climate is far less stable than is our understanding of the underlying science.

The cost of injecting aerosols into the stratosphere was analyzed by the U.S. National Academy of Sciences in 1992; it examined several delivery methods including high-altitude aircraft and naval guns, and found that annual costs of greater than $100 billion would be sufficient to produce a 1 percent reduction in effective insolation (average solar radiation) reaching the lower atmosphere.[7] While this cost may sound high, it is roughly a factor of ten lower than the cost to achieve an equivalent reduction in climate change through reductions in CO_2 emissions.[8] The amount of sulfate that would need to be injected would be about twenty to fifty times smaller than the amount of sulfur now added to the lower atmosphere by fossil fuel combustion, so the contribution to acid rain might be negligible. Moreover, later analysis has shown that it is technically possible to design aerosols that are far more effective per unit mass at scattering light, which could reduce costs by more than a factor of ten.[9]

Costs are unlikely to be a deciding factor in the implementation of geoengineering. Using engineered high-scattering-efficiency aerosols, it is conceivable that the cost of climate engineering could be within reach of the world's richest individuals or private foundations. Decisions about implementation should balance the reduction in climate risk against the direct risks of geoengineering; cost would be a minor factor in this risk-risk decision.

The use of sulfate aerosols poses serious risks, including the alteration of atmospheric chemistry that might further deplete stratospheric ozone. The role of natural aerosols in forming the Antarctic ozone hole serves as a warning about the sensitivity of ozone concentrations to aerosols. However, Paul Crutzen (who received a Nobel Prize for work on stratospheric ozone) has argued that ozone depletion due to aerosol geoengineering might be acceptably small and could be made smaller still. While increasing CO_2 warms the lower atmosphere, it paradoxically cools the stratosphere, which can lead to increased ozone depletion.[10] Crutzen points out that if absorbing aerosols were used (black carbon in addition to sulfate), it would be possible to increase stratospheric temperatures, offsetting the current stratospheric cooling and partially or entirely offsetting the ozone depletion due to aerosol geoengineering.[11]

While expensive, space-based sunshields have side effects that would be both less significant and more predictable than would be the case with aerosols. Assuming that the shields were steerable, their effect could be eliminated at will. Additionally, steerable shields might be used to direct radiation at specific areas, offering the possibility of weather control. In recent decades, proposals have focused on space-based systems that would be located in stable orbits on a line between the Earth and the sun, well beyond the moon's orbit. Edward Teller and collaborators have found that such a shield could be made with much lower mass than was previously thought, implying that costs might be dramatically reduced.[12] While little technical analysis has been done, it seems certain that the cost and technical challenges of creating space-based sunshields are far larger than the costs in injecting aerosols into the stratosphere.

Regardless of how it is achieved, a reduction of solar input cannot perfectly compensate for CO_2-induced warming. While insolation could presumably be adjusted so that a geoengineered climate matched the preindustrial mean surface temperature, the result would still be significantly different than from the preindustrial climate. Several climate model experiments have shown that albedo geoengineering may nevertheless reproduce preindustrial climate with reasonable fidelity.[13]

Controlling the Weather

Just as growing knowledge of the role of aerosols in the atmosphere might enable more efficient and precise geoengineering, advances in the science of weather prediction are inadvertently producing tools that enable more effective weather control. The key tool is the development of specialized numerical models that are able to efficiently predict the impacts of small changes in the atmospheric state (temperatures, winds, and so forth) on the evolution of weather systems.[14] These tools are used in advanced weather-prediction systems to estimate the effect of errors in current observations of atmospheric conditions on the accuracy of weather forecasts a few days later.

This ability might be used to build a system for weather control by exploiting a paradoxical feature of chaotic systems. We often assume that chaos makes systems hard to control. The hallmark of chaotic systems is their extreme sensitivity to initial conditions, the proverbial flapping of a butterfly's wings that alters the global weather. It is this sensitivity that makes it hard to predict the future state of a chaotic system, because errors in one's knowledge of the system's initial state are rapidly amplified. Sensitivity to initial conditions can, however, facilitate dynamic control or guidance of the system's evolution because small control inputs are subject to the same amplification. Given sufficiently accurate models and observations, it is possible to steer the time evolution of chaotic systems with surprisingly small control inputs. Ross Hoffman and collaborators have shown, for example, that this strategy might be used to steer hurricanes.[15]

If atmospheric models and measurements are the software of weather control, the hardware is the tools used to manipulate atmospheric conditions. At the simplest, manipulation of atmospheric conditions might be accomplished by perturbing the altitude or course of commercial aircraft, which already effect atmospheric heating by generating cirrus clouds. Alternatively, manipulation might be accomplished by cloud seeding or, most extravagantly, by the use of space-based systems that could direct solar infrared radiation

to selectively heat the atmosphere or the surface. Better measurement of atmospheric conditions and better models of the global atmosphere together allow the use of smaller levers to achieve a given degree of weather control. Better software allows use of less hardware.

The most obvious utility of weather control is the ability to minimize the impact of severe storms on human welfare; sustained and large-scale use of weather control is, however, a form of climate control. Like other means of geoengineering, such power might be used to alter the climate to suit human desires or counteract climatic changes arising from other causes.

Should We Engineer the Planet?

The postwar growth of the Earth sciences has been fueled, in part, by a drive to quantify environmental insults in order to support arguments for their reduction. Paradoxically, our growing understanding of the dynamics of the Earth system increasingly grants us leverage that may be used to manipulate the Earth system and deliberately engineer environmental processes on a planetary scale. The manipulation of solar flux using stratospheric scatterers is the best example of leverage: we could reduce solar input sufficiently to initiate an ice age at an annual cost of less than 1 percent of global economic output.

How should we use our growing ability to engineer the planet? There is no immediate prospect that geoengineering will be employed as a tool for managing the CO_2-climate problem, but looking further ahead the question is far less easily answered. Should geoengineering substitute, even partially, for mitigation? In my view, a crucial part of the answer turns on the ultimate objectives of climate policy. Why should we spend money to reduce climate change? What consequences concern us most? Is human welfare the sole consideration, or do we have a duty to protect natural systems independent of their utility to us?

Just as safer cars may encourage more aggressive driving, the mere knowledge that geoengineering is possible may reduce the incentive to cut emissions by reducing (or appearing to reduce) the worst-case consequences of climate change.

Geoengineering may nevertheless be needed even if we pursue an aggressive mitigation strategy: suppose that several decades hence real collective action is underway to reduce CO_2 emissions under a robust international agreement. Suppose further that the climate's sensitivity to CO_2, or the sensitivity of natural systems to changed climate and increased CO_2, turn out to be higher than we now anticipate. Finally, suppose that because of the long lifetime of CO_2 in the atmosphere, even strong action to abate emissions is insuf-

ficient to prevent rapid deglaciation and consequent sea-level rise. Under such conditions, temporary albedo modification to limit climate impacts during the period of peak CO_2 concentrations might be warranted to control climate risk, not to substitute for mitigation.

Figure 49.1 illustrates the distinction between geoengineering as a substitute for mitigation and geoengineering as a means to reduce the risks of climate change while mitigation is ongoing. If geoengineering were used as a substitute, as in the left panel of the figure, the scale of the engineered compensation for CO_2-driven warming would have to grow to offset growing CO_2 concentrations. The risks of unanticipated side effects would therefore grow without bound. In this case, one might view mitigation as a strategy to minimize the risks of the side effects of geoengineering. On the other hand, geoengineering might be used in conjunction with mitigation to reduce the risks of climate change during the period of peak CO_2 concentrations.

It is tempting to discount geoengineering because of the risk of unintended conse-

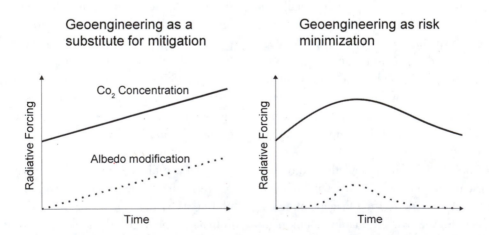

Geoengineering as a substitute for mitigation

Geoengineering as risk minimization

FIGURE 49.1. Schematic illustration of the distinction between geoengineering as a substitute for mitigation (left panel) and geoengineering as a supplement to mitigation used as a means to reduce the risks of climate change during the period of the peak radiative forcing (right panel).

quences. For example, Jeff Kiehl asserts that "a basic assumption to this approach [geo-engineering] is that we, humans, understand the Earth system sufficiently to modify it and 'know' how the system will respond."[16] If geo-engineering is used temporarily to reduce impacts of peak CO_2 concentrations, however, then it is misleading to argue against it solely because of the impossibility of predicting the system's response. Consider the choice between enduring a period in which CO_2 concentrations exceed 600 parts per million (ppm) and living with the same CO_2 concentration in conjunction with geoengineering that reduces insolation by 1 percent, as illustrated schematically in the right panel of figure 49.1. It is impossible to predict exactly how the planet will respond to either case, yet it is hard to argue that the risks of 600 ppm alone would be larger than the risks of 600 ppm with a little geoengineering to reduce peak temperatures.

Climate policy is often framed as a choice among various energy technologies and policy instruments. Beyond this choice of tools, however, lie hard choices about the objectives of planetary management. Should the planet be managed using all available tools so as to maximize human benefit, or should we seek to minimize human interference with nature? Advocates of active management argue that simple minimization of impacts is naive because the Earth is already so transformed by human actions that it is, in effect, a human artifact. According to this view, the proper goal of planetary management is the maximization of the planet's functionality to humans.[17] A strategy of active management might freely employ a mixture of responses, including the reduction of CO_2 emissions, geoengineering, and strategic adaptation to changing climate.[18] In this view, it makes lit-

tle sense to minimize impacts in order to let nature run free if there is no free nature left to protect.

If human utility is our sole concern, then active management seems an appropriate strategy. We may sensibly argue against geoengineering because it is too risky, too expensive, or too uncertain; but if methods of planetary engineering are proposed that are demonstrably less risky and more cost-effective than alternative measures, then, under this interpretation, we should use them.

An alternative view demands that we attribute intrinsic value to natural systems independent of their utility. According to this view, we should minimize our impact on the natural world—for its own sake—not solely to reduce the risk that manipulation of natural systems poses for humanity. Accepting such rights does not require that they trump all others—humans have rights, too—but attributing rights to nature does provide a basis for arguing that concerns other than pure human utility ought to enter into climate politics, and therefore that minimizing our impact on natural systems is a legitimate goal of climate policy.

Accepting minimization as a goal does not rule out geoengineering. What it does rule out is the use of geoengineering simply because it provides an expedient way of advancing human interests. Minimization (arguably) allows the use of geoengineering as a temporary measure if it provides an efficient method of minimizing impacts on the natural world.

As a thought experiment, imagine that alien visitors arrive and give us technology for climate and weather control. For illustration, imagine a box with knobs that allow independent control of global temperature and CO_2 concentration. Any adjustment of the knobs would inevitably benefit some and harm others. We do not yet possess a system of global governance that would allow a robust, let alone democratic, decision about how to set the knobs. One might readily imagine conflict arising from disputes about how the knobs should be set. Absent a credible system of global governance, perhaps the only robust decision would be to return the knobs to their preindustrial settings, that is, to minimize human influence rather than actively manipulating the planetary environment.

While a climate-control box is fiction, the ability to control nature on a planetary scale is not. Such powers are being gradually accumulated by the evolution of scientific knowledge and technologic ability. Unless a global war or other catastrophe should dramatically arrest or reverse technological progress, it seems inevitable that we will soon have such abilities.

Debate about deliberate modification of the global climate dates back at least a century. In 1908, Arrhenius, who was the first to analyze the role of CO_2 in regulating climate, suggested that warming resulting from fossil fuel combustion could increase food supply by allowing agriculture to extend northward. His contemporary, Eckhom, went further by suggesting that extra CO_2 could be injected into the atmosphere (by setting fire to shallow coal beds) to prevent the onset of ice ages and to enhance agricultural productivity through the fertilizing effect of CO_2. In the century since Arrhenius and Eckhom first considered these questions, our ability to manipulate the planet has grown in concert with knowledge of the global impacts of human activities. As remedies for the CO_2-climate problem, all proposed geoengineering schemes have serious flaws. Nevertheless, I judge it likely that this century will see serious debate about— and perhaps implementation of—deliberate

planetary-scale engineering. The continued acceleration of anthropogenic emissions coupled with growing concern about the possibility of dangerous nonlinear responses to climate forcing argue for more systemic exploration of the feasibility and risks of geoengineering. Active planetary management may be an inevitable step in the evolution of a technological society, but I urge caution. We would be wise to practice walking before we try to run, to learn to minimize impacts before we try our hand at planetary engineering.

Notes

1. Keith, D. W. (2000a). Geoengineering the climate: History and prospect. *Annual Review of Energy and Environment* 25: 245–84.

2. Allenby, B. (2000). Earth systems engineering and management. *IEEE Technology and Society Magazine* 19: 10–24; Friedman, R. M. (2000). When you find yourself in a hole, stop digging. *Journal of Industrial Ecology* 3: 15–19; Keith, D. W. (2000b). The Earth is not yet an artifact. *IEEE Technology and Society Magazine* 19: 25–28.

3. The average planetary reflectivity is called "albedo," so such methods are often called albedo modification.

4. Angel, R. (2006). Feasibility of cooling the Earth with a cloud of small spacecraft near the inner Lagrange point (L1). *Proceedings of the National Academy of Sciences* 103: 17184–89.

5. Budyko, M. I. (1982). *The Earth's Climate, Past and Future*. New York, Academic Press.

6. PSAC. (1965). President's Science Advisory Committee, *Restoring the Quality of Our Environment*. Washington, DC, Executive Office of the President.

7. Panel on Policy Implications of Greenhouse Warming, Committee on Science, Engineering, and Public Policy, National Academy of Sciences, 1992. *Policy Implications of Greenhouse Warming: Mitigation, Adaptation, and the Science Base*. Washington, DC: National Academy Press.

8. An atmospheric loading of around 10 g S offsets the effect of 1 ton of carbon, a S:C mass ratio of 1:105 (NAS, 1992 op. cit.; Crutzen, P. J. [2006].

Albedo enhancement by stratospheric sulfur injections: A contribution to resolve a policy dilemma? *Climatic Change* 77: 211–19). The NAS estimated a $20-per-kilogram cost to place aerosols in the stratosphere using naval rifles. Assuming a one-century CO_2 lifetime with a CO_2 atmospheric fraction of 0.5 and a two-year lifetime for stratospheric aerosols, and assuming that one can use elemental sulfur, which is oxidized in the stratosphere, the undiscounted cost of offsetting CO_2 emissions is around $5 per ton of carbon (in 2009 dollars per metric ton of carbon). In comparison, the cost of making large reductions in emissions by use of low-emission technologies is of order $100 per ton of carbon or larger.

9. Teller, E., L. Wood, and R. Hyde. (1997). *Global Warming and Ice Ages: I. Prospects for Physics Based Modulation of Global Change.* Livermore, CA, Lawrence Livermore National Laboratory, p. 20.

10. Kirk-Davidoff, D. B., E. J. Hintsa, J. G. Anderson, and D. W. Keith. (1999). The effect of climate change on ozone depletion through changes in stratospheric water vapor. *Nature* 402: 399–401.

11. Crutzen, 2006 op. cit.

12. Teller et al., 1997 op. cit.

13. Govindasamy, B., and K. Caldeira. (2000). Geoengineering Earth's radiation balance to mitigate CO_2-induced climate change. *Geophysical Research Letters* 27: 2141–44.

14. So-called tangent linear adjoint models enable one to efficiently run forecast models backward in time, allowing computation of the perturbation in the initial state required to produce some specified perturbation in the final state some days later. The full model is not actually run backward in time; instead a linearized model is generated that is valid only for small perturbations to the forward evolution of the atmospheric state.

15. Hoffman, R. N. (2004). Controlling hurricanes. *Scientific American* 291: 68–75; Henderson, J. M., R. N. Hoffman, S. M. Leidner, T. Nehrkorn, and C. R. Grassotti. (2005). A 4D-Var study on the potential of weather control and exigent weather forecasting. *Quarterly Journal of the Royal Meteorological Society* 131: 3037–51.

16. Kiehl, J. T. (2006). Geoengineering climate change: Treating the symptom over the cause? *Climatic Change* 77: 227–28.

17. Allenby, 2000 op. cit.

18. Ibid.

Contributors

Neil Adger, Tyndall Centre for Climate Change Research, University of East Anglia, Norwich, UK.

Philippe Ambrosi, The World Bank, Washington, DC.

André R. Aquino, Global Environmental Facility, The World Bank, Washington, D.C.

Vicki Arroyo, Executive Director, Georgetown State and Federal Climate Resource Center housed at Georgetown University Law Center, Washington, DC.

Christian Azar, Physical Resource Theory, Chalmers University of Technology, Sweden.

Paul Baer, Georgia Tech University, GA.

Jon Barnett, Melbourne School of Land and Environment, University of Melbourne, Australia.

Maxwell T. Boykoff, PhD, Assistant Professor, Environmental Studies and Center for Science and Technology Policy Research, University of Colorado, CO.

Tom Burns, Department of Sociology, University of Uppsala, Sweden and Woods Institute, Stanford, CA.

Josh Bushinsky, University of Chicago Law School (formerly at the Pew Center).

Audrey Chang, Director of the California Climate Program, Natural Resources Defense Council, CA.

Eileen Claussen, President of the Pew Center on Global Climate Change and Strategies for the Global Environment, Washington, DC.

Whitney Colella, Energy, Resources & Systems Analysis Center, Sandia National Laboratories, NM.

Russell Conklin, Master of Public Policy, 2008, University of Maryland, MD.

Danny Cullenward, PhD candidate, Interdisciplinary Program in Environment and Resources, Stanford University, CA.

Michael B. Cummings, ClearPeak Advisors, Inc., San Francisco, CA.

Judith A. Curry, School of Earth and Atmospheric Sciences, Georgia Tech University, GA.

Paul Dickinson, Carbon Disclosure Project, CA.

Kristie Ebi, Exponent Engineering and Scientific Consulting, U.S. and IPCC Working Group II member.

Heidi Ellemor, Melbourne School of Land and Environment, University of Melbourne, Australia.

Phil Fearnside, National Institute for Research in the Amazon (INPA), Brazil.

Martin Fussel, Working Group on Sustainable Solutions, Potsdam Institute for Climate Impact Research, Germany.

Ashok Gadgil, Environmental Energy Technologies Division, Lawrence Berkeley National Lab, CA.

Peter Gleick, Pacific Institute for Studies in Development, Environment, and Security, CA.

Stephane Hallegatte, Centre International de Recherche sur l'Environnement et le Développement & Ecole Nationale de la Météorologie, France.

Michael Hanemann, Department of Agricultural & Resource Economics, University of California Berkeley, CA.

Paul G. Harris, Department of Social Sciences, Hong Kong Institute of Education.

David Hawkins, Director of the Climate Center, Natural Resources Defense Council, Washington, DC.

Jim Hawley, Elfenworks Center for the Study of Fiduciary Capitalism, St. Mary's University, CA.

John Holdren, Kennedy School of Government and Department of Earth and Planetary Sciences, Harvard University, and Woods Hole Research Center (at time of writing); now Director of the White House Office of Science and Technology Policy (OSTP) and Assistant to the President for Science and Technology.

Chris Hotham, Conservation Council of South Australia.

Dan Kammen, Transportation Sustainability Research Center, University of California Berkeley, CA.

David Keith, ISEEE Energy and Environmental Systems Group, University of Calgary, Canada.

Klaus Lackner, The Earth Engineering Center, Columbia University, NY.

Rik Leemans, Environmental Systems Analysis Group, Wageningen University, The Netherlands.

Anthony Leiserowitz, Yale School of Forestry & Environmental Studies, CT.

Sharad Lele, Asoka Trust for Research in Ecology and the Environment, Bangalore, India.

Joanna Lewis, Assistant Professor, Science, Technology and International Affairs, Edmund A. Walsh School of Foreign Service, Georgetown University, Washington, DC.

David Lobell, Program on Food Security and Environment, Stanford University, CA.

Jeffrey Logan, Strategic Energy Analysis and Applications Center, National Renewable Energy Laboratory, CO.

Amy Luers, Environment Program, Google, CA.

M. J. Mace, lawyer and consultant, negotiator for the Federated States of Micronesia; former head of Climate Change & Energy Programme, Foundation for International Environmental Law and Development's Climate Change and Energy Programme, UK.

Jason Mark, Transportation Program, The Energy Foundation, CA.

Michael Mastrandrea, Consulting Assistant Professor at the Stanford University Woods Institute for the Environment.

Patrick K. McAuliffe.

Aaron McCright, Lyman Briggs College and Department of Sociology, Michigan State University, MI.

Frederick Meyerson, Department of Natural Resources Science, University of Rhode Island.

Fran Pavley, California State Senator.

Janet Peace, Vice President for Markets and Business Strategy, Pew Center on Global Climate Change, Washington, DC.

Burt Richter, Freeman Spogli Institute for International Studies, Stanford University, CA.

Mikael Roman, Stockholm Environmental Institute, Sweden.

Armin Rosencranz, SAIS-Johns Hopkins Bologna Center.

Art Rosenfeld, member, California Energy Commission.

Ambuj Sagar, Department of Humanities and Social Sciences, Indian Institute of Technology Delhi, India.

Ben Santer, Program for Climate Model Diagnosis and Intercomparison, Lawrence Livermore National Lab, CA.

Jayant Sathaye, International Energy Studies Group, Lawrence Berkeley National Lab, CA.

Stephen Schneider, Biological Sciences, Stanford.

Rachael Shwom, Department of Human Ecology, Rutgers University, NJ.

Truman Semans, Principal, GreenOrder, Washington, DC.

Carol Turley, Marine Biogeochemistry, Plymouth Marine Laboratory, UK.

Michael Wara, Stanford Law School, Stanford University, CA.

Robert Watson, Director of Strategic Development, Tyndall Centre for Climate Change Research, UK.

Peter J. Webster, School of Earth and Atmospheric Sciences, Georgia Tech, GA.

Anthony Westerling, Sierra Nevada Research Institute, University of California, Merced, CA.

Tom Wigley, Climate and Global Dynamics Division, University Corporation for Atmospheric Research, CO.

Andrew T. Williams, Graduate Business Programs of the School of Economics and Business Administration, St. Mary's University, CA.

Dale Willman, Executive Editor, Field Notes Productions; environmental journalist.

David Wolfowitz, Business Development Manager at EcoSecurities, Indonesia.

Index

About the Authors

Stephen H. Schneider is Melvin and Joan Lane Professor for Interdisciplinary Environmental Studies, professor of biology, and senior fellow in the Woods Institute for the Environment at Stanford University. He focuses on climate change science, integrated assessment of ecological and economic impacts of human-induced climate change, and identifying viable climate policies and technological solutions in a risk-management framework. He has consulted for federal agencies and White House staff in seven administrations. Involved since 1988 with the Intergovernmental Panel on Climate Change, which was awarded the Nobel Peace Prize in 2007, he was a core writer for the IPCC 2007 Fourth Assessment Report. Dr. Schneider was a MacArthur Fellow in 1992 and was elected to the U.S. National Academy of Sciences in 2002. He is actively engaged in improving public understanding of science and the environment through extensive media communications, television appearances, documentaries, and public outreach.

Armin Rosencranz, a lawyer and political scientist, taught a variety of environmental policy courses at Stanford from 1995 to 2006. He received three teaching awards at Stanford, including "Teacher of the Year" in 2005. In 1998 he organized a course on climate change policy that he and Steve Schneider cotaught five times. They also coedited *Climate Change Policy* (Island Press, 2002). Armin has received two Fulbrights to India and one to Australia. He taught one of India's first courses on environmental law, and his book, *Environmental Law and Policy in India* (2nd edition, 2001), is widely used throughout India. Until 1996, Armin was president of Pacific Environment, an international NGO that he founded in 1987. Since 2006 he has taught climate-related courses at Vermont and Georgetown Law Schools, the University of Maryland School of Public Policy, the University of Virginia's Semester at Sea, Stanford in Washington, and the Johns Hopkins School of Advanced International Studies in Bologna.

Michael D. Mastrandrea is Consulting Assistant Professor at the Stanford University Woods Institute for the Environment and a lecturer for the Stanford University interdisciplinary graduate program in environment and resources. His research focuses on the physical, biological, and societal impacts of climate change, policy strategies for reducing climate risks, and their accurate and effective translation for the general public, policy makers, and the business community. His work has been published in several journals, including *Science Magazine* and *Proceedings of the National Academy of Sciences*, and he is a coauthor of chapters on key vulnerabilities and climate risks and long-term mitigation strategies for the 2007 IPCC Fourth Assessment Report. He also serves on the editorial board for the journal *Climatic Change*.

Kristin Kuntz-Duriseti combines her professional interest in climate change policy as managing editor at *Climatic Change* with civic engagement as an environmental quality commissioner, business cochair of the Green Ribbon Citizens' Committee to reduce local carbon emissions, and parent mentor to student environment groups. Her research interests include economic evaluation of climate policies and distributional consequences of climate change and climate policy. More broadly, she is interested in environmentally sustainable lifestyles, including building construction, alternative transportation, water conservation, and home gardening.